W9-AQL-833

Network Science

Networks are everywhere, from the Internet, to social networks, and the genetic networks that determine our biological existence. Illustrated throughout in full color, this pioneering textbook, spanning a wide range of topics from physics to computer science, engineering, economics and social sciences, introduces network science to an interdisciplinary audience.

From the origins of the six degrees of separation to explaining why networks are robust to failures and fragile to attacks, the author explores how viruses like Ebola and H1N1 spread, and why it is that our friends have more friends than we do. Using numerous real-world examples, this innovative text includes clear delineation between undergraduate- and graduate-level material. The mathematical formulas and derivations are included within Advanced Topics sections, enabling use at a range of levels. Extensive online resources, including films and software for network analysis, make this a multifaceted companion for anyone with an interest in network science.

Albert-László Barabási is Robert Gray Dodge Professor of Network Science and Director of the Center for Complex Network Research at Northeastern University, with appointments at the Harvard Medical School and the Central European University in Budapest. His work in network science has led to the discovery of scale-free networks and elucidated many key network properties, from robustness to control.

Network Science

Albert-László Barabási

with

Márton Pósfai

Data analysis and simulations

CAMBRIDGE
UNIVERSITY PRESS

CAMBRIDGE
UNIVERSITY PRESS

University Printing House, Cambridge CB2 8BS, United Kingdom

Cambridge University Press is part of the University of Cambridge.

It furthers the University's mission by disseminating knowledge in the pursuit of education, learning and research at the highest international levels of excellence.

www.cambridge.org
Information on this title: www.cambridge.org/9781107076266

First published 2016
3rd printing 2017

Printed in the United Kingdom by Bell & Bain Ltd, Glasgow

A catalog record for this publication is available from the British Library

ISBN 978-1-107-07626-6 Hardback

Contents

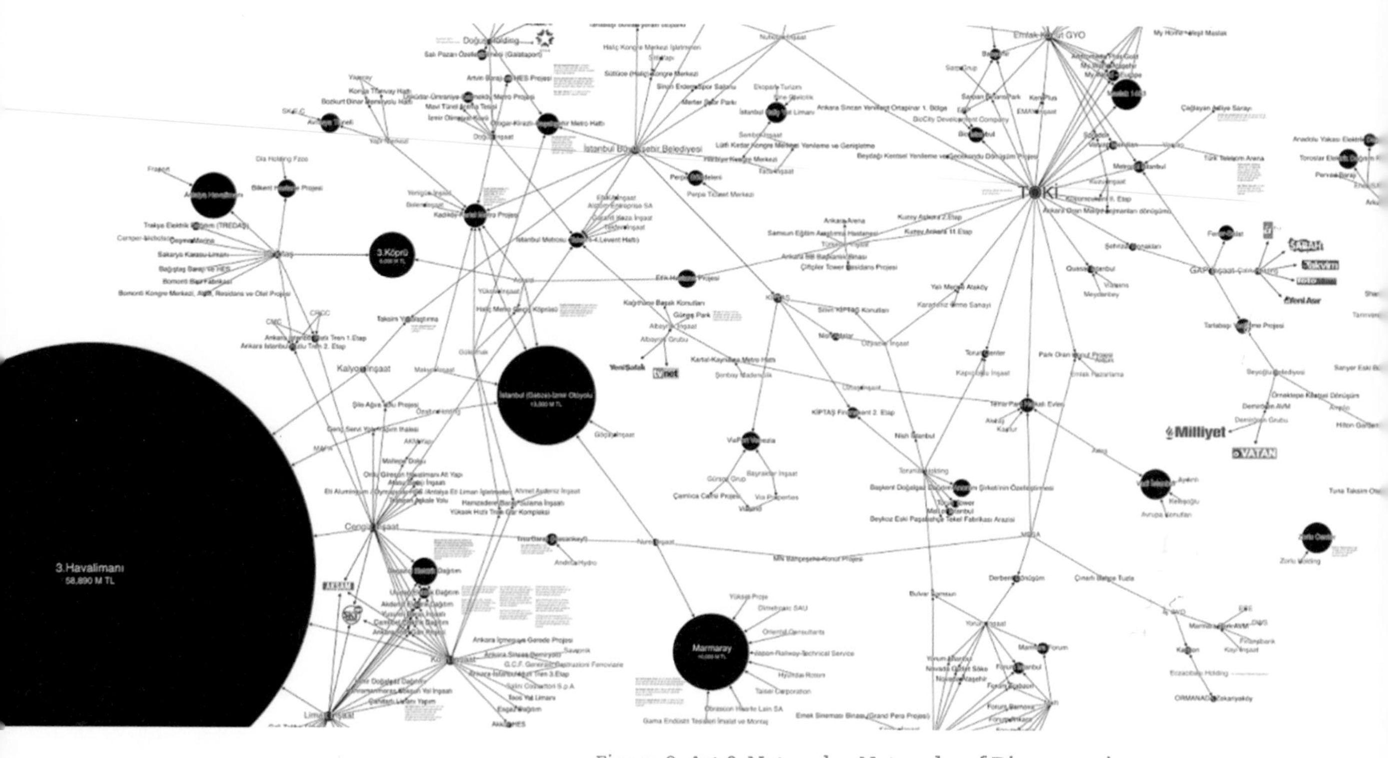

Figure 0 **Art & Networks: Networks of Dispossession**
In 2013 a popular protest movement shook Turkey, prompting thousands of activists and protesters to decamp at Gezi Park. The protests were accompanied by online campaigns, using Twitter or the WWW to mobilize supporters. A central component of this campaign was *Networks of Dispossession*, generated by a coalition of artists, lawyers, activists and journalists that mapped the complex financial relationships behind Istanbul's political and business elite. First exhibited at the Istanbul Biennial in 2013, the map reproduced here shows "dispossession" projects as black circles. The size of each circle represents the monetary value of the project. Corporations and media outlets, shown in blue, are linked directly to their projects. Work-related crimes are noted in red and supporters of Turkey's Olympic bid are shown in purple, while the sponsors of the Istanbul Biennial are in turquoise. The map was developed by Yaşar Adanalı, Burak Arıkan, Özgül Şen, Zeyno Üstün, Özlem Zıngıland and anonymous participants using Graph Commons (http://graphcommons.com/).

Preface

Teaching Network Science

The perspective offered by networks is indispensable for those who wish to understand today's interlinked world. This textbook is the best avenue I have found to share this perspective, offering anyone the opportunity to become a bit of a network scientist. Many of the choices I made in selecting the topics and in presenting the material were guided by the desire to offer a quantitative yet easy-to-follow introduction to the field. At the same time I have tried to pass on the many insights networks offer about the many complex systems that surround us. To resolve these often-conflicting desires, I have paired the technical advances with historical notes and boxes that illustrate the roots and the applications of the key discoveries.

This Preface has two puposes. On the one hand, by describing the class that motivated this text, it offers some practical tips on how to best use the textbook. On the other hand, and equally importantly, it acknowledges the long list of individuals who helped move this textbook forward.

Online Compendium

Network science is rich in content and knowledge that is best appreciated online. Therefore, throughout the chapters we

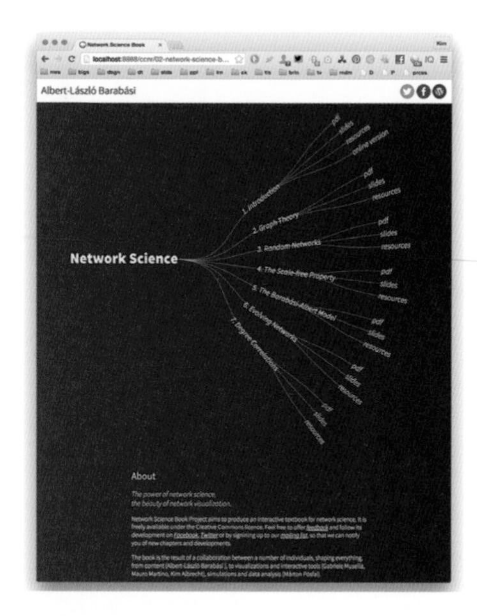

Online Resource 1
barabasi.com/NetworkScienceBook
The website offers online access to the textbook, the videos, software and interactive tools mentioned in the chapters, the slides I use to teach network science and the datasets analyzed in the book.

encounter numerous **Online resources** that point to pertinent online material – videos, software, interactive tools, datasets and data sources. These resources are available on the http://barabasi.com/NetworkScienceBook website.

The website also contains the PowerPoint slides that I use to teach network science, mirroring the content of this textbook. Anyone teaching networks should feel free to use these slides and modify them as they see fit to offer the best classroom experience. There is no need to ask the author for permission to use these slides in educational settings.

Given the empirical roots of network science, the book has a strong emphasis on the analysis of real networks. We have therefore assembled ten network maps that are frequently used in the literature to test various network characteristics. They were chosen to represent the diversity of the networks explored in network science, describing social, biological, technological and informational systems. The Online Compendium offers access to these datasets, which are used throughout the book to illustrate the tools of network science.

Finally, for those teaching the book in different languages, the website also mirrors the ongoing translation projects.

Network Science Class

I have taught network science in two different settings. The first is a full-semester class that attracts graduate and advanced undergraduate students with a physics, computer science and engineering background. The second is a three-week, two-credit class for students with an economics and social science background. The textbook builds on both teaching experiences. In the full-semester class I cover the full text, integrating into the lectures the proofs and derivations contained in the Advanced topics. In the shorter class I only cover the content of the main sections, omitting the Advanced topics and the chapter on degree correlation (**Chapter** 7).

In both settings key components of the class are assignments and the research project described below.

Homework Problems

For the longer class we assign as homework a subset of the problems listed at the end of each chapter, testing the technical proficiency of the students with the material, and their problem-solving ability. Two rounds of homework cover the material as we progress with the class.

Wiki Assignment

We ask each student to select a concept or a term related to network science and write a Wikipedia page on it (**Figure 1**). What makes this assignment somewhat challenging is that the topic must not already be covered by Wikipedia, yet must be sufficiently notable to be covered. The Wiki assignment tests the students' ability to synthesize and distill material in an easy-to-understand encyclopedic style, potentially turning them into regular Wikipedia contributors. At the same time, the assignment enriches Wikipedia with network science content, offering a service to the whole community. Those teaching network science in other languages should consider contributing to Wikipedia in their native language.

Social Network Analysis

As a warmup to network analysis, students are asked to analyze the social network of the class. This requires a bit of preparation and the help of a teaching assistant. In the very first class the instructor hands out the class list and asks everyone to check that they are on the list or add their name if they are missing. The teaching assistant takes the final list, and during the class prints an accurate class list for each student. At the end of the class each student is asked to mark everyone they knew before coming to the class. To help students match the faces with the names, each student is asked to briefly introduce themselves – also offering a chance for the instructor to learn more about the students in the class. These lists are then compiled to generate a social network of the class, enriching the nodes with gender and the name of the program the students are engaged in. The anonymized version of this network is returned to the class halfway through the course, the assignment being to analyze its properties using the network science tools the students have acquired up to that point. This allows them to explore a relatively small network that they are invested in and understand. The assignment offers a preparation for the more extensive network analysis they will perform for their final research project. This homework is assigned after

WIKI ASSIGNMENT

1. Select a keyword related to network science and check that it is not already covered in Wikipedia. "Related" is defined widely – you can select a technical concept (degree distribution), a network-related concept (terrorist networks), an application of networks (networks in finance), you can write about a network scientist, or anything else that you can convincingly relate to networks.

2. You are not expected to generate original material. Instead you need to identify 2-5 sources on the subject (research papers, books, etc.) and write a succinct, self-contained encyclopedic-style summary with references, graphs, tables, images, photos, as required to best cover the material. Observe Wikipedia's copyright and notability guidelines.

3. Upload your page on Wikipedia and send us the link. You will need to sign up for an account in Wikipedia, as anonymous editors cannot add new pages. Make sure that the page is not deleted by the Wikipedia administrators, which happens when the concept is not well documented or referenced, or is not written in an encyclopedic style.

4. The grade will reflect how understandable, pertinent, self-contained and accurate the content of your page is.

Figure 1 Wikipedia Assignment Guidelines

PRELIMINARY PROJECT

Present 5 slides in no more than five minutes:

- Introduce your network, discussing its nodes and links.
- Tell us how you will collect the data and estimate size of the network (N, L). Make sure that $N > 100$.
- Tell us what questions you are planning to ask. We understand that they may change as you advance with your project and the class.
- Tell us why you care about your network.

Figure 2 **Preliminary Project Guidelines**

the hands-on class on software, so that the students are already familiar with the online tools available for network analysis.

Final Research Project

The final project is the most rewarding part of the class, offering students the opportunity to combine and utilize all the knowledge they have acquired. Students are asked to select a network of interest to them, map it out and analyze it. Some procedural details enrich this assignment.

(a) The project is carried out in pairs. If the class composition allows, the students are asked to form professionally heterogenous pairs: undergraduate students are asked to pair up with graduate students, or students from different programs are asked to work together, like a physics student with a biology student. This forces the students to collaborate outside their expertise level and comfort zone, a common ingredient of interdisciplinary research. The instructor does not do the pairing, but students are encouraged to find their own partners.

(b) A few weeks into the course one class is devoted to *preliminary project presentations*. Each group is asked to offer a five-minute presentation with no more than five slides, offering a preview of the dataset they selected (**Figure 2**). Students are advised to collect their own data – simply downloading a dataset already prepared for network analysis is not acceptable. Indeed, one of the goals of the project is to experience the choices and compromises one must make in network mapping. Manual mapping is allowed, like looking up the ingredients of recipes in a cookbook or the interaction of characters in a novel or historical text. Digital mapping is encouraged, like scrapping data from a website or a database that is not explicitly organized as a network map, but the students must reinterpret and clean the data to make it amenable for network analysis. For example, one can systematically scrap data from Wikipedia to identify relationships between writers and scientists or concepts.

(c) It is important to always emphasize that the purpose of the final project is to test a student's ability to analyze a network. Consequently, students must stay focused on exploring the network aspect of the data, and avoid being carried away by other tempting questions their dataset poses that would take them away from this goal.

(d) The course ends with the final project presentations. Depending on the size of the class, we devote one or two classes to this (**Figure 3**).

The choice of Wikipedia keywords, partner selection for the research project and choice of topic for the final project require repeated feedback from the instructor, making sure that all students are on track. To achieve this, the last ten minutes of each class is devoted to asking everyone: Have you chosen a network that you wish to analyze? What are your nodes and your links? Do you know how to get the data? Do you have a partner for your final project? What is your Wiki word? Did you check if it is already covered by Wikipedia? Did you collect literature pertaining to it? The answers range from "Not yet," to firm or vague ideas the students are entertaining. Providing public feedback about the appropriateness and the feasibility of their plans helps those who are behind to crystallize their ideas, and to identify potential partners with common interests. After a few classes typically everyone finds a partner and identifies a research project and a Wikipedia keyword, at which point this end-of-class ritual finishes.

Software

We devote one class to various network analysis and visualization software, like Gephi, Cytoscape or NetworkX. In the longer course we devote another class to other numerical procedures, like fitting, log-binning or network visualization. We ask students to bring their laptops to these classes, so that they can try out these tools immediately.

Movie Night

We devote one night, typically outside class time, to a movie night, where we screen the documentary *Connected* by Annamaria Talas. The one-hour documentary features many contributors to network science, and offers a compelling narrative of the field's importance. Movie night is advertised university wide, offering a chance to reach out to the wider community.

FINAL PROJECT

Each group has 10 minutes to present their final project. Time limit is strictly enforced. On the first slide, give your title, name and program.

Tell us about your data and the data collection method. Show an entry of the data source to offer a sense of where you started from.

Measure: N, L, and their time dependence if you have a time-dependent network; degree distribution, average path length, clustering coefficient $C(k)$, the weight distribution $P(w)$ if you have a weighted network. Visualize communities; discuss network robustness and spreading, degree correlations, whichever is appropriate for your project.

It is not sufficient to simply measure things - you need to discuss the insights you gained, always asking:

- What was your expectation?
- What is the proper random reference?
- How do the results compare to your expectation?
- What did you learn from each quantity?

Grading criteria:

- Use of network tools (completeness/correctness);
- Ability to extract information/insights from your data using the network tools;
- Overall quality of the project/presentation.

No need to write a report - email us the presentation as a pdf file.

Figure 3 Final Project Guidelines

COMPLEX NETWORKS: SYLLABUS

Week 1
- **Class 1** *Ch. 1: Introduction*
- **Class 2** *Ch. 2: Graph Theory*

Week 2
- **Class 1** *Ch. 3: Random Networks*
- **Class 2** *Ch. 3: Random Networks*

Week 3
- **Class 1** *Ch. 4: The Scale-Free Property*
- **Class 2** *Ch. 4: The Scale-Free Property*

Hand-out Assignment 1 (Problems for Chapters 1-5)

Week 4
- **Class 1** *Ch. 5: The Barabási-Albert model*
- **Class 2** *Ch. 5: The Barabási-Albert model*

Week 5
- **Class 1** *Preliminary Project Presentations*
- **Class 2** *Hands-on Class: Graph representation, binning, fitting*

Week 6
- **Class 1** *Hands-on Class: Gephi and Python*

Collect Assignment 1;
Hand-out Assignment 2: Class Network Analysis
- **Class 2** *Guest Speaker*

Week 7
- Class 1 *Ch. 6: Evolving Networks*
- Class 2 *Ch. 6: Evolving Networks*

Week 8
- **Class 1** *Guest Speaker*

Collect Assignment 2
- **Class 2** *Ch. 7: Degree Correlations*

Hand out Assignment 3 (Problems for Chapters 6-10)

Week 9
- **Class 1** *Ch. 8: Network Robustness*

Hand out Assignment 4: Wikipedia Page
- **Class 2** *Ch. 8: Network Robustness*

Week 10
- **Class 1** *Ch. 9: Communities*
- **Class 2** *Ch. 9: Communities*

Movie Night: *Connected*, by Annamaria Talas

Week 11
- **Class 1** *Ch. 10: Spreading Phenomena*
- **Class 2** *Ch. 10: Spreading Phenomena*

Week 12
- **Class 1** *Guest Speaker*
- **Class 2** *Ch. 10: Spreading Phenomena*

Collect Assignment 4

Week 13
- **Class 1** *Guest Speaker*
- **Class 2** *Open-Door class (Research Project Discussions)*

Collect Assignment 3

Week 14
- **Exam Week** *Final Project Presentations (10 min per group)*

Figure 4 The Syllabus
The week-by-week schedule of the four-credit network science class, which meets twice a week.

Guest Speakers

In the full-semester class we invite researchers from the area to give research seminars about their work pertaining to networks. This offers the students a sense of what cutting-edge research looks like in this area. This is typically (but not always) done toward the end of the class, by which point most theoretical tools have been covered and the students are focusing on their final project. Such talks, advertised and open to the local research community, often inspire additional perspectives and ideas for the final project.

Figure 5 details the grading system used for the one-semester class. To aid in planning, **Figure 4** offers the schedule of the full-semester class I co-taught before this book went to print.

GRADE DISTRIBUTION

(1) Assignment 1 (Homework 1): 15%
(2) Assignment 2 (Homework 2): 15%
(3) Assignment 3 (Class Network): 15%
(4) Assignment 4 (Wikipedia): 15%
(5) Preliminary Project Presentation: No grading, only feedback.
(6) Final Project: 40%

Figure 5 **Grading**
The grading system used in the one-semester class.

Acknowledgments

Writing a book, any book, is an exercise in lonely endurance. This project was no different, dominating all my free time between 2011 and 2015. It was mostly time spent alone, working in one of the many coffeehouses I frequent in Boston and Budapest, or wherever in the world the morning found me. Despite this, the book is far from being a lonely achievement: during these four years a number of individuals donated their time and expertise to help move the project forward, offering me the opportunity to discuss the subject with colleagues, friends and lab members. I also shared the chapters on the Internet for everyone to use, receiving valuable feedback from many individuals. In this section I wish to acknowledge the professional network that stepped in to help at various stages of this long journey.

Formulas, Graphs, Simulations

A textbook must ensure that everything works as promised: that one can derive the key formulas, and that the measures described in the text – when applied to real data – work as the theory predicts. There is only one way to achieve this: one must check and repeat each calculation, measurement and simulation. This was a heroic job, most of it done by *Márton Pósfai* (**Figure 6**), who joined the project when he was a visiting

Figure 6 **The Math Team**
Márton Pósfai was responsible for the calculations, simulations and measurements in the textbook.

student in my lab in Boston and stayed with it throughout his PhD work in Budapest. He checked all derivations, if necessary helped re-derive key formulas, performed all the simulations and measurements and prepared the book's figures and tables. Many figures and tables amounted to small research projects, their outcome forcing us to de-emphasize some quantities because they did not work as promised, or helping us appreciate and understand the importance of others. His deep understanding of the network science literature and his careful work offered many subtle insights that enriched the book. There is no way I could have achieved this depth and reliability if it wasn't for Márton's tireless dedication to the project.

The Design

The ambition to create a book that had a clear aesthetic and visual appeal was planted by *Mauro Martino*, a data visualization expert in my lab. He created the first face of the chapters and many visual elements designed by him stayed with us until the end. After Mauro moved on to lead a team of designers at IBM Research, *Gabriele Musella* took over the design. He standardized the color palette and designed the basic elements of the info-graphics appearing throughout the book, also redrawing most images. He worked with us until the fall of 2014, when he too had to return to London to take up his dream job. At that time the design was taken over by *Nicole Samay*, who tirelessly and gently retouched the whole book as we neared the finish line (**Figure 7**). The website for the book was designed by *Kim Albrecht*, who is currently collaborating with Mauro to design the online experience that trails the book.

An important component of the visual design are the images included at the beginning of each chapter, illustrating the

Figure 7 **The Design Team**
Mauro Martino, Gabriele Musella and Nicole Samay developed the look and feel of the chapters and the figures, offering the book an elegant and consistent style. Kim Albrecht designed the online experience that trails the book.

interplay between networks and art. In selecting these images I have benefited from advice and discussions with several artists and designers, academics and practicing artists alike. Many thanks go to *Isabel Meirelles* and *Dietmar Offenhuber* from the Art and Design Department at Northeastern, *Mathew Ritchie* from Columbia University and *Meredith Tromble* from the San Francisco Art Institute, for helping me navigate the boundaries of art, data and network science.

The Daily Drill: Typing, Editing

I remain an old-fashioned writer, who writes with a pencil rather than a computer. I am lost, therefore, without editors and typists, who integrate my hand-written notes, corrections and recommendations into each chapter. *Sabrina Rabello* and *Galen Wilkerson* helped get this project started, yet the bulk of editing fell on the shoulders of three individuals (**Figure 8**): *Payam Parsinejad* worked with me during the first year of the project; after he had to refocus on his research, *Amal Al-Husseini*, a former student from my network science class, joined us and stayed until the very end; equally defining was the help of *Sarah Morrison*, my former assistant, who joined the project after she moved to Lucca, Italy. Her timely and accurate editing were essential in finishing the book.

Each chapter, before being released on our webpage, has undergone a final check by *Philipp Hoevel* (**Figure 9**), who

Figure 8 **The Editorial Team**
Payam Parsinejad, Amal Al-Hussieni and Sarah Morrison worked on the book on a daily basis, editing and correcting it.

Figure 9 **Accuracy and Rights**
Philipp Hoevel acted as our first reader and last editor. The rights were obtained and managed by Brett Common.

Figure 10 **Homework**
Roberta Sinatra conceived and compiled the homework at the end of each chapter in the textbook.

joined the project while visiting my lab and continued to work with us even after he returned to Berlin to run his own lab. Philipp methodically reviewed everything, from the science to notations, becoming our first reader and final filter.

Brett Common has worked tirelessly to secure all the permissions for the visual materials used throughout the textbook. This was a major project on its own, whose magnitude and difficulty was hard to anticipate.

Homework

The homework at the end of each chapter was conceived and curated by *Roberta Sinatra* (**Figure 10**). As a research faculty affiliated with my lab, Roberta co-taught the network science class with me in the fall of 2014, helping also to catch and correct many typos and misunderstandings that surfaced while teaching the material.

Science Input

Throughout the project I have received comments, recommendations, advice, clarifications and key material from numerous scientists and students. It is impossible to recall them all, but I will try.

Chaoming Song helped estimate the degree exponent of scale-free networks and helped me uncover the literature pertaining to cascading failures. The mathematician *Endre Csóka* helped clarify the subtle details of the Bollobás model. I benefited from a great discussion with *Raissa D'Souza* on optimization models, with *Ginestra Bianconi* on the fitness model and with *Erzsébet Ravasz Reagan* on the Ravasz algorithm. *Alex Vespignani* was a great resource on spreading processes and degree correlations. *Marian Boguña* snapped the picture for the Karate Club Trophy. *Huawei Shen* calculated the future citations of research papers. *Gergely Palla* and *Tamás Vicsek* helped me understand the CFinder algorithm and *Martin Rosvall* pointed us to some key material on the InfoMap algorithm. *Gergely Palla, Sune Lehmann* and *Santo Fortunato* offered critical comments on the community detection chapter. *Yong-Yeol Ahn* helped me develop an early version of the material on spreading phenomena and *Kate Coronges* helped improve the clarity of the first four chapters. *Ramis Movassagh, Hiroki Sayama* and *Sid Redner* provided careful feedback on several chapters.

Finally, *Sarah Sugar, R. Bharath, Susanne Nies, Harsha Gwalani, Jörg Franke*, who by communicating a series of typos, have helped make this print as error free as feasible.

Publishing

Simon Capelin, my longtime editor at Cambridge University Press, had been encouraging this project even before I was ready to write it. He also had the patience to see the book to its completion, through many missed deadlines. *Róisín Munnelly* helped move the book through production.

Institutions

This book would not have been possible if several institutions had not offered inspiring environments and a supporting infrastructure. First and foremost I need to thank the leadership of *Northeastern University*: President *Joseph Aoun*, Provost *Steve Director*, my deans *Murray Gibson* and *Larry Finkelstein* and my department chair *Paul Champion*, who were true champions of network science, turning it into a major cross-disciplinary topic within Northeastern. Their relentless support has led to the hiring of several superb faculty focusing on networks, spanning all domains of inquiry, from physics and mathematics to social, political, computer and health sciences, turning Northeastern into the leading institution in this area. They have also urged and supported the creation of a network science PhD program and helped found the *Network Science Institute* led by *Alessandro Vespignani.*

My appointment at *Harvard Medical School*, through the *Network Medicine Division* at *Brigham and Women's Hospital* and the *Center for Cancer Systems Biology* at *Dana Farber Cancer Institute (DFCI)*, offered a window on the applications of network science in cell biology and medicine. Many thanks to *Marc Vidal* from DFCI and *Joe Loscalzo* from Brigham who, as colleagues and mentors, have defined my work in this area – an experience that found its way into this book as well.

My visiting appointment at *Central European University (CEU)*, and the network science class I teach there in the summer, have exposed me to a student body with an economics and social science background, an experience that has shaped this textbook. *Balázs Vedres* had the vision to bring network science to CEU, *George Soros* convinced me to get involved with the university and President *John Shattuck* and Provosts *Farkas Katalin* and *Liviu Matei*, with their relentless support, have smoothed the path toward CEU's superb program in this area, giving birth to CEU's PhD program in network science.

Thanks to the place where it all began: when I was a young Assistant Professor, the *University of Notre Dame* offered me the support and the serene environment to think about something different. And big thanks to *Suzanne Aleva*, who followed my lab from Notre Dame to Northeastern, and worked tirelessly for over a decade to foster an environment where I can focus, uninterrupted, on science.

Finally, I am truly indebted to my children, Dániel, Izabella and Lénárd, and my wife, Janet, who accepted that I devote countless hours to this book. Without their understanding and patience I would never have finished *Network Science*.

Figure 0.0 **Art & Networks: Chiharu Shiota**
Chiharu Shiota is a Japanese installation artist working in Berlin. She creates interwoven webs of wool that engulf ordinary objects. The apparent randomness of these networks embraces the inherent tension between order and disorder, the very topic of network science.

CHAPTER 0

Personal Introduction

Introduction

Today, when each year a dozen or so conferences, workshops and schools focus on networks, when over a hundred books and four journals are devoted to the field, when most universities offer network science courses and one can get a PhD in network science on three continents, and when funding agencies have earmarked hundreds of millions of dollars for the subject, it is tempting to see this decade-old field's evolution as a straight path to success. But blinded by this cumulative impact, we may miss the most fascinating question: How could the field grow up this fast?

I call this chapter a *personal introduction* for the simple reason that I have no intention of offering an unbiased answer to this question. On the contrary, I plan to recall the emergence of network science from the perspective of a participant whose story I best know, which happens to be me. This is not a victory march, but my goal is to recall the winding and convoluted journey that I experienced, with its numerous setbacks and bursts. Instead of a bird's eye perspective, I will focus on those hard-to-forget trees that I repeatedly bumped into as I attempted to cross the forest. It is a reminder that scientific discovery is not as straightforward and smooth as textbooks, like ours here, may occasionally insinuate.

My First Network Paper (1994)

My fascination with networks started in December 1994, a few months into my brief postdoctoral position at IBM's legendary

Figure 0.1 **1994–1995: My First Take on Networks**
Conceived over the winter break at the end of 1994, my first network paper [1] mapped the minimal spanning-tree problem, a well-known algorithm in computer science, into invasion percolation, a much-studied problem in statistical physics. It marked the beginning of my long engagement with network science.

VOLUME 76, NUMBER 20 PHYSICAL REVIEW LETTERS 13 MAY 1996

Invasion Percolation and Global Optimization

Albert-László Barabási

Department of Physics, University of Notre Dame, Notre Dame, Indiana 46556
and T.J. Watson Research Center, IBM, P.O. Box 218, Yorktown Heights, New York 10598
(Received 24 February 1995)

Invasion bond percolation (IBP) is mapped exactly into Prim's algorithm for finding the shortest spanning tree of a weighted random graph. Exploring this mapping, which is valid for arbitrary dimensions and lattices, we introduce a new IBP model that belongs to the same universality class as IBP and generates the minimal energy tree spanning the IBP cluster. [S0031-9007(96)00106-8]

PACS numbers: 47.55.Mh

Flow in a porous medium, a problem with important practical applications, has motivated a large number of theoretical and experimental studies [1]. Aiming to understand the complex interplay between the dynamics of flow processes and randomness characterizing the porous medium, a number of models have been introduced that capture different aspects of various experimental situations. One of the most investigated models in this respect is invasion percolation [2], which describes low flow rate drainage experiments or secondary migration of oil during the formation of underground oil reservoirs [1,3].

this graph is a connected graph of n vertices and $n-1$ bonds. Of the many possible spanning trees one wants to find the one for which the sum of the weights p_{ij} is the smallest. A well known example is designing a network that connects n cities with direct city-to-city links (whose length is p_{ij}) and shortest possible total length. This is a problem of major interest in the planning of large scale communication networks and is one of the few problems in graph theory that can be considered completely solved. Since for a fully connected graph with n vertices there are n^{n-2} spanning trees [5], designing an algorithm that finds the shortest one in nonexponential time steps is a formidable global optimization problem.

T. J. Watson Research Center. As the approaching holidays had brought a predictable halt to life at Watson, I decided to use the break to learn a bit more about my employer. Back then IBM was synonymous with computers, so I went to Watson's library looking for an introduction to computer science.

Curious about the field's intellectual challenges, I walked away with a book covering an array of problems, from algorithms to Boolean logic and NP-completeness. One chapter, focusing on the minimal spanning-tree problem, particularly piqued my interest. For good reason: I realized that the Kruskal algorithm described in the book mapped into a well-known model of statistical physics, called invasion percolation. So exactly two months after Christmas, on February 24, 1995, I submitted my first paper on networks to *Physical Review Letters* [1], demonstrating the equivalence of two much-studied network problems of physics and computer science (**Figure 0.1**). While a single-author paper in this prestigious physics journal was undoubtedly a smart career move, its true impact was more far-reaching: the hidden intellectual floodgates the paper unlocked laid the ground for my subsequent decades-long love affair with networks.

Fail 1: The Second Paper (1995)

The more I learned, the more puzzled I was about how little we knew about real networks. Living in New York City, I imagined the remarkable complexity of the millions of electric, telephone and Internet cables cramped under Manhattan's pavements. Graph theory envisioned that these networks were wired randomly. That didn't make much sense to me. There must be some organizing principles governing the numerous networks that we depend on. Finding these principles was a fitting challenge for a statistical physicist trained at the border of order and randomness.

So, I devoted the subsequent months to Béla Bollobás' excellent book on random graphs [2], which introduced me to the classical work of Erdős and Rényi [3]. At the same time, Stuart Kaufmann's visionary writing made me appreciate the importance of networks in biology [4]. Two very different perspectives collided in these books: the dry, theorem-driven world of mathematics and the wandering imagination of Stu, which saw no mathematical bounds (**Figure 0.2**).

Eight months into my postdoctoral position I accepted a faculty position at the University of Notre Dame, allowing me to devote the remaining four months at IBM to my second network paper. Entitled "Dynamics of random networks: Connectivity and first order phase transitions" [5], it was my first attempt to probe the implications of altering the topology of a network. The paper merged the world of Bollobás and Kaufmann, asking how changes in the network structure affect the dynamics of a Boolean system (**Figure 0.3**). The underlying observation was simple: if we alter the average degree of a random network, the Boolean system undergoes a dynamic phase transition. Hence we cannot interpret a system's behavior without fully accounting for the structure of the network behind it.

The paper was motivated by a mixture of ideas rooted in cellular networks, the Internet and the World Wide Web (WWW), yet these topics were largely absent from the physics journals that normally published my work. I struggled, therefore, to find some tangible applications within my own domain. At the end I put the results in the context of neural networks, a much-studied problem among physicists. This community, I thought, should be inclined to think positively about networks. I was wrong, of course, and this decision marked the first of a series of failures that

Figure 0.2 **1995: Order or Randomness?**
Two of the three books that inspired my early journey toward network science. I could never track down the first book, whose title (or subtitle) was something like *Fifty Problems in Computer Science*, the one I borrowed in 1994 from the library of IBM's Watson Research Center.

Figure 0.3 **1995–1997: The Never-Published Network Paper**
My second take on networks, and the first paper in which I explored the role of the network topology. It was posted on the online server Arxiv in November 1995, after it was rejected by four journals. I eventually gave up trying to get it published in a journal.

Dynamics of Random Networks: Connectivity and First Order Phase Transitions

Albert-László Barabási
Department of Physics, University of Notre Dame, Notre Dame, IN 46556.
(February 1, 2008)

Abstract

The connectivity of individual neurons of large neural networks determine both the steady state activity of the network and its answer to external stimulus. Highly diluted random networks have zero activity. We show that increasing the network connectivity the activity changes discontinuously from zero to a finite value as a critical value in the connectivity is reached. Theoretical arguments and extensive numerical simulations indicate that the origin of this discontinuity in the activity of random networks is a first order phase transition from an inactive to an active state as the connectivity of the network is increased.

trailed my journey toward network science for the next four years.

On November 10, 1995 I mailed the finished manuscript to *Science* and returned to Boston for the annual meeting of the Materials Research Society. Philipp Ball, a *Nature* editor with an interest in interdisciplinary subjects, was at the meeting, giving me the opportunity to tell him about my fascination with my new subject, networks. So when a few weeks later *Science* rejected the paper without review, I sent it to Philipp, hoping that *Nature* would show more interest. And it did, sending the paper for review.

The referees were much less fascinated, however. One of them put this bluntly, writing in the referee report that:

1. **It is badly motivated.**
2. **It is technically very constrained.**
3. **The speculations (about evolution and the Internet) do not materialize.**

The referee was right, of course: I failed to explain why we care about networks in the first place. It was all in my head. But barely a year after my PhD, relying on a language (English) that I had acquired only four years earlier, I could not yet translate my ideas into a story that sticked.

Disappointed, on April 25, 1996 I resubmitted the paper to *Physical Review Letters*. It did not fare much better there either, being rejected after a lengthy review. When, on November 21, 1997, two years after its first submission, I resubmitted the paper to *Europhysics Letters*, I was already experiencing the second major failure of my network-bound journey.

Fail 2: Mapping the Web (1996)

While struggling to get my second paper published, I became increasingly convinced that to move forward I would need to abandon the graph-theoretical path I had pursued thus far. I should instead do what physicists are good at: look at the real world for inspiration. That is, I decided that I needed maps. Maps of real networks, to be precise.

Five years after Tim Berners-Lee unleashed the code behind the WWW and two years before Google was founded, the Web just started humming. An odd collection of search engines – going by names like JumpStation, RBSE Spider or Webcrawler – hacked together in research labs, were trying to map its link structure. In February 1996 I sent an email to several researchers running such crawlers (**Figure 0.4**), hoping to get a sample of their data. A full map would have been ideal. Short of that it would have been sufficient to get the number of links each node had. "I wish to make a simple histogram of the previous data," I wrote, asking for something that we would name only three years later: the degree distribution of the WWW.

No one said no. But no one bothered to answer either. And as I waited for a reply, my second network paper got its final blow, being rejected by *Europhysics Letters* as well.

By that point my journey into networks was quite disappointing. My second network paper had been seen by four journals and three rounds of referees. No one said that it was wrong. The referees' message was simple: Who cares? Then my Plan B to access real data had slowly reached a dead end. Disappointed and under pressure to publish and obtain grants, I gradually replaced networks with a safer line of research on quantum dots.

I had no choice, really. Two years into my assistant professorship, my startup funds were dwindling and my prospects of tenure looked thin. As much as I believed in networks, all I could show for the past three years was one publication and

Figure 0.4 **1996: Begging for Data**
One of the emails I sent to computer scientists building Web crawlers in the mid-1990s, hoping to convince them to share data on the Web's topology. In hindsight, not a very convincing letter. No wonder no one responded. I had to wait two more years, until Hawoong Jeong joined my research lab and built our own crawler, to get the data that allowed us to discover scale-free networks. Had we gotten the data in 1996, as I originally hoped to, we might have discovered it three years earlier.

Date: Fri, 09 Feb 1996 10:34:17 -0500
From: Albert-Laszlo Barabasi <alb@nd.edu>
To: .gov
Cc: alb@nd.edu
Subject: Robots
X-Url: .html

Dear ,

I am doing some research on random networks and their statistical mechanical properties. The best available real word example of such networks is the WWW with its almost random links. To try out my approach I need some data that that Robots could provide without much difficulty. I friend of mine (who knows much more about the dangers of writing and operating a poorly working- robot) convinced me that instead of attempting to write my own robot, I should rather check if somebody with an already running robot could either (i) help me with the data I need or (ii) allow me to use his/her robot for this purpose.

I wonder if you are willing to give me a help in this direction? Of course, any help will be carefully acknowledged when the results of this research will be published (this is all-academic, non-profit basic research).

When a robot visits a new site, it finds a number of external links (pointing to other home pages). I need statistics regarding this number. Robots regularly collect this information, since this is how they assemble their database. Thus the only thing I need is to have the robot write this info into a file in a structured form, that would
allow me to extract this information. Maybe some of the robots do save the obtained data in a format that would allow me to simply collect these numbers.

For example, if the robot visits the home page http://www.new.homepage.edu/bbb.html it finds that there are for example four
links there, pointing to the addresses:
http://www.aps.org/xxx.html
http://www.my.best.friend/home.html
http://www.my.hobby/joke.html
http://www.my.preffered.newspaper/news.html

So the type of list I could most use is this one (or something equivalent):
http://www.new.homepage.edu/bbb.html
HAS LINKS TO:
http://www.aps.org/xxx.html
http://www.my.best.friend/home.html
http://www.my.hobby/joke.html
http://www.my.preffered.newspaper/news.html

Moreover, to start with it would be enough less information, for example a just listing the number of links he found:

4

After visiting a fair number of home pages the table would look like this:

4
2
3
2
0
19
10
1
0
1

How many datapoints do I need? Well, I wish to make a simple histogram of the previous data, thus I need enough data to obtain a smooth histogram. This histogram will be the starting point of my investigation.

I hope you are willing to help me to obtain this information. If you are not running your robot currently, but are willing to lend me your code so that I can run from my computer to collect this data (I have an IBM RISC 6000 that I could use for this purpose), that is also a solution. Again, I do not plan to use the robot for any other purpose than collecting this (and similar) statistics on the connectivity of the web. If you are interested in more details regarding the nature of the scientific questions I am investigating, I am happy to provide it to you.

laszlo

Albert-Laszlo Barabasi
Assistant Professor

a string of failures. The transition to the more conventional topic paid off, however: by the end of 1997 I was awarded two research grants, allowing me to hire several students and a postdoctoral researcher.

Reboot (1998)

In 1997 I was living in Chicago, commuting every second day to Notre Dame. To kill the boredom of the two-hour drive, I started to listen to books on tape. One day I picked up from the library Asimov's *Foundation*, a book that I had devoured as a child (**Figure 0.5**). As I slipped into the magical world of the *Second Foundation*, I was captivated by Harry Seldon's ability to forecast the fate of humanity hundreds of years into the future. It was the best of science fiction: fascinating, out of reach, but still plausible in some abstract dimension.

The monolithic cornfields that surround Route 90, connecting Notre Dame to Chicago, allowed my mind to contemplate a whole range of quixotic questions: What would it take to turn Asimov's fiction into reality? Could one indeed formulate a set of equations that could predict the future of a system as complex as society? Is there anything I could do to help achieve this? As my research on quantum dots blossomed, Asimov kept pulling my mind back to the questions that never stopped fascinating me – despite the many setbacks I had experienced earlier: networks and complex systems.

By early 1998 I was ready to try it again. I started by sketching out a new network-related research project and in March I invited Réka Albert to lunch at *Sorins*, the most elegant restaurant Notre Dame had to offer. Réka, a year and a half into her graduate studies, was on to a stellar career. Her paper on granular media had just made the cover of *Nature* and the preliminary results of her ongoing projects were just as promising. Hence, my purpose with the lunch defied all wisdom: I wanted to persuade her to give up the research she had been so successful at. I wanted her to explore networks instead.

As I asked my best student to join me on my network crusade, I could offer little encouragement. I had to tell her that my second paper on the subject had been rejected by four journals and that I could never get it published [5]. Networks had no community, no journal and no funding. I had to be

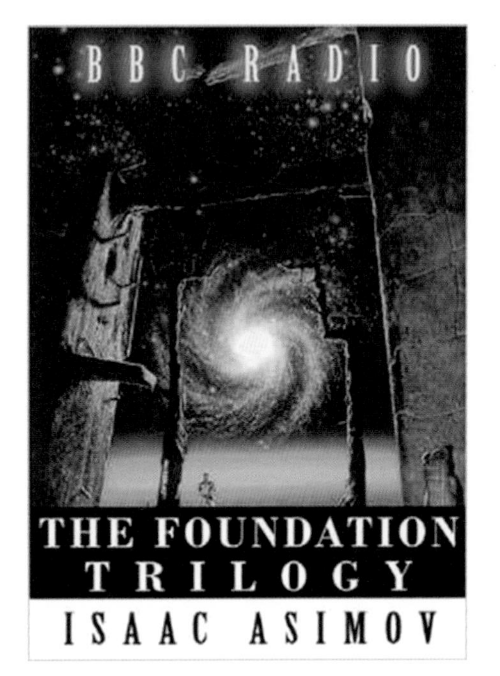

Figure 0.5 **1997: Reboot**
Isaac Asimov's science fiction trilogy that inspired my return to networks.

honest, confessing to her that no one seemed to care about the subject. She was therefore risking a sudden end to the success story she had so far experienced.

Yet, I also told her that to succeed we must take risks. And that in my view networks were worth the gamble.

At the end of the lunch I gave Réka a densely typed document, my early vision of network science. I estimated that it would take us about six months to quantify the network topology and another six months to understand the impact of the topology on network dynamics. Then we could move on to the real problem, exploring the joint evolution of network topology and dynamics.

I was completely off the mark, of course: I could not foresee the fantastic richness the topology had to offer. But that was besides the point back then. What mattered was that in her quiet and gracious manner, Réka agreed to join me on this risky network-bound journey.

Fail 3: Small Worlds (1998)

I still find it puzzling how disjoint the communities were in thinking about networks prior to 1999. On the one hand there was a small but active social network community, whose roots went back to the 1940s. Indeed, much of what we know today about the small-world problem is contained in a little-known paper written around 1960 by the social scientist Ithiel de Sola Pool and the mathematician Manfred Kochen. While their work remained unpublished until 1978 [6], its preprint was widely circulated in the social network community, inspiring Stanley Milgram's 1967 small-world experiment [7]. And it was Milgram's work that a quarter of a century later inspired the playwright John Guare to invent the "six degrees of separation" phrase.

While Pool and Kochen relied on the same models that the graph theorists Erdős and Rényi explored in parallel, no sociology paper showed even the faintest evidence that they were aware of the massive mathematical literature emerging on random graphs. On the other hand there was the extensive random graph literature inspired by Erdős and Rényi's pioneering work. Yet, no one in graph theory had any awareness of the social network community, nor did they make any reference to small worlds.

This disciplinary gap was reflected in the different questions the two communities asked: the graph theorists worried about phase transitions, subgraphs and giant components; the social scientists were fascinated by small worlds, weak ties and communities. While for social scientists networks with a hundred nodes were beyond comprehension, mathematicians got excited only in the $N \to \infty$ limit.

When the Watts and Strogatz paper about small-world networks was published in *Nature* in 1998 [8], it first brought back memories of my failed attempt to publish my second network paper in the same journal three years earlier. The roots of my failure became painfully obvious as I read their paper: I had a massive framing problem. Both papers used the random network paradigm, yet I asked questions of interest to physicists, while directing the paper to neuroscientists. In contrast, the questions asked by Duncan and Steve were deeply rooted in sociology, six degrees offering a brilliant narrative for their manuscript.

At the same time the small-world model appeared to be a dead end for the questions Réka and I were pursuing. As physicists we cared about patterns that could *not* be produced by randomness. Hence we were searching for phenomena that went beyond both regular lattices, the over-explored bread-and-butter of solid state physics, and the purely random network model of Erdős and Rényi. The Watts–Strogatz model interpolated between a regular and a random network, precisely the two limits we sought to avoid. So we set the paper aside, seeing it as a distraction from the path we had embarked on. I pulled it out again only months later, when the small-world framing offered some unexpected help on our journey.

Mapping the Web (1998)

When Hawoong Jeong joined my group as a postdoctoral researcher in 1998, Réka and I were already deeply immersed in networks. A graduate of Korea's prestigious Seoul National University, Hawoong's knowledge of computers was prodigious. One night, in the fall of 1998, I dropped by his office to chat about his progress on quantum dots, his main project at that time. Somehow we slipped into networks, prompting me to tell him about my failures to access real data on the topology of the WWW. I asked if he knew how to build a robot, the colloquial term for a Web crawler. He responded that he had never built one, but was willing to give it a try. And try he did: a few weeks later Hawoong's robot was busily crawling the Web, reviving my failed Plan B to explore the structure of the WWW.

We decided to use the data collected by Hawoong to continue where I left off in 1996 (**Figure 0.4**), measuring the degree distribution of the WWW. We were motivated by a simple question: Had the WWW reached its percolation threshold? Erdős and Rényi predicted that under a critical link density a network is fragmented into many isolated clusters. Yet once the density reaches a critical threshold a giant component, something that we would perceive as a network, emerges.

Could the WWW still be broken into many disconnected components? Or was it already one big network, as everyone perceived it back then? These were intriguing questions, no matter what the outcome. To answer them we needed the Web's degree distribution, which was now being provided by Hawoong's robot. The data granted us our first real surprise: we did not see the Poisson distribution that random network theory predicted. A power law greeted us instead.

Hawoong's data was a shocking departure from everything I had learned during my four-year journey into networks. There was no trace in the literature of a network with a power-law degree distribution. In fact, no one seemed to care much about the degree distribution up to that point: both the random graph and the social network literature took the Poisson form for granted. The power law observed by us predicted that the Web has hubs, nodes with a huge number of links, outliers forbidden in a random universe. None of the existing models could account for these.

According to a surviving email I sent to Hawoong, I started writing my third network paper on March 30, 1999, my 32nd birthday. It was tempting to focus on the true discovery, which was simple: the WWW represents a new type of network, a previously unrecognized form of organization. I sensed, however, that this would be a mistake. By then I was convinced that the failure of my second network paper had little to do with its science, but was a framing problem. Focusing on the inherent, scientific value of the observation was too dry, unlikely to excite the editors of *Nature*. So I decided to use a Trojan horse instead: I hid the discovery under six degrees (**Figure 0.6**). We entitled the paper "The diameter of the World Wide Web," trumped up the fact that six is really nineteen on the Web, and mailed it off to *Nature* [9].

Figure 0.6 **1999: The 19 Degrees Team**
A photo taken for *Business 2.0* magazine in 2000, showing Réka Albert, Hawoong Jeong and the author, soon after our publication of the paper on the topology of the WWW.

The Discovery (1999)

Shortly after the submission I started a two-week-long trip to Spain and Portugal, whose last leg was a workshop at the University of Porto. As I drove across the Iberian peninsula, I could not help carrying with me the questions: Why hubs? Why power laws?

To understand what makes the WWW so special, we needed to learn more about other networks. Therefore, before boarding my flight to Europe, I embarked on an aggressive pursuit of additional network maps. The first map came from Jay Brockman, a computer science professor at Notre Dame, who gave us the wiring diagram of a computer chip manufactured by IBM. Duncan Watts sent us the map of the power grid, and Brett Tjaden shared the Hollywood actor database. I left them all with Réka, to analyze them while I was traveling.

On June 14, 1999 I was already in Porto when Réka sent a message detailing some ongoing activities. At the end of her email, more like an afterthought, she added a line: "I looked at the connectivity distribution too, and in almost all systems (IBM, actors, power grid), the tail of the distribution follows a power law."

Réka's sentence hit me like a thunderbolt. I could not pay attention to the talks any longer. My mind was spinning around the implications. If networks as different as the Web and Hollywood display the same power-law degree distribution, then the property we saw earlier on the WWW is universal! Hence, some common law or mechanism must be responsible for its emergence! And if it applies to systems as different as actors, a computer chip and the Web, the explanation needs to be fundamental and simple.

I needed a quiet spot to think about this, so I left the workshop to withdraw to *Casa Diocesana*, the seminary that housed us during the conference.

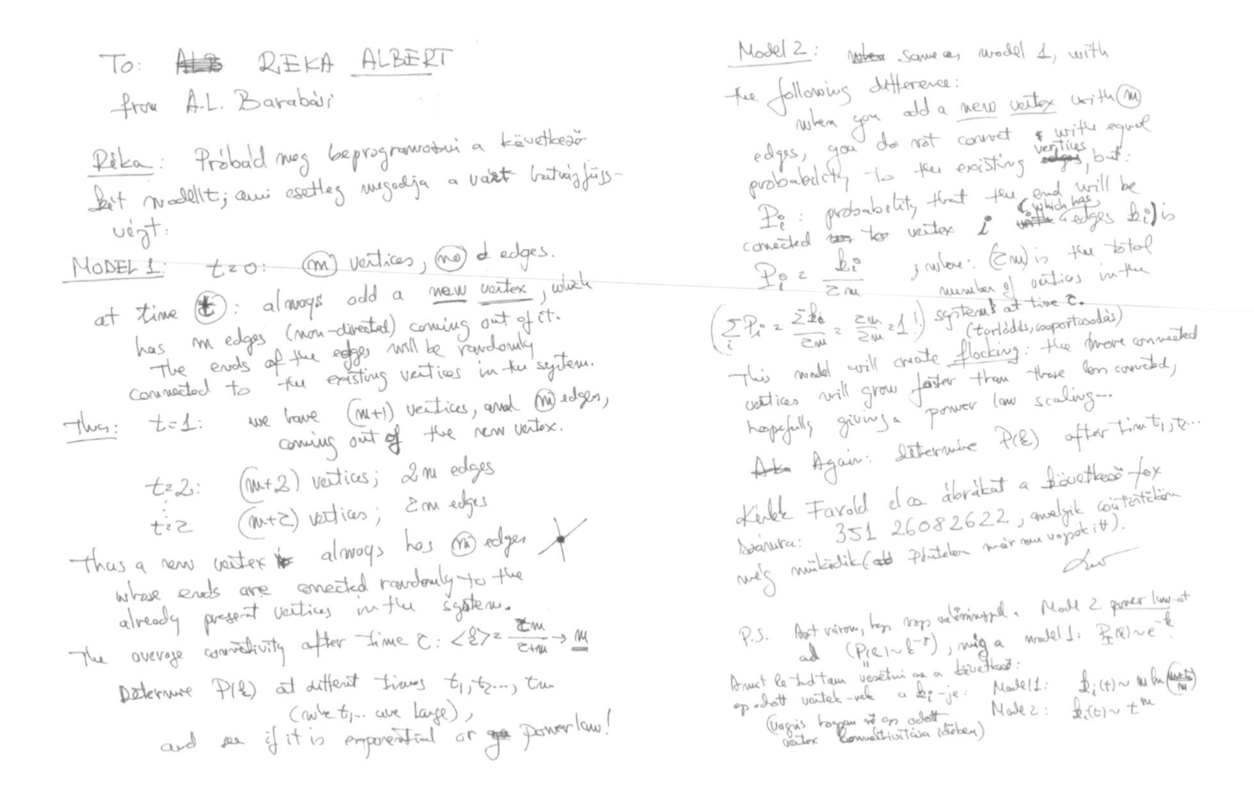

To: RÉKA ALBERT
from A.L. Barabási

Réka: Próbáld meg beprogramozni a következő két modellt; ami esetleg megadja a várt hatványfüggvényt:

MODEL 1: $t=0$: m vertices, m_0 edges.

at time t: always add a new vertex, which has m edges (non-directed) coming out of it. The ends of the edges will be randomly connected to the existing vertices in the system.

Thus: $t=1$: we have $(m+1)$ vertices, and m edges, coming out of the new vertex.

$t=2$: $(m+2)$ vertices; $2m$ edges

$t=\tau$: $(m+\tau)$ vertices; τm edges

Thus a new vertex always has m edges whose ends are connected randomly to the already present vertices in the system.

The average connectivity after time τ: $\langle k \rangle = \frac{\tau m}{\tau + m} \to m$

Determine $P(k)$ at different times $t_1, t_2, \ldots, t_m$ (when $t_1 \ldots$ are large), and see if it is exponential or power law!

Model 2: same as model 1, with the following difference: when you add a new vertex with m edges, you do not connect with equal probability to the existing vertices, but:

P_i: probability that the end will be connected to vertex i (which has k_i edges) is

$P_i = \frac{k_i}{\tau m}$, where: (τm) is the total number of vertices in the systems at time τ.

$\left(\sum_i P_i = \frac{\sum k_i}{\tau m} = \frac{\tau m}{\tau m} = 1!\right)$

This model will create flocking (torlódás, csoportosodás): the more connected vertices will grow faster than those less connected, hopefully giving a power law scaling...

Again: determine $P(k)$ after time $t_1, t_2 \ldots$

Kérlek Faxold el az ábrákat a következő fax számra: 351 26082622, amelyik csütörtökön még működik (a pénteken már nem kapok itt).

Laci

P.S. Azt várom, hogy nagy valószínűséggel Model 2 power law-t ad $(P(k) \sim k^{-\gamma})$, míg a model 1: $P_I(k) \sim e^{-k}$.

Amit le tudtam vezetni az a következő: egy adott vertex-nek a k_i-je: Model 1: $k_i(t) \sim m \ln\left(\frac{m+t}{m}\right)$ Model 2: $k_i(t) \sim t^{m}$

(vagyis hogyan nő egy adott vertex konnektivitása időben)

Figure 0.7 **1999: Scale-Free Fax**
The fax sent to Réka Albert on June 14, 1999 from Porto, Portugal. In a mixture of Hungarian and English, the fax describes the algorithm called today the Barabási–Albert model, which explains the origins of scale-free networks and outlines the continuum theory to calculate the degree exponent (**Chapter 5**).

I did not get far, however. During the fifteen-minute walk from the university to the seminary I found the explanation. I made a few frantic calculations in my room to formulate the idea in mathematical terms and wrote a fax to Réka, asking her to perform numerical simulations to verify my quick conclusions (**Figure 0.7**).

A few hours later her answer arrived by email. To my great astonishment, it worked. A simple model, relying on only two ingredients, *growth* and *preferential attachment*, could explain the power laws we had spotted on the Web and Hollywood.

The Rush (1999)

After Portugal I had only seven days at Notre Dame before my month-long vacation in Transylvania. Yet I could not see myself sitting for a full month on the discovery, which meant that I had two more days in Portugal and seven in the USA to write the paper.

I wanted to get started right away. My girlfriend reminded me, however, of my promise that there would be no work during the last two days of the trip. We had planned to take a vacation in Lisbon. So I postponed the writing until the eight-hour flight from Lisbon to New York, joining her to explore the city instead. My brain could not free itself of networks, however. The paper was slowly taking shape in my mind as my feet mapped out the narrow streets of Santa Cruz.

Once the plane took off, I started typing frantically. I had just finished the manuscript's introduction when a flight attendant, handing a Coke to the passenger next to me, poured the contents of the glass over my keyboard. With that she ruined both my brand new laptop and my dreams of finishing the first draft of the paper on the plane.

I could not give up, however. So I ended up writing the manuscript on a pad that the remorseful flight attendant gave me following the accident. By the end of the week, the paper had been submitted to *Science*.

I was paranoid, however. At that point we had two manuscripts under submission. The first reported the discovery of scale-free networks in the context of the WWW, and was under review at *Nature*. The second, just submitted to *Science*, showed that the scale-free property was universal and proposed a theory to explain its origin. To be sure, *Nature* and *Science* were the most prestigious journals out there. They are equally infamous, however, for their huge rejection rate. With less than 10% of submissions published, the likelihood that both papers would be rejected was over 81%. More disturbing, the odds that *both* papers would be accepted was less than 1%.

My previous network paper had languished for two years before I gave up on it. What if these two manuscripts suffered a similar fate? And what if, in the meantime, someone else made the same discovery? The phenomenon was so robust and obvious that an independent discovery was rather plausible. I needed a backup plan.

There is an expectation in the physics community that short publications, like a *Nature*, *Science* or *Physical Review Letters* paper, should be followed by a "long paper," offering a detailed exposition of the results. So in those seven days Réka, Hawoong and I wrote the long paper as well. I called Gene Stanley, the Editor-in-Chief of *Physica A*, telling him that I would send him our hottest paper if he was willing to make a quick decision on it. I doubt that he believed me on the "hot" issue. He promised to act on it quickly, nevertheless.

It took only a few days to learn that my paranoia had been well founded. I had barely arrived in Csíkszereda, my hometown in Transylvania, when the rejection from *Science* arrived. Disappointed, but convinced that the paper was important, I did something that I had never done before: I called the editor who had rejected the paper in a desperate attempt to change his mind.

To my astonishment, I succeeded and he sent the paper for review. A few months later the 1% scenario came true: *Nature*, *Science* and *Physica A* all accepted our papers [9, 10, 11]. It was an unexpected but delightful payoff after five years of setbacks.

Leap of Faith (1999)

With four students and one postdoc, my research group was modest back then. All but Réka were working on surfaces and quantum dots. A few days after the *Science* paper was accepted I called a group meeting to make an announcement, one that promised to be shocking to several group members. I told them that I was quitting materials science. The reason was simple: I did not want to divide my time and attention any longer between my passion and the topic that paid the bills. So, with three years left on my tenure clock, I decided to switch fields, from quantum dots to networks. I offered each group member a choice: join me in this new journey, or leave.

Two students bailed ship. The rest followed me on this new, untested voyage.

Fail 4: Funding (1999)

If you enter an established field and do good work, it is only a matter of time before you get funding. Entering a nonexistent field, however, creates some puzzling difficulties.

The good news was that I had just been awarded a new grant by the National Science Foundation (NSF) to explore quantum dots. The bad news was that I had no interest in pursuing that topic any longer. Of course, I could have just let the funds pour in and pretended that I was carrying on with the subject. I was not comfortable with that option, however. So I called the NSF program manager, asking if I could use the funds to pursue networks instead.

He said no. I either did what I had promised, or I needed to return the money to NSF. This was a Catch 22: I had money to pursue an incremental project that I had completely lost interest in, but I had no funds to pursue a problem that was potentially transformative.

In the end I followed my dream and returned the funds. That, however, left my group in a precarious situation: we desperately needed a new grant but no funding agency considered networks a legitimate research subject. That is until I noticed a call by the Defense Advanced Research Projects Agency (DARPA) for technologies that would "allow the networks of the future to be resistant to attacks and continue to provide network services."

In hindsight it is obvious that the call was intended for networking experts, an active branch of computer science. I was convinced, however, that no one can build fault-tolerant networks without first understanding the network topology and its inherent vulnerabilities. The scale-free property we had just discovered, with its accompanying hubs, must impact any technology DARPA would develop. With the insights we had just reported in *Nature* and *Science* we could nail this, I decided, and immersed myself in proposal writing.

I felt, however, that we needed a "smoking gun": something that would convince the program manager of the key role that network topology plays in robustness. Therefore, Réka started to randomly remove nodes from a scale-free network, mimicking component failures. She compared the impact of these failures with an attack, when only the hubs were removed. The difference was dramatic: scale-free networks proved to be surprisingly resistant to random failures but shockingly sensitive to attacks. We quickly incorporated this discovery into the proposal, convinced that we had now demonstrated beyond doubt the key role that network topology plays in fault tolerance (**Figure 0.8**).

After submitting the proposal by the November 1st deadline, I suggested to Réka and Hawoong that the questions we had formulated in the proposal were too exciting to wait for the DARPA funding. We must pursue them right away. So, we

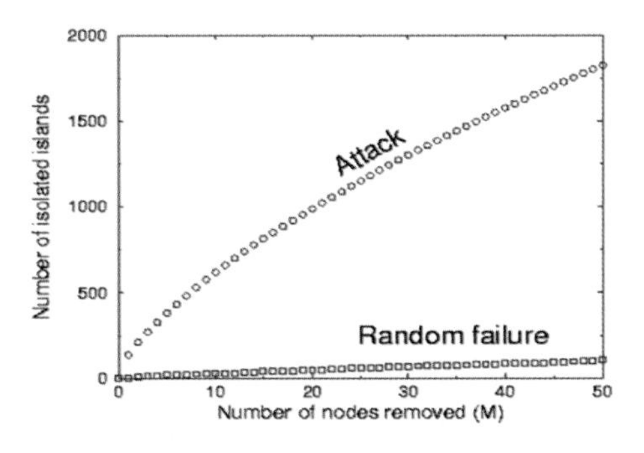

Figure 0.8 **1999: Funding Robustness**
A figure from the proposal submitted to DARPA on November 1, 1999, illustrating the impact of the network topology on a network's error and attack tolerance. The original caption foreshadows our error-tolerance paper published a year later in *Nature* [12]:

"The effect of attacks or accidental failures on the connectivity of power networks. We constructed a power network with 40,000 nodes such that the probability that a node has k links follows $P(k) \sim k^{-3}$. Attacks typically target the most connected nodes in the system.

To investigate their effect, we removed from the network the M nodes with the largest number of links. The upper curve shows the number of isolated islands as a function of M, indicating for example that the removal of only 10 of the largest nodes broke the system into 500 islands, which do not communicate with each other. Random failures uniformly affect any node in the system, and since most nodes have only a few links, failures rarely have a large effect. Indeed, removing randomly M nodes breaks the system only into a few clusters, leaving the overall connectivity practically unchanged."

Figure 0.9 **2000: Achilles' Heel**
The cover of the July 27, 2000 issue of *Nature*, highlighting our paper "Attack and error tolerance of complex networks" inspired by the (unsuccessful) DARPA proposal [12].

expanded our findings and sent the manuscript to *Science*, only to be rejected without review again.

I called the editor once more, to be told that in his view the paper offered few advances over our previous *Science* paper. I was flabbergasted, but failed to convince him otherwise. So we resubmitted it to *Nature*.

A few months later DARPA rejected our proposal. *Nature*, however, accepted the paper and put it on its cover (**Figure 0.9**) [12].

Fail 5: "Comically Wrong"

Each time there is a major scientific discovery, some researchers will feel that their mission in life is to restore the balance of the Universe by doing everything in their power to erase the topic from the face of the Earth. If network science were to live up to its transformative power, it had to have its own nemesis.

"Comically wrong" was the phrase John Doyle, a control theorist from Caltech and a self-proclaimed expert on networks, used in one of his numerous interviews that dotted his decade-long crusade against network science. The small-world property was surprising at first, but it was easy to derive and had decades of research backing it. The scale-free property was a different story altogether, raising a host of questions that originally we could not answer. If it was so universal, how did we miss it for decades? Aren't growth and preferential attachment just one of many potential explanations? We had known about power laws since Pareto, why was this any different? Béla Bollobás was even more blunt, telling me during our first meeting in the Buda Castle that absent an exact mathematical proof, the scale-free property "does not exist."

These were legitimate questions with consequences: only researchers with a strong mathematical training could build on the scale-free concept. The resulting vacuum of understanding gave room for confusion and misinformation. This void was filled by the bombastic statements John Doyle made each time a journalist waved a mike in front of him.

Then, slowly, the tide turned. José Mendes, Sergey Dorogovtsev and Sid Redner used a rate-equation approach to put the continuum theory of scale-free networks on a firm mathematical footing [13,14]. Béla Bollobás and several colleagues, in a landmark paper, offered an exact proof of the scale-free property [15]. Shlomo Havlin and his students connected robustness with percolation theory [16], and Bollobás and Oliver Riordan weighed in with an exact proof [17]. A string of discoveries started to document how deeply the scale-free property alters a network's behavior, like Romualdo Pastor-Satorras and Alessandro Vespignani's classic result on the disappearance of the epidemic threshold [18]. With that the community started to appreciate the central role the degree distribution plays in networks. As we will see throughout this textbook, virtually all network characteristics, from six degrees to robustness and community structure, must be interpreted in this context. With hundreds of researchers attracted to networks by the many fundamental questions the field poses, gradually network science has taken shape.

Summary

One could easily view the events recounted above as a string of successes. In the next decade our 1999 *Science* paper on scale-free networks became the most-cited paper in physical sciences. The 2000 *Nature* paper on error and attack tolerance not only made the journal's cover [12], but also had a profound impact on our understanding of network robustness. Réka and I spent the subsequent year writing a review on networks, which formalized the field's intellectual foundations, eventually becoming the most-cited paper in *Review of Modern Physics* [19]. Then the National Research Council report of 2005, published by the US National Academies, coined the term *network science* and persuaded the US government to spend hundreds of millions of dollars to support this new research field as a new and separate discipline. Eventually the most venerable scientific publishers, like Cambridge University Press, Oxford University Press and Springer, and the top engineering society, IEEE, launched journals to cover the field's advances. By all measures a new discipline was born, supported by a vibrant interdisciplinary community.

Science rarely follows a straight path to success (**Figure 0.10**). New ideas require years of gestation. One could see the theory of scale-free networks as an exception, a spark that took only

Figure 0.10 **Different Paths to Success**
A cartoonist's take on success that vividly captures the convoluted path, punctuated by failures and dead ends, that characterized my early research in networks.

ten days from idea to paper submission. Yet had it not been preceded by five years of apparently fruitless work on the problem, the spark could never have started a fire.

Network science offers a reminder of the important role collaboration and mentorship plays in science. Before Réka and Hawoong joined me on this journey, all I could produce was a string of ideas and failures. Without Hawoong's skill in mapping out the WWW we would never have discovered the scale-free property. Réka's ability to roll the math was essential in developing the theory behind the scale-free model. Our subsequent research on biological networks would never have happened if Zoltán Oltvai, a medical doctor and researcher at Northwestern University, had not convinced us to apply networks to cell biology, patiently guiding us through the maze of proteins and metabolites [20, 21, 22]. These were not the work of one individual, but truly joint discoveries.

Today, many fields consider network science their own. Mathematicians rightly claim ownership and priority through graph theory; the exploration of social networks by sociologists goes back decades; physics lent the universality concept and infused many analytical tools that are now unavoidable in the study of networks; biology invested hundreds of millions of dollars into mapping subcellular networks; computer science offered an algorithmic perspective, allowing us to explore very large networks; engineering invested considerable effort in the exploration of infrastructural networks. It is remarkable how these many disparate pieces managed to fit together, giving birth to a new discipline.

This textbook is a testimony to the amazing progress that this community has achieved on this fascinating journey. The continued success of network science depends on our ability to maintain its multi-disciplinary nature, allowing each scientist to infuse a unique perspective. This clash of ideas and viewpoints remains the field's strength and its intellectual engine.

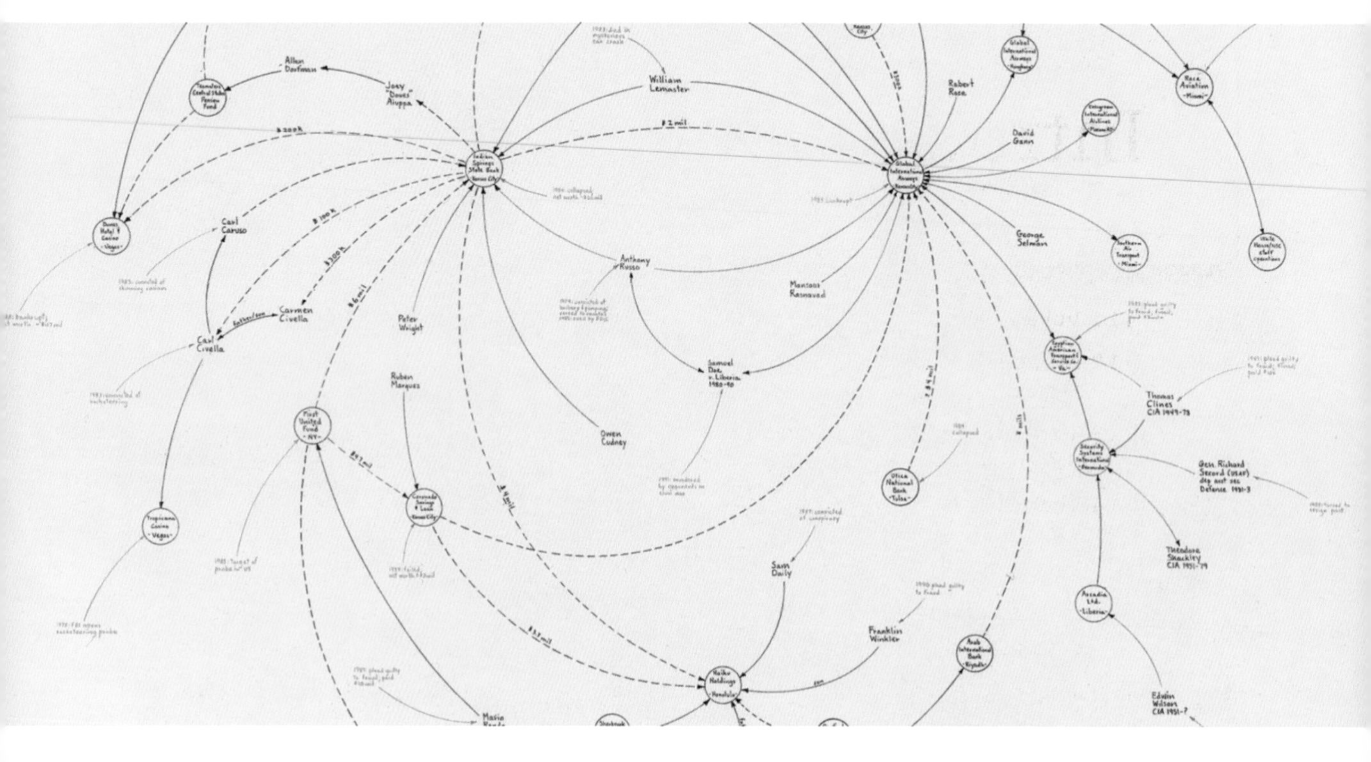

Figure 1.0 **Art & Networks: Mark Lombardi**
Mark Lombardi (1951–2000) was an American artist who documented "the uses and abuses of power." His work was preceded by careful research, resulting in thousands of index cards, whose number began to overwhelm his ability to deal with them. Hence Lombardi began assembling them into hand-drawn diagrams, intended to focus his work. Eventually these diagrams became a form of art on their own [23]. The image shows one such drawing, titled *Global International Airway and Indian Spring State Bank*, created between 1977 and 1983 in colored pencil and graphite on paper.

CHAPTER 1

Introduction

1.1 Vulnerability Due to Interconnectivity

At first glance the two satellite images of **Figure 1.1** are indistinguishable, showing lights shining brightly in highly populated areas and dark spaces that mark vast uninhabited forests and oceans. Yet, upon closer inspection, we notice differences: Toronto, Detroit, Cleveland, Columbus and Long Island, bright and shining in **Figure 1.1a**, have gone dark in **Figure 1.1b**. This is not a doctored shot from the next *Armageddon* movie but represents a real image of the US Northeast on August 14, 2003, before and after the blackout that left an estimated 45 million people without power in eight US states and another 10 million in Ontario.

The 2003 blackout is a typical example of a cascading failure. When a network acts as a transportation system, a local failure shifts loads to other nodes. If the extra load is negligible, the system can seamlessly absorb it and the failure goes unnoticed. If, however, the extra load is too much for the neighboring nodes, they too will tip and redistribute the load to their neighbors. In no time, we are faced with a cascading event,

(a)

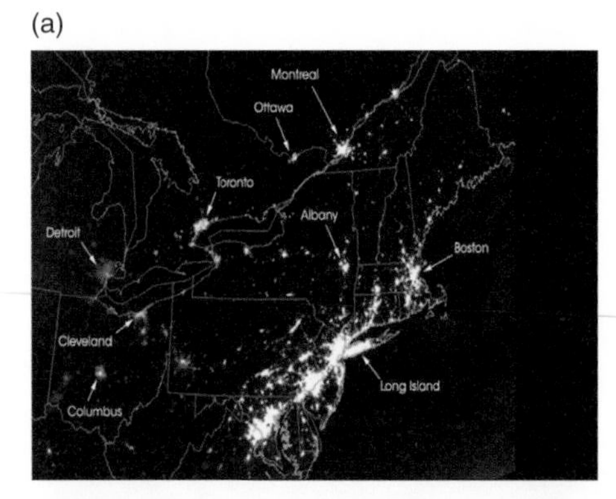

(b)

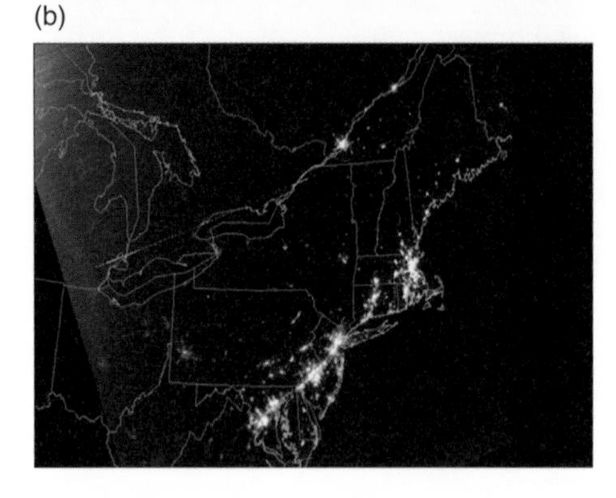

Figure 1.1 **2003 North American Blackout**
(**a**) Satellite image of the US Northeast on August 13, 2003 at 9: 29pm (EDT), 20 hours *before* the blackout.
(**b**) Similarly, but 5 hours *after* the blackout.

whose magnitude depends on the position and the capacity of the nodes that failed initially.

Cascading failures have been observed in many complex systems. They take place on the Internet, when traffic is rerouted to bypass malfunctioning routers. This routine operation can occasionally create denial-of-service attacks, which make fully functional routers unavailable by overwhelming them with traffic. We witness cascading events in financial systems, like in 1997, when the International Monetary Fund pressured the central banks of several Pacific nations to limit their credit, which defaulted multiple corporations, eventually resulting in stockmarket crashes worldwide. The 2009–2011 financial meltdown is often seen as a classic example of a cascading failure, the US credit crisis paralyzing the economy of the globe, leaving behind scores of failed banks, corporations and even bankrupt states. Cascading failures can also be induced artificially. An example is the worldwide effort to dry up the money supply of terrorist organizations, aimed at crippling their ability to function. Similarly, cancer researchers aim to induce cascading failures in our cells to kill cancer cells.

The US Northeast blackout illustrates several important themes of this book. First, to avoid damaging cascades, we must understand the structure of the network on which the cascade propagates. Second, we must be able to model the dynamical processes taking place on these networks, like the flow of electricity. Finally, we need to uncover how the interplay between the network structure and dynamics affects the robustness of the whole system. Although cascading failures may appear random and unpredictable, they follow reproducible laws that can be quantified and even predicted using the tools of network science.

The blackout also illustrates a bigger theme: *vulnerability due to interconnectivity*. Indeed, in the early years of electric power each city had its own generators and electric network. Electricity cannot be stored, however. Once produced, electricity must be consumed immediately. It made economic sense, therefore, to link neighboring cities up, allowing them to share the extra production and borrow electricity if needed. We owe the low price of electricity today to the power grid, the network that emerged through these pairwise connections, linking all producers and consumers in a single network. It allows cheaply produced power to be transported anywhere instantly.

Electricity hence offers a wonderful example of the huge positive impact networks have on our lives.

Being part of a network has its catches, however. Local failures, like the breaking of a fuse somewhere in Ohio, may not stay local any longer. Their impact can travel along the network's links and affect other nodes, consumers and individuals – apparently removed from the original problem. In general, interconnectivity induces a remarkable non-locality: it allows information, memes, business practices, power, energy and viruses to spread on their respective social or technological networks, reaching us no matter what our distance from the source. Hence, networks carry both benefits and vulnerabilities. Uncovering the factors that can enhance the spread of traits deemed positive, and limit others that make networks weak or vulnerable, is one of the goals of this book.

1.2 Networks at the Heart of Complex Systems

"I think the next century will be the century of complexity."

Stephen Hawking

We are surrounded by systems that are hopelessly complicated. Consider, for example, the society that requires cooperation between billions of individuals, or communications infrastructures that integrate billions of cell phones with computers and satellites. Our ability to reason and comprehend our world requires the coherent activity of billions of neurons in our brain. Our biological existence is rooted in seamless interactions between thousands of genes and metabolites within our cells.

These systems are collectively called *complex systems* (**Box 1.1**), capturing the fact that it is difficult to derive their collective behavior from a knowledge of the system's components. Given the important role complex systems play in our daily life, in science and in the economy, their understanding, mathematical description, prediction and eventually control is one of the major intellectual and scientific challenges of the twenty-first century.

The emergence of network science at the dawn of the twenty-first century is a vivid demonstration that science can live up to this challenge. Indeed, behind each complex system

Box 1.1 Complex

[adj., v. kuh m-pleks, kom-pleks; n. kom-pleks]

(1) composed of many interconnected parts; compound; composite: a complex highway system
(2) characterized by a very complicated or involved arrangement of parts, units, etc.: complex machinery
(3) so complicated or intricate as to be hard to understand or deal with: a complex problem

Source: *Dictionary.com*

there is an intricate network that encodes the interactions between the system's components:

(a) The network encoding the interactions between genes, proteins and metabolites integrates these components into live cells. The very existence of this *cellular network* is a prerequisite of life.
(b) The wiring diagram capturing the connections between neurons, called the *neural network*, holds the key to our understanding of how the brain functions and to our consciousness.
(c) The sum of all professional, friendship and family ties, often called the *social network*, is the fabric of society and determines the spread of knowledge, behavior and resources.
(d) *Communication networks*, describing which communication devices interact with each other, through wired Internet connections or wireless links, are at the heart of the modern communication system.
(e) The *power grid*, a network of generators and transmission lines, supplies virtually all modern technology with energy.
(f) *Trade networks* maintain our ability to exchange goods and services, being responsible for the material prosperity that the world has enjoyed since World War II (**Figure 1.2**).

Figure 1.2 **Subtle Networks Behind the Economy**
A credit card selected as the 99th object in the *History of the World in 100 Objects* exhibit at the British Museum. This card is a vivid demonstration of the highly interconnected nature of the modern economy, relying on subtle economic and social connections that normally go unnoticed.

The card was issued in the United Arab Emirates in 2009 by the Hong Kong and Shanghai Banking Corporation, known as HSBC, a London-based bank. The card functions through protocols provided by VISA, a US-based credit association. Yet the card adheres to Islamic banking principles, operating in accordance with Fiqhal-Muamalat (Islamic rules of transactions), most notably eliminating interest or *riba*. The card is not limited to Muslims in the United Arab Emirates, but is offered in non-Muslim countries as well, to anyone who agrees with its strict ethical guidelines.

Networks are also at the heart of some of the most revolutionary technologies of the twenty-first century, empowering everything from Google to Facebook, Cisco and Twitter. In the end, networks permeate science, technology, business and nature to a much higher degree than may be evident on casual inspection. Consequently, *we will never understand complex systems unless we develop a deep understanding of the networks behind them.*

The exploding interest in network science during the first decade of the twenty-first century is rooted in the discovery that despite the obvious diversity of complex systems, the structure and the evolution of the networks behind each system is driven by a common set of fundamental laws and principles. Therefore, notwithstanding the amazing differences in form, size, nature, age and scope of real networks, most networks are driven by common organizing principles. Once we disregard the nature of the components and the precise nature of the interactions between them, the networks obtained are more similar to than different from each other. In the following sections we discuss the forces that have led to the emergence of this new research field and its impact on science, technology and society.

1.3 Two Forces that Helped Network Science

Network science is a new discipline. One may debate its precise beginning, but by all accounts the field has emerged as a separate discipline only in the twenty-first century.

Why didn't we have network science two hundred years earlier? After all, many of the networks that the field explores are by no means new: metabolic networks date back to the origins of life, with a history of four billion years, and the social network is as old as humanity. Furthermore, many disciplines, from biochemistry to sociology and brain science, have been dealing with their own networks for decades. Graph theory, a prolific subfield of mathematics, has explored graphs since 1735. Is there a reason, therefore, to call network science *the science of the twenty-first century*?

Something special happened at the dawn of the twenty-first century that transcended individual research fields and catalyzed the emergence of a new discipline (**Figure 1.3**). To understand why this happened now and not two hundred years earlier, we need to discuss the two forces that have contributed to the emergence of network science.

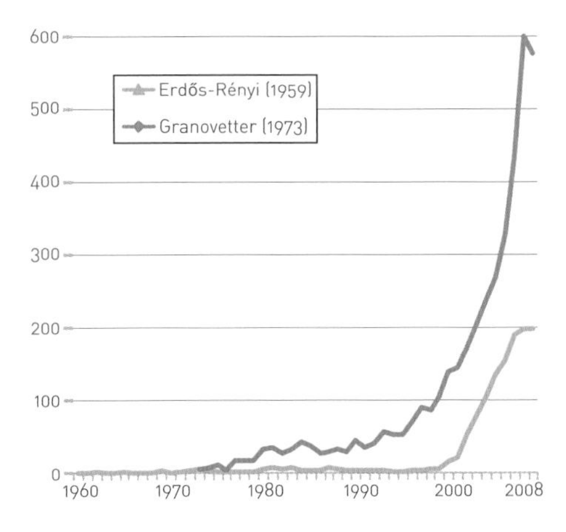

Figure 1.3 **The Emergence of Network Science**

While the study of networks has a long history, with roots in graph theory and sociology, the modern chapter of network science emerged only during the first decade of the twenty-first century.

The explosive interest in networks is well documented by the citation pattern of two classic papers, the 1959 paper by Paul Erdős and Alfréd Rényi that marks the beginning of the study of random networks in graph theory [3] and the 1973 paper by Mark Granovetter, the most-cited social network paper [24]. The figure shows the yearly citations each paper has acquired since their publication. Both papers were highly regarded within their discipline, but had only limited impact outside their field. The explosive growth in citations of these papers in the twenty-first century is a consequence of the emergence of network science, drawing new interdisciplinary attention to these classic publications.

1.3.1 The Emergence of Network Maps

To describe the detailed behavior of a system consisting of hundreds to billions of interacting components, we need a map of the system's wiring diagram. In a social system this would require an accurate list of your friends, your friends' friends, and so on. In the WWW this map tells us which webpages link to each other. In a cell the map corresponds to a detailed list of binding interactions and chemical reactions involving genes, proteins and metabolites.

> **Box 1.2 The Origins of Network Maps**
>
> A few of the maps studied today by network scientists were generated with the purpose of studying networks. Most are the byproduct of other projects and morphed into maps only in the hands of network scientists.
>
> (a) The lists of chemical reactions in a cell were discovered – one by one – over a 150-year period by biochemists. In the 1990s they were collected in central databases, offering the first chance to assemble the biochemical networks within a cell.
> (b) The lists of actors playing in any given movie were traditionally scattered over newspapers, books and encyclopedias. With the advent of the Internet, these data were assembled into central databases, like imdb.com, feeding the curiosity of movie aficionados. This database allowed network scientists to reconstruct the affiliation network behind Hollywood.
> (c) The lists of authors of millions of research papers were traditionally scattered in the tables of content of thousands of journals. Recently, Web of Science, Google Scholar and other services have assembled them into comprehensive databases, allowing network scientists to reconstruct accurate maps of scientific collaboration networks.
>
> Much of the early history of network science relied on the investigators' ingenuity to recognize and extract networks from pre-existing databases. Network science changed that: today, well-funded research collaborations focus on map making, capturing accurate wiring diagrams of biological, communication and social systems.

In the past, we lacked the tools to map these networks. It was equally difficult to keep track of the huge amount of data behind them. The digital revolution, offering effective and fast data-sharing methods and cheap digital storage, fundamentally changed our ability to collect, assemble, share and analyze data pertaining to real networks.

Thanks to these technological advances, at the turn of the millennium we witnessed an explosion of map making (**Box 1.2**). Examples range from the CAIDA or DIMES projects that offered the first large-scale maps of the Internet; to the hundreds of millions of dollars spent by biologists to experimentally map out protein–protein interactions in human cells; the efforts made by social network companies, like Facebook, Twitter or LinkedIn, to develop accurate depositories of our friendships and professional ties; and the Connectome project of the US National Institute of Health that aims to systematically trace the neural connections in mammalian brains. The sudden availability of these maps at the end of the twentieth century has catalyzed the emergence of network science.

1.3.2 The Universality of Network Characteristics

It is easy to list the differences between the various networks we encounter in nature or society: the nodes of the metabolic network are tiny molecules and the links are chemical reactions governed by the laws of chemistry and quantum mechanics; the nodes of the WWW are web documents and the links are URLs guaranteed by computer algorithms; the nodes of the social network are individuals and the links represent family, professional, friendship and acquaintance ties.

The processes that generated these networks also differ greatly: metabolic networks were shaped by billions of years of evolution; the WWW is built on the collective actions of millions of individuals and organizations; social networks are shaped by social norms whose roots go back thousands of years. Given this diversity in size, nature, scope, history and evolution, one would not be surprised if the networks behind these systems differed greatly.

A key discovery of network science is that *the architectures of networks emerging in various domains of science, nature and technology are similar to each other, a consequence of being governed by the same organizing principles. Consequently, we can use a common set of mathematical tools to explore these systems.*

This universality is one of the guiding principles of this book: we will not only seek to uncover specific network properties, but each time ask how widely they apply. We will also aim to understand their origins, uncovering the laws that shape network evolution and their consequences on network behavior.

In summary, while many disciplines have made important contributions to network science, the emergence of a new field was partly made possible by data availability, offering accurate maps of networks encountered in different disciplines. These diverse maps allowed network scientists to identify the universal properties of various network characteristics. This universality offers the foundation of the new discipline of network science.

1.4 The Characteristics of Network Science

Network science is defined not only by its subject matter, but also by its methodology. In this section we discuss the key characteristics of the approach network science adopts to understand complex systems.

1.4.1 Interdisciplinary Nature

Network science offers a language through which different disciplines can seamlessly interact with each other. Indeed, cell biologists, brain scientists (**Figure 1.4**) and computer scientists alike are faced with the task of characterizing the wiring diagram behind their system, extracting information from incomplete and noisy datasets and understanding their systems' robustness to failures or attacks.

To be sure, each discipline brings a different set of goals, technical details and challenges, which are important on their own. Yet, the common nature of many issues these fields struggle with has led to a cross-disciplinary fertilization of tools and ideas. For example, the concept of betweenness centrality that emerged in the social network literature of the 1970s today plays a key role in identifying high-traffic nodes on the

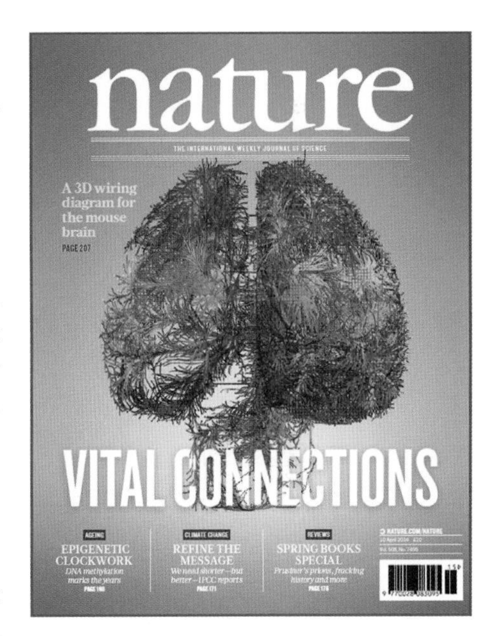

Figure 1.4 **Mapping the Brain**
An exploding application area for network science is brain research. The wiring diagram of a complete nervous system has long been available for *C. elegans*, a small roundworm, but neuronal connectivity data for larger animals has been missing until recently. That is changing, thanks to major efforts by the scientific community to develop technologies that can map out the brain's wiring diagram. The image shows the cover of the April 10, 2014 issue of *Nature*, reporting an extensive map of the laboratory mouse [25] generated by researchers at the Allen Institute in Seattle.

Internet. Similarly, algorithms developed by computer scientists for graph partitioning have found novel applications in identifying disease modules in medicine or detecting communities within large social networks.

1.4.2 Empirical, Data-Driven Nature

Several key concepts of network science have their roots in graph theory, a fertile field of mathematics. What distinguishes network science from graph theory is its empirical nature, that is its focus on data, function and utility. As we will see in the coming chapters, in network science we are never satisfied with developing abstract mathematical tools to describe a certain network property. Each tool we develop is tested on real data and its value is judged by the insights it offers about a system's properties and behavior.

1.4.3 Quantitative and Mathematical Nature

To contribute to the development of network science and to properly use its tools, it is essential to master the mathematical formalism behind it. Network science borrowed the formalism to deal with graphs from graph theory and the conceptual framework to deal with randomness and seek universal organizing principles from statistical physics. Lately, the field has benefited from concepts borrowed from engineering (like control and information theory), allowing us to understand the control principles of networks and from statistics, helping us to extract information from incomplete and noisy datasets.

The development of network analysis software has made the tools of network science available to a wider community, even those who may not be familiar with the intellectual foundations and the full mathematical depths of the discipline. Yet, to further the field and to efficiently use its tools, we need to master its theoretical formalism.

1.4.4 Computational Nature

Given the size of many of the networks of practical interest, and the exceptional amount of auxiliary data behind them, network scientists are regularly confronted by a series of formidable computational challenges. Hence, the field has a strong computational character, actively borrowing from algorithms, database management and data mining. A series of software tools are available to address these computational problems, enabling practitioners with diverse computational skills to analyze the networks of interest to them.

In summary, a mastery of network science requires familiarity with each of these aspects of the field. It is their combination that offers the multi-faceted tools and perspectives necessary to understand the properties of real networks.

1.5 Societal Impact

The impact of a new research field is measured both by its intellectual achievements as well as by its societal impact, indicated by the reach and the potential of its applications. While network science is a young field, its impact is everywhere.

1.5.1 Economic Impact: From Web Search to Social Networking

The most successful companies of the twenty-first century, from Google to Facebook, Twitter, LinkedIn, Cisco, Apple and Akamai, base their technology and business model on networks. Indeed, Google not only runs the biggest network mapping operation that humanity has ever built, generating a comprehensive and constantly updated map of the WWW, but its search technology is deeply interlinked with the network characteristics of the Web.

Networks have gained particular popularity with the emergence of Facebook, the company with the ambition to map out the social network of the whole planet. Facebook was not the first social networking site and it will likely not be the last either: an impressive ecosystem of social networking tools, from Twitter to LinkedIn, are fighting for the attention of millions of users. Algorithms conceived by network scientists fuel these sites, aiding everything from friend recommendation to advertising.

1.5.2 Health: From Drug Design to Metabolic Engineering

Completed in 2001, the human genome project offered the first comprehensive list of all human genes [26, 27]. Yet, to fully

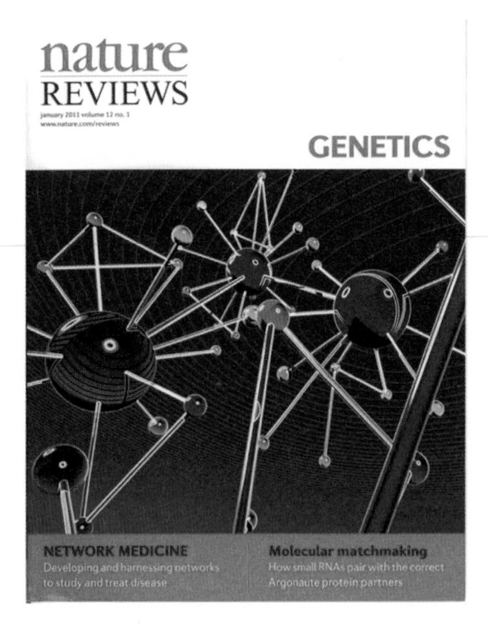

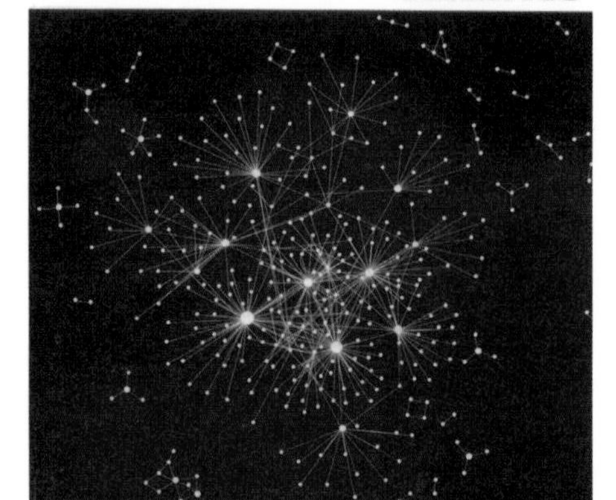

Figure 1.5 **Network Biology and Medicine**
The cover of two issues of *Nature Reviews Genetics*, the leading review journal in genetics. The journal has devoted exceptional attention to the impact of networks: the 2004 cover focuses on *network biology* [22] (top), the 2011 cover is inspired by *network medicine* [29] (bottom).

understand how our cells function, and the origin of disease, a full list of genes is not sufficient: we also need an accurate map of how genes, proteins, metabolites and other cellular components interact with each other. Indeed, most cellular processes, from food processing to sensing changes in the environment, rely on molecular networks. The breakdown of these networks is responsible for human diseases.

The increasing awareness of the importance of molecular networks has led to the emergence of *network biology*, a new subfield of biology that aims to understand the behavior of cellular networks. A parallel movement within medicine, called *network medicine*, aims to uncover the role of networks in human disease (**Figure 1.5**). The importance of these advances is illustrated by the fact that Harvard University in 2012 started its Division of Network Medicine, which employs researchers and medical doctors who apply network-based ideas to understanding human disease.

Networks play a particularly important role in drug development. The ultimate goal of *network pharmacology* [28] is to develop drugs that can cure diseases without significant side-effects. This goal is pursued at many levels, from the millions of dollars invested in mapping out cellular networks to the development of tools and databases to store, curate and analyze patient and genetic data.

Several new companies take advantage of the opportunities offered by networks for health and medicine. For example, GeneGo collects maps of cellular interactions from the scientific literature and Genomatica uses the predictive power behind metabolic networks to identify drug targets in bacteria and humans. Recently, major pharmaceutical companies, like Johnson & Johnson, have made significant investments in network medicine, seeing it as a path toward future drugs.

1.5.3 Security: Fighting Terrorism

Terrorism is a malady of the twenty-first century requiring significant resources to combat it worldwide. Network thinking is increasingly present in the arsenal of various law-enforcement agencies in charge of responding to terrorist activities. It is used to disrupt the financial network of terrorist

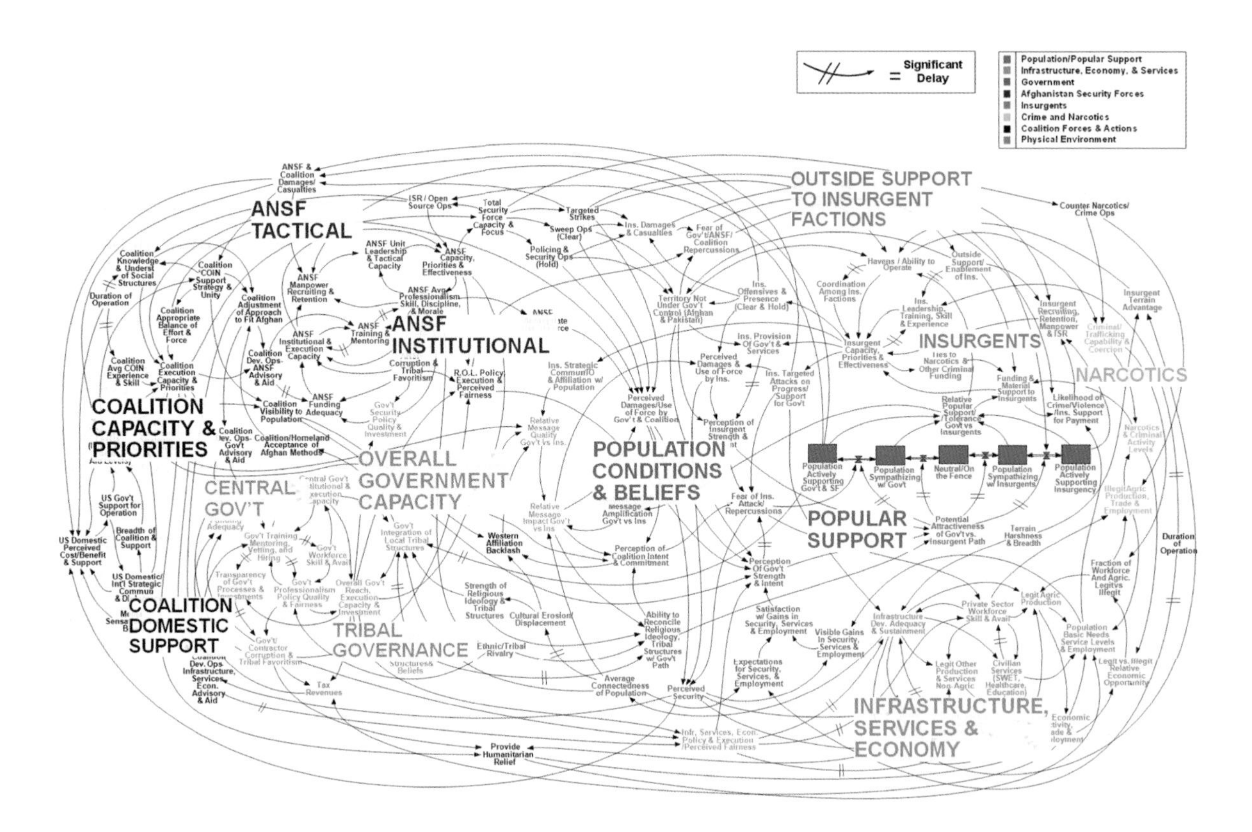

Figure 1.6 **The Network Behind a Military Engagement**
This diagram was designed during the Afghan war in 2012 to portray the American operational plans in Afghanistan. While it has been ridiculed in the press for displaying too much complexity and detail in one chart, it vividly illustrates the interconnected nature of a modern military engagement. Today this example is studied by officers and military students to demonstrate the power and utility of network models for decision making and operational coordination. Indeed, the job of military generals is not limited to ensuring the necessary military capacities, but must also factor in the beliefs and the living conditions of the local population or the impact of the narcotics trade that finances the operations of the insurgents. Image from the *New York Times*.

organizations and to map adversarial networks, helping to uncover the roles of their members and their capabilities. While much of the work in this area is classified, several well-documented case studies have been made public. Examples include the use of social networks to find Saddam Hussein [30] or those responsible for the March 11, 2004 Madrid train bombings through examination of the mobile call network. Network concepts have impacted military doctrine as well, leading to the concept of *network-centric warfare*, aimed at fighting low-intensity conflicts against terrorist and criminal networks that employ decentralized flexible network organization [31] (**Figure 1.6**).

Given the numerous potential military applications, it is perhaps not surprising that one of the first academic programs in network science was started at West Point, the US Army Military Academy. Furthermore, starting in 2009 the Army Research Lab devoted over $300 million to support network science centers across the USA.

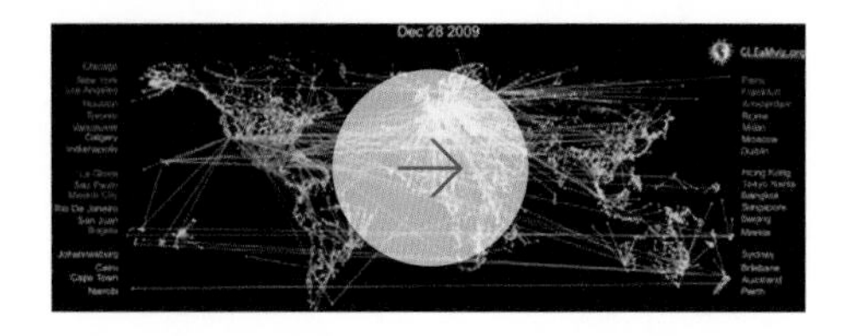

Online Resource 1.1

Predicting the H1N1 Epidemic

The predicted spread of the H1N1 pandemic during 2009, representing the first successful real-time prediction of a pandemic [33]. The project, relying on data describing the structure and the dynamics of the worldwide transportation network, foresaw that H1N1 would peak out in October 2009, in contrast with the expected January–February peak of influenza. This meant that the vaccines timed for November 2009 were too late, eventually having little impact on the outcome of the epdemic. The success of this project shows the power of network science in facilitating advances in areas of key importance for humanity.

The knowledge and the capabilities offered by networks can also be abused. Such misuses were well illustrated by the indiscriminate network mapping operation by the National Security Agency (NSA) [32]. Under the pretext of stopping future terrorist attacks, the NSA monitored the communications of hundreds of millions of individuals, from the USA and abroad, rebuilding their social networks. With that, network scientists have awoken to a new social responsibility: to ensure the ethical use of our tools and knowledge.

1.5.4 Epidemics: From Forecasting to Halting Deadly Viruses

While the H1N1 pandemic was not as devastating as feared at the beginning of the outbreak in 2009, it gained a special role in the history of epidemics: it was the first pandemic whose course and time evolution was accurately predicted months before the pandemic reached its peak (**Online resource 1.1**) [33]. This was possible thanks to fundamental advances in understanding the role of transportation networks in the spread of viruses.

Before 2000, epidemic modeling was dominated by compartment-based models, assuming that everyone can infect everyone else in the same socio-physical compartment. The emergence of a network-based framework has brought a fundamental change, offering a new level of predictability. Today, epidemic prediction is one of the most active applications of network science [33, 34], being used to foresee the spread of influenza or to contain Ebola. It is also the source of several fundamental results covered in this book, allowing us to model and predict the spread of biological, digital and social viruses (memes).

The impact of these advances is felt beyond epidemiology. Indeed, in January 2010 network science tools predicted the conditions necessary for the emergence of viruses spreading through mobile phones [35]. The first major mobile epidemic outbreak, which started in the fall of 2010 in China, infected over 300,000 phones each day and closely followed the predicted scenario.

1.5.5 Neuroscience: Mapping the Brain

The human brain, consisting of hundreds of billions of interlinked neurons, is one of the least-understood networks from the perspective of network science. The reason is simple: we lack maps telling us which neurons are linked together. The only fully mapped brain available for research is that of the *C. elegans* worm, consisting of only 302 neurons. Detailed maps of mammalian brains could lead to a revolution in brain science, allowing us to understand and cure numerous neurological and brain diseases. With that, brain research could turn into one of the most prolific application areas of network science [36]. Driven by the potential transformative impact of such maps, in 2010 the National Institutes of Health in the USA initiated the *Connectome* project, aimed at developing technologies that could provide accurate neuron-level maps of mammalian brains (**Figure 1.4**).

1.5.6 Management: Uncovering the Internal Structure of an Organization

While management tends to rely on the official chain of command, it is increasingly evident that the informal network, capturing who really communicates with whom, plays the most important role in the success of an organization. Accurate maps of such *organizational networks* can expose the potential lack of interactions between key units, help identify individuals who play an important role in bringing different departments and products together, and help higher management diagnose diverse organizational issues. Furthermore, there is increasing evidence in the management literature that the productivity of an employee is determined by his/her position in this informal organizational network [37].

Therefore, numerous companies, like Maven 7, Activate Networks or Orgnet, offer tools and methodologies to map out the true structure of an organization. These companies offer a host of services, from identifying opinion leaders to reducing employee churn, optimizing knowledge and product diffusion and designing teams with the diversity, size and expertise to be the most effective for specific tasks (**Figure 1.7**). Established firms, from IBM to SAP, have added social networking capabilities to their business. Overall, network science tools are indispensable in management and business, enhancing productivity and boosting innovation within an organization.

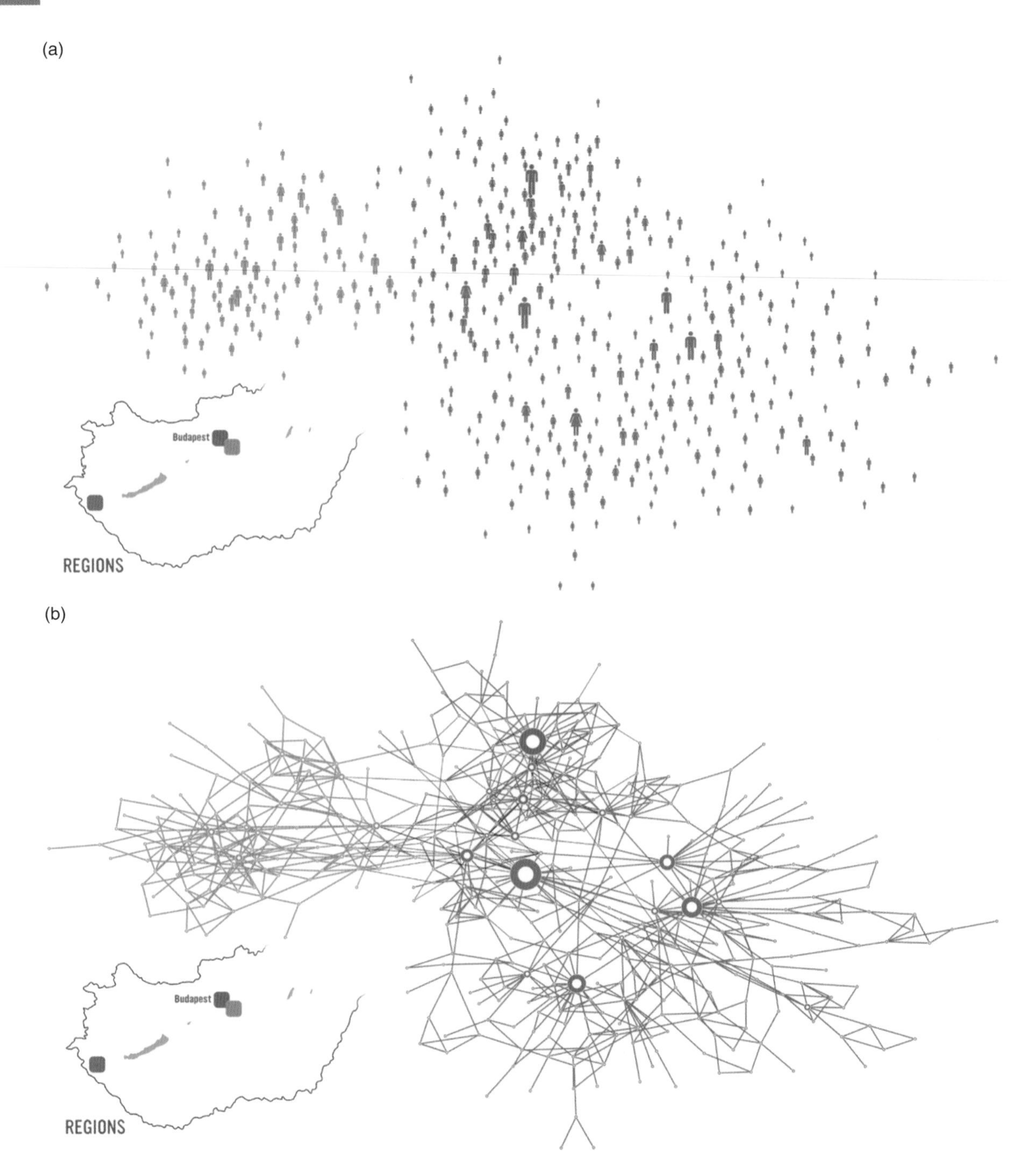

Figure 1.7 **Mapping Organizations**

(a) Employees of a Hungarian company with three main locations (purple, yellow and blue). The management realized that information reaching the workers about the intentions of the higher management often had nothing to do with their real plans. Seeking to enhance information flow within the company, they turned to Maven 7, a company founded by the author that applies network science in an organizational setting.

(b) Maven 7 developed an online platform to ask each employee who they turned to for advice when it comes to decisions impacting the company. This platform provided the map shown, where two individuals are connected if one nominated the other as his/her source of information on organizational and professional issues. The map identifies several highly influential individuals, appearing as large hubs.

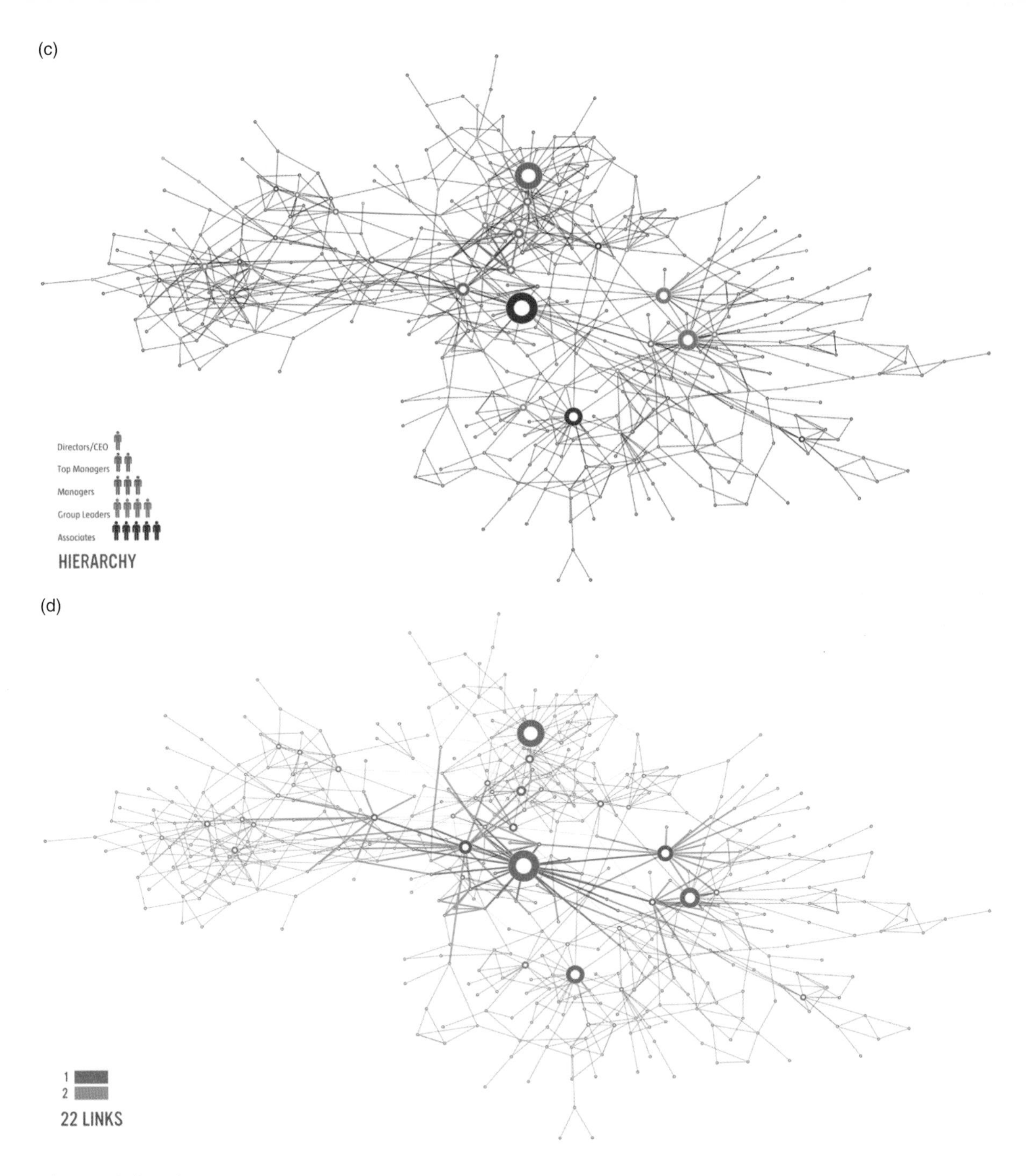

Figure 1.7 *(cont.)*
(c) The position of the leadership within the company's informal network, nodes being colored based on their rank within the company. Note that none of the directors, shown in red, are hubs. Nor are the top managers, shown in blue. The hubs come from lower ranks: they are managers, group leaders and associates. The biggest hub, hence the most influential individual, is an ordinary employee, appearing as a gray node.
(d) The links of the largest hub (red) and those two links away from this hub (orange) demonstrate that a significant fraction of employees are at most two links from this hub. But who is this hub? He is the employee in charge of safety and environmental issues. Hence he regularly visits each location and talks with the employees. He is connected to everyone, except the top management. With little knowledge of the true intentions of the management, he passes on information that he collects along his trail, effectively running a gossip center.

Should they fire or promote the biggest hub? What is the best solution to this problem?

Figure 1.8 **Complex Systems and Networks**
Special issue of *Science* magazine devoted to networks, published on July 24, 2009, on the 10th anniversary of the 1999 discovery of scale-free networks.

1.6 Scientific Impact

Nowhere is the impact of network science more evident than in the scientific community. The most prominent scientific journals, from *Nature* to *Science*, *Cell* and *PNAS*, have devoted reviews and editorials addressing the impact of networks on various topics, from biology to social sciences. For example, *Science* has published a special issue on networks, marking the ten-year anniversary of the discovery of scale-free networks (**Figure 1.8**).

During the past decade, each year about a dozen international conferences, workshops, summer and winter schools have focused on network science. A highly successful network science conference series, called NetSci, has attracted the field's practitioners since 2005. Several general-interest books have made bestseller lists in many countries, bringing network science to the general public. Most major universities offer network science courses, attracting a diverse student body, and in 2014 both Northeastern University in Boston and the Central European University in Budapest launched PhD programs in network science.

To see the impact of networks on the scientific community it is useful to inspect the citation patterns of the most-cited papers in the area of complex systems. Each of these papers are citation classics, reporting discoveries like the butterfly effect, renormalization groups, spin glasses, fractals and neural networks, and cumulatively amass anywhere between 2000 and 5000 citations. To see how the interest in network science compares with the impact of these foundational papers, in **Figure 1.9** we compare their citation patterns with the citations of the two most-cited network science papers: the 1998 paper on small-world phenomena and the 1999 *Science* paper reporting the discovery of scale-free networks. As one can see, the rapid rise in yearly citations of these two papers is without precedent in the area of complex systems.

Several other metrics indicate that network science has an impact in a defining manner on numerous disciplines. For example, in several research fields network papers became the most-cited papers in their leading journals:

(a) The 1998 paper by Watts and Strogatz in *Nature* on small-world phenomena and the 1999 paper by Barabási and Albert in *Science* on scale-free networks were identified by Thompson-Reuters as being among the top ten

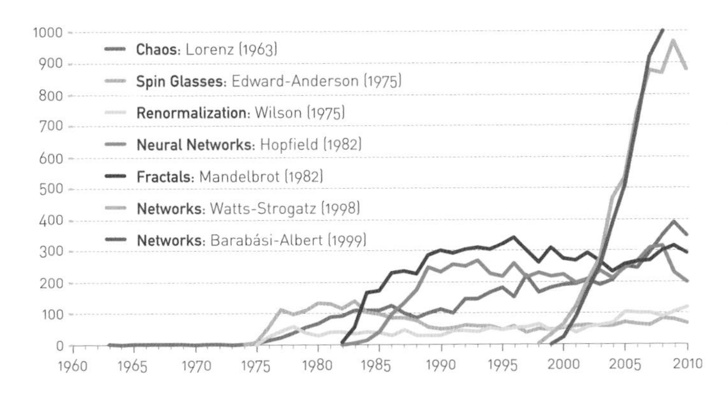

Figure 1.9 **Complexity and Network Science**

The scientific impact of network science, as seen through citation patterns, compared with the citations of the most-cited papers in complexity. The study of complex systems in the 1960s and 1970s was dominated by Edward Lorenz's 1963 classic work on chaos [38], Kenneth G. Wilson's renormalization group [39], and Samuel F. Edwards and Philip W. Anderson's work on spin glasses [40]. In the 1980s the community shifted its focus to pattern formation, following Benoit Mandelbrot's book on fractals [41] and Thomas Witten and Len Sander's introduction of the diffusion-limited aggregation model [42]. Equally influential was John Hopfield's paper on neural networks [43] and Per Bak, Chao Tang and Kurt Wiesenfeld's work on self-organized criticality [44]. These papers continue to define our understanding of complex systems. The figure compares the yearly citations of these landmark papers with the citations of the two most-cited papers in network science, the paper by Watts and Strogatz on small-world networks and by Barabási and Albert, reporting the discovery of scale-free networks.

most-cited papers in physical sciences during the decade after their publication. As of 2011 the Watts–Strogatz paper was the second most cited of all papers published in *Nature* in 1998 and the Barabási–Albert paper was the most-cited paper among all papers published in *Science* in 1999.

(b) Four years after its publication, the *SIAM* review by Mark Newman on network science became the most-cited paper of any journal published by the Society of Industrial & Applied Mathematics [45].

(c) *Reviews of Modern Physics*, published since 1929, is the physics journal with the highest impact factor. Until 2012 the most-cited paper of that journal was written by Nobel Prize winner Subrahmanyan Chandrasekhar, his classic 1944 review entitled "Stochastic problems in physics and astronomy" [46]. During the 70+ years since its publication, the paper has gathered over 5000 citations. Yet, in 2012 it was taken over by the first review of network science published in 2001, entitled "Statistical mechanics of complex networks" [19].

(d) The paper reporting the discovery that in scale-free networks the epidemic threshold vanishes, by Pastor-Satorras and Vespignani, is the most-cited paper among all papers published in 2001 by *Physical Review Letters*, shared with a paper on quantum computing.

(e) The paper by Michelle Girvan and Mark Newman on community discovery in networks [47] is the most-cited paper published in 2002 by *Proceedings of the National Academy of Sciences, USA*.

(f) The 2004 review entitled "Network biology" [22] is the second most-cited paper in the history of *Nature Reviews Genetics*, the top review journal in genetics.

Prompted by this extraordinary enthusiasm within the scientific community, network science was examined by the National Research Council (NRC), the arm of the US National Academies in charge of offering policy recommendations to the US government. The NRC has assembled two panels, resulting in recommendations summarized in two NRC reports [48, 49], defining the field of network science (**Figure 1.10**). These reports not only document the emergence of a new research field, but also highlight the field's role for science, national competitiveness and security. Following these reports, the National Science Foundation (NSF) in the USA established a network science directorate and several network science centers were funded at US universities by the Army Research Labs.

Network science has excited the public as well. This was fueled by the success of several general-audience books, like *Linked*, *Nexus*, *Six Degrees* and *Connected* (**Figure 1.11**). *Connected*, an award-winning documentary by Australian film-maker Annamaria Talas, has brought the field to our TV screen, being broadcast all over the world and winning several prestigious prizes (**Online resource 1.2**).

Networks have inspired artists as well, leading to a wide range of network-related art projects and an annual symposium series that brings together artists and network scientists [54]. Fueled by successful movies like *The Social Network* or *Six Degrees of Separation*, and a series of science fiction novels and short stories exploiting the network paradigm, today networks are deeply ingrained in popular culture.

Figure 1.10 **National Research Council**
Two NRC reports on network science have documented the emergence of the new discipline and highlighted its long-term impact on research and national competitiveness [48, 49]. They have recommended dedicated support for the field, prompting the establishment of network science centers at US universities and a network science program within the NSF.

1.7 Summary

While the emergence of network science may appear to have been rather sudden (**Figures 1.3 and 1.9**), the field was responding to a wider social awareness of the role and importance of networks. This is illustrated in **Figure 1.12**, which shows the usage frequency of words that capture two important scientific revolutions of the past two centuries: *evolution*, the most common term referring to Darwin's theory of evolution and *quantum*, the most frequently used term when one refers

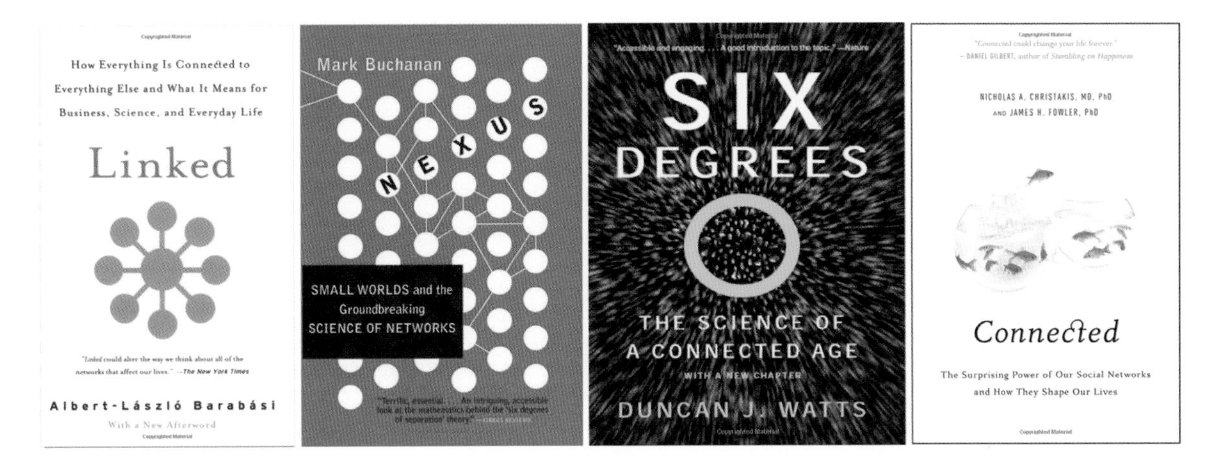

Figure 1.11 **Wide Impact**
Four widely read books, translated into over twenty languages, have brought network science to the general public [50, 51, 52, 53].

to quantum mechanics. As expected, the use of *evolution* increases after the 1859 publication of Darwin's *On the Origin of Species*. The word *quantum*, first used in 1902, remained virtually absent until the 1920s, when quantum mechanics gained acceptance among physicists and reached the public consciousness.

The figure compares these words with the usage of *network*, which enjoyed a spectacular increase following the 1980s, surpassing both *evolution* and *quantum*. While the term *network* has many uses (as do *evolution* and *quantum*), its dramatic rise captures the increasing societal awareness of networks.

There is something in common between the advances facilitated by evolutionary theory, quantum mechanics and network science: they are not only important scientific fields with their own intellectual core and body of knowledge, but also enabling platforms. Indeed, the current revolution in genetics is built on evolutionary theory and quantum mechanics offers a platform for a wide range of advances in contemporary science, from chemistry to electronics. In a similar fashion, network science is an *enabling platform*, offering novel tools and perspectives for a wide range of scientific problems, from social networking to drug design.

Given this exceptional impact networks have both in science and in society, we must master the tools to study and quantify them. The rest of this book is devoted to this worthy subject.

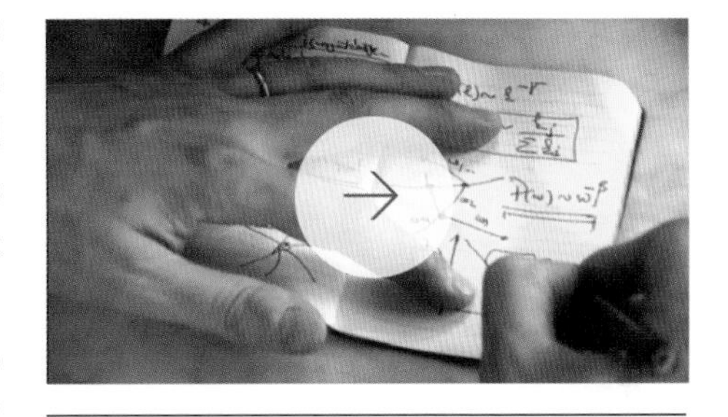

Online Resource 1.2
Connected
The trailer of the award-winning documentary *Connected*, directed by Annamaria Talas, offering an introduction to network science. It features the actor Kevin Bacon and several well-known network scientists.

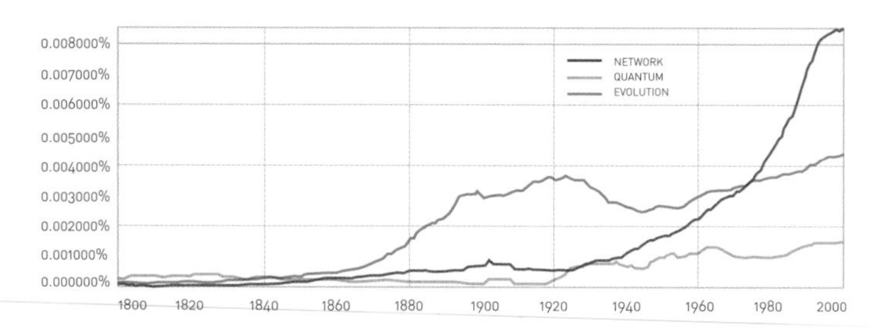

Figure 1.12 **The Rise of Networks**
The frequency of use of the words *evolution, quantum* and *networks* in books since 1880. The plot indicates the exploding societal awareness of networks in the last decades of the twentieth century, laying the ground for the emergence of network science. The plots were generated by Google's *ngram* platform, calculating the fraction of books published in a year that mention *evolution, quantum* or *networks.*

1.8 Homework

1.8.1 Networks Everywhere

List three different real networks and state the nodes and links for each of them.

1.8.2 Your Interest

Tell us of the network you are personally most interested in. Address the following questions:

(a) What are its nodes and links?
(b) How large is it?
(c) Can it be mapped out?
(d) Why do you care about it?

1.8.3 Impact

In your view, in what area could network science have the biggest impact in the next decade? Explain your answer.

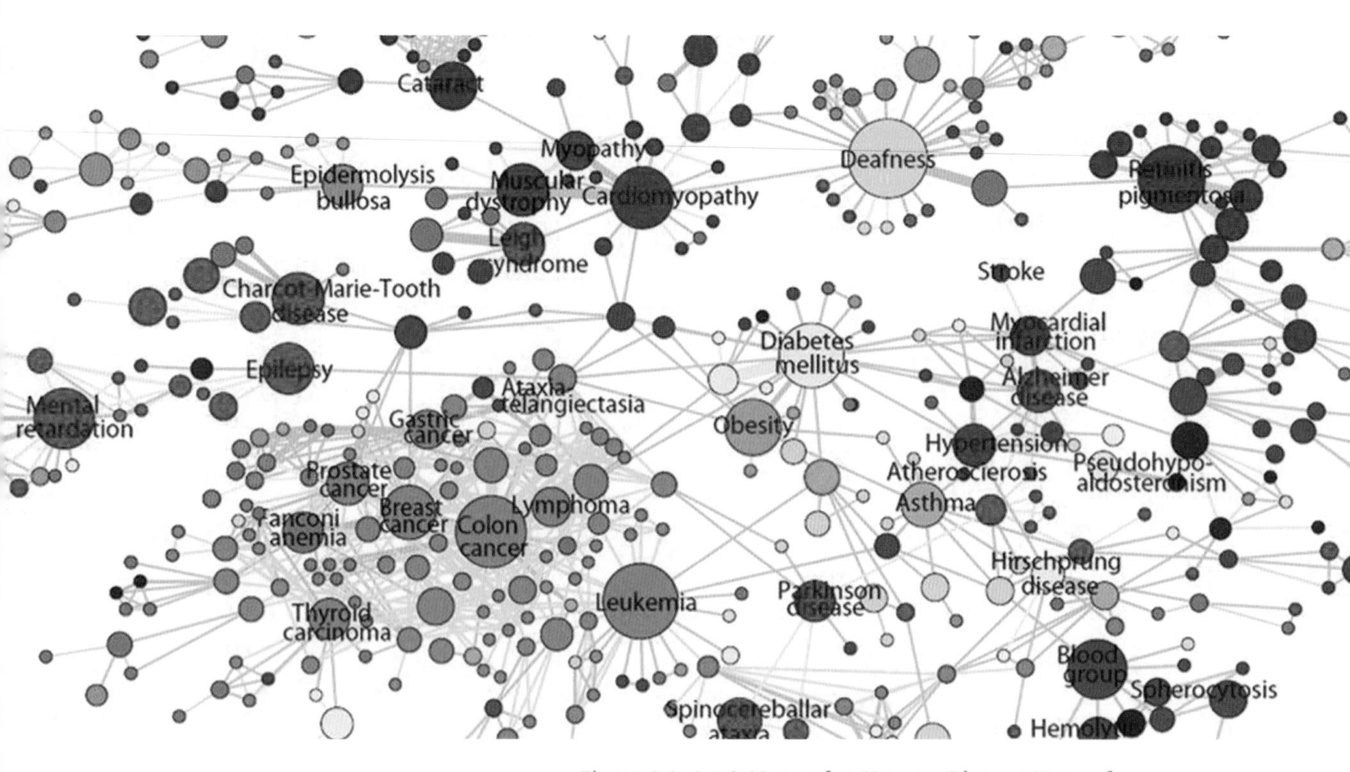

Figure 2.0 **Art & Networks: Human Disease Network**
The human disease network, whose nodes are diseases connected if they have a common genetic origin. Published as a supplement to the *Proceedings of the National Academy of Sciences, USA* [55], the map was created to illustrate the genetic interconnectedness of apparently distinct diseases. With time it crossed disciplinary boundaries, taking up a life of its own. The *New York Times* created an interactive version of the map and the London-based Serpentine Gallery, one of the top contemporary art galleries in the world, exhibited it as part of their focus on networks and maps [56]. It is also featured in numerous books on design and maps [57, 58, 59].

CHAPTER 2

Graph Theory

2.1 The Bridges of Königsberg

Few research fields can trace their birth to a single moment and place in history. Graph theory, the mathematical scaffold behind network science, can. Its roots go back to 1735 in Königsberg, the capital of Eastern Prussia, a thriving merchant city of its time. The trade supported by its busy fleet of ships allowed city officials to build seven bridges across the River Pregel that surrounded the town. Five of these connected to the mainland the elegant island of Kneiphof, caught between the two branches of the Pregel. The remaining two crossed the two branches of the river (**Online resource 2.1**, **Figure 2.1**). This peculiar arrangement gave birth to a contemporary puzzle: Can one walk across all seven bridges and never cross the same one twice? Despite many attempts, no one could find such a path. The problem remained unsolved until 1735, when Leonard Euler, a

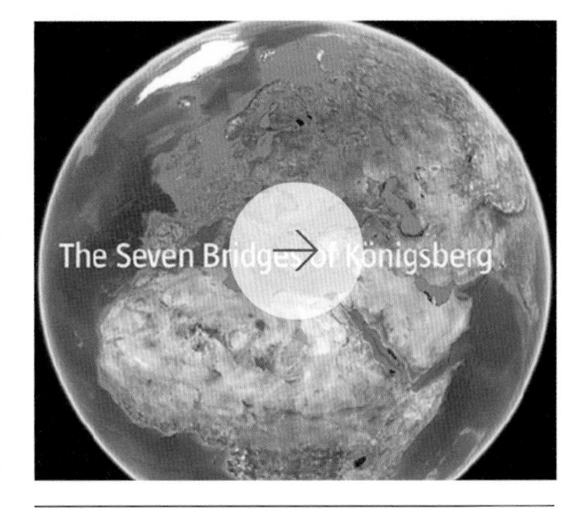

Online Resource 2.1
The Bridges of Königsberg
Watch a short video introducing the Königsberg problem and Euler's solution.

(a)

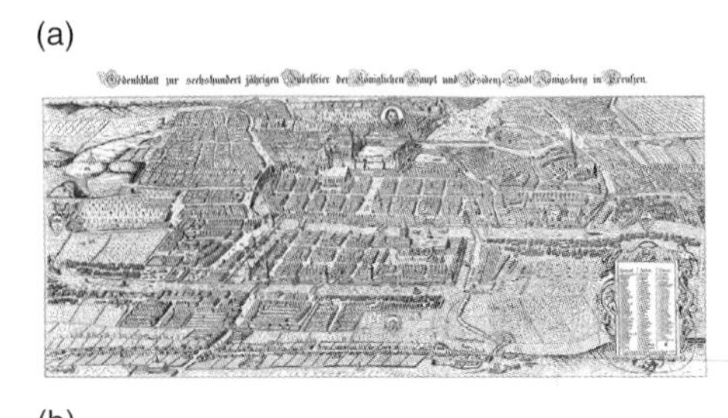

(b)

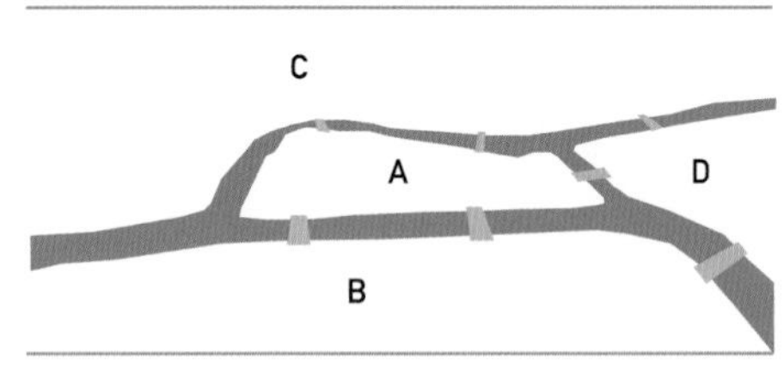

(c)

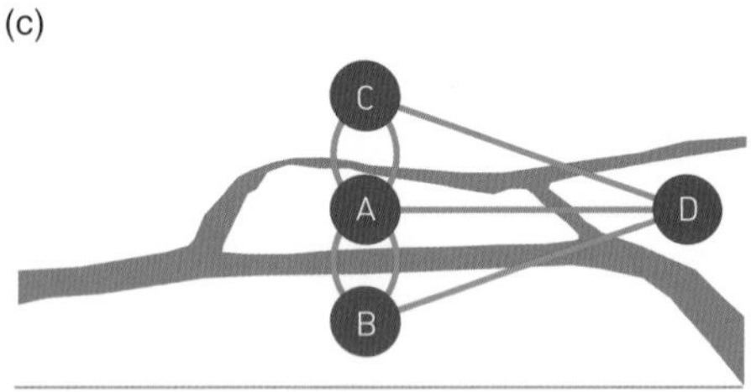

Figure 2.1 **The Bridges of Königsberg**

(**a**) A contemporary map of Königsberg (now Kaliningrad, Russia) during Euler's time.

(**b**) A schematic illustration of Königsberg's four land pieces and the seven bridges across them.

(**c**) Euler constructed a graph that has four nodes (A, B, C, D), each corresponding to a patch of land, and seven links, each corresponding to a bridge. He then showed that there is no continuous path that would cross the seven bridges while never crossing the same bridge twice. The people of Königsberg gave up their fruitless search and in 1875 built a new bridge between B and C, increasing the number of links of these two nodes to four. Now only two nodes were left with an odd number of links. Consequently, we should be able to find the desired path. Can you find one yourself?

Swiss-born mathematician, offered a rigorous mathematical proof that such a path does not exist [60, 61].

Euler represented each of the four land areas separated by the river with letters A, B, C and D (**Figure 2.1**). Next he connected with lines all pieces of land that had a bridge between them. He thus built a graph, whose nodes were pieces of land and whose links were the bridges. Then Euler made a simple observation: if there is a path crossing all bridges, but never the same bridge twice, then nodes with an odd number of links must be either the starting point or the end point of this path. Indeed, if you arrive at a node with an odd number of links, you may find yourself having no unused link to leave it by.

A walking path that goes through all bridges can have only one starting and one end point. Thus, such a path cannot exist on a graph that has more than two nodes with an odd number of links. The Königsberg graph had four nodes with an odd number of links, A, B, C and D, so no path could satisfy the problem (**Online resource 2.1**).

Euler's proof was the first time someone had solved a mathematical problem using a graph. For us, the proof has two important messages. The first is that some problems become simpler and more tractable if they are represented as a graph. The second is that the existence of the path does not depend on our ingenuity in finding it. Rather, it is a property of the graph. Indeed, given the structure of the Königsberg graph, no matter how smart we are, we will never find the desired path. In other words, networks have properties encoded in their structure that limit or enhance their behavior.

To understand the many ways that networks can affect the properties of a system, we need to become familiar with graph theory, a branch of mathematics that grew out of Euler's proof. In this chapter we learn how to represent a network as a graph and introduce the elementary characteristics of networks, from degrees to degree distributions, from paths to distances, and learn to distinguish weighted, directed and bipartite networks. We will introduce a graph-theoretic formalism and language that will be used throughout this book.

2.2 Networks and Graphs

If we want to understand a complex system, we first need to know how its components interact with each other. In other words, we need a map of its wiring diagram. A network is a catalog of a system's components – often called *nodes* or *vertices* – and the direct interactions between them, called *links* or *edges* (**Box 2.1**). This network representation offers a common language to study systems that may differ greatly in nature, appearance or scope. Indeed, as shown in **Figure 2.2**, three rather different systems have exactly the same network representation.

Figure 2.2 introduces two basic network parameters.

- *The number of nodes*, or N, represents the number of components in the system. We will often call N the *size of the network*. To distinguish the nodes, we label them $i = 1, 2, \ldots, N$.
- *The number of links*, which we denote by L, represents the total number of interactions between the nodes. Links are rarely labeled, as they can be identified through the nodes they connect. For example, the (2, 4) link connects nodes 2 and 4.

The networks shown in **Figure 2.2** have $N = 4$ and $L = 4$.

The links of a network can be *directed* or *undirected*. Some systems have directed links, like the WWW, whose uniform resource locators (URLs) point from one web document to another, or phone calls, where one person calls another. Other systems have undirected links, like romantic ties: if I date Janet, Janet also dates me, or like transmission lines on the power grid, where electric current can flow in both directions.

A network is called directed (or a *digraph*) if all of its links are directed; it is called undirected if all of its links are undirected. Some networks simultaneously have directed and undirected links. For example, in the metabolic network some reactions are reversible (i.e., bidirectional or undirected) and others are irreversible, taking place in only one direction (directed).

The choices we make when we represent a system as a network will determine our ability to use network science

Box 2.1 Networks or Graphs?

In the scientific literature the terms *network* and *graph* are used interchangeably:

Network Science	Graph Theory
Network	Graph
Node	Vertex
Link	Edge

Yet there is a subtle distinction between the two terminologies. The {*network, node, link*} combination often refers to real systems: the WWW is a network of web documents linked by URLs; society is a network of individuals linked by family, friendship or professional ties; the metabolic network is the sum of all chemical reactions that take place in a cell. In contrast, we use the terms {*graph, vertex, edge*} when we discuss the mathematical representation of these networks: we talk about the Web graph, the social graph (a term made popular by Facebook) or the metabolic graph. Yet this distinction is rarely made, so these two terminologies are often synonymous.

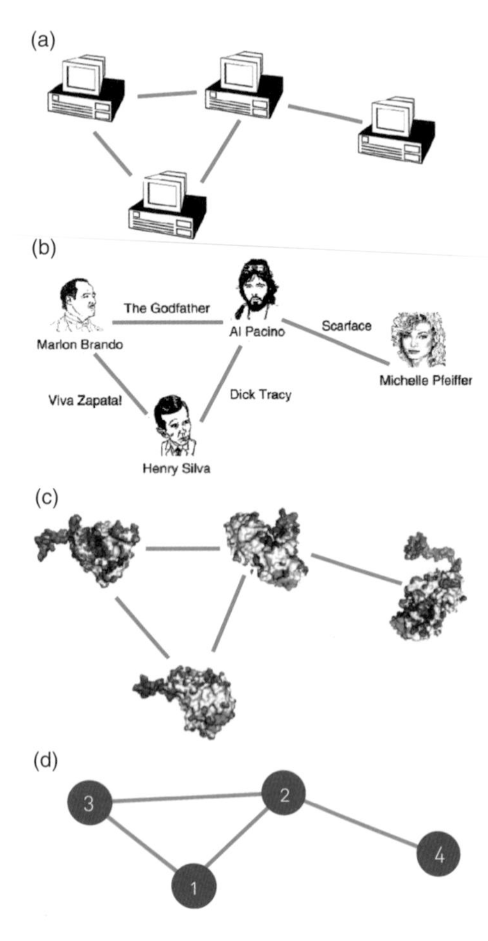

Figure 2.2 **Different Networks, Same Graph**
The figure shows a small subset of
(**a**) the Internet, where routers (specialized computers) are connected to each other;
(**b**) the Hollywood actor network, where two actors are connected if they played in the same movie; and
(**c**) a protein–protein interaction network, where two proteins are connected if there is experimental evidence that they can bind to each other in the cell. While the nature of the nodes and the links differs, these networks have the same graph representation, consisting of $N = 4$ nodes and $L = 4$ links, shown in (**d**).

successfully to solve a particular problem. For example, the way we define the links between two individuals dictates the nature of the questions we can explore:

(a) By connecting individuals who regularly interact with each other in the context of their work, we obtain the *organizational* or *professional network* that plays a key role in the success of a company or an institution, and is of major interest to organizational research (**Figure 1.7**).
(b) By linking friends to each other, we obtain the *friendship network* that plays an important role in the spread of ideas, products and habits and is of major interest to sociology, marketing and health sciences.
(c) By connecting individuals that have an intimate relationship, we obtain the *sexual network* that is of key importance for the spread of sexually transmitted diseases, like AIDS, and is of major interest for epidemiology.
(d) By using phone and email records to connect individuals that call or email each other, we obtain the *acquaintance network* that captures a mixture of professional, friendship and intimate links, and is of importance to communications and marketing.

While many links in these four networks overlap (some coworkers may be friends or may have an intimate relationship), these networks have different uses and purposes.

We can also build networks that may be valid from a graph-theoretic perspective, but may have little practical utility. For example, if we link all individuals with the same first name, Johns with Johns and Marys with Marys, we obtain a well-defined graph whose properties can be analyzed with the tools of network science. Its utility is questionable, however. Hence, in order to apply network theory to a system, careful consideration must precede our choice of nodes and links, ensuring their significance to the problem we wish to explore.

Throughout this book we will use ten networks to illustrate the tools of network science. These *reference networks*, listed in **Table 2.1**, span social systems (mobile call graph or email network), collaboration and affiliation networks (science collaboration network, Hollywood actor network), information systems (WWW), technological and infrastructural systems (Internet and power grid), biological systems (protein interaction and metabolic

Table 2.1 **Canonical Network Maps**

The basic characteristics of ten networks used throughout this book to illustrate the tools of network science. The table lists the nature of their nodes and links, indicating if links are directed or undirected, the number of nodes (N) and links (L), and the average degree for each network. For directed networks the average degree shown is the average number of in- or out-degrees $\langle k \rangle = \langle k_{in} \rangle = \langle k_{out} \rangle$ (see equation (**2.5**)).

Network	Nodes	Links	Directed or undirected	N	L	$\langle k \rangle$
Internet	Routers	Internet connections	Undirected	192,244	609,066	6.34
WWW	Webpages	Links	Directed	325,729	1,497,134	4.60
Power grid	Power plants, transformers	Cables	Undirected	4,941	6,594	2.67
Mobile phone calls	Subscribers	Calls	Directed	36,595	91,826	2.51
Email	Email addresses	Emails	Directed	57,194	103,731	1.81
Science collaboration	Scientists	Co-authorship	Undirected	23,133	93,439	8.08
Actor network	Actors	Co-acting	Undirected	702,388	29,397,908	83.71
Citation network	Papers	Citations	Directed	449,673	4,689,479	10.43
***E. coli* metabolism**	Metabolites	Chemical reactions	Directed	1,039	5,802	5.58
Protein interactions	Proteins	Binding interactions	Undirected	2,018	2,930	2.90

network) and reference networks (citations). They differ widely in their size, from as few as $N = 1{,}039$ nodes in the *E. coli* metabolism to almost half a million nodes in the citation network. They cover several areas where networks are actively applied, representing "canonical" datasets frequently used by researchers to illustrate key network properties. As we indicate in **Table 2.1**, some of them are directed, others are undirected. In the coming chapters we will discuss in detail the nature and the characteristics of each of these datasets, turning them into the guinea pigs of our journey to understand complex networks.

2.3 Degree, Average Degree and Degree Distribution

A key property of each node is its *degree*, representing the number of links it has to other nodes. The degree can represent the number of mobile phone contacts an individual has in the call graph (i.e., the number of different individuals the person has talked to), or the number of citations a research paper gets in the citation network.

2.3.1 Degree

We denote by k the degree of the ith node in the network. For example, for the undirected networks shown in **Figure 2.2**, we have $k_1 = 2$, $k_2 = 3$, $k_3 = 2$, $k_4 = 1$. In an undirected network the *total number of links*, L, can be expressed as the sum of the node degrees:

$$L = \frac{1}{2}\sum_{i=1}^{N} k_i. \tag{2.1}$$

Here the factor $\frac{1}{2}$ corrects for the fact that in the sum (**2.1**) each link is counted twice. For example, the link connecting nodes 2 and 4 in **Figure 2.2** will be counted once in the degree of node 1 and once in the degree of node 4.

2.3.2 Average Degree

An important property of a network is its *average degree* (**Box 2.2**), which for an undirected network is

$$\langle k \rangle = \frac{1}{N}\sum_{i=1}^{N} k_i = \frac{2L}{N}. \tag{2.2}$$

In directed networks we distinguish between the *incoming degree*, k_i^{in}, representing the number of links that point to node i and the *outgoing degree*, k_i^{out}, representing the number of links that point from node i to other nodes. Finally, a node's *total degree*, k_i, is given by

$$k_i = k_i^{\text{in}} + k_i^{\text{out}}. \tag{2.3}$$

For example, on the WWW the number of pages a given document points to represents its outgoing degree, k_i^{out} and the number of documents that point to it represents its incoming degree, k_i^{in}. The total number of links in a directed network is

$$L = \sum_{i=1}^{N} k_i^{\text{in}} = \sum_{i=1}^{N} k_i^{\text{out}}. \tag{2.4}$$

The factor $\frac{1}{2}$ seen in (**2.1**) is now absent, as for directed networks the two sums in (**2.4**) separately count the outgoing and the incoming degrees. The average degree of a directed network is

Box 2.2 Brief Statistics Review

Four key quantities characterize a sample of N values $x_1, \ldots, x_N$:

Average (mean)

$$\langle x \rangle = \frac{x_1 + x_2 + \ldots + x_N}{N} = \frac{1}{N}\sum_{i=1}^{N} x_i$$

The *n*th moment

$$\langle x^n \rangle = \frac{x_1^n + x_2^n + \ldots + x_N^n}{N} = \frac{1}{N}\sum_{i=1}^{N} x_i^n$$

Standard deviation

$$\sigma_x = \sqrt{\frac{1}{N}\sum_{i=1}^{N} (x_i - \langle x \rangle)^2}$$

Distribution of *x*

In directed networks we distinguish the *incoming degree*, k_i:

$$p_x = \frac{1}{N}\sum_i \delta_{x,x_i}$$

where p_x follows

$$\sum_i p_x = 1 \left(\int p_x dx = 1 \right).$$

$$\langle k^{in} \rangle = \frac{1}{N} \sum_{i=1}^{N} k_i^{in} = \langle k^{out} \rangle = \frac{1}{N} \sum_{i=1}^{N} k_i^{out} = \frac{L}{N}. \qquad (2.5)$$

2.3.3 Degree Distribution

The *degree distribution*, p_k, provides the probability that a randomly selected node in the network has degree k. Since p_k is a probability, it must be normalized:

$$\sum_{k=1}^{\infty} p_k = 1. \qquad (2.6)$$

For a network with N nodes, the degree distribution is the normalized histogram (**Figure 2.3**) given by

$$p_k = \frac{N_k}{N}, \qquad (2.7)$$

where N_k is the number of degree-k nodes (**Figure 2.4**). Hence, the number of degree-k nodes can be obtained from the degree distribution as $N_k = Np_k$.

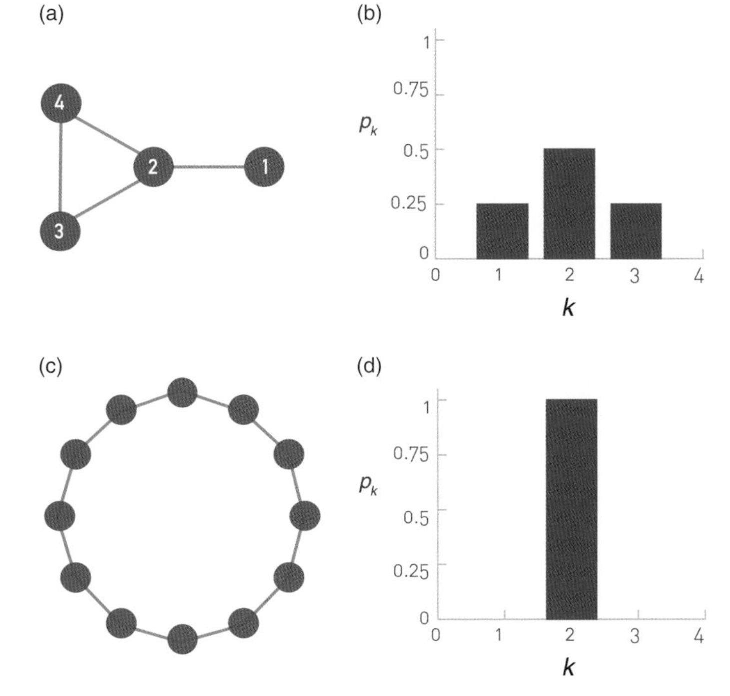

Figure 2.3 **Degree Distribution**
The degree distribution of a network is provided by the ratio (**2.7**).
(**a**) For the network in (a) with $N = 4$, the degree distribution is shown in (b).
(**b**) We have $p_1 = 1/4$ (one of the four nodes has degree $k_1 = 1$), $p_2 = 1/2$ (two nodes have $k_3 = k_4 = 2$) and $p_3 = 1/4$ (as $k_2 = 3$). As we lack nodes with degree $k > 3$, $p_k = 0$ for any $k > 3$.
(**c**) A one-dimensional lattice for which each node has the same degree $k = 2$.
(**d**) The degree distribution of (c) is a Kronecker delta function, $p_k = \delta(k - 2)$.

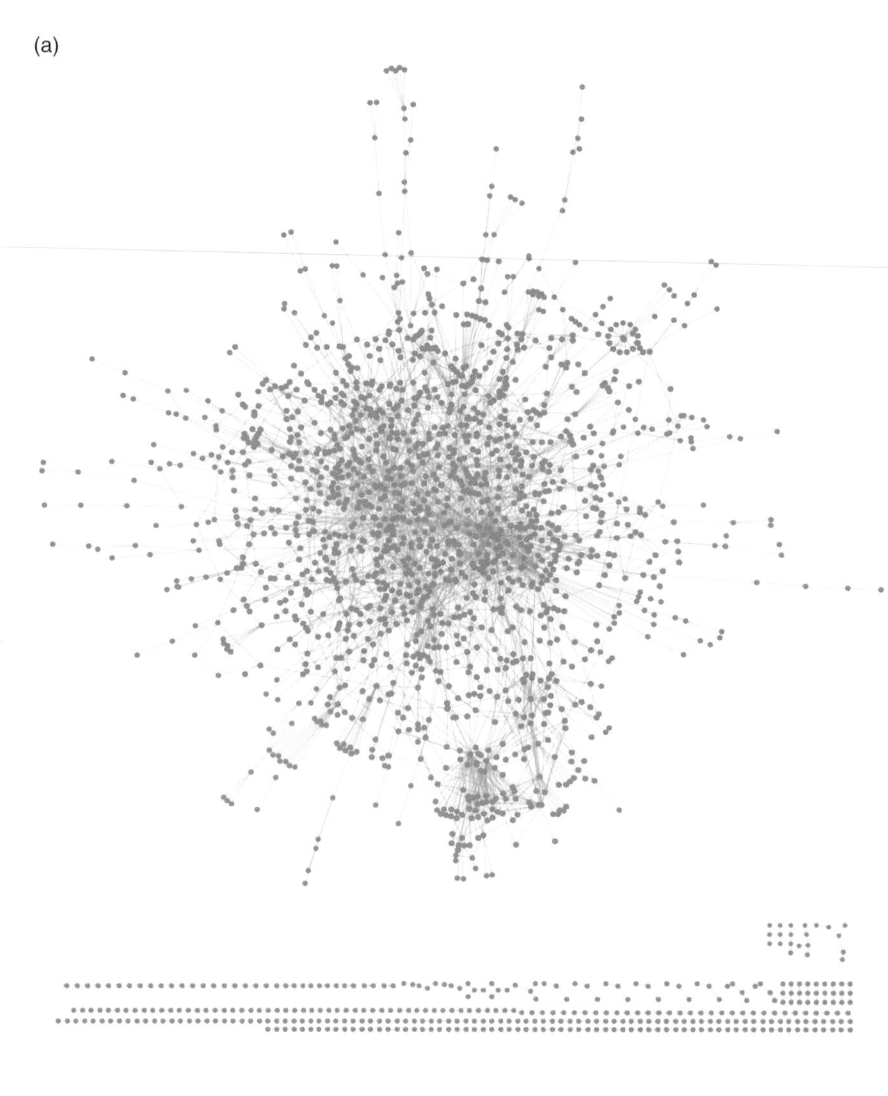

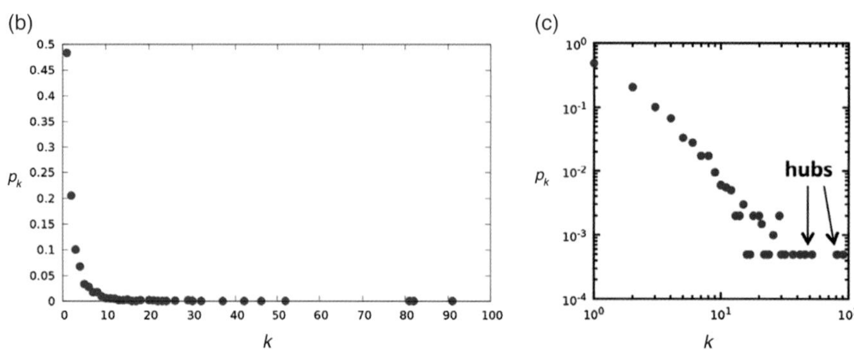

Figure 2.4 **Degree Distribution of a Real Network**
In real networks the node degrees can vary widely.
(**a**) A layout of the protein interaction network of yeast (**Table 2.1**). Each node corresponds to a yeast protein and links correspond to experimentally detected binding interactions. Note that the proteins shown on the bottom have self-loops, hence for them $k = 2$.
(**b**) The degree distribution of the protein interaction network shown in (a). The observed degrees vary between $k = 0$ (isolated nodes) and $k = 92$, which is the degree of the most connected node, called a *hub*. There are also wide differences in the number of nodes with different degrees: almost half of the nodes have degree one (i.e., $p_1 = 0.48$), while we have only one copy of the biggest node (i.e., $p_{92} = 1/N = 0.0005$).
(**c**) The degree distribution is often shown on a log–log plot, in which we either plot $\log p_k$ as a function of $\ln k$ or we use logarithmic axes. The advantages of this representation are discussed in **Chapter 4**.

The degree distribution has assumed a central role in network theory following the discovery of scale-free networks [10]. One reason is that the calculation of most network properties requires us to know p_k. For example, the average degree of a network can be written as

$$\langle k \rangle = \sum_{k=0}^{\infty} k p_k. \tag{2.8}$$

The other reason is that the precise functional form of p_k determines many network phenomena, from network robustness to the spread of viruses.

2.4 Adjacency Matrix

A complete description of a network requires us to keep track of its links. The simplest way to achieve this is to provide a complete list of the links. For example, the network of **Figure 2.2** is uniquely described by listing its four links: {(1, 2), (1, 3), (2, 3), (2, 4)}. For mathematical purposes we often represent a network through its adjacency matrix. The *adjacency matrix* of a directed network of N nodes has N rows and N columns, its elements being:

- $A_{ij} = 1$ if there is a link pointing from node j to node i;
- $A_{ij} = 0$ if nodes i and j are not connected to each other.

The adjacency matrix of an undirected network has two entries for each link, for example link (1, 2) is represented as $A_{12} = 1$ and $A_{21} = 1$. Hence, the adjacency matrix of an undirected network is symmetric, $A_{ij} = A_{ji}$ (**Figure 2.5b**).

The degree k_i of node i can be obtained directly from the elements of the adjacency matrix. For undirected networks a node's degree is a sum over either the rows or the columns of the matrix:

$$k_i = \sum_{j=1}^{N} A_{ji} = \sum_{i=1}^{N} A_{ji}. \tag{2.9}$$

For directed networks the sums over the adjacency matrix's rows and columns provide the incoming and outgoing degrees, respectively:

$$k_i^{\text{in}} = \sum_{j=1}^{N} A_{ij}, \ k_i^{\text{out}} = \sum_{j=1}^{N} A_{ji}. \tag{2.10}$$

Figure 2.5 **The Adjacency Matrix**
(**a**) The labeling of the elements of the adjacency matrix.
(**b**) The adjacency matrix of an *undirected network*. The figure shows that the degree of a node (in this case node 2) can be expressed as the sum over the appropriate column or row of the adjacency matrix. It also shows a few basic network characteristics, like the total number of links, L, and the average degree, $\langle k \rangle$, expressed in terms of the elements of the adjacency matrix.
(**c**) The same as (b) but for a *directed network*.

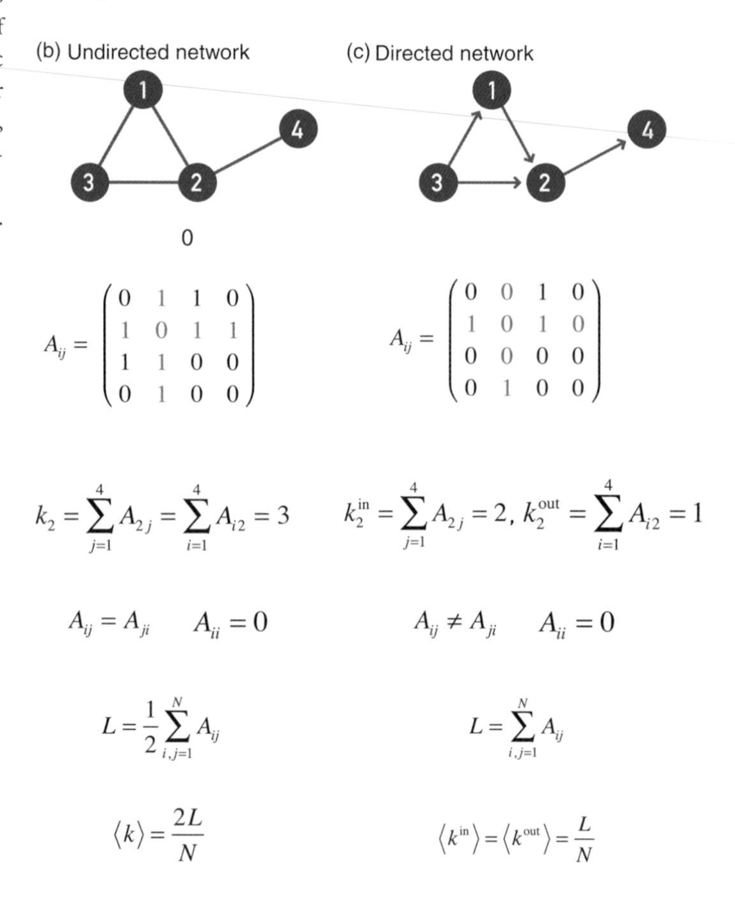

Given that in an undirected network the number of outgoing links equals the number of incoming links, we have

$$2L = \sum_{i=1}^{N} k_i^{in} = \sum_{i=1}^{N} k_i^{out} = \sum_{ij}^{N} A_{ij}. \tag{2.11}$$

The number of nonzero elements of the adjacency matrix is $2L$, or twice the number of links. Indeed, an undirected link connecting nodes i and j appears in two entries: $A_{ij} = 1$, a link pointing from node j to node i, and $A_{ji} = 1$, a link pointing from node i to node j (**Figure 2.5b**).

2.5 Real Networks are Sparse

In real networks the number of nodes (N) and links (L) can vary widely. For example, the neural network of the worm *C. elegans*, the only fully mapped nervous system of a living organism, has $N = 302$ neurons (nodes). In contrast, the human brain is estimated to have about a hundred billion ($N \approx 10^{11}$) neurons. The genetic network of a human cell has about 20,000 genes as nodes; the social network consists of seven billion individuals ($N \approx 7 \times 10^9$) and the WWW is estimated to have over a trillion web documents ($N > 10^{12}$).

These wide differences in size are noticeable in **Table 2.1**, which lists N and L for several network maps. Some of these maps offer a complete wiring diagram of the system they describe (like the actor network or the *E. coli* metabolism), while others are only samples, representing a subset of the full network (like the WWW or the mobile call graph).

Table 2.1 indicates that the number of links also varies widely. In a network of N nodes, the number of links can change between 0 and $L_{\max}$, where

$$L_{\max} = \binom{N}{2} = \frac{N(N-1)}{2} \tag{2.12}$$

is the total number of links present in a *complete graph* of size N (**Figure 2.6**). In a complete graph each node is connected to every other node.

In real networks L is much smaller than $L_{\max}$, reflecting the fact that most real networks are sparse. We call a network *sparse* if $L \ll L_{\max}$. For example, the WWW graph in **Table 2.1** has about 1.5 million links. Yet, if the WWW were to be a complete graph, it should have $L_{\max} \approx 5 \times 10^{10}$ links according to (**2.12**). Consequently, the WWW graph has only a fraction 3×10^{-5} of the links it could have. This is true for all the networks in **Table 2.1**: One can check that their number of links is only a tiny fraction of the expected number of links for a complete graph of the same number of nodes.

The sparsity of real networks implies that the adjacency matrices are also sparse. Indeed, a complete network has $A_{ij} = 1$ for all (i, j), that is each of its matrix elements is equal to one. In contrast, in real networks only a tiny fraction

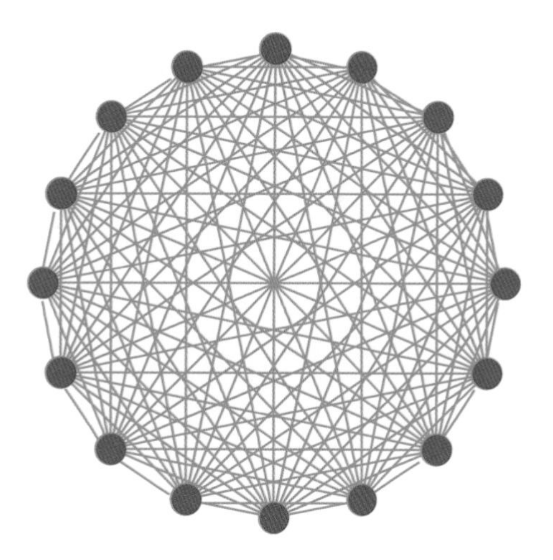

Figure 2.6 **Complete Graph**
A complete graph with $N = 16$ nodes and $L_{\max} = 120$ links, as predicted by (**2.12**). The adjacency matrix of a complete graph is $A_{ij} = 1$ for all $i, j = 1, \ldots, N$ and $A_{ii} = 0$. The average degree of a complete graph is $\langle k \rangle = N - 1$. A complete graph is often called a *clique*, a term frequently used in community identification, a problem discussed in **Chapter 9**.

Figure 2.7 **The Adjacency Matrix is Sparse**
The adjacency matrix of the yeast protein–protein interaction network, consisting of 2018 nodes, each representing a yeast protein (**Table 2.1**). A dot is placed at each position of the adjacency matrix for which $A_{ij} = 1$, indicating the presence of an interaction. There are no dots for $A_{ij} = 0$. The small fraction of dots illustrates the sparse nature of the protein–protein interaction network.

of the matrix elements are nonzero. This is illustrated in **Figure 2.7**, which shows the adjacency matrix of the protein–protein interaction network listed in **Table 2.1** and shown in **Figure 2.4a**. One can see that the matrix is nearly empty.

Sparseness has important consequences for the way we explore and store real networks. For example, when we store a large network in our computer, it is better to store only the list of links (i.e., elements for which $A_{ij} \neq 0$) rather than the full adjacency matrix, as an overwhelming fraction of the A_{ij} elements are zero. Hence, the matrix representation will block a huge chunk of memory, filled mainly with zeros (**Figure 2.7**).

2.6 Weighted Networks

So far we have discussed only networks for which all links have the same weight, that is $A_{ij} = 1$. In many applications we need to study *weighted networks*, where each link (i, j) has a unique weight w_{ij}. In mobile call networks the weight can represent the total number of minutes two individuals talk with each other on the phone; on the power grid the weight is the amount of current flowing through a transmission line.

For *weighted networks* the elements of the adjacency matrix carry the weight of the link as

$$A_{ij} = w_{ij}. \tag{2.13}$$

Most networks of scientific interest are weighted, but we cannot always measure the appropriate weights. Consequently, we often approximate these networks with an unweighted graph. In this book we predominantly focus on unweighted networks, but whenever appropriate we discuss how the weights alter the corresponding network property (**Box 2.3**).

2.7 Bipartite Networks

A *bipartite graph* (or *bigraph*) is a network whose nodes can be divided into two disjoint sets U and V such that each link connects a U-node to a V-node. In other words, if we color the U-nodes green and the V-nodes purple, then each link must connect nodes of different colors (**Figure 2.9**).

Box 2.3 Metcalfe's law: The Value of a Network

Metcalfe's law states that the *value of a network* is proportional to the square of the number of its nodes (i.e., N^2). Formulated around 1980 in the context of communication devices by Robert M. Metcalfe [62], the idea behind Metcalfe's law is that the more individuals use a network, the more valuable it becomes. Indeed, the more of your friends use email, the more valuable the service is to you.

During the Internet boom of the late 1990s, Metcalfe's law was frequently used to offer a quantitative valuation for Internet companies. It suggested that the value of a service is proportional to the number of connections it can create, which is the square of the number of its users. In contrast, the cost grows only linearly with N. Hence, if the service attracts a sufficient number of users, it will inevitably become profitable, as N^2 will surpass N at some large N (**Figure 2.8**). Metcalfe's law therefore supports a "build it and they will come" mentality [63], offering credibility to growth over profits.

Metcalfe's law is based on (**2.12**), telling us that if *all links* of a communication network with N users are equally valuable, the total value of the network is proportional to $N(N-1)/2$, that is, roughly N^2. If a network has $N = 10$ consumers, there are $L_{max} = 45$ different possible connections between them. If the network doubles in size to $N = 20$, the number of connections doesn't merely double but roughly quadruples to 190, a phenomenon called *network externality* in economics.

Two issues limit the validity of Metcalfe's law:

(a) Most real networks are sparse, which means that only a very small fraction of the links are present. Hence, the value of the network does not grow like N^2 but increases only linearly with N.

(b) As the links have weights, not all links are of equal value. Some links are used heavily while the vast majority of links are rarely utilized.

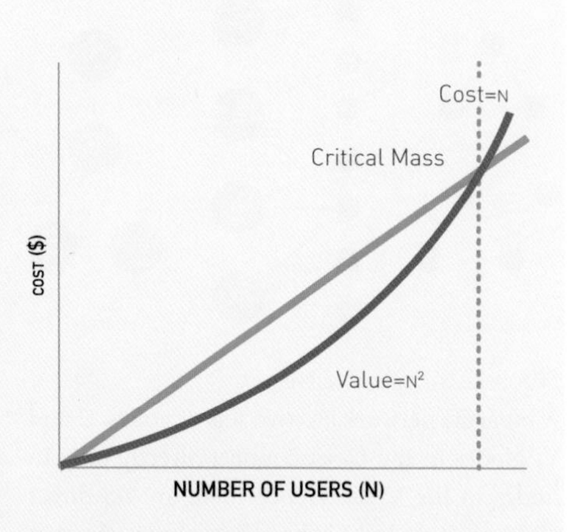

Figure 2.8 **Metcalfe's Law**
According to Metcalfe's law the *cost* of network-based services increases linearly with the number of nodes (users or devices). In contrast, the *benefits* or *income* are driven by the number of links L_{max} the technology makes possible, which grows like N^2 according to (**2.12**). Hence, once the number of users or devices exceeds some *critical mass*, the technology becomes profitable.

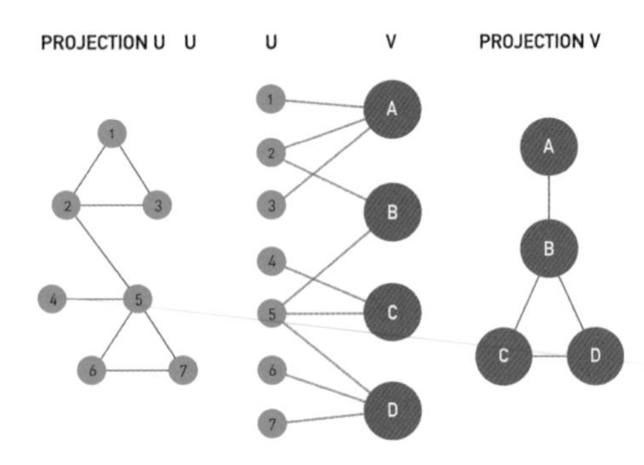

Figure 2.9 **Bipartite Network**
A bipartite network has two sets of nodes, U and V. Nodes in the U-set connect directly only to nodes in the V-set. Hence there are no direct U–U or V–V links. The figure shows the two projections we can generate from any bipartite network. Projection U is obtained by connecting two U-nodes to each other if they link to the same V-node in the bipartite representation. Projection V is obtained by connecting two V-nodes to each other if they link to the same U-node in the bipartite network.

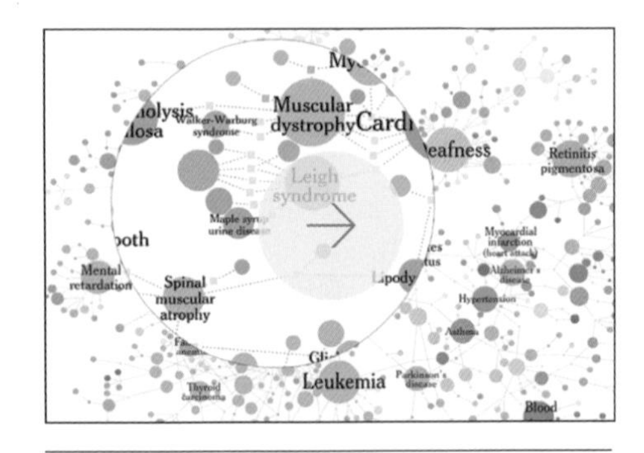

Online Resource 2.2
Human Disease Network
Download the high-reslution version of the human disease network [55], or explore it using the online interface built by the *New York Times*.

We can generate two *projections* for each bipartite network. The first projection connects two U-nodes by a link if they are linked to the same V-node in the bipartite representation. The second projection connects the V-nodes by a link if they connect to the same U-node (**Figure 2.9**).

In network theory we encounter numerous bipartite networks. A well-known example is the Hollywood actor network, in which one set of nodes corresponds to movies (U) and the other to actors (V). A movie is connected to an actor if the actor plays in that movie. One projection of this bipartite network is the *actor network*, in which two nodes are connected to each other if they played in the same movie. This is the network listed in **Table 2.1**. The other projection is the *movie network*, in which two movies are connected if they share at least one actor in their cast.

Medicine offers another prominent example of a bipartite network: the *human disease network* connects diseases to the genes whose mutations are known to cause or effect the corresponding disease (**Figure 2.10**).

Finally, one can also define *multipartite networks*, like the *tripartite* recipe–ingredient–compound network shown in **Figure 2.11**.

2.8 Paths and Distances

Physical distance plays a key role in determining the interactions between the components of physical systems. For example, the distance between two atoms in a crystal or between two galaxies in the Universe determines the forces that act between them.

In networks, distance is a challenging concept. Indeed, what is the distance between two webpages, or between two individuals who do not know each other? The physical distance is not relevant here: two webpages could be sitting on computers on the opposite sides of the globe, yet have a link to each other. At the same time two individuals who live in the same building may not know each other.

In networks, physical distance is replaced by *path length*. A *path* is a route that runs along the links of the network. A path's *length* represents the number of links the path contains (**Figure 2.12a**). Note that some texts require that each node a path visits is distinct.

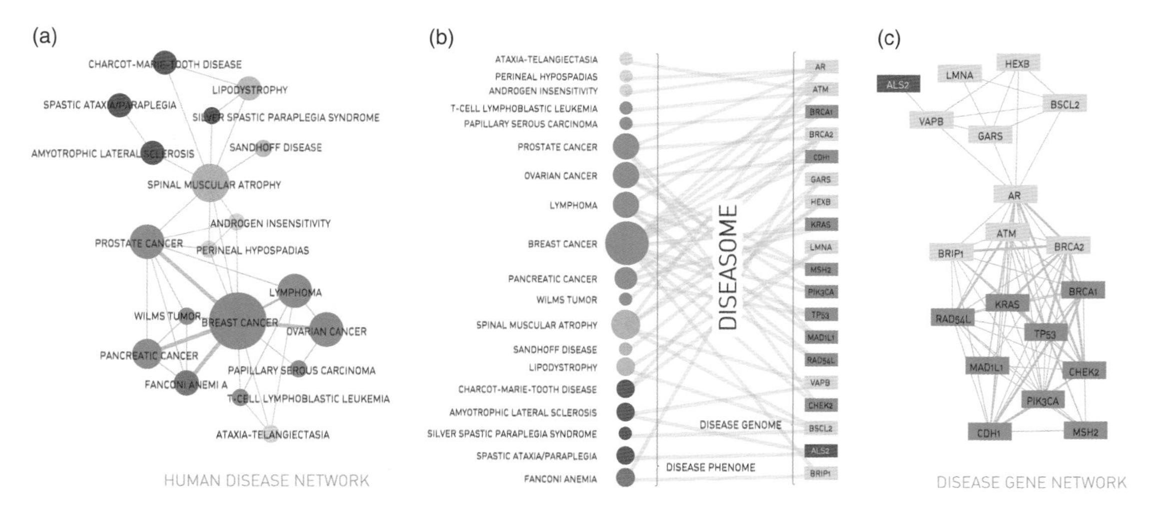

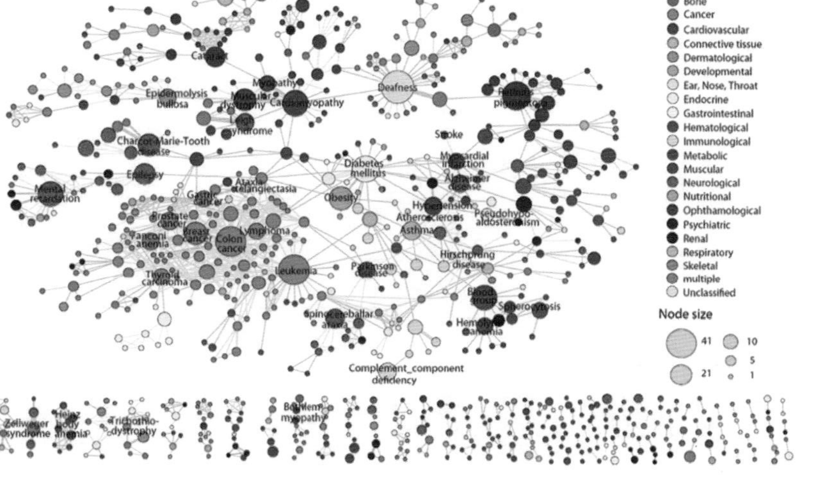

Figure 2.10 **Human Disease Network**

(**a**) One projection of the diseaseome is the *disease network*, whose nodes are diseases. Two diseases are connected if the same genes are associated with them, indicating that the two diseases have a common genetic origin.

(**b**) The human disease network (or *diseaseome*) is a bipartite network, whose nodes are diseases (U) and genes (V). A disease is connected to a gene if mutations in that gene are known to effect the particular disease [58].

(**c**) The second projection is the *gene network*, whose nodes are genes, and where two genes are connected if they are associated with the same disease. Figures (a)–(c) show a subset of the diseaseome, focusing on cancers.

(**d**) The full diseaseome, connecting 1,283 disorders via 1,777 shared disease genes. After [55]. See **Online resource 2.2** for the detailed map.

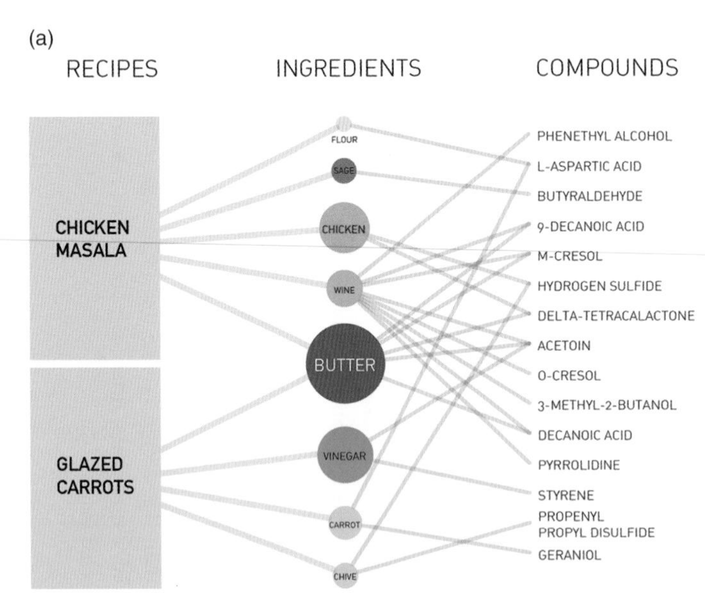

Figure 2.11 **Tripartite Network**

(a) The construction of the tripartite recipe–ingredient–compound network, in which one set of nodes are recipes, like Chicken Marsala; the second set corresponds to the ingredients each recipe has (like flour, sage, chicken, wine and butter for Chicken Marsala); the third set captures the flavor compounds, or chemicals, that contribute to the taste of each ingredient.

(b) The *ingredient* or the *flavor network* represents a projection of the tripartite network. Each node denotes an ingredient; the node color indicates the food category and the node size indicates the ingredient's prevalence in recipes. Two ingredients are connected if they share a significant number of flavor compounds. Link thickness represents the number of shared compounds.

After [64].

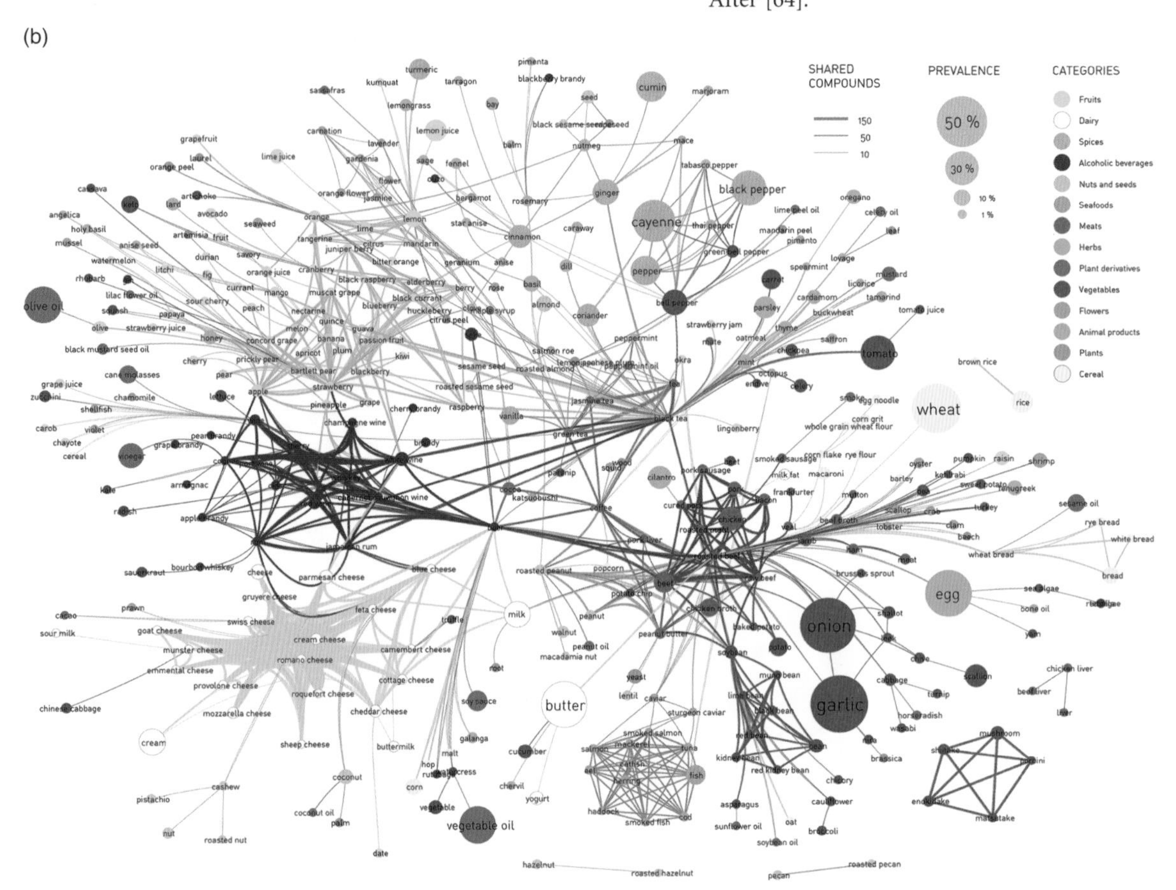

In network science, paths play a central role. Next we discuss some of their most important properties, many more being summarized in **Figure 2.13**.

2.8.1 Shortest Path

The shortest path between nodes i and j is the path with the fewest number of links (**Figure 2.12b**). The shortest path is often called the distance between nodes i and j, and is denoted by d_{ij} or simply d. We can have multiple shortest paths of the same length d between a pair of nodes (**Figure 2.12b**). The shortest path never contains loops or intersects itself.

In an undirected network $d_{ij} = d_{ji}$, that is the distance between node i and node j is the same as the distance between node j and node i. In a directed network often $d_{ij} \neq d_{ji}$. Furthermore, in a directed network the existence of a path from node i to node j does not guarantee the existence of a path from node j to node i.

In real networks we often need to determine the distance between two nodes. For a small network, like the one shown in **Figure 2.12**, this is an easy task. For a network with millions of nodes, finding the shortest path between two nodes can be rather time consuming. The length of the shortest path and the number of such paths can be formally obtained from the adjacency matrix (**Box 2.4**). In practice, we use the breadth-first-search (BFS) algorithm discussed in **Box 2.5** for this purpose.

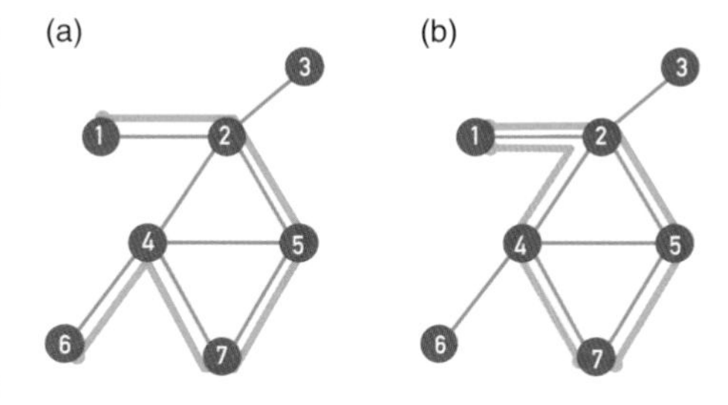

Figure 2.12 **Paths**

(a) A path between nodes i_0 and i_n is an ordered list of n links $P = \{(i_0, i_1), (i_1, i_2), (i_2, i_3), \ldots, (i_n - 1, i_n)\}$. The length of this path is n. The path shown in orange follows the route $1 \to 2 \to 5 \to 7 \to 4 \to 6$, hence its length is $n = 5$.

(b) The shortest paths between nodes 1 and 7, or the distance d_{17}, correspond to the path with the fewest number of links that connect nodes 1 to 7. There can be multiple paths of the same length, as illustrated by the two paths shown in orange and gray. The network diameter is the largest distance in the network, being $d_{max} = 3$ here.

2.8.2 Network Diameter

The *diameter* of a network, denoted by d_{max}, is the maximum shortest path in the network. In other words, it is the largest distance recorded between *any* pair of nodes. One can verify that the diameter of the network shown in **Figure 2.13** is $d_{max} = 3$. For larger networks the diameter can be determined using the BFS algorithm described in **Box 2.5**.

2.8.3 Average Path Length

The *average path length*, denoted by $\langle d \rangle$, is the average distance between all pairs of nodes in the network. For a directed network of N nodes, $\langle d \rangle$ is

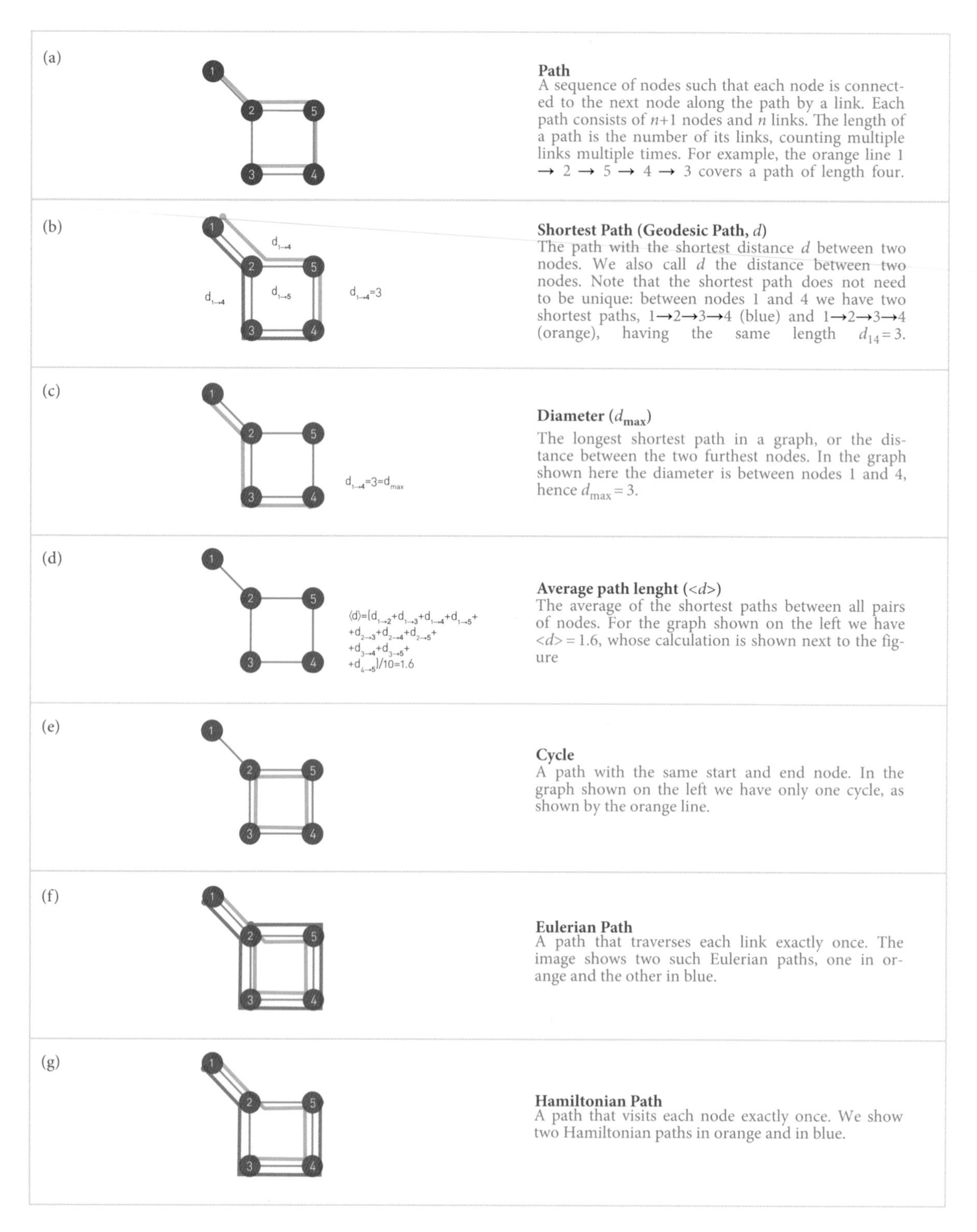

Figure 2.13 **Pathology**

$$d = \frac{1}{N(N-1)} \sum_{\substack{i,j=1,N \\ i \neq j}} d_{i,j}. \quad (2.14)$$

Note that **(2.14)** is measured only for node pairs that are in the same component (**Section 2.9**). We can use the BFS algorithm to determine the average path length for a large network. For this we first determine the distances between the first node and all other nodes in the network using the algorithm described in **Box 2.5**. We then determine the distances between the second node and all other nodes but the first one (if the network is undirected). We then repeat this procedure for all nodes.

2.9 Connectedness

A phone would be of limited use as a communication device if we could not call any valid phone number; email would be rather useless if we could send emails to only certain email addresses and not to others. From a network perspective this means that the network behind the phone or the Internet must be capable of establishing a path between *any* two nodes. This is in fact the key utility of most networks: they ensure *connectedness*. In this section we discuss the graph-theoretic formulation of connectedness.

In an undirected network nodes i and j are *connected* if there is a path between them. They are *disconnected* if such a path does not exist, in which case we have $d_{ij} = \infty$. This is illustrated in **Figure 2.15a**, which shows a network consisting of two disconnected clusters. While there are paths between any two nodes on the same cluster (e.g., nodes 4 and 6), there are no paths between nodes that belong to different clusters (e.g., nodes 1 and 6).

A *network is connected* if all pairs of nodes in the network are connected. A *network is disconnected* if there is at least one pair with $d_{ij} = \infty$. Clearly the network shown in **Figure 2.15a** is disconnected, and we call its two subnetworks *components* or *clusters*. A *component* is a subset of nodes in a network, so that there is a path between any two nodes that belong to the component, but one cannot add any more nodes to it that would have the same property.

Box 2.4 Number of Shortest Paths Between two Nodes

The number of shortest paths, N_{ij} and the distance, d_{ij} between nodes i and j can be calculated directly from the adjacency matrix A_{ij}:

- $d_{ij} = 1$ if there is a direct link between i and j, then $A_{ij} = 1$ (otherwise $A_{ij} = 0$).
- $d_{ij} = 2$ if there is a path of length two between i and j, then $A_{ik}A_{kj} = 1$ (otherwise $A_{ik}A_{kj} = 0$). The number of $d_{ij} = 2$ paths between i and j is

$$N_{ij}^{(2)} = \sum_{k=1}^{N} A_{ik}A_{kj} = A_{ij}^2$$

where $[\ldots]_{ij}$ denotes the (ij)th element of a matrix.

- $d_{ij} = d$ *if there is a path of length* d *between* i *and* j, *then* $A_{ik} \ldots A_{lj} = 1$ (otherwise $A_{ik} \ldots A_{lj} = 0$). The number of paths of length d between i and j is

$$N_{ij}^{(d)} = A_{ij}^d.$$

These equations hold for directed and undirected networks. The *distance* between nodes i and j is the path with the smallest d for which $N_{ij}^{(d)} > 0$. Despite the elegancy of this approach, faced with a large network it is more efficient to use the breadth-first-search algorithm described in **Box 2.5**.

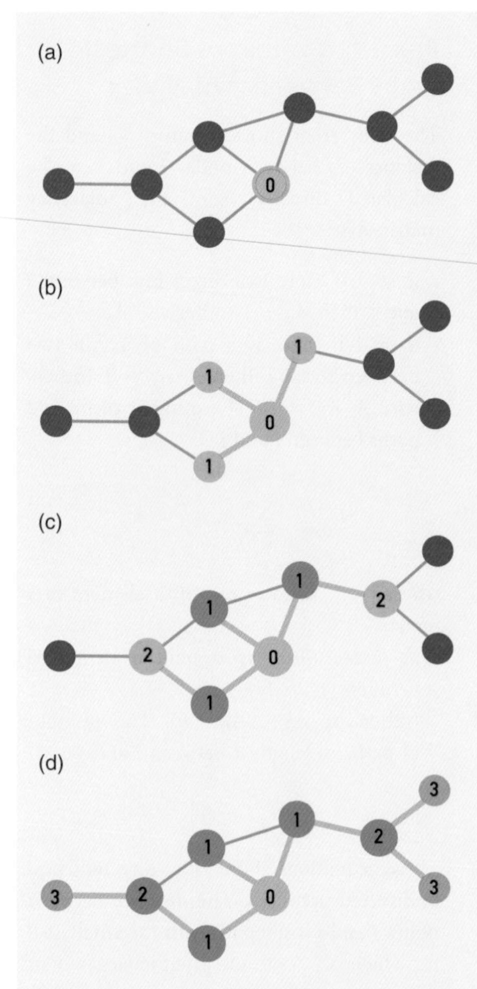

Figure 2.14 **Applying the BFS Algorithm**
(a) Starting from the orange node, labeled "0," we identify all its neighbors, labeling them "1."
(b)–(d) Next we label "2" the unlabeled neighbors of all nodes labeled "1," and so on, in each iteration increasing the label number, until no node is left unlabeled. The length of the shortest path or the distance d_{0i} between node 0 and any other node i in the network is given by the label of node i. For example, the distance between node 0 and the leftmost node is $d = 3$.

Box 2.5 **Breadth-First-Search (BFS) Algorithm**

BFS is a frequently used algorithm in network science. Similar to throwing a pebble in a pond and watching the ripples spread from it, BFS starts from a node and labels its neighbors, then the neighbors' neighbors, until it reaches the target node. The number of "ripples" needed to reach the target provides the distance.

The identification of the shortest path between node i and node j follows the following steps (**Figure 2.14**):

1. Start at node i, which we label "0."
2. Find the nodes directly linked to node i. Label them distance "1" and put them in a queue.
3. Take the first node, labeled n, out of the queue ($n = 1$ in the first step). Find the unlabeled nodes adjacent to it in the graph. Label them and put them in the queue.
4. Repeat step 3 until you find the target node j or there are no more nodes in the queue.
5. The distance between i and j is the label of j. If j does not have a label, then $d_{ij} = \infty$.

The computational complexity of the BFS algorithm, representing the approximate number of steps the computer needs to find d_{ij} on a network of N nodes and L links, is$O(N + L)$. It is linear in N and L as each node needs to be entered and removed from the queue at most once, and each link has to be tested only once.

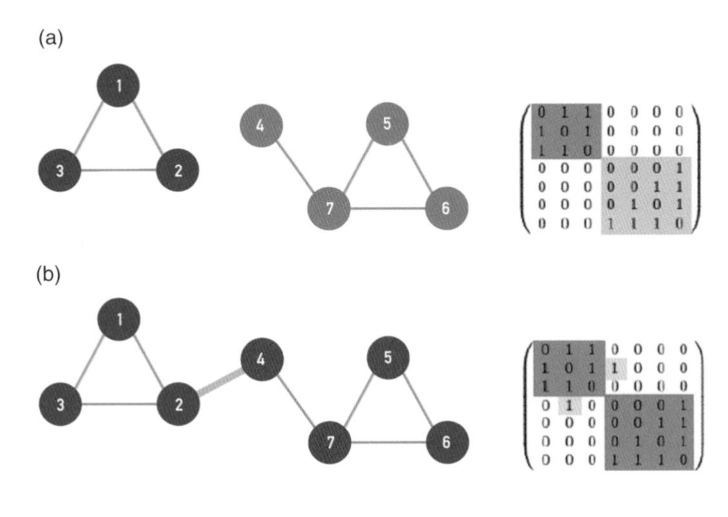

Figure 2.15 **Connected and Disconnected Networks**

(a) A small network consisting of two disconnected components. Indeed, there is a path between any pair of nodes in the (1,2,3) component, as well as in the (4,5,6,7) component. However, there are no paths between nodes that belong to different components.

The right panel shows the adjacently matrix of the network. If the network has disconnected components, the adjacency matrix can be rearranged into a block diagonal form, such that all nonzero elements of the matrix are contained in square blocks along the diagonal of the matrix and all other elements are zero.

(b) The addition of a single link, called a *bridge*, shown in gray, turns a disconnected network into a single connected component. Now there is a path between every pair of nodes in the network. Consequently, the adjacency matrix cannot be written in a block diagonal form.

If a network consists of two components, a properly placed single link can connect them, making the network connected (**Figure 2.15b**). Such a link is called a *bridge*. In general a bridge is any link that, if cut, disconnects the network.

While for a small network visual inspection can help us decide if it is connected or disconnected, for a network consisting of millions of nodes connectedness is a challenging question. Mathematical and algorithmic tools can help us identify the connected components of a graph. For example, for a disconnected network the adjacency matrix can be rearranged into a block diagonal form, such that all nonzero elements in the matrix are contained in square blocks along the matrix's diagonal and all other elements are zero (**Figure 2.15a**). Each square block corresponds to a component. We can use the tools of linear algebra to decide if the adjacency matrix is block diagonal, helping us to identify the connected components.

In practice, for large networks the components are more efficiently identified using the BFS algorithm (**Box 2.6**).

Box 2.6 Finding the Connected Components of a Network

Step 1. Start from a randomly chosen node i and perform a BFS (**Box 2.5**). Label all nodes reached in this way $n = 1$.

Step 2. If the total number of labeled nodes equals N, then the network is connected. If the number of labeled nodes is smaller than N, then the network consists of several components. To identify them, proceed to step 3.

Step 3. Increase the label $n \to n + 1$. Choose an unmarked node j, label it n. Use BFS to find all nodes reachable from j, label them all n. Return to step 2.

2.10 Clustering Coefficient

The clustering coefficient captures the degree to which the neighbors of a given node link to each other. For a node i with degree k_i, the *local clustering coefficient* is defined as [8]

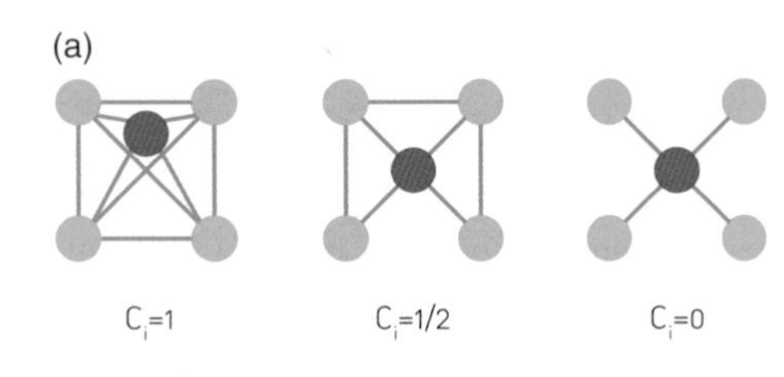

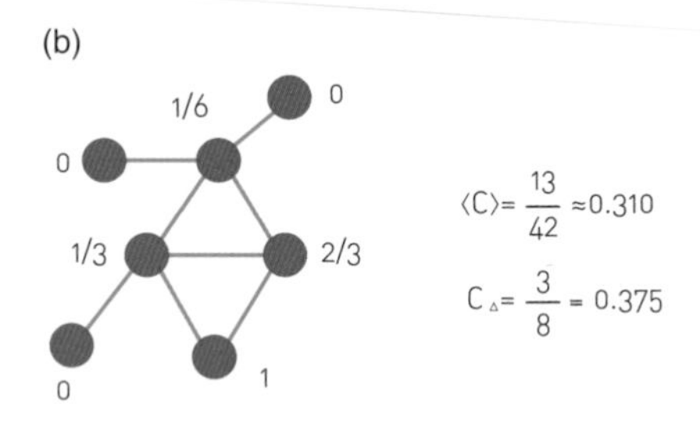

Figure 2.16 **Clustering Coefficient**
(**a**) The local clustering coefficient, C_i, of the central node with degree $k_i = 4$ for three different configurations of its neighborhood. The local clustering coefficient measures the local density of links in a node's vicinity.
(**b**) A small network, with the local clustering coefficient of each node shown next to it. We also list the network's average clustering coefficient $\langle C \rangle$, according to (**2.16**), and its global clustering coefficient C_Δ, defined in **Section 2.12**, equation (**2.17**). Note that for nodes with degree $k_i = 0, 1$, the clustering coefficient is zero.

$$C_i = \frac{2L_i}{k_i(k_i - 1)}, \tag{2.15}$$

where L_i represents the number of links between the k_i neighbors of node i. Note that C_i is between 0 and 1 (**Figure 2.16a**):

- $C_i = 0$ if none of the neighbors of node i link to each other.
- $C_i = 1$ if the neighbors of node i form a complete graph, that is they all link to each other.
- *C_i is the probability that two neighbors of a node link to each other. Consequently, $C = 0.5$ implies that there is a 50% chance that two neighbors of a node are linked.*

In summary, C_i measures the network's local link density: the more densely interconnected the neighborhood of node i, the higher is its local clustering coefficient.

The degree of clustering of a whole network is captured by the *average clustering coefficient* $\langle C \rangle$, representing the average of C_i over all nodes $i = 1, \ldots, N$ [8]

$$\langle C \rangle = \frac{1}{N} \sum_{i=1}^{N} C_i. \tag{2.16}$$

In line with the probabilistic interpretation, $\langle C \rangle$ is the probability that two neighbors of a randomly selected node link to each other.

While (**2.16**) is defined for undirected networks, the clustering coefficient can be generalized to directed and weighted [65, 66, 67, 68] networks as well. In the network literature we may encounter the *global clustering coefficient* as well, discussed in **Advanced topics 2.A**.

2.11 Summary

The crash course offered in this chapter has introduced some of the basic graph-theoretical concepts and tools used in network science. The set of elementary network characteristics, summarized in **Figure 2.17**, offers a formal language through which we can explore networks.

Many of the networks we study in network science consist of thousands or even millions of nodes and links (**Table 2.1**). To explore them, we need to go beyond the small graphs shown in **Figure 2.17**. A glimpse of what we are about to encounter is offered by the protein–protein interaction network of yeast

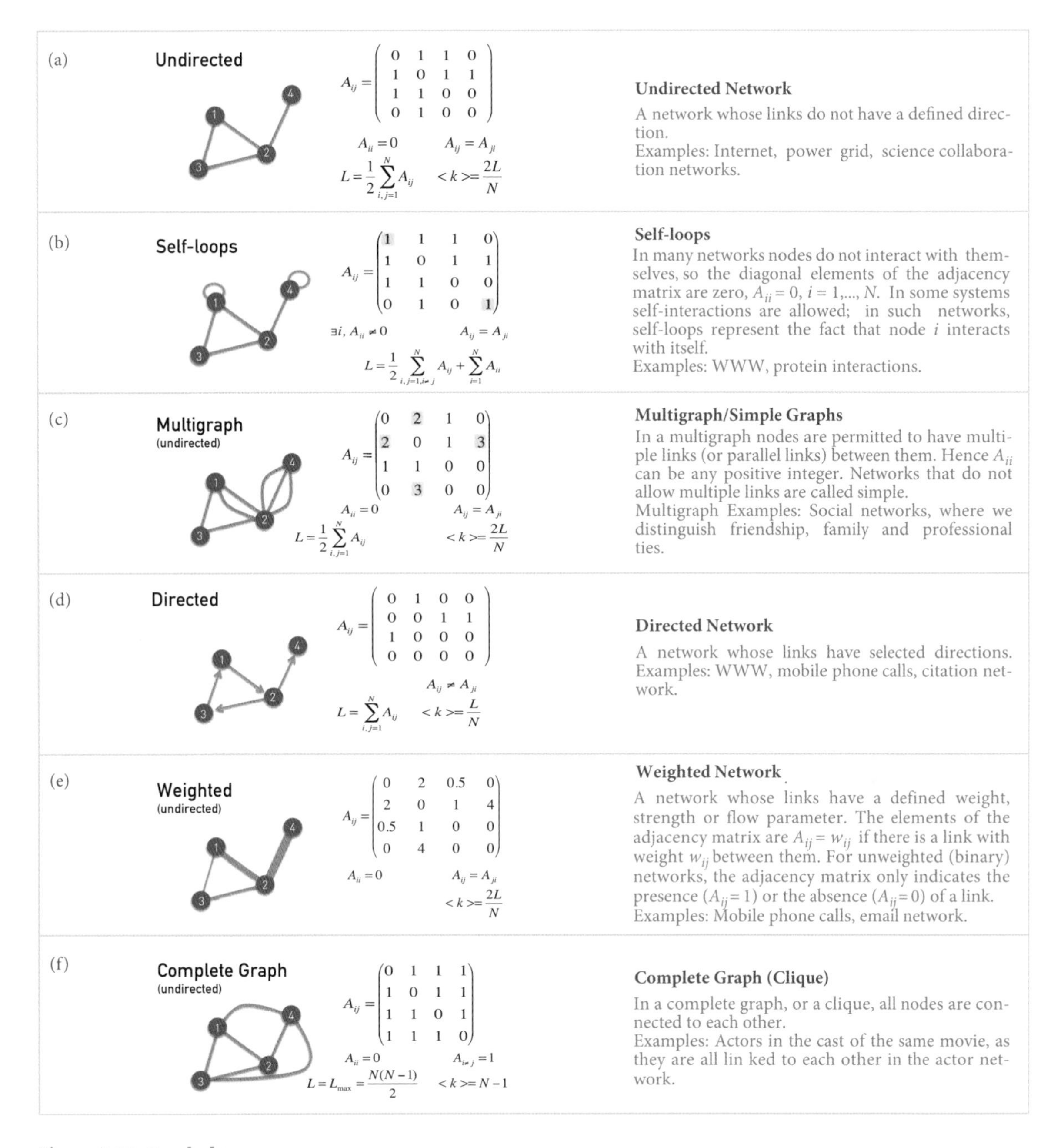

Figure 2.17 **Graphology**

In network science we often distinguish networks by some elementary property of the underlying graph. Here we summarize the most commonly encountered network types. We also list real systems that share the particular property. Note that many real networks combine several of these elementary network characteristics. For example, the WWW is a directed multi-graph with self-interactions; the mobile call network is directed and weighted, without self-loops.

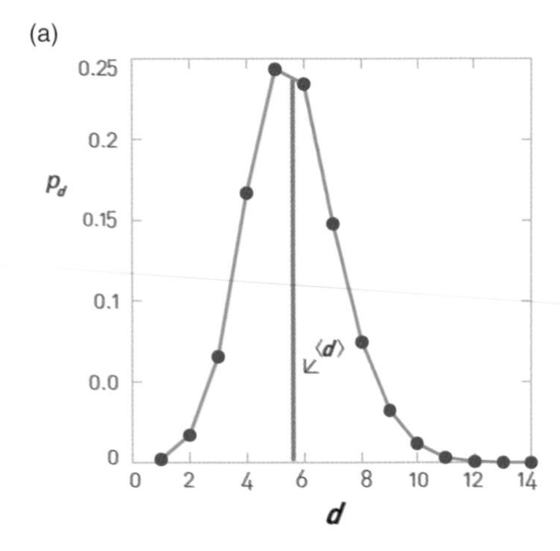

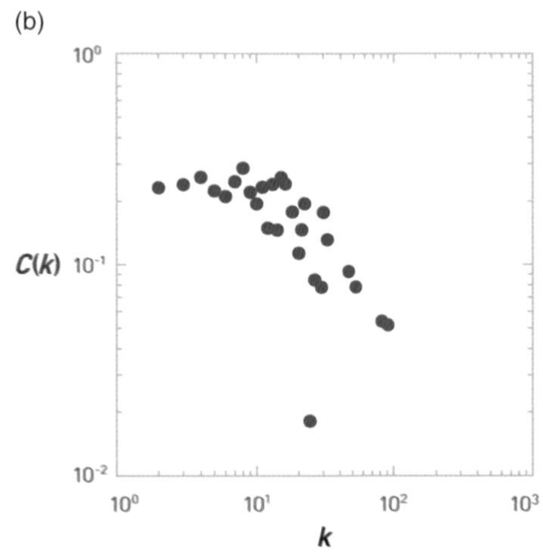

Figure 2.18 **Characterizing a Real Network**
The protein–protein interaction (PPI) network of yeast is frequently studied by biologists and network scientists. The detailed wiring diagram is shown in **Figure 2.4a**, indicating, that the network has a large component that connects 81% of the proteins and several smaller components.
(a) The distance distribution p_d for the PPI network, providing the probability that two randomly chosen nodes have a distance d between them (shortest path). The gray vertical line shows the average path length, which is $\langle d \rangle = 5.61$.
(b) The dependence of the average local clustering coefficient on the node's degree, k, obtained by averaging over the local clustering coefficient of all nodes with the same degree k.

(**Figure 2.4a**). The network is too complex to understand its properties through a visual inspection of its wiring diagram. We therefore need to turn to the tools of network science to characterize its topology.

Let us use the measures introduced so far to explore some basic characteristics of this network. The undirected network, shown in **Figure 2.4a**, has $N = 2{,}018$ proteins as nodes and $L = 2{,}930$ binding interactions as links. Hence its average degree, according to (**2.2**), is $\langle k \rangle = 2.90$, suggesting that a typical protein interacts with approximately two to three other proteins. Yet this number is somewhat misleading. Indeed, the degree distribution p_k shown in **Figure 2.4b,c** indicates that the vast majority of nodes have only a few links. To be precise, in this network 69% of nodes have fewer than three links, that is for these $k < \langle k \rangle$. These numerous nodes with few links coexist with a few highly connected nodes, or hubs, the largest having as many as 92 links. Such wide differences in node degree is a consequence of the network's scale-free property, discussed in **Chapter 4**. We will see that the shape of the degree distribution determines a wide range of network properties, from the network's robustness to the spread of viruses.

The BFS algorithm (**Box 2.5**) helps us determine the network's diameter, finding $d_{\max} = 14$. We might be tempted to expect wide variations in d, as some nodes are close to each other and others are quite far away. The distance distribution (**Figure 2.18a**) indicates otherwise: p_d has a prominent peak between 5 and 6, telling us that most distances are rather short, being in the vicinity of $\langle d \rangle = 5.61$. Also, p_d decays fast for large d, suggesting that large distances are absent. Indeed, the variance of the distances is $\sigma_d = 1.64$, indicating that most path lengths are in the close vicinity of $\langle d \rangle$. These are manifestations of the small-world property discussed in **Chapter 3**.

The BFS algorithm also tells us that the protein interaction network is not connected, but consists of 185 components, shown as isolated clusters and nodes in **Figure 2.4a**. The largest, called the giant component, contains 1,647 of the 2,018 nodes; all other components are tiny. As we will see in the coming chapters, such fragmentation is common in real networks.

The average clustering coefficient of the protein interaction network is $\langle C \rangle = 0.12$, which, as we will come to appreciate in the following chapters, indicates a significant degree of local clustering. A further caveat is provided by the dependence of the clustering coefficient on the node's degree, or the $C(k)$ function (**Figure 2.18b**). The fact that $C(k)$ decreases for large k indicates that the local clustering coefficient of the small nodes is significantly higher than the local clustering coefficient of the hubs. Hence, the small degree nodes are located in dense local network neighborhoods, while the neighborhood of the hubs is much sparser. This is a consequence of *hierarchy*, a network property discussed in **Chapter 9**.

Finally, a visual inspection reveals an interesting pattern: hubs have a tendency to connect to small nodes, giving the network a hub-and-spoke character (**Figure 2.4a**). This is a consequence of degree correlations, discussed in **Chapter 7**. Such correlations influence a number of network-based processes, from spreading phenomena to the number of driver nodes needed to control a network.

Taken together, **Figures 2.4** and **2.18** illustrate that the quantities we introduced in this chapter can help us diagnose several key properties of real networks. The purpose of the coming chapters is to study systematically these network characteristics and understand what they tell us about a particular complex system.

2.12 Homework

2.12.1 Königsberg Problem

Which of the icons in **Figure 2.19** can be drawn without raising your pencil from the paper, and without drawing any line more than once? Why?

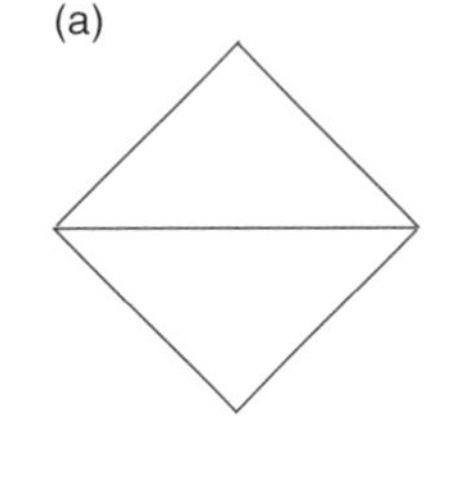

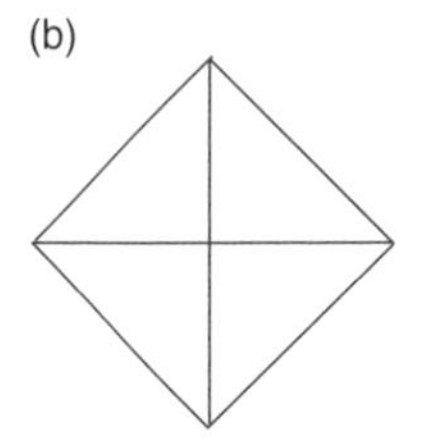

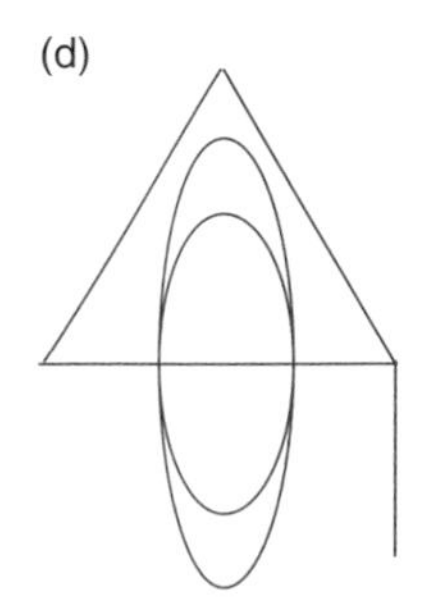

Figure 2.19 **Königsberg Problem**

2.12.2 Matrix Formalism

Let A be the $N \times N$ adjacency matrix of an undirected unweighted network, without self-loops. Let **1** be a column vector of N elements, all equal to 1. In other words, $\mathbf{1} = (1, 1, ..., 1)^T$, where the superscript T indicates the *transpose* operation. Use the matrix formalism (multiplicative constants, multiplication row by column, matrix operations like transpose and trace, etc. but avoid the sum symbol $\sum$) to write expressions for:

(a) The vector $\mathbf{k}$ whose elements are the degrees k_i of all nodes $i = 1, 2, \ldots, N$.
(b) The total number of links, L, in the network.
(c) The number of triangles T present in the network, where a triangle means three nodes, each connected by links to the other two (*Hint:* you can use the trace of a matrix).
(d) The vector $\mathbf{k}_{nn}$ whose element i is the sum of the degrees of node i's neighbors.
(e) The vector $\mathbf{k}_{nnn}$ whose element i is the sum of the degrees of node i's second neighbors.

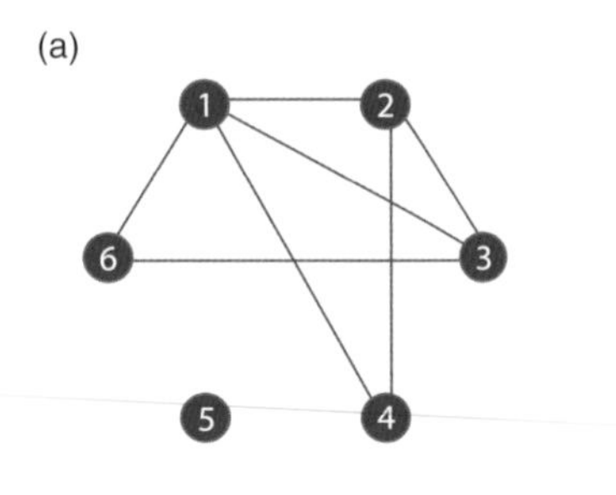

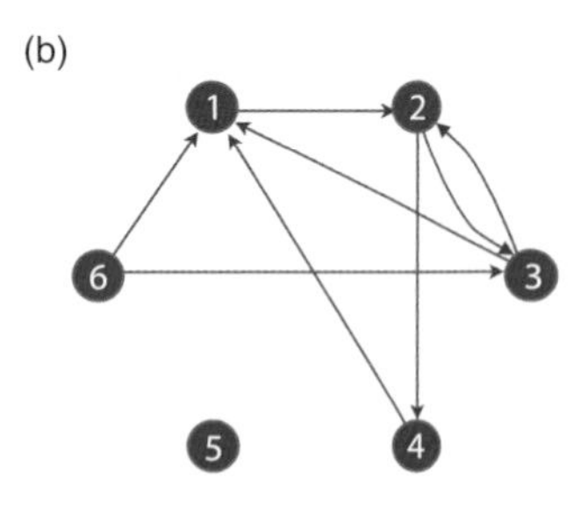

Figure 2.20 **Graph Representation**
(**a**) Undirected graph of 6 nodes and 7 links.
(**b**) Directed graph of 6 nodes and 8 directed links.

2.12.3 Graph Representation

The adjacency matrix is a useful graph representation for many analytical calculations. However, when we need to store a network in a computer, we can save computer memory by offering the list of links in an $L \times 2$ matrix, whose rows contain the starting and end point i and j of each link.

Construct for the networks (a) and (b) in **Figure 2.20**:

(a) The corresponding adjacency matrices.
(b) The corresponding link lists.
(c) Determine the average clustering coefficient of the network shown in **Figure 2.20a**.
(d) If you switch the labels of nodes 5 and 6 in **Figure 2.20a**, how does that move change the adjacency matrix? And the link list?
(e) What kind of information can you not infer from the link list representation of the network that you can infer from the adjacency matrix?
(f) In network (a), how many paths (with possible repetition of nodes and links) of length 3 exist starting from node 1 and ending at node 3? And in (b)?
(g) With the help of a computer, count the number of cycles of length 4 in both networks.

2.12.4 Degree, Clustering Coefficient and Components

(a) Consider an undirected network of size N in which each node has degree $k = 1$. Which condition does N have to satisfy? What is the degree distribution of this network? How many components does the network have?

(b) Consider now a network in which each node has degree $k = 2$ and clustering coefficient $C = 1$. How does the network look now? What condition does N satisfy in this case?

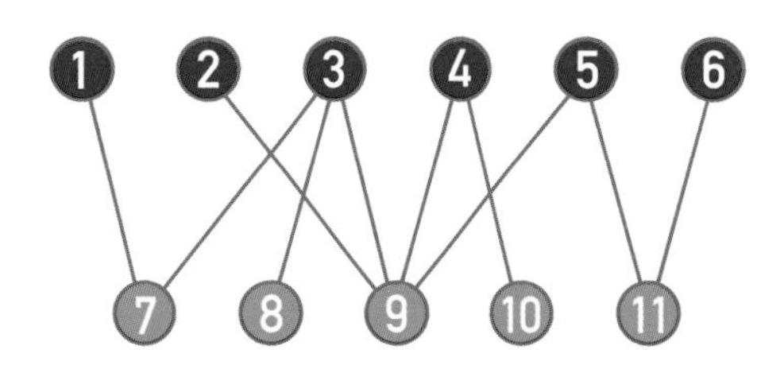

Figure 2.21 **Bipartite Network**
Bipartite network with 6 nodes in one set and 5 nodes in the other, connected by 10 links.

2.12.5 Bipartite Networks

Consider the bipartite network of **Figure 2.21**:

(a) Construct its adjacency matrix. Why is it a block-diagonal matrix?

(b) Construct the adjacency matrix of its two projections, on the purple and on the green nodes, respectively.

(c) Calculate the average degree of the purple nodes and the average degree of the green nodes in the bipartite network.

(d) Calculate the average degree in each of the two network projections. Is it surprising that the values are different from those obtained in (c)?

2.12.6 Bipartite Networks – General Considerations

Consider a bipartite network with N_1 and N_2 nodes in the two sets:

(a) What is the maximum number of links L_{max} the network can have?

(b) How many links cannot occur compared with a non-bipartite network of size $N = N_1 + N_2$?

(c) If $N_1 \ll N_2$, what can you say about the network density, that is the total number of links over the maximum number of links L_{max}?

(d) Find an expression connecting N_1, N_2 and the average degree for the two sets in the bipartite network, $\langle k_1 \rangle$ and $\langle k_2 \rangle$.

2.13 Advanced Topics 2.A Global Clustering Coefficient

In the network literature we occasionally encounter the *global clustering coefficient*, which measures the total number of closed triangles in a network. Indeed, L_i in (**2.15**) is the number of triangles that node i participates in, as each link between two neighbors of node i closes a triangle (**Figure 2.17**). Hence, the degree of a network's global clustering can also be captured by the *global clustering coefficient*, defined as

$$C_\Delta = \frac{3 \times \text{number of triangles}}{\text{number of connected triples}}, \tag{2.17}$$

where a *connected triplet* is an ordered set of three nodes ABC such that A connects to B and B connects to C. For example, an A, B, C triangle is made up of three triplets ABC, BCA and CAB. In contrast, a chain of connected nodes A, B, C, in which B connects to A and C but A does not link to C, forms a single open triplet ABC. The factor three in the numerator of (**2.17**) is due to the fact that each triangle is counted three times in the triplet count. The roots of the global clustering coefficient go back to the social network literature of the 1940s [69, 70], where C_Δ is often called the *ratio of transitive triplets.*

Note that the average clustering coefficient $\langle C \rangle$ defined in (**2.16**) and the global clustering coefficient (**2.17**) are not equivalent. Indeed, take a network that is a double star, consisting of N nodes, where nodes 1 and 2 are joined to each other and to all other nodes, and there are no other links. Then the local clustering coefficient C_i is 1 for $i \geq 3$ and $2/(N-1)$ for $i = 1, 2$. It follows that the average clustering coefficient of the network is $\langle C \rangle = 1 - O(1)$, while the global clustering coefficient is $C_\Delta \sim 1/N$. In less extreme networks the two definitions will give more comparable values, but they still differ from each other [71]. For example, for the network in **Figure 2.16b** we have $\langle C \rangle = 0.31$ and $C_\Delta = 0.375$.

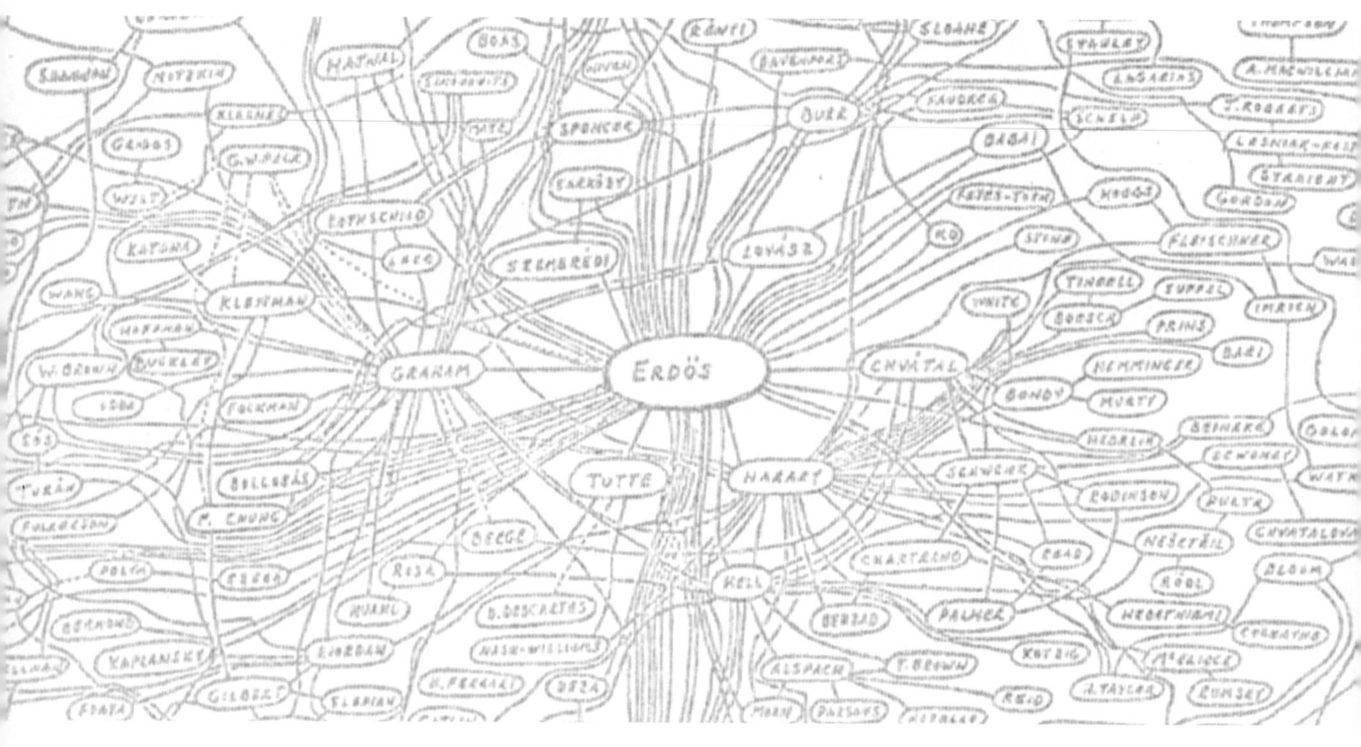

Figure 3.0 **Art & Networks: Erdős Number**

The Hungarian mathematician Pál Erdős authored hundreds of research papers, many of them in collaboration with other mathematicians. His relentless collaborative approach to mathematics inspired the *Erdős number*, which works like this: Erdős' Erdős number is 0. Erdős' co-authors have Erdős number 1. Those who have written a paper with someone with Erdős number 1 have Erdős number 2, and so on. If there is no chain of co-authorships connecting someone to Erdős, then that person's Erdős number is infinite. Many famous scientists have low Erdős numbers: Albert Einstein has Erdős number 2 and Richard Feynman has 3. The image shows the collaborators of Pál Erdős, as drawn in 1970 by Ronald Graham, one of Erdős' close collaborators. As Erdős' fame rose, this image has achieved iconic status.

CHAPTER 3

Random Networks

3.1 Introduction

Imagine organizing a party for a hundred guests who initially do not know each other [50]. Offer them wine and cheese and you will soon see them chatting in groups of two to three. Now mention to Mary, one of your guests, that the red wine in the unlabeled dark green bottles is a rare vintage, much better than the one with the fancy red label. If she shares this information only with her acquaintances, your expensive wine appears to be safe, as she has only had time to meet a few others so far.

The guests will continue to mingle, however, creating subtle paths between individuals who may still be strangers to each other. For example, while John has not yet met Mary, they have

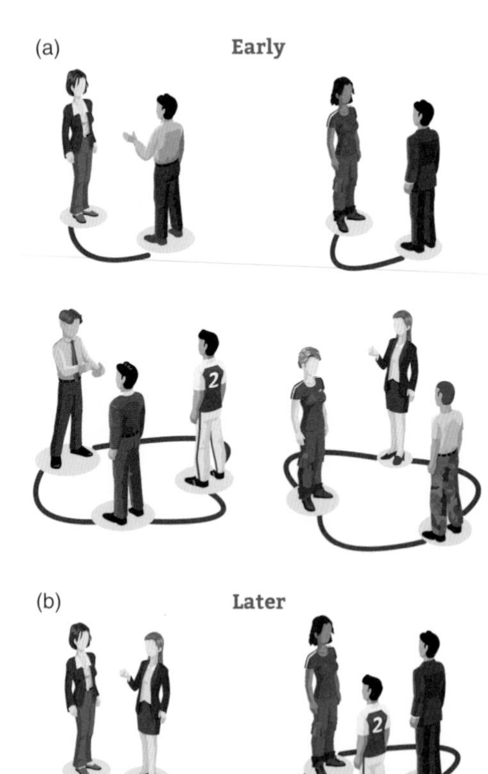

Figure 3.1 From a Cocktail Party to Random Networks
The emergence of an acquaintance network through random encounters at a cocktail party. (**a**) Early on the guests form isolated groups. (**b**) As individuals mingle, changing groups, an invisible network emerges that connects all of them into a single network.

both met Mike, so there is an invisible path from John to Mary through Mike. As time goes on, the guests will be increasingly interwoven by such elusive links. With that the secret of the unlabeled bottle will pass from Mary to Mike and from Mike to John, escaping into a rapidly expanding group (**Figure 3.1**).

To be sure, when all the guests had gotten to know each other, everyone would be pouring the superior wine. But if each encounter took only ten minutes, meeting all ninety-nine others would take about sixteen hours. Thus, you could reasonably hope that a few drops of your fine wine would be left for you to enjoy once the guests are gone.

Yet you would be wrong. In this chapter we show you why. We will see that the party maps into a classic model in network science called the random network model. And random network theory tells us that we do not have to wait until *all* individuals get to know each other for our expensive wine to be in danger. Rather, soon after each person meets at least one other guest, an invisible network will emerge allowing the information to reach all of them. Hence in no time everyone will be enjoying the better wine.

3.2 The Random Network Model

Network science aims to build models that reproduce the properties of real networks. Most networks we encounter do not have the comforting regularity of a crystal lattice or the predictable radial architecture of a spider web. Rather, at first inspection they look as if they were spun randomly (**Figure 2.4**). Random network theory embraces this apparent randomness by constructing and characterizing networks that are *truly random*.

From a modeling perspective a network is a relatively simple object, consisting of only nodes and links. The real challenge, however, is to decide where to place the links between the nodes so that we reproduce the complexity of a real system. In this respect the philosophy behind a random network is simple: we assume that this goal is best achieved by placing the links randomly between the nodes. That takes us to the definition of a random network (**Box 3.1**):

A random network consists of N nodes where each node pair is connected with probability p.

To construct a random network we follow these steps:

1. Start with N isolated nodes.
2. Select a node pair and generate a random number between 0 and 1. If the random number is less than p, connect the selected node pair with a link, otherwise leave them disconnected.
3. Repeat step 2 for each of the $N(N-1)/2$ node pairs.

The network obtained after this procedure is called a *random graph* or a *random network*. Two mathematicians, Pál Erdős and Alfréd Rényi, have played an important role in understanding the properties of these networks. In their honor, a random network is called an *Erdős–Rényi network* (**Box 3.2**).

3.3 Number of Links

Each random network generated with the same parameters N, p looks slightly different (**Figure 3.3**). Not only does the detailed wiring diagram change between realizations, but so does the number of links L. It is useful, therefore, to determine how many links we expect for a particular realization of a random network with fixed N and p.

The probability that a random network has exactly L links is the product of three terms:

1. The probability that L of the attempts to connect the $N(N-1)/2$ pairs of nodes have resulted in a link, which is p^L.
2. The probability that the remaining $N(N-1)/2 - L$ attempts have not resulted in a link, which is $(1-p)^{N(N-1)/2-L}$.
3. A combinational factor

$$\binom{\frac{N(N-1)}{2}}{L} \tag{3.0}$$

counting the number of different ways we can place L links among $N(N-1)/2$ node pairs.

We can therefore write the probability that a particular realization of a random network has exactly L links as

Box 3.1 Defining Random Networks

There are two definitions of a random network:

$G(N, L)$ Model

N labeled nodes are connected with L randomly placed links. Erdős and Rényi used this definition in their string of papers on random networks [3, 72, 73, 74, 75, 76, 77, 78].

$G(N, p)$ Model

Each pair of N labeled nodes is connected with probability p, a model introduced by Gilbert [79].

Hence, the $G(N,p)$ model fixes the probability p that two nodes are connected and the $G(N,L)$ model fixes the total number of links L. While in the $G(N,L)$ model the average degree of a node is simply $\langle k \rangle = 2L/N$, other network characteristics are easier to calculate in the $G(N,p)$ model. Throughout this book we will explore the $G(N,p)$ model, not only for the ease with which it allows us to calculate key network characteristics, but also because in real networks the number of links rarely stays fixed.

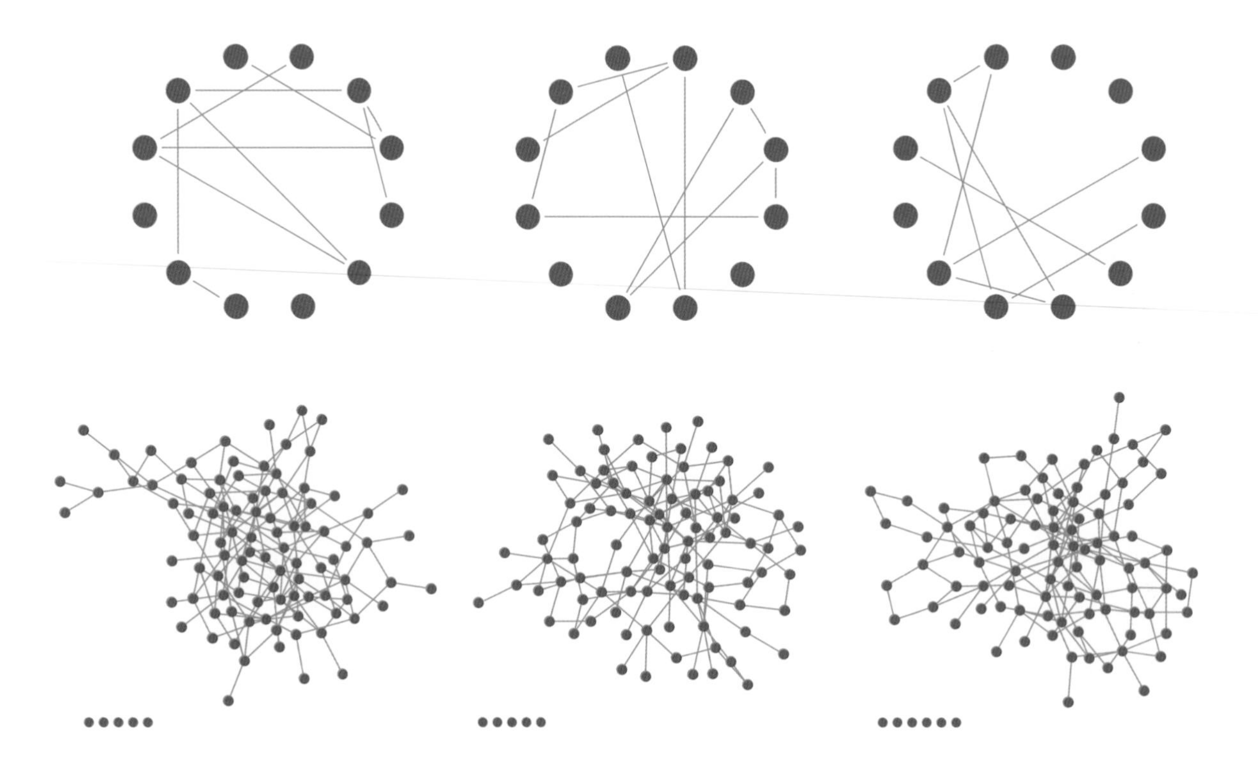

Figure 3.3 **Random Networks are Truly Random**

Top row

Three realizations of a random network generated with the same parameters $p = 1/6$ and $N = 12$. Despite the identical parameters, the networks not only look different but also have a different number of links as well ($L = 10$, 10, 8).

Bottom row

Three realizations of a random network with $p = 0.03$ and $N = 100$. Several nodes have degree $k = 0$, shown as isolated nodes at the bottom.

$$p_L = \begin{pmatrix} \frac{N(N-1)}{2} \\ L \end{pmatrix} p^L (1-p)^{\frac{N(N-1)}{2}-L}. \qquad (3.1)$$

As **(3.1)** is a binomial distribution (**Box 3.3**), the expected number of links in a random graph is

$$\langle L \rangle = \sum_{L=0}^{\frac{N(N-1)}{2}} L p_L = p\frac{N(N-1)}{2}. \qquad (3.2)$$

Hence, $\langle L \rangle$ is the product of the probability p that two nodes are connected and the number of pairs we attempt to connect, which is $L_{max} = N(N-1)/2$ (**Chapter 2**).

Using **(3.2)** we obtain the average degree of a random network

$$\langle k \rangle = \frac{2\langle L \rangle}{N} = p(N-1). \qquad (3.3)$$

Hence, $\langle k \rangle$ is the product of the probability p that two nodes are connected and $(N-1)$, which is the maximum number of links a node can have in a network of size N.

Box 3.2 Random Networks: A Brief History

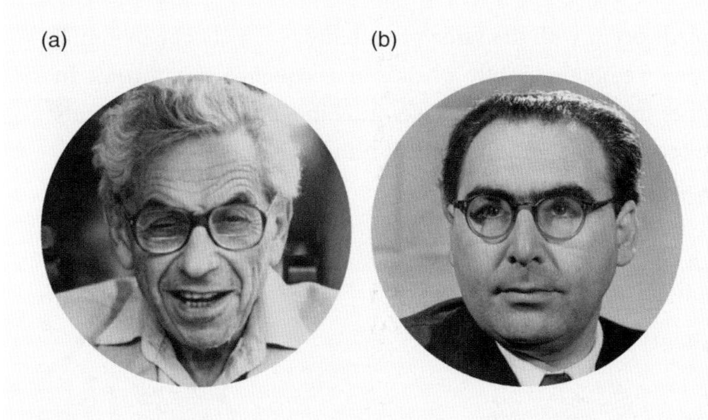

Figure 3.2

(a) **Pál Erdős (1913–1996)**

Hungarian mathematician known for both his exceptional scientific output and his eccentricity. Indeed, Erdős published more papers than any other mathematician in the history of mathematics. He co-authored papers with over five hundred mathematicians, inspiring the concept of the Erdős number. His legendary personality and profound professional impact inspired two biographies [81, 82] and a documentary [83] (**Online resource 3.1**).

(b) **Alfréd Rényi (1921–1970)**

Hungarian mathematician with fundamental contributions to combinatorics, graph theory and number theory. His impact goes beyond mathematics: Rényi entropy is widely used in chaos theory and the random network theory he co-developed is at the heart of network science. He is remembered through that hotbed of Hungarian mathematics, the Alfred Rényi Institute of Mathematics in Budapest.

Anatol Rapoport (1911–2007), a Russian immigrant to the USA, was the first to study random networks. Rapoport's interest turned to mathematics after realizing that a successful career as a concert pianist would require a wealthy patron. He focused on mathematical biology at a time when mathematicians and biologists hardly spoke to each other. In a paper written with Ray Solomonoff in 1951 [80], Rapoport demonstrated that if we increase the average degree of a network, we observe an abrupt transition from disconnected nodes to a graph with a giant component.

The study of random networks reached prominence thanks to the fundamental work of Pál Erdős and Alfréd Rényi (**Figure 3.2**). In a sequence of eight papers published between 1959 and 1968 [3, 72, 73, 74, 75, 76, 77, 78], they merged probability theory and combinatorics with graph theory, establishing *random graph theory*, a new branch of mathematics.

The random network model was independently introduced by Edgar Nelson Gilbert (1923–2013) [79], the same year Erdős and Rényi published their first paper on the subject. Yet the impact of Erdős and Rényi's work is so overwhelming that they are rightly considered the founders of random graph theory.

"A mathematician is a device for turning coffee into theorems"
Alfréd Rényi (a quote often attributed to Erdős)

Online Resource 3.1

N is a Number: A Portrait of Paul Erdős

The 1993 biographical documentary of Pál Erdős, directed by George Paul Csicsery, offers a glimpse into Erdős' life and scientific impact [83].

Box 3.3 Binomial Distribution: Mean and Variance

If we toss a fair coin N times, tails and heads occur with the same probability $p = 1/2$. The binomial distribution provides the probability p_x that we obtain exactly x heads in a sequence of N throws. In general, the binomial distribution describes the number of successes in N independent experiments with two possible outcomes, in which the probability of one outcome is p and that of the other is $1 - p$.

The binomial distribution has the form

$$p_x = \binom{N}{x} p^x (1-p)^{N-x}.$$

The mean of the distribution (first moment) is

$$\langle x \rangle = \sum_{x=0}^{N} x p_x = Np. \tag{3.4}$$

Its second moment is

$$\langle x^2 \rangle = \sum_{x=0}^{N} x^2 p_x = p(1-p)N + p^2 N^2, \tag{3.5}$$

providing its standard deviation as

$$\sigma_x = \left(\langle x^2 \rangle - \langle x \rangle^2 \right)^{\frac{1}{2}} = [p(1-p)N]^{\frac{1}{2}}. \tag{3.6}$$

Equations **(3.4)–(3.6)** are used repeatedly as we characterize random networks.

In summary, the number of links in a random network varies between realizations. Its expected value is determined by N and p. If we increase p, a random network becomes denser: the average number of links increases linearly from $\langle L \rangle = 0$ to $L_{\max}$ and the average degree of a node increases from $\langle k \rangle = 0$ to $\langle k \rangle = N - 1$.

3.4 Degree Distribution

In a given realization of a random network some nodes gain numerous links, while others acquire only a few or no links (**Figure 3.3**). These differences are captured by the degree distribution p_k, which is the probability that a randomly chosen node has degree k. In this section we derive p_k for a random network and discuss its properties.

3.4.1 Binomial Distribution

In a random network the probability that node i has exactly k links is the product of three terms [2]:

- The probability that k of its links are present, or p^k.
- The probability that the remaining $(N-1-k)$ links are missing, or $(1-p)^{N-1-k}$.
- The number of ways we can select k links from the $N-1$ potential links a node can have, or

$$\binom{N-1}{k}.$$

Consequently, the degree distribution of a random network follows the binomial distribution:

$$p_k = \binom{N-1}{k} p^k (1-p)^{N-1-k}. \tag{3.7}$$

The shape of this distribution depends on the system size N and the probability p (**Figure 3.4**). The binomial distribution (**Box 3.3**) allows us to calculate the network's average degree $\langle k \rangle$, recovering **(3.3)**, as well as its second moment $\langle k^2 \rangle$ and variance σ (**Figure 3.4**).

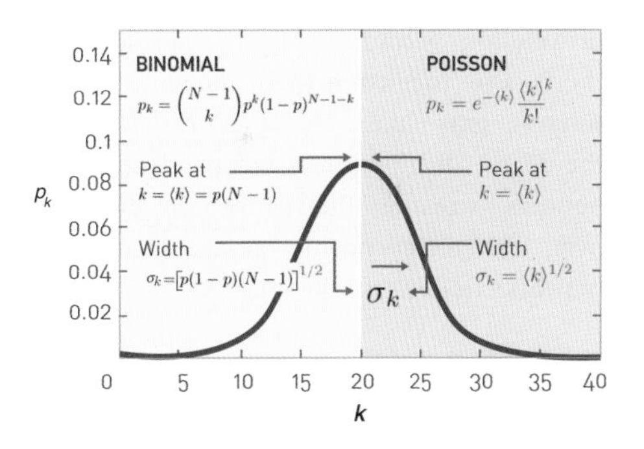

Figure 3.4 **Binomial vs. Poisson Degree Distribution**

The exact form of the degree distribution of a random network is the binomial distribution (left half). For $N \gg \langle k \rangle$ the binomial is well approximated by a Poisson distribution (right half). As both formulas describe the same distribution, they have identical properties but are expressed in terms of different parameters: the binomial distribution depends on p and N while the Poisson distribution has only one parameter, $\langle k \rangle$. It is this simplicity that makes the Poisson form preferred in calculations.

3.4.2 Poisson Distribution

Most real networks are sparse, meaning that $\langle k \rangle \ll N$ (**Table 2.1**). In this limit the degree distribution **(3.7)** is well approximated by the Poisson distribution (**Advanced topics 3.A**):

$$p_k = e^{-\langle k \rangle} \frac{\langle k \rangle^k}{k!}, \tag{3.8}$$

which is often called, together with **(3.7)**, the *degree distribution of a random network*.

The binomial and the Poisson distribution describe the same quantity, hence they have similar properties (**Figure 3.4**):

- Both distributions have a peak around $\langle k \rangle$. If we increase p, the network becomes denser, increasing $\langle k \rangle$ and moving the peak to the right.

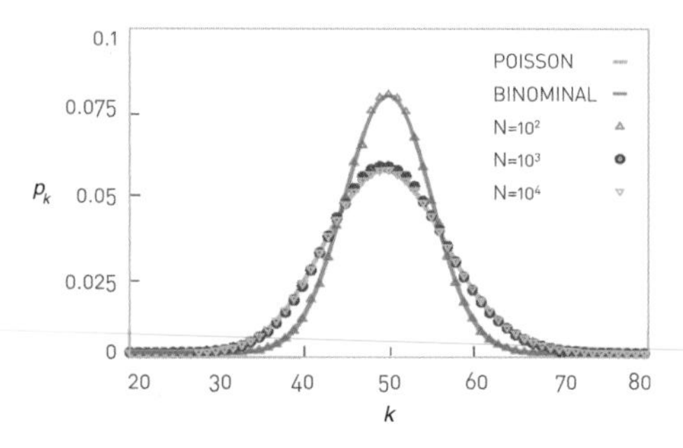

Figure 3.5 **Degree Distribution is Independent of the Network Size**

The degree distribution of a random network with $\langle k \rangle = 50$ and $N = 10^2, 10^3, 10^4$.

Small Networks: Binomial

For a small network ($N = 10^2$) the degree distribution deviates significantly from the Poisson form **(3.8)**, as the condition for the Poisson approximation $N \gg \langle k \rangle$ is not satisfied. Hence, for small networks one needs to use the exact binomial form **(3.7)** (green line).

Large Networks: Poisson

For larger networks ($N = 10^3, 10^4$) the degree distribution becomes indistinguishable from the Poisson prediction **(3.8)**, shown as a continuous gray line. Therefore for large N the degree distribution is independent of the network size. In the figure we averaged over 1,000 independently generated random networks to decrease the noise.

- The width of the distribution (dispersion) is also controlled by p or $\langle k \rangle$. The denser the network, the wider is the distribution, hence the larger are the differences in the degrees.

When we use the Poisson form **(3.8)**, we need to keep in mind that:

- The exact result for the degree distribution is the binomial form **(3.7)**, thus **(3.8)** represents only an approximation to **(3.7)** valid in the $\langle k \rangle \ll N$ limit. As most networks of practical importance are sparse, this condition is typically satisfied.
- The advantage of the Poisson form is that key network characteristics, like $\langle k \rangle$, $\langle k^2 \rangle$ and σ, have a much simpler form (**Figure 3.4**), depending on a single parameter, $\langle k \rangle$.
- The Poisson distribution in **(3.8)** does not depend explicitly on the number of nodes N. Therefore, **(3.8)** predicts that the degree distributions of networks of different sizes but the same average degree $\langle k \rangle$ are indistinguishable from each other (**Figure 3.5**).

In summary, while the Poisson distribution is only an approximation to the degree distribution of a random network, thanks to its analytical simplicity, it is the preferred form for p_k. Hence throughout this book, unless noted otherwise, we will refer to the Poisson form **(3.8)** as the degree distribution of a random network. Its key feature is that its properties are independent of the network size and depend on a single parameter, the average degree $\langle k \rangle$.

3.5 Real Networks are not Poisson

As the degree of a node in a random network can vary between 0 and $N - 1$, we must ask how big the differences are between the node degrees in a particular realization of a random network. That is, can high-degree nodes coexist with small-degree nodes? We address these questions by estimating the size of the largest and the smallest node in a random network.

Let us assume that the world's social network is described by the random network model. This random society may not be as far fetched as it first sounds: there is significant randomness in whom we meet and whom we choose to become acquainted with.

Sociologists estimate that a typical person knows about 1,000 individuals on a first-name basis, prompting us to assume $\langle k \rangle \approx 1{,}000$. Using the results obtained so far about random networks, we arrive at a number of intriguing conclusions about a random society of $N \simeq 7 \times 10^9$ individuals (**Advanced topics 3.B**):

- The most connected individual (the largest-degree node) in a random society is expected to have $k_{\max} = 1{,}185$ acquaintances.
- The degree of the least-connected individual is $k_{\min} = 816$, not that different from $k_{\max}$ or $\langle k \rangle$.
- The dispersion of a random network is $\sigma_k = \langle k \rangle^{1/2}$, which for $\langle k \rangle = 1{,}000$ is $\sigma_k = 31.62$. This means that the number of friends a typical individual has is in the $\langle k \rangle \pm \sigma_k$ range, or between 968 and 1,032, a rather narrow window.

Taken together, in a random society all individuals are expected to have a comparable number of friends. Hence, if people are randomly connected to each other, we lack outliers: there are no highly popular individuals and no one is left behind, having only a few friends. This surprising conclusion is a consequence of an important property of random networks: *in a large random network the degree of most nodes is in the narrow vicinity of* $\langle k \rangle$ (**Box 3.4**).

This prediction blatantly conflicts with reality. Indeed, there is extensive evidence of individuals who have considerably more than 1,185 acquaintances. For example, US President Franklin Delano Roosevelt's appointment book has about 22,000 names, all individuals he met personally [84, 85]. Similarly, a study of the social network behind Facebook has documented numerous individuals with 5,000 Facebook friends, the maximum allowed by the social networking platform [86]. To understand the origin of these discrepancies we must compare the degree distribution of real and random networks.

Box 3.4 Why are Hubs Missing?

To understand why hubs (nodes with a very large degree) are absent in random networks, we turn to the degree distribution **(3.8)**.

We first note that the $1/k!$ term in **(3.8)** significantly decreases the chances of observing large-degree nodes. Indeed, the Stirling approximation

$$k! \sim [\sqrt{2\pi k}] \left(\frac{k}{e}\right)^k$$

allows us to rewrite **(3.8)** as

$$p_k = \frac{e^{-\langle k \rangle}}{\sqrt{2\pi k}} \left(\frac{e\langle k \rangle}{k}\right)^k. \qquad (3.9)$$

For degrees $k > e\langle k \rangle$ the term in parentheses is smaller than one, hence for large k both k-dependent terms in **(3.9)**, that is $1/\sqrt{k}$ and $(e\langle k \rangle/k)^k$, decrease rapidly with increasing k. Overall, **(3.9)** predicts that in a random network the chance of observing a hub decreases faster than exponentially.

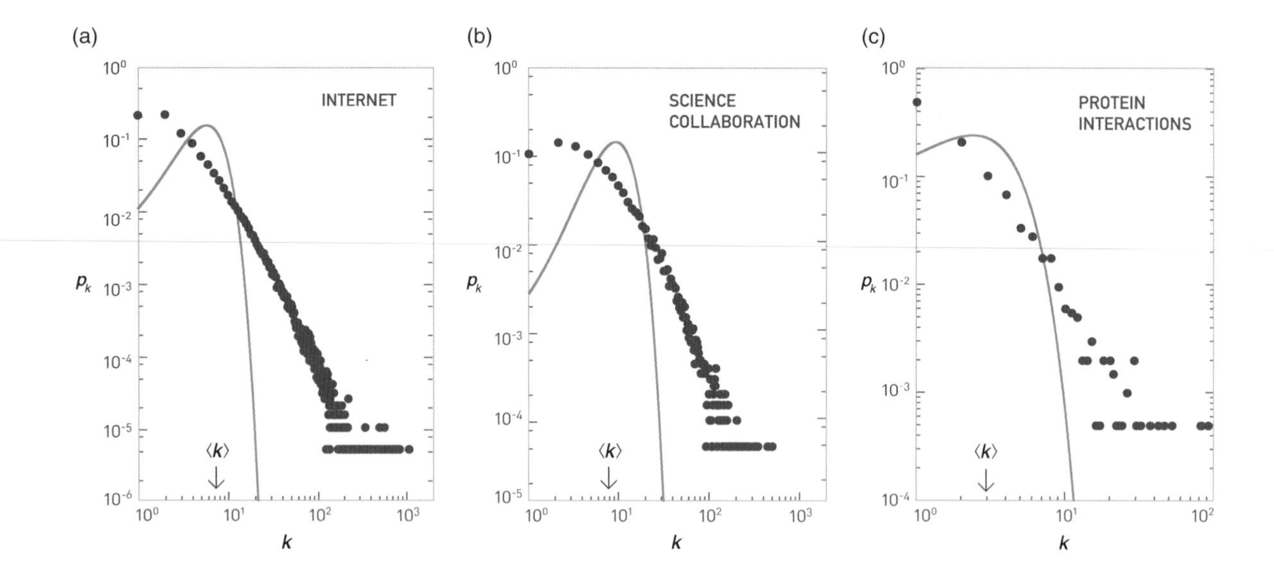

Figure 3.6 **Degree Distribution of Real Networks**
The degree distribution of
(**a**) the Internet,
(**b**) the science collaboration network and
(**c**) the protein interaction network (**Table 2.1**).
The green line corresponds to the Poisson prediction obtained by measuring $\langle k \rangle$ for the real network and then plotting (**3.8**). The significant deviation between the data and the Poisson fit indicates that the random network model underestimates the size and the frequency of the high-degree nodes, as well as the number of low-degree nodes. Instead, the random network model predicts a larger number of nodes in the vicinity of $\langle k \rangle$ than seen in real networks.

In **Figure 3.6** we show the degree distribution of three real networks, together with the corresponding Poisson fit. The figure documents systematic differences between the random network predictions and the real data:

- The Poisson form significantly underestimates the number of high-degree nodes. For example, according to the random network model the maximum degree of the Internet is expected to be around 20. In contrast, the data indicates the existence of routers with degrees close to 103.
- The spread in the degrees of real networks is much wider than expected in a random network. This difference is captured by the dispersion σ_k (**Figure 3.4**). If the Internet were to be random, we would expect $\sigma = 2.52$. The measurements indicate $\sigma_{\text{Internet}} = 14.14$, significantly higher than the random prediction. These differences are not limited to the networks shown in **Figure 3.6**; all the networks listed in **Table 2.1** share this property.

In summary, comparison with the real data indicates that the random network model does not capture the degree distribution of real networks. In a random network most nodes have comparable degrees, forbidding hubs. In contrast, in real networks we observe a significant number of highly connected nodes and there are large differences in node degrees. We will resolve these differences in **Chapter 4**.

3.6 The Evolution of a Random Network

The cocktail party we encountered at the beginning of this chapter captures a dynamical process: starting with N isolated nodes, the links are added gradually through random encounters between the guests. This corresponds to a gradual increase of p, with striking consequences on the network topology (**Online resource 3.2**). To quantify this process, we first inspect how the size of the largest connected cluster within the network, N_G, varies with $\langle k \rangle$. Two extreme cases are easy to understand:

- For $p = 0$ we have $\langle k \rangle = 0$, hence all nodes are isolated. Therefore, the largest component has size $N_G = 1$ and $N_G/N \to 0$ for large N.
- For $p = 1$ we have $\langle k \rangle = N - 1$, hence the network is a complete graph and all nodes belong to a single component. Therefore, $N_G = N$ and $N_G/N = 1$.

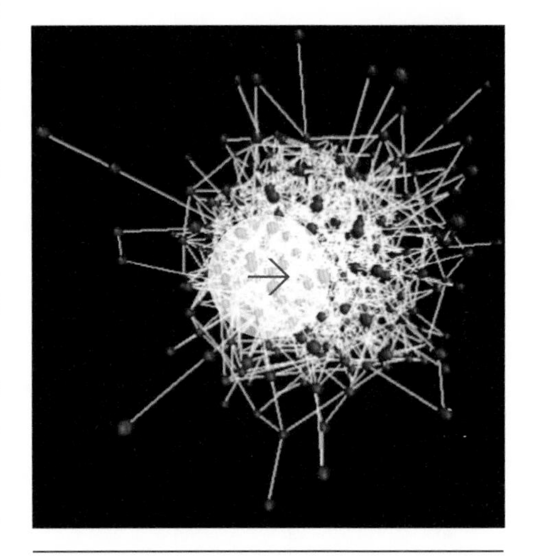

Online Resource 3.2
Evolution of a Random Network
A video showing the change in structure of a random network with increasing p. It vividly illustrates the absence of a giant component for small p and its sudden emergence once p reaches a critical value.

One would expect that the largest component grows gradually from $N_G = 1$ to $N_G = N$ if $\langle k \rangle$ increases from 0 to $N - 1$. Yet, as **Figure 3.7a** indicates, this is not the case: N_G/N remains zero for small $\langle k \rangle$, indicating the lack of a large cluster. Once $\langle k \rangle$ exceeds a critical value, N_G/N increases, signaling the rapid emergence of a large cluster that we call the *giant component*. Erdős and Rényi, in their classical 1959 paper, predicted that the condition for the emergence of the giant component is [3]

$$\langle k \rangle = 1. \tag{3.10}$$

In other words, we have a giant component if and only if each node has on average more than one link (**Advanced topics 3.C**).

The fact that we need at least one link per node to observe a giant component is not unexpected. Indeed, for a giant component to exist, each of its nodes must be linked to at least one other node. It is somewhat counterintuitive, however, that one link is *sufficient* for its emergence.

We can express **(3.10)** in terms of p using **(3.3)**, obtaining

$$p_c = \frac{1}{N-1} \approx \frac{1}{N}. \tag{3.11}$$

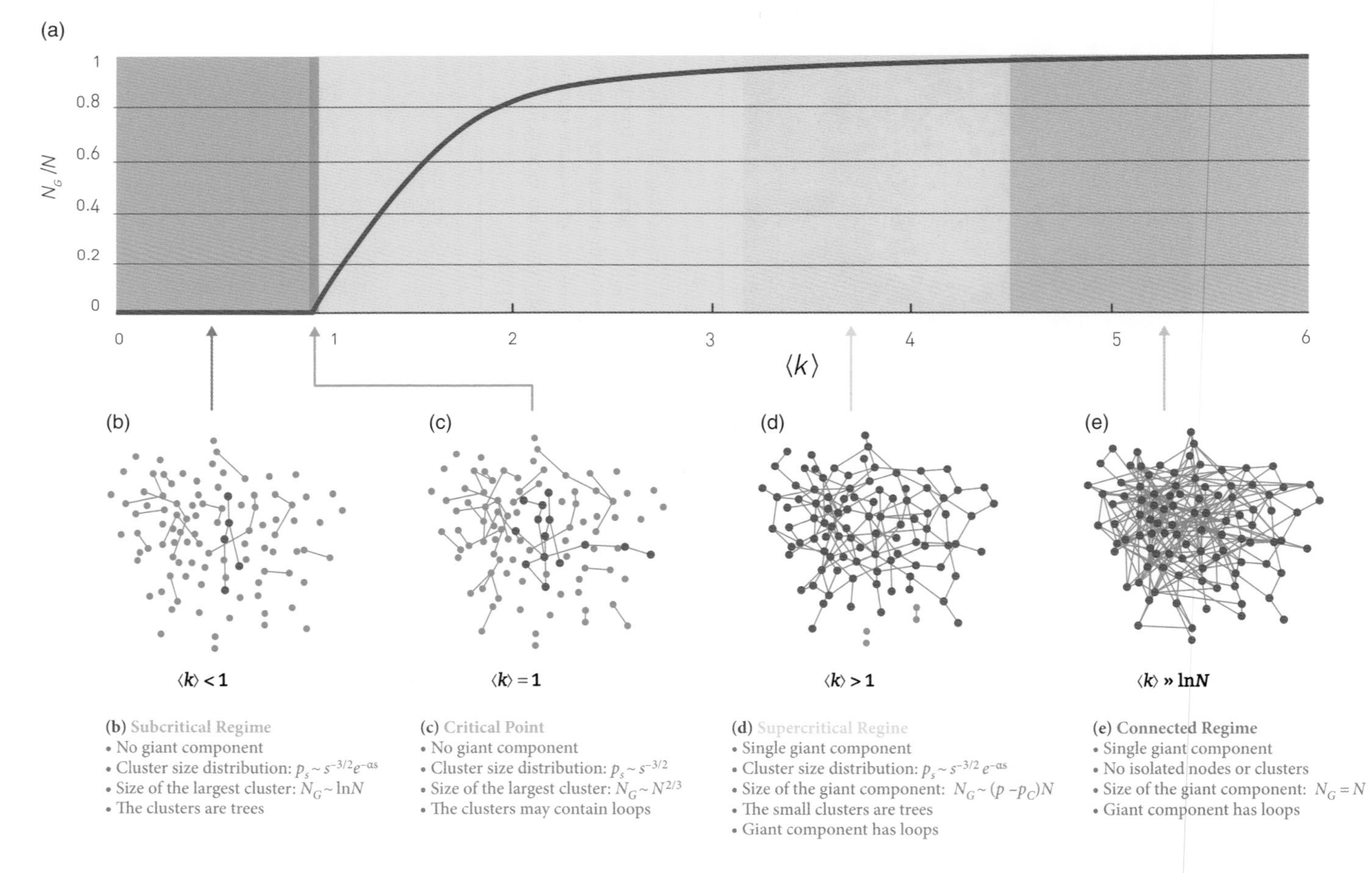

Figure 3.7 **Evolution of a Random Network**

(a) The relative size of the giant component as a function of the average degree $\langle k \rangle$ in the Erdős–Rényi model. The figure illustrates the phase tranisition at $\langle k \rangle = 1$, marking the emergence of a giant component with nonzero N_G.

(b)–(e) A sample network and its properties in the four regimes that characterize a random network.

Therefore, the larger a network, the smaller the value of p that is sufficient for the giant component.

The emergence of the giant component is only one of the transitions characterizing a random network as we change $\langle k \rangle$. We can distinguish four topologically distinct regimes (**Figure 3.7a**), each with unique characteristics:

Subcritical regime: $0 < \langle k \rangle < 1 \left(p < \frac{1}{N} \right.$, **Figure 3.7b**$\left. \right)$

For $\langle k \rangle = 0$ the network consists of N isolated nodes. Increasing $\langle k \rangle$ means that we are adding $N\langle k \rangle = pN(N-1)/2$ links to the network. Yet, given that $\langle k \rangle < 1$, we have only a small number of links in this regime, hence we mainly observe tiny clusters (**Figure 3.7b**).

We can designate at any moment the largest cluster to be the giant component. Yet in this regime the relative size of the largest cluster, N_G/N, remains zero. The reason is that for $\langle k \rangle < 1$ the largest cluster is a tree with size $N_G \sim \ln N$, hence its size increases much slower than the size of the network. Therefore, $N_G/N \simeq \ln N/N \to 0$ in the $N \to \infty$ limit.

In summary, in the subcritical regime the network consists of numerous tiny components, whose size follows the exponential distribution (**3.35**). Hence these components have comparable sizes, lacking a clear winner that we could designate as a giant component.

Critical point: $\langle k \rangle = 1$ ($p = 1/N$, **Figure 3.7c**)

The critical point separates the regime where there is not yet a giant component ($\langle k \rangle < 1$) from the regime where there is one ($\langle k \rangle > 1$). At this point the relative size of the largest component is still zero (**Figure 3.7c**). Indeed, the size of the largest component is $N_G \sim N^{2/3}$. Consequently, N grows much slower than the network's size, so its relative size decreases as $N/N \sim N^{-1/3}$ in the $N \to \infty$ limit.

Note, however, that in absolute terms there is a significant jump in the size of the largest component at $\langle k \rangle = 1$. For example, for a random network with $N = 7 \times 10^9$ nodes, comparable with the globe's social network, for $\langle k \rangle < 1$ the largest cluster is of the order of $N_G \simeq \ln N = \ln(7 \times 10) \simeq 22.7$. In contrast, at $\langle k \rangle = 1$ we expect $N_G \sim N = (7 \times 10^9)^{2/3} \simeq 3 \times 10^6$, a jump of about five orders of magnitude. Yet, both in the subcritical regime and at the critical point, the largest component contains only a vanishing fraction of the total number of nodes in the network.

In summary, at the critical point most nodes are located in numerous small components, whose size distribution follows **(3.36)**. The power law form indicates that components of rather different sizes coexist. These numerous small components are mainly trees, while the giant component may contain loops. Note that many properties of the network at the critical point resemble the properties of a physical system undergoing a phase transition (**Advanced topics 3.F**).

Supercritical regime: $\langle k \rangle > 1 \left(p > \frac{1}{N} \right.$, **Figure 3.7d**$\left. \right)$

This regime has the most relevance to real systems, as for the first time we have a giant component that looks like a network. In the vicinity of the critical point the size of the giant component varies as

$$N_G/N \sim \langle k \rangle - 1, \qquad (3.12)$$

or

$$N_G \sim (p - p_c)N, \qquad (3.13)$$

where p_c is given by **(3.11)**. In other words, the giant component contains a finite fraction of the nodes. The further we move from the critical point, the larger the fraction of nodes that will belong to it. Note that **(3.12)** is valid only in the vicinity of $\langle k \rangle = 1$. For large $\langle k \rangle$ the dependence between N_G and $\langle k \rangle$ is nonlinear (**Figure 3.7a**).

In summary, in the supercritical regime numerous isolated components coexist with the giant component, their size distribution following **(3.35)**. These small components are trees, while the giant component contains loops and cycles. The supercritical regime lasts until all nodes are absorbed by the giant component.

Connected regime: $\langle k \rangle > \ln N \left(p > \frac{\ln N}{N} \right.$, **Figure 3.7e**$\left. \right)$

For sufficiently large p the giant component absorbs all nodes and components, hence $N_G \simeq N$. In the absence of isolated nodes, the network becomes connected. The average degree at which this happens depends on N (**Advanced topics 3.E**):

$$\langle k \rangle = \ln N. \qquad (3.14)$$

Note that when we enter the connected regime the network is still relatively sparse, as $\ln N/N \rightarrow 0$ for large N. The network turns into a complete graph only at $\langle k \rangle = N - 1$.

In summary, the random network model predicts that the emergence of a network is not a smooth, gradual process: the isolated nodes and tiny components observed for small $\langle k \rangle$ collapse into a giant component through a phase transition (**Advanced topics 3.F**). As we vary $\langle k \rangle$ we encounter four topologically distinct regimes (**Figure 3.7**).

The discussion offered above follows an empirical perspective, fruitful if we wish to compare a random network with real systems. A different perspective, with its own rich behavior, is offered by the mathematical literature (**Box 3.5**).

3.7 Real Networks are Supercritical

Two predictions of random network theory are of direct importance for real networks:

1. Once the average degree exceeds $\langle k \rangle = 1$, a giant component should emerge that contains a finite fraction of all nodes. Hence, only for $\langle k \rangle > 1$ do the nodes organize themselves into a recognizable network.
2. For $\langle k \rangle > \ln N$ all components are absorbed by the giant component, resulting in a single connected network.

Do real networks satisfy the criteria for the existence of a giant component, that is $\langle k \rangle > 1$? And will this giant component contain all nodes for $\langle k \rangle > \ln N$, or will we continue to see some disconnected nodes and components? To answer these questions we compare the structure of a real network for a given $\langle k \rangle$ with the theoretical predictions discussed above.

The measurements indicate that real networks extravagantly exceed the $\langle k \rangle = 1$ threshold. Indeed, sociologists estimate that an average person has around 1,000 acquaintances; a typical neuron in the human brain has about 7,000 synapses; in our cells each molecule takes part in several chemical reactions.

This conclusion is supported by **Table 3.1**, which lists the average degree of several undirected networks, in each case finding $\langle k \rangle > 1$. Hence the average degree of real networks is well beyond the $\langle k \rangle = 1$ threshold, implying that they all have a giant component. The same is true for the reference networks listed in **Table 3.1**.

Box 3.5 Network Evolution in Graph Theory

In the random graph literature it is often assumed that the connection probability $p(N)$ scales as N^z, where z is a tunable parameter between $-\infty$ and 0 [2]. In this language Erdős and Rényi discovered that as we vary z, key properties of random graphs appear quite suddenly.

A graph has a given property Q if the probability of having Q approaches 1 as $N \to \infty$. That is, for a given z either almost every graph has the property Q or almost no graph has it. For example, for z less than $-3/2$ almost all graphs contain only isolated nodes and pairs of nodes connected by a link. Once z exceeds $-3/2$, most networks will contain paths connecting three or more nodes (**Figure 3.8**).

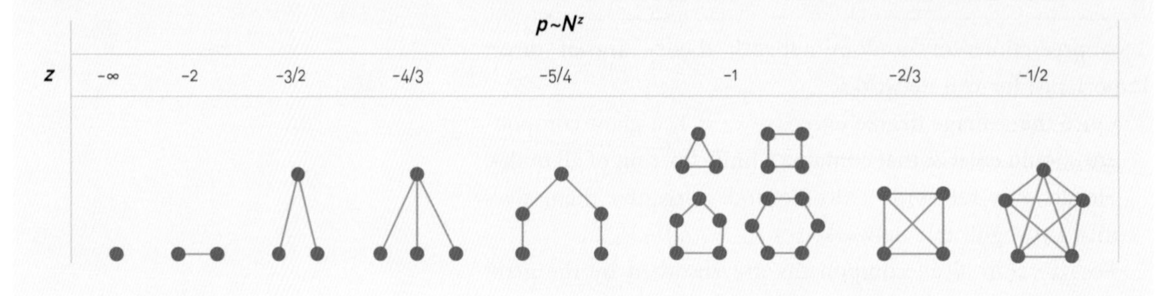

Figure 3.8 **Evolution of a Random Graph**
The threshold probabilities at which different subgraphs appear in a random graph, as defined by the exponent z in the $p(N) \sim N^z$ relationship. For $z < -3/2$ the graph consists of isolated nodes and edges. When z passes $-3/2$ trees of order 3 appear, while at $z = -4/3$ trees of order 4 appear. At $z = -1$ trees of all orders are present, together with cycles of all orders. Complete subgraphs of order 4 appear at $z = -2/3$, and as z increases further, complete subgraphs of larger and larger order emerge. After [19].

Let us now turn to the second prediction, inspecting if we have a single component (i.e., $\langle k \rangle > \ln N$) or if the network is fragmented into multiple components (i.e., $\langle k \rangle < \ln N$). For social networks the transition between the supercritical and the fully connected regime should be at $\langle k \rangle > \ln(7 \times 10^9) \approx 22.7$. That is, if the average individual has more than two dozen acquaintances, then a random society must have a single component, leaving no individual disconnected. With $\langle k \rangle \approx 1{,}000$ this condition is clearly satisfied. Yet, according to **Table 3.1**, many real networks do not obey the fully connected criterion. Consequently, according to random network theory these networks should be fragmented into several disconnected components. This is a disconcerting prediction for the Internet, indicating that some routers should be disconnected from the giant component, being unable to communicate with other routers. It is equally problematic for the power grid, indicating that some consumers should not get power. These predictions are clearly at odds with reality.

In summary, we find that most real networks are in the supercritical regime (**Figure 3.9**). Therefore these networks are expected to have a giant component, which is in agreement with observations. Yet this giant component should coexist with many disconnected components, a prediction that fails for several real networks. Note that these predictions should be valid only if real networks are accurately described by the Erdős–Rényi model, that is if real networks are random. In the coming chapters, as we learn more about the structure of real networks, we will understand why real networks can stay connected despite failing the $\langle k \rangle > \ln N$ criterion.

Table 3.1 **Are Real Networks Connected?**
The number of nodes N and links L for the undirected networks of our reference network list, shown together with $\langle k \rangle$ and $\ln N$. A giant component is expected for $\langle k \rangle > 1$ and all nodes should join the giant component for $\langle k \rangle > \ln N$. While for all networks $\langle k \rangle > 1$, for most networks $\langle k \rangle$ is below the $\ln N$ threshold (see also **Figure 3.9**).

Network	N	L	$\langle k \rangle$	$\ln N$
Internet	192,244	609,066	6.34	12.17
Power grid	4,941	6,594	2.67	8.51
Science collaboration	23,133	94,439	8.08	10.05
Actor network	702,388	29,397,908	83.71	13.46
Protein interactions	2,018	2,930	2.90	7.61

3.8 Small Worlds

The *small-world phenomenon*, also known as *six degrees of separation*, has long fascinated the general public. It states that if you choose any two individuals anywhere on Earth, you will find a path of at most six acquaintances between them (**Figure 3.10**). The fact that individuals who live in the same city are only a few handshakes from each other is by no means surprising. The small-world concept states, however, that even individuals who are on opposite sides of the globe can be connected to us via a few acquaintances.

In the language of network science the small-world phenomenon implies that *the distance between two randomly chosen nodes in a network is short.* This statement raises two questions: What does short (or small) mean – short compared with what? How do we explain the existence of these short distances?

Both questions are answered by a simple calculation. Consider a random network with average degree $\langle k \rangle$. A node in this network has on average:

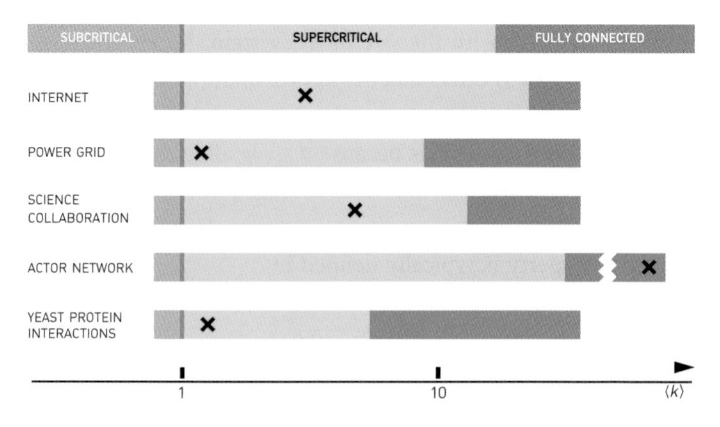

Figure 3.9 **Most Real Networks are Supercritical**
The four regimes predicted by random network theory, marking with a cross the location of $\langle k \rangle$ for the undirected networks listed in **Table 3.1**. The diagram indicates that most networks are in the supercritical regime, hence they are expected to be broken into numerous isolated components. Only the actor network is in the connected regime, meaning that all nodes are part of a single giant component. Note that while the boundary between the subcritical and the supercritical regime is always at $\langle k \rangle = 1$, the boundary between the supercritical and the connected regime is at $\ln N$, which varies from system to system.

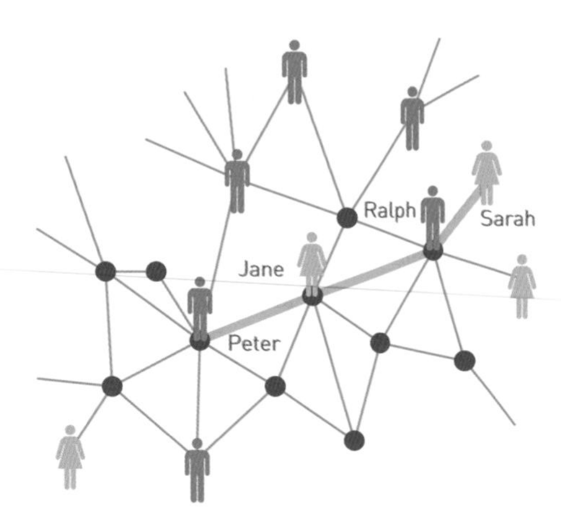

Figure 3.10 **Six Degrees of Separation**
According to six degrees of separation, two individuals anywhere in the world can be connected through a chain of six or fewer acquaintances. This means that while Sarah does not know Peter, she knows Ralph, who knows Jane and who in turn knows Peter. Hence Sarah is three handshakes, or three degrees, from Peter. In the language of network science six degrees, also called the small-world property, means that the distance between any two nodes in a network is unexpectedly small.

$\langle k \rangle$ nodes at distance one ($d = 1$)
$\langle k \rangle^2$ nodes at distance two ($d = 2$)
$\langle k \rangle^3$ nodes at distance three ($d = 3$)
...
$\langle k \rangle^d$ nodes at distance d.

For example, if $\langle k \rangle \approx 1{,}000$, which is the estimated number of acquaintances an individual has, we expect 10^6 individuals at distance two and about a billion, almost the whole of Earth's population, at distance three from us.

To be precise, the expected number of nodes up to distance d from our starting node is

$$N(d) \approx 1 + \langle k \rangle + \langle k \rangle^2 + \cdots + \langle k \rangle^d = \frac{\langle k \rangle^{d+1} - 1}{\langle k \rangle - 1}. \quad (3.15)$$

$N(d)$ must not exceed the total number of nodes, N, in the network. Therefore, the distances cannot take up arbitrary values. We can identify the maximum distance, d_{max}, or the network's diameter by setting

$$N(d_{max}) \approx N. \quad (3.16)$$

Assuming that $\langle k \rangle \gg 1$, we can neglect the (-1) term in the numerator and the denominator of **(3.15)**, obtaining that the diameter of a random network follows

$$\langle k \rangle^{d_{max}} \approx N. \quad (3.17)$$

Hence

$$d_{max} \approx \frac{\ln N}{\ln \langle k \rangle}, \quad (3.18)$$

which represents the mathematical formulation of the small-world phenomenon. The key, however, lies in its interpretation:

- As derived, **(3.18)** predicts the scaling of the network diameter, d_{max}, with the size of the system, N. Yet for most networks, **(3.18)** offers a better approximation to the average distance between two randomly chosen nodes, $\langle d \rangle$, than to d_{max} (**Table 3.2**). This is because d_{max} is often dominated by a few extreme paths, while $\langle d \rangle$ is averaged over all node pairs, a process that suppresses the fluctuations. Hence, the small-world property is typically defined by

$$\langle d \rangle \approx \frac{\ln N}{\ln \langle k \rangle}, \quad (3.19)$$

Table 3.2 **Six Degrees of Separation**

The average distance $\langle d \rangle$ and the maximum distance d_{max} for the ten reference networks. The last column provides $\langle d \rangle$ predicted by (3.19), indicating that it offers a reasonable approximation to the measured $\langle d \rangle$. Yet the agreement is not perfect-we will see in the next chapter that for many real networks (3.19) needs to be adjusted. For directed networks the average degree and the path lengths are measured along the direction of the links.

Network	N	L	$\langle k \rangle$	$\langle d \rangle$	d_{max}	$\frac{\ln N}{\ln \langle k \rangle}$
Internet	192,244	609,066	6.34	6.98	26	6.58
WWW	325,729	1,497,134	4.60	11.27	93	8.31
Power grid	4,941	6,594	2.67	18.99	46	8.66
Mobile phone calls	36,595	91,826	2.51	11.72	39	11.42
Email	57,194	103,731	1.81	5.88	18	18.4
Science collaboration	23,133	93,439	8.08	5.35	15	4.81
Actor network	702,388	29,397,908	83.71	3.91	14	3.04
Citation network	449,673	4,707,958	10.43	11.21	42	5.55
***E. coli* metabolism**	1,039	5,802	5.58	2.98	8	4.04
Protein interactions	2,018	2,930	2.90	5.61	14	7.14

describing the dependence of the average distance in a network on N and $\langle k \rangle$.

- In general $\ln N \ll N$, hence the dependence of $\langle d \rangle$ on $\ln N$ implies that the distances in a random network are *orders of magnitude smaller than the size of the network*. Consequently, by *small* in the "small-world phenomenon" we mean that *the average path length or the diameter depends logarithmically on the system size*. Hence, "small" means that $\langle d \rangle$ is proportional to $\ln N$, rather than N or some power of N (**Figure 3.11**).
- The $1/\ln \langle k \rangle$ term implies that the denser the network, the smaller is the distance between the nodes.
- In real networks there are systematic corrections to **(3.19)**, rooted in the fact that the number of nodes at distance $d > \langle d \rangle$ drops rapidly (**Advanced topics 3.G**).

Let us illustrate the implications of **(3.19)** for social networks. Using $N \approx 7 \times 10^9$ and $\langle k \rangle \approx 10^3$, we obtain

$$\langle d \rangle \approx \frac{\ln(7 \times 10^9)}{\ln(10^3)} = 3.28. \qquad (3.20)$$

Therefore, all individuals on Earth should be within three to four handshakes of each other. The estimate **(3.20)** is probably closer to the real value than the frequently quoted six degrees (**Box 3.6**).

(a)

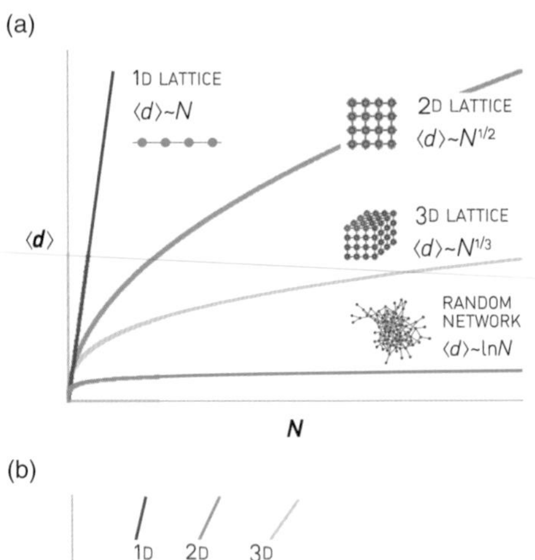

(b)

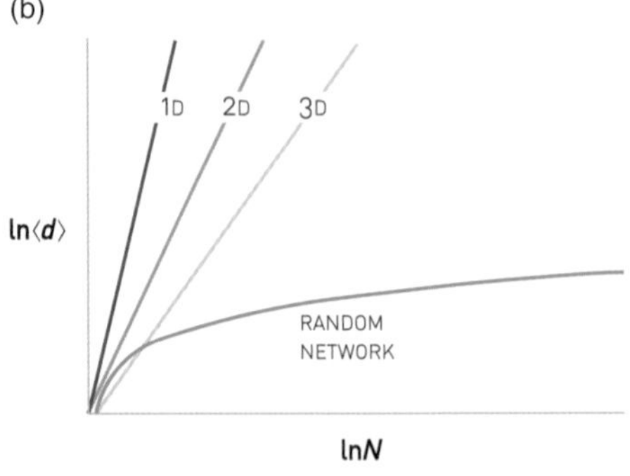

Figure 3.11 **Why are Small Worlds Surprising?**
Much of our intuition about distance is based on our experience with regular lattices, which do not display the small-world property.
1D. For a one-dimensional lattice (a line of length N), the diameter and the average path length scale linearly with N: $d_{\max} \sim \langle d \rangle \sim N$.
2D. For a square lattice, $d_{\max} \sim \langle d \rangle \sim N^{1/2}$.
3D. For a cubic lattice, $d_{\max} \sim \langle d \rangle \sim N^{1/3}$.
4D. In general, for a d-dimensional lattice, $d_{\max} \sim \langle d \rangle \sim N^{1/d}$.
These polynomial dependences predict a much faster increase with N than (**3.19**), indicating that in lattices the path lengths are significantly longer than in a random network. For example, if the social network were to form a square lattice (2D), where each individual knows only its neighbors, the average distance between two individuals would be roughly $(7 \times 10^9)^{1/2} = 83,666$. Even if we correct for the fact that a person has about 1000 acquaintances, not four, the average separation will be orders of magnitude larger than predicted by (**3.19**).
(**a**) The figure shows the predicted N-dependence of $\langle d \rangle$ for regular and random networks on a linear scale.
(**b**) The same as in (a), but shown on a log–log scale.

Much of what we know about the small-world property in random networks, including the result (**3.19**), is in a little-known paper by Manfred Kochen and Ithiel de Sola Pool, in which they mathematically formulated the problem and discussed in depth its sociological implications. This paper inspired the well-known Milgram experiment (**Box 3.7**), which in turn inspired the *six degrees of separation* phrase.

While discovered in the context of social systems, the small-world property applies beyond social networks (**Box 3.6**). To demonstrate this, in **Table 3.2** we compare the prediction of (**3.19**) with the average path length $\langle d \rangle$ for several real networks, finding that despite the diversity of these systems and the significant differences between them in terms of N and $\langle k \rangle$, (**3.19**) offers a good approximation to the empirically observed $\langle d \rangle$.

In summary, the small-world property has not only ignited the public's imagination (**Box 3.8**), but also plays an important role in network science as well. The small-world phenomenon can be reasonably well understood in the context of the random network model: it is rooted in the fact that the number of nodes at distance d from a node increases exponentially with d. In the coming chapters we will see that in real networks we encounter systematic deviations from (**3.19**), forcing us to replace it with more accurate predictions. Yet the intuition offered by the random network model on the origin of the small-world phenomenon remains valid.

3.9 Clustering Coefficient

The degree of a node contains no information about the relationship between the node's neighbors. Do they all know each other, or are they perhaps isolated from each other? The answer is provided by the local clustering coefficient C_i, which measures the density of links in node i's immediate neighborhood: $C_i = 0$ means that there are no links between i's neighbors; $C_i = 1$ implies that each of i's neighbors link to each other (**Section 2.10**).

To calculate C_i for a node in a random network, we need to estimate the expected number of links L_i between the node's k_i neighbors. In a random network the probability that two of i's neighbors link to each other is p. As there are $k_i(k_i - 1)/2$ possible links between the k_i neighbors of node i, the expected value of L_i is

$$\langle L_i \rangle = p\frac{k_i(k_i - 1)}{2}.$$

Thus the local clustering coefficient of a random network is

$$C_i = \frac{2\langle L_i \rangle}{k_i(k_i - 1)} = p = \frac{\langle k \rangle}{N}. \tag{3.21}$$

Equation **(3.21)** makes two predictions:

(1) For fixed $\langle k \rangle$, the larger the network, the smaller is a node's clustering coefficient. Consequently, a node's local clustering coefficient C_i is expected to decrease as $1/N$. Note that the network's average clustering coefficient, $\langle C \rangle$, also follows **(3.21)**.

(2) The local clustering coefficient of a node is independent of the node's degree.

To test the validity of **(3.21)**, we plot $\langle C \rangle / \langle k \rangle$ as a function of N for several undirected networks (**Figure 3.13a**). We find that $\langle C \rangle / \langle k \rangle$ does not decrease as N^{-1} but is largely independent of N, in violation of the prediction **(3.21)** and point **(1)** above. In **Figure 3.13b–d** we also show the dependency of C on the node's degree k_i for three real networks, finding that $C(k)$ systematically decreases with the degree, again in violation of **(3.21)** and point **(2)**.

Box 3.6 19 Degrees of Separation

How many clicks do we need to reach a randomly chosen document on the Web? The difficulty in addressing this question is rooted in the fact that we lack a complete map of the WWW – we only have access to small samples of the full map. We can start, however, by measuring the WWW's average path length in samples of increasing size, a procedure called *finite size scaling*. The measurements indicate that the average path length of the WWW increases with the size of the network as [9]

$$\langle d \rangle \approx 0.35 + 0.89 \ln N.$$

In 1999 the WWW was estimated to have about 800 million documents [87], in which case the above equation predicts $\langle d \rangle \approx 18.69$. In other words, in 1999 two randomly chosen documents were on average 19 clicks from each other, a result that became known as *19 degrees of separation*. Subsequent measurements on a sample of 200 million documents found $\langle d \rangle \approx 16$ [88], in good agreement with the $\langle d \rangle \approx 17$ prediction. Currently the WWW is estimated to have about a trillion nodes ($N \sim 10^{12}$), in which case the formula predicts $\langle d \rangle \approx 25$. Hence $\langle d \rangle$ is not fixed but as the network grows, so does the distance between two documents.

The average path length of 25 is much larger than the proverbial six degrees (**Box 3.7**). The difference is easy to understand: the WWW has smaller average degree and larger size than the social network. According to **(3.19)**, both of these differences increase the Web's diameter.

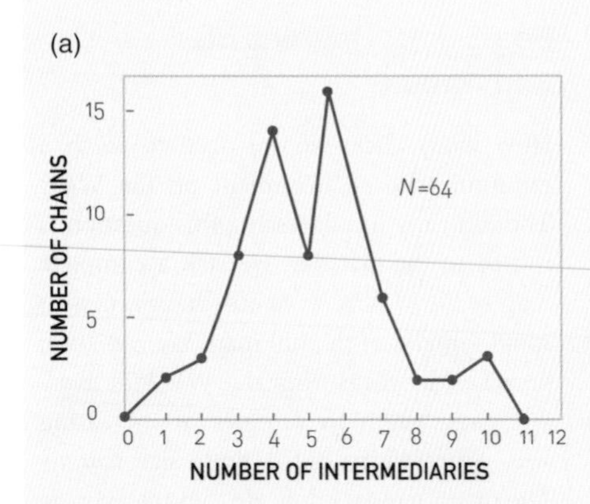

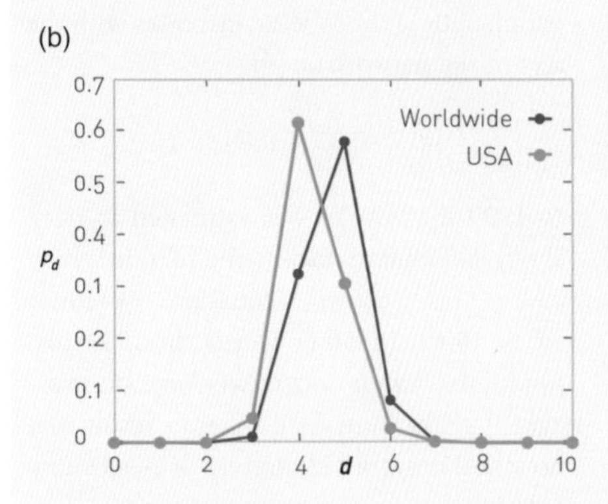

Figure 3.12 **Six Degrees? From Milgram to Facebook**

(a) In Milgram's experiment, 64 of the 296 letters made it to the recipient. The figure shows the length distribution of the completed chains, indicating that some letters required only one intermediary, while others required as many as ten. The median of the distribution was 5.2, indicating that on average six "handshakes" were required to get a letter to its recipient. The playwright John Guare renamed this "six degrees of separation" two decades later. After [89].

(b) The distance distribution, p_d, for all pairs of Facebook users worldwide and within the USA only. Using Facebook's N and L, (**3.19**) predicts the average degree to be approximately 3.90, not far from the reported four degrees. After [86].

Box 3.7 **Six Degrees: Experimental Confirmation**

The first empirical study of the small-world phenomenon took place in 1967, when Stanley Milgram, building on the work of Pool and Kochen, designed an experiment to measure the distances in social networks [7, 89]. Milgram chose a stock broker in Boston and a divinity student in Sharon, Massachusetts as *targets*. He then randomly selected residents of Wichita and Omaha, sending them a letter containing a short summary of the study's purpose, a photograph, and the name, address and information about the target person. They were asked to forward the letter to a friend, relative or acquantance who is most likely to know the target person.

Within a few days the first letter arrived, passing through only two links. Eventually 64 of the 296 letters made it back, some, however, requiring close to a dozen intermediates [89]. These completed chains allowed Milgram to determine the number of individuals required to get the letter to the target (**Figure 3.12a**). He found that the median number of intermediates was 5.2, a relatively small number that was remarkably close to Frigyes Karinthy's 1929 insight (**Box 3.8**).

Milgram lacked an accurate map of the full acquaintance network, hence his experiment could not detect the true distance between his study's participants. Today, Facebook has the most extensive social network map ever assembled. Using Facebook's social graph of May 2011, consisting of 721 million active users and 68 billion symmetric friendship links, researchers found an average distance of 4.74 between users (**Figure 3.12b**). Therefore, the study detected only "four degrees of separation" [86], closer to the prediction of (**3.20**) than to Milgram's six degrees [7, 89].

"I asked a person of intelligence how many steps he thought it would take, and he said that it would require 100 intermediate persons, or more, to move from Nebraska to Sharon."

Stanley Milgram, 1969

Box 3.8 19 Degrees of the WWW

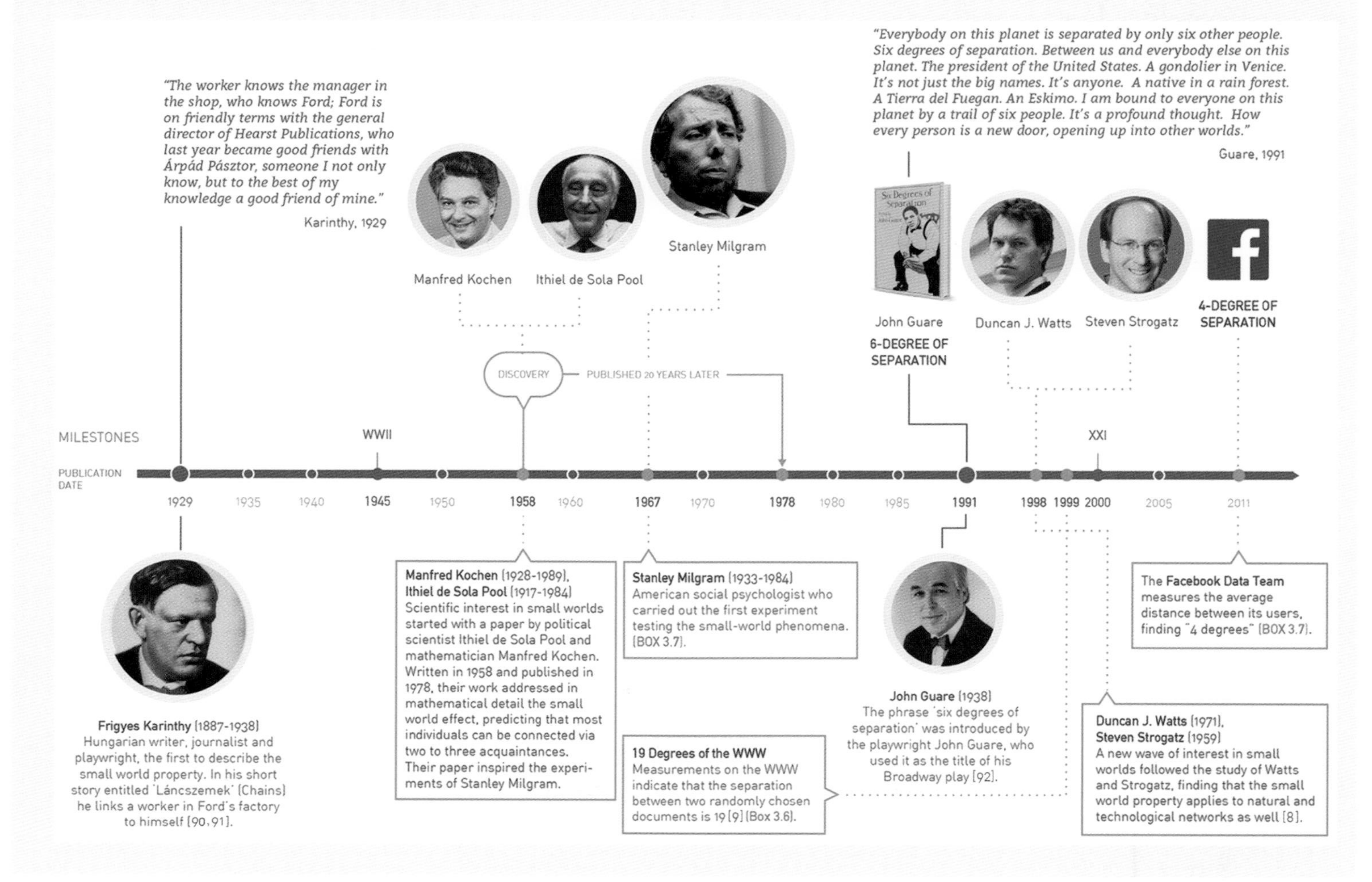

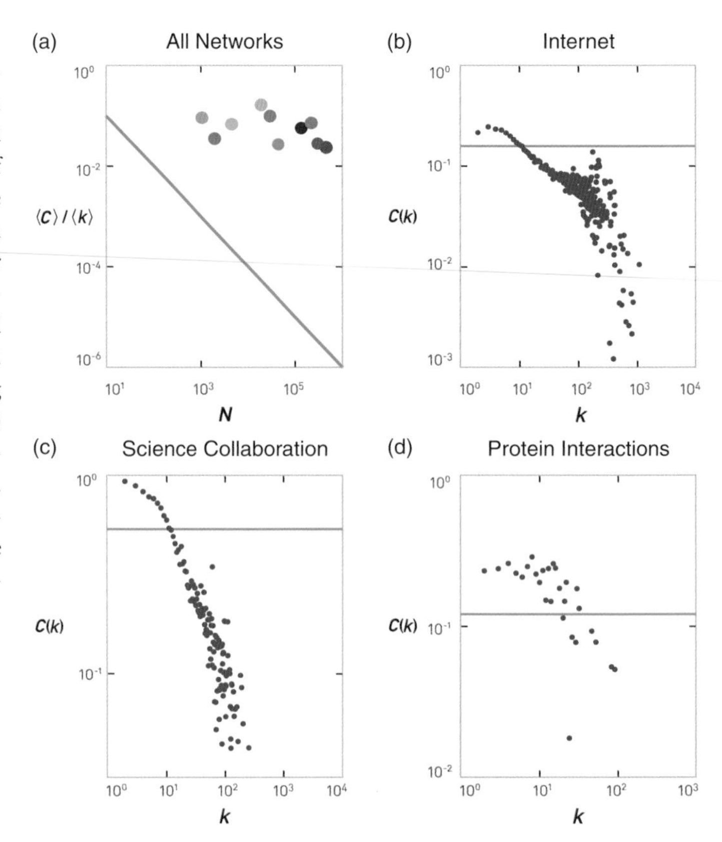

Figure 3.13 **Clustering in Real Networks**
(**a**) Comparing the average clustering coefficient of real networks with the prediction **(3.21)** for random networks. The circles and their colors correspond to the networks of **Table 3.2**. Directed networks were made undirected to calculate $\langle C \rangle$ and $\langle k \rangle$. The green line corresponds to **(3.21)**, predicting that for random networks the average clustering coefficient decreases as N^{-1}. In contrast, for real networks $\langle C \rangle$ appears to be independent of N.
(**b**)–(**d**) The dependence of the local clustering coefficient, $C(k)$, on the node's degree for (b) the Internet, (c) the science collaboration network and (d) the protein interaction network. $C(k)$ is measured by averaging the local clustering coefficient of all nodes with the same degree k. The green horizontal line corresponds to $\langle C \rangle$.

In summary, we find that the random network model does not capture the clustering of real networks. Instead, real networks have a much higher clustering coefficient than expected for a random network of similar N and L. An extension of the random network model proposed by Watts and Strogatz [8] addresses the coexistence of high $\langle C \rangle$ and the small-world property (**Box 3.9**). It fails to explain, however, why high-degree nodes have a smaller clustering coefficient than low-degree nodes. Models explaining the shape of $C(k)$ are discussed in **Chapter 9**.

3.10 Summary: Real Networks are not Random

Since its introduction in 1959, the random network model has dominated mathematical approaches to complex networks. The model suggests that the random-looking networks

Box 3.9 Watts–Strogatz Model

Duncan Watts and Steven Strogatz proposed an extension of the random network model (**Figure 3.14**) motivated by two observations [8]:

(a) Small-World Property
In real networks the average distance between two nodes depends logarithmically on N (**3.18**), rather than following a polynomial as expected for regular lattices (**Figure 3.11**).

(b) High Clustering
The average clustering coefficient of real networks is much higher than expected for a random network of similar N and L (**Figure 3.13a**).

The *Watts–Strogatz model* (also called the *small-world model*) interpolates between a *regular lattice*, which has high clustering but lacks the small-world phenomenon, and a *random network*, which has low clustering but displays the small-world property (**Figure 3.14a–c**). Numerical simulations indicate that for a range of rewiring parameters the model's average path length is low but the clustering coefficient is high, hence reproducing the coexistence of high clustering and small-world phenomenon (**Figure 3.14d**).

Being an extension of the random network model, the Watts–Strogatz model predicts a Poisson-like bounded degree distribution. Consequently high-degree nodes, like those seen in **Figure 3.6**, are absent from it. Furthermore, it predicts a k-independent $C(k)$, being unable to recover the k-dependence observed in **Figure 3.13b–d**. As we show in the next chapters, understanding the coexistence of the small-world property with high clustering must start from the network's correct degree distribution.

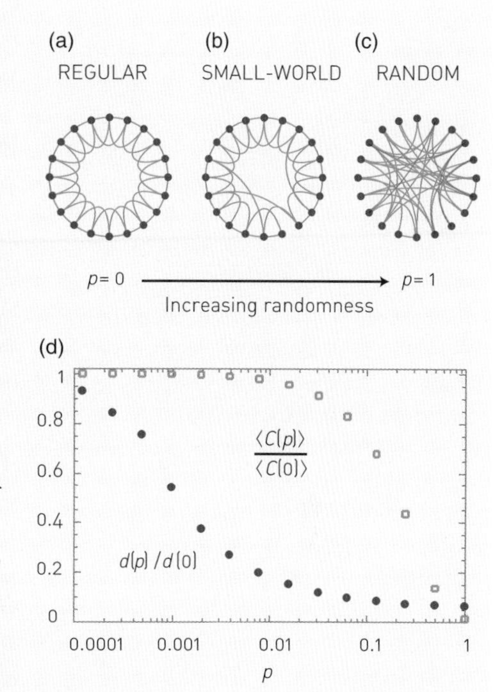

Figure 3.14 The Watts–Strogatz Model
(a) We start from a ring of nodes, each node being connected to their immediate and next neighbors. Hence, initially each node has $\langle C \rangle = 1/2 (p = 0)$.
(b) With probability p each link is rewired to a randomly chosen node. For small p the network maintains high clustering but the random long-range links can drastically decrease the distances between the nodes.
(c) For $p = 1$ all links have been rewired, so the network turns into a random network.
(d) The dependence of the average path length $d(p)$ and clustering coefficient $\langle C(p) \rangle$ on the rewiring parameter p. Note that $d(p)$ and $\langle C(p) \rangle$ have been normalized by $d(0)$ and $\langle C(0) \rangle$ obtained for a regular lattice (i.e., for $p = 0$ in (a)). The rapid drop in $d(p)$ signals the onset of the small-world phenomenon. During this drop, $\langle C(p) \rangle$ remains high. Hence, in the range $0.001 < p < 0.1$ short path lengths and high clustering coexist in the network. All graphs have $N = 1{,}000$ and $\langle k \rangle = 10$. After [8].

observed in complex systems should be described as purely random. With that it equated complexity with randomness. We must therefore ask:

Do we really believe that real networks are random?

The answer is clearly no. As the interactions between our proteins are governed by the strict laws of biochemistry, for the cell to function its chemical architecture cannot be random. Similarly, in a random society an American student would be as likely to have among his friends Chinese factory workers as one of his classmates.

In reality we suspect the existence of a deep order behind most complex systems. That order must be reflected in the structure of the network that describes their architecture, resulting in systematic deviations from a purely random configuration.

The degree to which random networks describe, or fail to describe, real systems must not be decided by epistemological arguments, but by a systematic quantitative comparison. We can do this by taking advantage of the fact that random network theory makes a number of quantitative predictions:

Degree Distribution

A random network has a binomial distribution, well approximated by a Poisson distribution in the $k \ll N$ limit. Yet, as shown in **Figure 3.5**, the Poisson distribution fails to capture the degree distribution of real networks. In real systems we have more highly connected nodes than the random network model could account for.

Connectedness

Random network theory predicts that for $\langle k \rangle > 1$ we should observe a giant component, a condition satisfied by all networks we examined. Most networks, however, do not satisfy the $\langle k \rangle > \ln N$ condition, implying that they should be broken into isolated clusters (**Table 3.1**). Some networks are indeed fragmented, most are not.

Average Path Length

Random network theory predicts that the average path length follows **(3.19)**, a prediction that offers a reasonable approximation for the observed path lengths. Hence the random network model can account for the emergence of small-world phenomena.

Box 3.10 Random Networks and Network Science

The lack of agreement between random and real networks raises an important question: How could a theory survive so long given its poor agreement with reality? The answer is simple: random network theory was never meant to serve as a model for real systems.

Erdős and Rényi, in their first paper [3], wrote that random networks "may be interesting not only from a purely mathematical point of view. In fact, the evolution of graphs may be considered as a rather simplified model of the evolution of certain communication nets (railways, road or electric network systems, etc.) of a country or some unit." Yet, in the string of eight papers authored by them on the subject [3, 72, 73, 74, 75, 76, 77, 78], this is the *only* mention of the potential practical value of their approach. The subsequent development of random graphs was driven by the problem's inherent mathematical challenges, rather than its applications.

It is tempting to follow Thomas Kuhn and view network science as a paradigm change from random graphs to a theory of real networks [93]. In reality, there was no network paradigm before the end of the 1990s. This period is characterized by a lack of systematic attempts to compare the properties of real networks with graph-theoretical models. The work of Erdős and Rényi gained prominence outside mathematics only after the emergence of network science (**Figure 3.15**).

Network theory does not lessen the contributions of Erdős and Rényi, but celebrates the unintended impact of their work. When we discuss the discrepancies between random and real networks, we do so mainly for pedagogical reasons: to offer a proper foundation on which we can understand the properties of real systems.

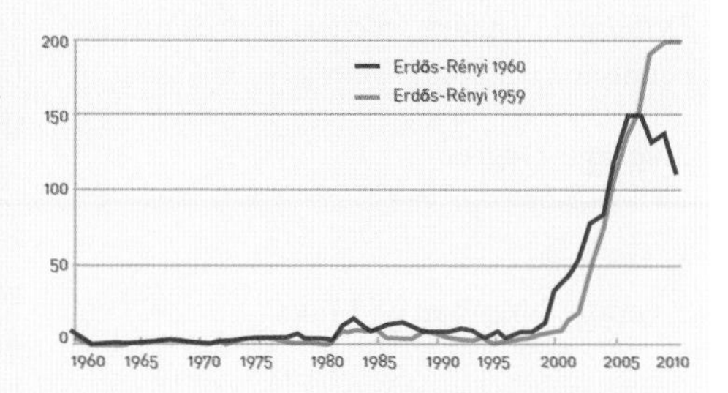

Figure 3.15 **Network Science and Random Networks**

While today we perceive the Erdős–Rényi model as the cornerstone of network theory, the model was hardly known outside a small subfield of mathematics. This is illustrated by the yearly citations of the first two papers by Erdős and Rényi, published in 1959 and 1960 [3, 72]. For four decades after their publication the papers gathered less than ten citations each year. The number of citations exploded after the first papers on scale-free networks [9, 10, 11] turned Erdős and Rényi's work into the reference model of network theory.

Box 3.11 At a Glance: Random Networks

Definition: N nodes, where each node pair is connected with probability p.

Average Degree

$$\langle k \rangle = p(N-1)$$

Average Number of Links

$$\langle L \rangle = \frac{pN(N-1)}{2}$$

Degree Distribution

Binomial form

$$p_k = \binom{N-1}{k} p^k (1-p)^{N-1-k}$$

Poisson form

$$p_k = e^{-\langle k \rangle} \frac{\langle k \rangle^k}{k!}$$

Giant Component (GC) (N_G)

$$\begin{aligned} \langle k \rangle < 1 &: \quad N_G \sim \ln N \\ 1 < \langle k \rangle < \ln N &: \quad N_G \sim N^{\frac{2}{3}} \\ \langle k \rangle > \ln N &: \quad N_G \sim (p - p_c)N \end{aligned}$$

Average Distance

$$\langle d \rangle \propto \frac{\ln N}{\ln \langle k \rangle}$$

Average Clustering Coefficient

$$\langle C \rangle = \frac{\langle k \rangle}{N}$$

Clustering Coefficient

In a random network the local clustering coefficient is independent of the node's degree and $\langle C \rangle$ depends on the system size as $1/N$. In contrast, measurements indicate that for real networks $C(k)$ decreases with the node degree and is largely independent of the system size (**Figure 3.13**).

Taken together, it appears that the small-world phenomenon is the only property reasonably explained by the random network model. All other network characteristics, from the degree distribution to the clustering coefficient, are significantly different in real networks. The extension of the Erdős–Rényi model proposed by Watts and Strogatz successfully predicts the coexistence of high C and low $\langle d \rangle$, but fails to explain the degree distribution and $C(k)$. In fact, the more we learn about real networks, the more we arrive at the startling conclusion that *we do not know of any real network that is accurately described by the random network model.*

This conclusion begs a legitimate question: If real networks are not random, why did we devote a full chapter to the random network model? The answer is simple: the model serves as an important reference as we proceed to explore the properties of real networks. Each time we observe some network property we will have to ask if it could have emerged by chance. For this we turn to the random network model as a guide. If the property is present in the model, it means that randomness can account for it. If the property is absent in random networks, it may represent some signature of order, requiring a deeper explanation. So, the random network model may be the wrong model for most real systems, but *it remains quite relevant for network science* (**Box 3.10**).

3.11 Homework

3.11.1 Erdős–Rényi Networks

Consider an Erdős–Rényi network with $N = 3{,}000$ nodes, connected to each other with probability $p = 10^{-3}$.

(a) What is the expected number of links, $\langle L \rangle$?
(b) In which regime is the network?

(c) Calculate the probability p_c so that the network is at the critical point.

(d) Given the linking probability $p = 10^{-3}$, calculate the number of nodes N^{cr} so that the network has only one component.

(e) For the network in (d), calculate the average degree $\langle k^{cr} \rangle$ and the average distance between two randomly chosen nodes $\langle d \rangle$.

(f) Calculate the degree distribution p_k of this network (approximate with a Poisson degree distribution).

3.11.2 Generating Erdős–Rényi Networks

Relying on the $G(N, p)$ model, generate with a computer three networks with $N = 500$ nodes and average degree (a) $\langle k \rangle = 0.8$, (b) $\langle k \rangle = 1$ and (c) $\langle k \rangle = 8$. Visualize these networks.

3.11.3 Circle Network

Consider a network with N nodes placed on a circle, so that each node connects to m neighbors on either side (consequently each node has degree $2m$). **Figure 3.14a** shows an example of such a network with $m = 2$ and $N = 20$. Calculate the average clustering coefficient $\langle C \rangle$ of this network and the average shortest path $\langle d \rangle$. For simplicity, assume that N and m are chosen such that $(n - 1)/2m$ is an integer. What happens to $\langle C \rangle$ if $N \gg 1$? And what happens to $\langle d \rangle$?

3.11.4 Cayley Tree

A Cayley tree is a symmetric tree, constructed starting from a central node of degree k. Each node at distance d from the central node has degree k, until we reach the nodes at distance P that have degree one and are called leaves (see **Figure 3.16** for a Cayley tree with $k = 3$ and $P = 5$).

(a) Calculate the number of nodes reachable in t steps from the central node.

(b) Calculate the degree distribution of the network.

(c) Calculate the diameter $d_{\max}$.

(d) Find an expression for the diameter $d_{\max}$ in terms of the total number of nodes N.

(e) Does the network display the small-world property?

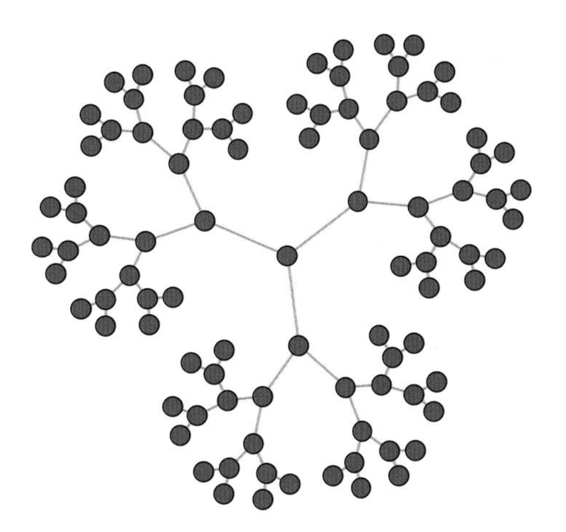

Figure 3.16 **Cayley Tree**
A Cayley tree with $k = 3$ and $P = 5$.

3.11.5 Snobbish Network

Consider a network of N red and N blue nodes. The probability that there is a link between nodes of identical color is p and the probability that there is a link between nodes of different color is q. A network is snobbish if $p > q$, capturing a tendency to connect to nodes of the same color. For $q = 0$ the network has at least two components, containing nodes with the same color.

(a) Calculate the average degree of the "blue" subnetwork made up of only blue nodes, and the average degree in the full network.
(b) Determine the minimal p and q required to have, with high probability, just one component.
(c) Show that for large N even very snobbish networks ($p \gg q$) display the small-world property.

3.11.6 Snobbish Social Networks

Consider the following variant of the model discussed above. We have a network of $2N$ nodes, consisting of an equal number of red and blue nodes, while a fraction f of the $2N$ nodes are purple. Blue and red nodes do not connect to each other ($q = 0$), while they connect with probability p to nodes of the same color. Purple nodes connect with the same probability p to both red and blue nodes.

(a) We call the red and blue communities *interactive* if a typical red node is just two steps away from a blue node and vice versa. Evaluate the fraction of purple nodes required for the communities to be interactive.
(b) Comment on the size of the purple community if the average degree of the blue (or red) nodes is $\langle k \rangle \gg 1$.
(c) What are the implications of this model for the structure of social (and other) networks?

3.12 Advanced Topics 3.A Deriving the Poisson Distribution

To derive the Poisson form of the degree distribution we start from the exact binomial distribution **(3.7)**

$$p_k = \binom{N-1}{k} p^k (1-p)^{N-1-k} \tag{3.22}$$

that characterizes a random graph. We rewrite the first term on the r.h.s. as

$$\begin{aligned}\binom{N-1}{k} &= \frac{(N-1)\ (N-1-1)\ (N-1-2)\cdots(N-1-k+1)}{k!} \\ &\approx \frac{(N-1)^k}{k!},\end{aligned} \tag{3.23}$$

where in the last term we have used that $k \ll N$. The last term of **(3.22)** can be simplified as

$$\ln[(1-p)^{(N-1)-k}] = (N-1-k)\ln\left(1 - \frac{\langle k \rangle}{N-1}\right)$$

and using the series expansion

$$\ln(1+x) = \sum_{n=1}^{\infty} \frac{(-1)^{n+1}}{n} x^n = x - \frac{x^2}{2} + \frac{x^3}{3} - \cdots, \forall |x| \leq 1$$

we obtain

$$\begin{aligned}\ln[(1-p)^{N-1-k}] &\approx (N-1-k)\frac{\langle k \rangle}{N-1} \\ &= -\langle k \rangle \left(1 - \frac{k}{N-1}\right) \approx -\langle k \rangle,\end{aligned}$$

which is valid if $N \gg k$. This represents the *small-degree approximation* at the heart of this derivation. Therefore, the last term of **(3.22)** becomes

$$(1-p)^{N-1-k} = e^{-\langle k \rangle}. \tag{3.24}$$

Combining **(3.22)**, **(3.23)** and **(3.24)**, we obtain the Poisson form of the degree distribution

$$\begin{aligned}p_k &= \binom{N-1}{k} p^k (1-p)^{(N-1)-k} = \frac{(N-1)}{k!} p^k e^{-\langle k \rangle} \\ &= \frac{(N-1)^k}{k!} \left(\frac{\langle k \rangle}{N-1}\right)^k e^{-\langle k \rangle},\end{aligned}$$

or

$$p_k = e^{-\langle k \rangle} \frac{\langle k \rangle^k}{k!}. \tag{3.25}$$

3.13 Advanced Topics 3.B Maximum and Minimum Degrees

To determine the expected degree of the largest node in a random network, called the network's *upper natural cutoff*, we define the degree k_{max} such that in a network of N nodes we have at most one node with degree higher than k_{max}. Mathematically this means that the area behind the Poisson distribution p_k for $k \geq k_{max}$ should be approximately one (**Figure 3.17**). Since the area is given by $1 - P(k_{max})$, where $P(k)$ is the cumulative degree distribution of p_k, the network's largest node satisfies

$$N[1 - P(k_{max})] \approx 1. \tag{3.26}$$

We write $\approx$ instead of $=$ because k_{max} is an integer, so in general the exact equation does not have a solution. For a Poisson distribution

$$1 - P(k_{max}) = 1 - e^{-\langle k \rangle} \sum_{k=0}^{k_{max}} \frac{\langle k \rangle^k}{k!} = e^{-\langle k \rangle} \sum_{k=k_{max}+1}^{\infty} \frac{\langle k \rangle^k}{k!}$$
$$\approx e^{-\langle k \rangle} \frac{\langle k \rangle^{k_{max}+1}}{(k_{max}+1)!}, \tag{3.27}$$

where in the last term we approximate the sum by its largest term.

For $N = 10^9$ and $\langle k \rangle = 1{,}000$, roughly the size and the average degree of the globe's social network, **(3.26)** and **(3.27)** predict $k_{max} = 1{,}185$, indicating that a random network lacks extremely popular individuals, or hubs.

We can use a similar argument to calculate the expected degree of the smallest node, k_{min}. By requiring that there should be at most one node with degree smaller than k_{min}, we can write

$$NP(k_{min} - 1) \approx 1. \tag{3.28}$$

For the Erdős–Rényi network we have

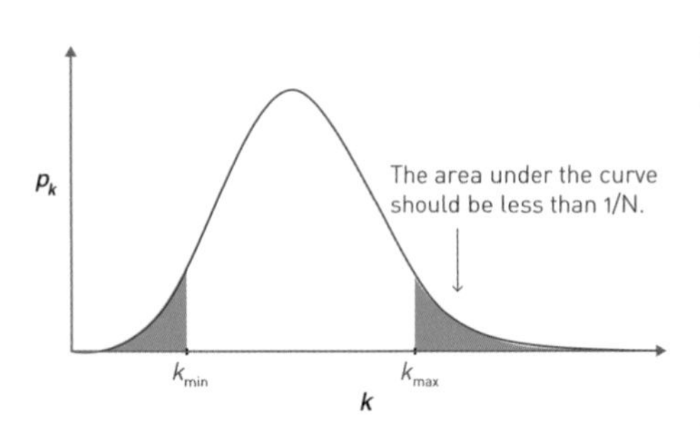

Figure 3.17 **Minimum and Maximum Degree**
The estimated maximum degree of a network, k_{max}, is chosen so that there is at most one maximum node whose degree is higher than k_{max}. This is often called the *natural upper cutoff* of a degree distribution. To calculate it, we need to set k_{max} such that the area under the degree distribution p_k for $k > k_{max}$ equals $1/N$, hence the total number of nodes expected in this region is exactly one. We follow a similar argument to determine the expected smallest degree, k_{min}.

$$P(k_{\min} - 1) = e^{-\langle k \rangle} \sum_{k=0}^{k_{\min}-1} \frac{\langle k \rangle^k}{k!}. \tag{3.29}$$

Solving **(3.28)** with $N = 10^9$ and $\langle k \rangle = 1{,}000$ we obtain $k_{\min} = 816$.

3.14 Advanced Topics 3.C Giant Component

In this section we introduce the argument, proposed independently by Solomonoff and Rapoport [80], and by Erdős and Rényi [3], for the emergence of a giant component at $\langle k \rangle = 1$ [94].

Let us denote by $u = 1 - N_G/N$ the fraction of nodes that are not in the giant component (GC), whose size we take to be N_G. If node i is part of the GC, it must link to another node j, which must also be part of the GC. Hence, if i is *not* part of the GC that could happen for two reasons:

- There is no link between i and j (the probability of this is $1 - p$).
- There is a link between i and j, but j is not part of the GC (the probability of this is pu).

Therefore, the total probability that i is not part of the GC via node j is $1 - p + pu$. The probability that i is not linked to the GC via any other node is therefore $(1 - p + pu)^{N-1}$, as there are $N - 1$ nodes that could serve as potential links to the GC for node i. As u is the fraction of nodes that do not belong to the GC, for any p and N the solution of the equation

$$u = (1 - p + pu)^{N-1} \tag{3.30}$$

provides the size of the giant component via $N_G = N(1 - u)$. Using $p = \langle k \rangle/(N - 1)$ and taking the logarithm of both sides, for $\langle k \rangle \ll N$ we obtain

$$\begin{aligned} \ln u &= (N-1)\ln\left[1 - \frac{\langle k \rangle}{N-1}(1-u)\right] \\ &\approx (N-1)\left[-\frac{\langle k \rangle}{N-1}(1-u)\right] = -\langle k \rangle \quad (1-u), \end{aligned} \tag{3.31}$$

where we have used the series expansion for $\ln(1 + x)$.

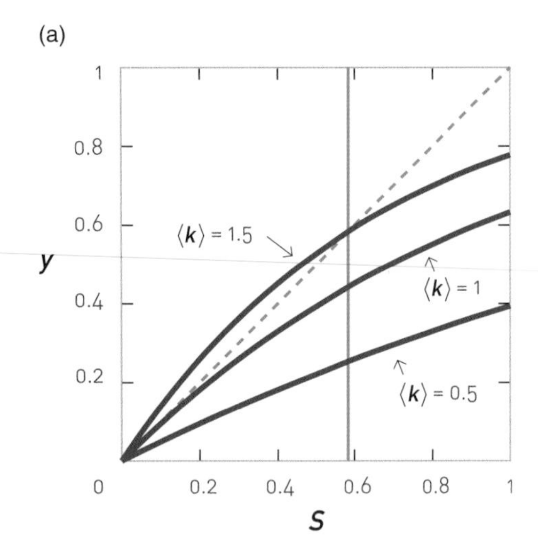

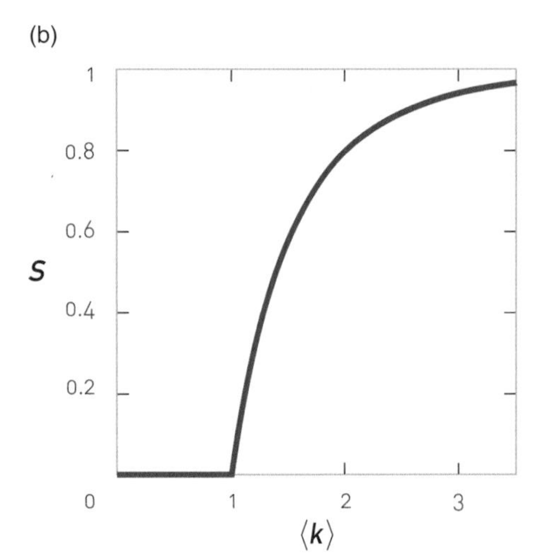

Figure 3.18 **Graphical Solution**
(a) The three purple curves correspond to $y = 1 - \exp[-\langle k\rangle S]$ for $\langle k\rangle = 0.5, 1, 1.5$. The green dashed diagonal corresponds to $y = S$ and the intersection of the dashed and purple curves provides the solution to **(3.32)**. For $\langle k\rangle = 0.5$ there is only one intersection at $S = 0$, indicating the absence of a giant component. The $\langle k\rangle = 1.5$ curve has a solution at $S = 0.583$ (green vertical line). The $\langle k\rangle = 1$ curve is precisely at the critical point, representing the separation between the regime where a nonzero solution for S exists and the regime where there is only the solution at $S = 0$.
(b) The size of the giant component as a function of $\langle k\rangle$, as predicted by **(3.32)**. After [94].

Taking the exponential of both sides leads to $u = \exp[-\langle k\rangle(1-u)]$. If we denote by S the fraction of nodes in the GC, $S = N_G/N$, then $S = 1 - u$ and **(3.31)** results in

$$S = 1 - e^{-\langle k\rangle S}. \tag{3.32}$$

This equation provides the size of the GC S as a function of $\langle k\rangle$ (**Figure 3.18**). While **(3.32)** looks simple, it does not have a closed solution. We can solve it graphically by plotting the r.h.s. of **(3.32)** as a function of S for various values of $\langle k\rangle$. To have a nonzero solution, the obtained curve must intersect with the dotted diagonal, representing the l.h.s. of **(3.32)**. For small $\langle k\rangle$ the two curves intersect each other only at $S = 0$, indicating that for small $\langle k\rangle$ the size of the GC is zero. Only when $\langle k\rangle$ exceeds a threshold value does a nonzero solution emerge.

To determine the value of $\langle k\rangle$ at which we start having a nonzero solution, we take a derivative of **(3.32)** at the phase transition point where the r.h.s. of **(3.32)** has the same derivative as the l.h.s. of **(3.32)**, that is when

$$\begin{gathered}\frac{d}{dS}\left(1 - e^{-\langle k\rangle S}\right) = 1, \\ \langle k\rangle e^{-\langle k\rangle S} = 1.\end{gathered} \tag{3.33}$$

Setting $S = 0$, we obtain that the phase transition point is at $\langle k\rangle = 1$ (see also **Advanced topics 3.F**).

3.15 Advanced Topics 3.D Component Sizes

In **Figure 3.7** we explored the size of the giant component, leaving an important question open: How many components do we expect for a given $\langle k\rangle$? What is their size distribution? The aim of this section is to discuss these topics.

3.15.1 Component Size Distribution

For a random network the probability that a randomly chosen node belongs to a component of size s (which is different from the giant component GC) is [94]

$$p_s \sim \frac{(s\langle k\rangle)^{s-1}}{s!} e^{-\langle k\rangle s}. \quad (3.34)$$

Replacing $\langle k\rangle^{s-1}$ with $\exp[(s-1)\ln\langle k\rangle]$ and using the Stirling formula for large s, we obtain

$$p_s \sim s^{-3/2} e^{-(\langle k\rangle -1)s+(s-1)\ln\langle k\rangle}. \quad (3.35)$$

Therefore the component size distribution has two contributions: a slowly decreasing power law term $s^{-3/2}$ and a rapidly decreasing exponential term $e^{-(<k>-1)s+(s-1)\ln<k>}$. Given that the exponential term dominates for large s, **(3.35)** predicts that large components are prohibited. At the *critical point*, $\langle k\rangle = 1$, all terms in the exponential cancel, hence p_s follows the power law

$$p_s \sim s^{-3/2}. \quad (3.36)$$

As a power law decreases relatively slowly, at the critical point we expect to observe clusters of widely different sizes, a property consistent with the behavior of a system during a phase transition (**Advanced topics 3.F**). These predictions are supported by the numerical simulations shown in **Figure 3.19**.

3.15.2 Average Component Size

The calculations also indicate that the average component size (once again, excluding the giant component) follows [94]

$$\langle s\rangle = \frac{1}{1-\langle k\rangle + \langle k\rangle N_G/N} \quad (3.37)$$

For $\langle k\rangle < 1$ we lack a giant component ($N_G = 0$), hence **(3.37)** becomes

(a)

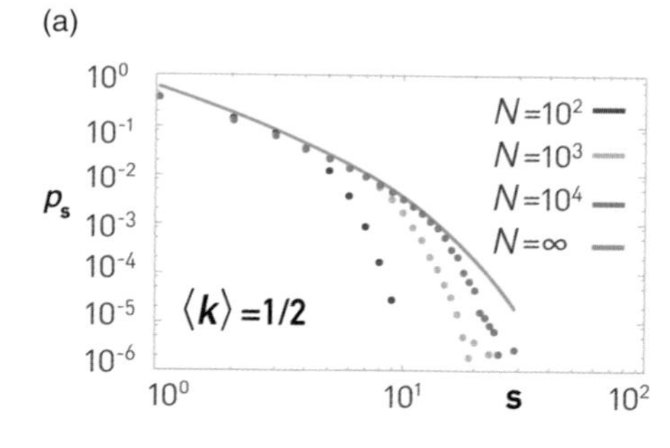

(b)

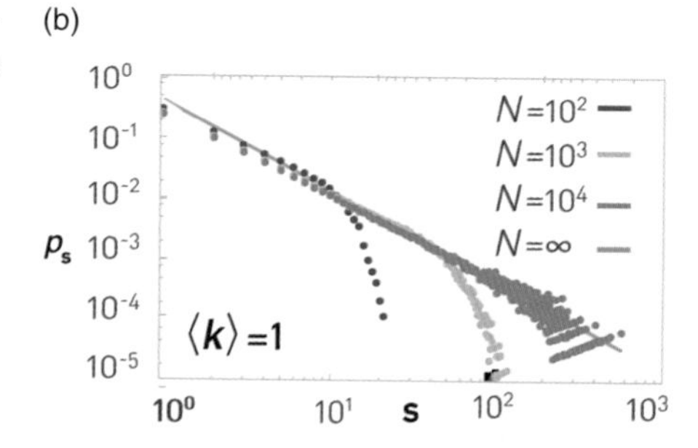

(c)

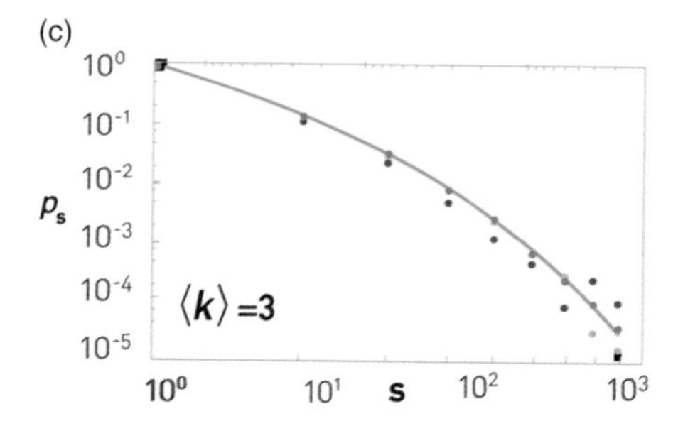

(d)

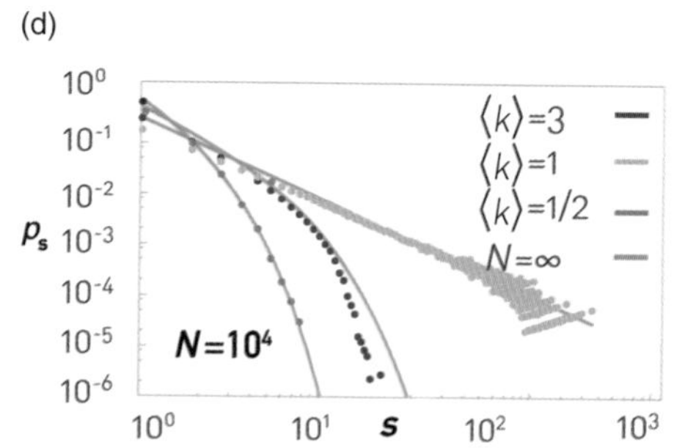

Figure 3.19 **Component Size Distribution**

Component size distribution p_s in a random network, excluding the giant component.

(a)–(c) p_s for different $\langle k\rangle$ values and N, indicating that p_s converges for large N to the prediction **(3.34)**.

(d) p_s for $N = 10^4$, shown for different $\langle k\rangle$. While for $\langle k\rangle < 1$ and $\langle k\rangle > 1$ the p_s distribution has an exponential form, right at the critical point $\langle k\rangle = 1$ the distribution follows the power law **(3.36)**. The continuous green lines correspond to **(3.35)**. The first numerical study of the component size distribution in random networks was carried out in 1998 [94], preceding the exploding interest in complex networks.

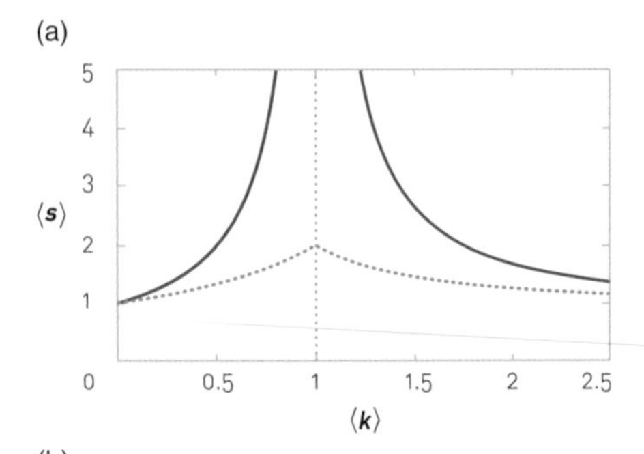

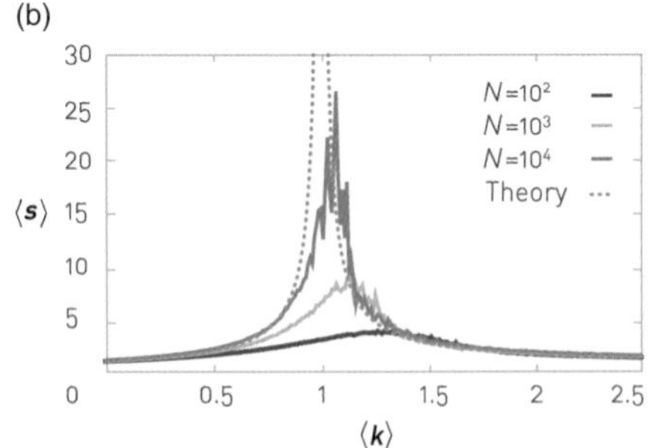

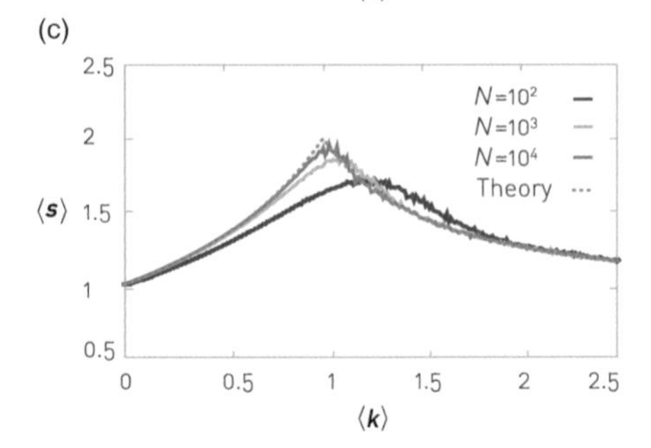

Figure 3.20 **Average Component Size**
(**a**) The average size $\langle s \rangle$ of a component to which a randomly chosen node belongs, as predicted by **(3.39)** (purple). The green curve shows the overall average size $\langle s' \rangle$ of a component as predicted by **(3.37).** After [94].
(**b**) The average cluster size in a random network. We chose a node and determined the size of the cluster it belongs to. This measure is biased, as each component of size s will be counted s times. The larger N becomes, the more closely the numerical data follows the prediction **(3.37).** As predicted, $\langle s \rangle$ diverges at the $\langle k \rangle = 1$ critical point, supporting the existence of a phase transition (**Advanced topics 3.F**).
(**c**) The average cluster size in a random network, where we corrected for the bias in (b) by selecting each component only once. The larger N becomes, the more closely the numerical data follows the prediction **(3.39).**

$$\langle s \rangle = \frac{1}{1 - \langle k \rangle}, \tag{3.38}$$

which diverges when the average degree approaches the critical point $\langle k \rangle = 1$. Therefore, as we approach the critical point, the size of the clusters increases, signaling the emergence of the giant component at $\langle k \rangle = 1$. Numerical simulations support these predictions for large N (**Figure 3.20**).

To determine the average component size for $\langle k \rangle > 1$ using **(3.37)**, we need to first calculate the size of the giant component. This can be done in a self-consistent manner, obtaining that the average cluster size decreases for $\langle k \rangle > 1$, as most clusters are gradually absorbed by the giant component.

Note that **(3.37)** predicts the size of the component to which a randomly chosen node belongs. This is a biased measure, as the chance of belonging to a larger cluster is higher than the chance of belonging to a smaller one. The bias is linear in the cluster size s. If we correct for this bias, we obtain the average size of the small components that we would get if we were to inspect each cluster one by one and then measure their average size [94]

$$\langle s' \rangle = \frac{2}{2 - \langle k \rangle + \langle k \rangle N_G / N}. \tag{3.39}$$

Figure 3.20 offers numerical support for **(3.39).**

3.16 Advanced Topics 3.E Fully Connected Regime

To determine the value of $\langle k \rangle$ at which most nodes became part of the giant component, we calculate the probability that a randomly selected node does not have a link to the giant component, which is $(1-p)^{N_G} \approx (1-p)^N$, as in this regime $N_G \simeq N$. The expected number of such isolated nodes is

$$I_N = N(1-p)^N = N\left(1 - \frac{N \cdot p}{N}\right)^N \approx Ne^{-Np}, \tag{3.40}$$

where we used $\left(1 - \frac{x}{n}\right)^n \approx e^{-x}$, an approximation valid for large n. If we make p sufficiently large, we arrive at the point where only one node is disconnected from the giant component. At this point $I_N = 1$, hence according to **(3.40)** p needs to satisfy $Ne^{-Np} = 1$. Consequently, the value of p at which we are about to enter the fully connected regime is

$$p = \frac{\ln N}{N}, \tag{3.41}$$

which leads to **(3.14)** in terms of $\langle k \rangle$.

3.17 Advanced Topics 3.F Phase Transitions

The emergence of the giant component at $\langle k \rangle = 1$ in the random network model is reminiscent of a *phase transition*, a much-studied phenomenon in physics and chemistry [96]. Consider two examples:

(i) **Water–Ice Transition (Figure 3.21a).** At high temperatures the H_2O molecules engage in a diffusive motion, forming small groups and then breaking apart to group up with other water molecules. If cooled, at 0°C the molecules suddenly stop this diffusive dance, forming an ordered rigid ice crystal.

(ii) **Magnetism (Figure 3.21b).** In ferromagnetic metals like iron, at high temperatures the spins point in randomly

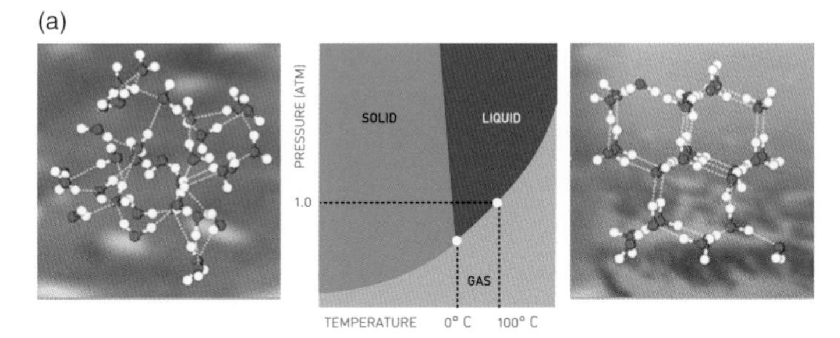

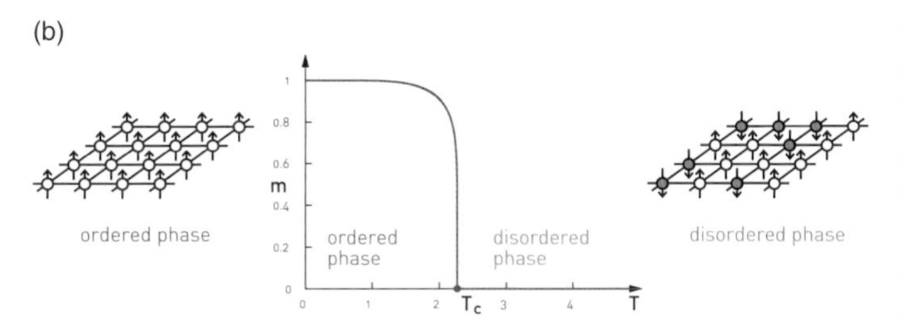

Figure 3.21 Phase Transitions

(a) Water–Ice Phase Transition

The hydrogen bonds that hold the water molecules together (dotted lines) are weak, constantly breaking up and reforming, maintaining partially ordered local structures (left panel). The temperature–pressure phase diagram indicates (center panel) that by lowering the temperature, the water undergoes a phase transition, moving from a liquid (purple) to a frozen solid (green) phase. In the solid phase each water molecule binds rigidly to four other molecules, forming an ice lattice (right panel). After http://www.lbl.gov/Science-Articles/Archive/sabl/2005/February/water-solid.html.

(b) Magnetic Phase Transition

In ferromagnetic materials the magnetic moments of the individual atoms (spins) can point in two different directions. At high temperatures they choose randomly their direction (right panel). In this *disordered state* the system's total magnetization ($m = \Delta M/N$, where ΔM is the number of up spins minus the number of down spins) is zero. The phase diagram (middle panel) indicates that by lowering the temperature T, the system undergoes a phase transition at $T = T_c$, when a nonzero magnetization emerges. Lowering T further allows m to converge to one. In this *ordered phase* all spins point in the same direction (left panel).

chosen directions. Under some critical temperature T_c, all atoms orient their spins in the same direction and the metal turns into a magnet.

The freezing of a liquid and the emergence of magnetization are examples of phase transitions, representing *transitions from disorder to order*. Indeed, relative to the perfect order of the crystalline ice, liquid water is rather disordered. Similarly, the randomly oriented spins in a ferromagnet take up the highly ordered common orientation under T_c.

Many properties of a system undergoing a phase transition are *universal*. This means that the same quantitative patterns are observed in a wide range of systems, from magma freezing into rock to a ceramic material turning into a superconductor. Furthermore, near the phase transition point, called the critical point, many quantities of interest follow power laws.

The phenomenon observed near the critical point $\langle k \rangle = 1$ in a random network in many ways is similar to a phase transition:

- The similarity between **Figure 3.7a** and the magnetization diagram of **Figure 3.21b** is not accidental – they both show a transition from disorder to order. In random networks this corresponds to the emergence of a giant component when $\langle k \rangle$ exceeds $\langle k \rangle = 1$.
- As we approach the freezing point, ice crystals of widely different sizes are observed, and so are domains of atoms with spins pointing in the same direction. The size distribution of the ice crystals or magnetic domains follows a power law. Similarly, while for $\langle k \rangle < 1$ and $\langle k \rangle > 1$ the cluster sizes follow an exponential distribution, right at the phase transition point p_s follows a power law **(3.36)**, indicating the coexistence of components of widely different sizes.
- At the critical point the average size of the ice crystals or of the magnetic domains diverges, ensuring that the whole system turns into a single frozen ice crystal or that all spins point in the same direction. Similarly, in a random network the average cluster size $\langle s \rangle$ diverges as we approach $\langle k \rangle = 1$ (**Figure 3.20**).

3.18 Advanced Topics 3.G Small-World Corrections

Equation **(3.18)** offers only an approximation to the network diameter, valid for very large N and small d. Indeed, as soon as $\langle k \rangle^d$ approaches the system size N, the $\langle k \rangle^d$ scaling must break down, as we do not have enough nodes to continue the $\langle k \rangle^d$ expansion. Such finite size effects result in corrections to **(3.18)**. For a random network with average degree $\langle k \rangle$, the network diameter is better approximated by [97]

$$d_{\max} = \frac{\ln N}{\ln\langle k \rangle} + \frac{2 \ln N}{\ln[-W(\langle k \rangle \exp - \langle k \rangle)]}, \tag{3.42}$$

where the Lambert W-function $W(z)$ is the principal inverse of $f(z) = z\exp(z)$. The first term on the r.h.s. is **(3.18)**, while the second is the correction that depends on the average degree. The correction increases the diameter, accounting for the fact that when we approach the network's diameter the number of nodes must grow slower than $\langle k \rangle$. The magnitude of the correction becomes more obvious if we consider the various limits of **(3.42)**.

In the $\langle k \rangle \to 1$ limit we can calculate the Lambert W-function, finding for the diameter [97]

$$d_{\max} = 3\frac{\ln N}{\ln\langle k \rangle}. \tag{3.43}$$

Hence in the moment when the giant component emerges the network diameter is three times our prediction **(3.18)**. This is due to the fact that at the critical point $\langle k \rangle = 1$ the network has a tree-like structure, consisting of long chains with hardly any loops, a configuration that increases $d_{\max}$.

In the $\langle k \rangle \to \infty$ limit, corresponding to a very dense network, **(3.42)** becomes

$$d_{\max} = \frac{\ln N}{\ln\langle k \rangle} + \frac{2 \ln N}{\langle k \rangle} + \ln N\left(\frac{\ln \langle k \rangle}{\langle k \rangle^2}\right). \tag{3.44}$$

Hence, if $\langle k \rangle$ increases, the second and third terms vanish and the solution **(3.42)** converges to the result **(3.18)**.

Figure 4.0 **Art & Networks: Tomás Saraceno**
Tomás Saraceno creates art inspired by spider webs and neural networks. Trained as an architect, he deploys insights from engineering, physics, chemistry, aeronautics and materials science, using networks as a source of inspiration and metaphor.
The image shows his work displayed in the Miami Art Museum, an example of the artist's take on complex networks.

CHAPTER 4

The Scale-Free Property

4.1 Introduction

The WWW is a network whose nodes are documents and whose links are the URLs that allow us to "surf" with a click from one web document to another. With an estimated size of over one trillion documents ($N \approx 10^{12}$), the Web is the largest network humanity has ever built. It exceeds in size even the human brain ($N \approx 10^{11}$ neurons).

It is difficult to overstate the importance of the WWW in our daily life. Similarly, we cannot exaggerate the role the WWW played in the development of network theory: it facilitated the discovery of a number of fundamental network characteristics and became a standard testbed for most network measures.

We can use software called a *crawler* to map out the Web's wiring diagram. A crawler can start from any web document,

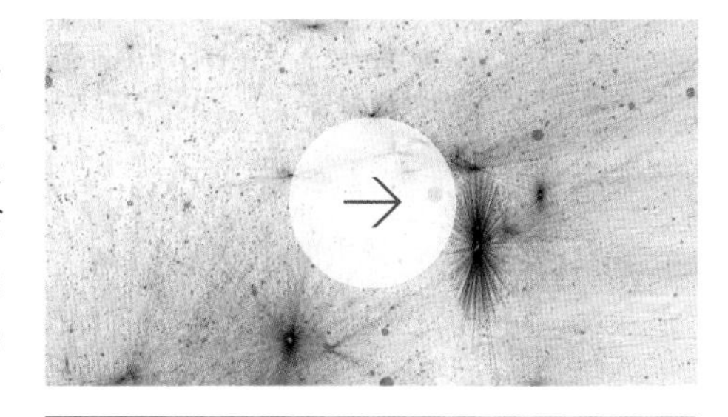

Online Resource 4.1

Zooming into the World Wide Web

Watch an online video that zooms into the WWW sample, which has led to the discovery of the scale-free property [9]. This is the network featured in **Table 2.1** and shown in **Figure 4.1**, whose characteristics are tested throughout this book.

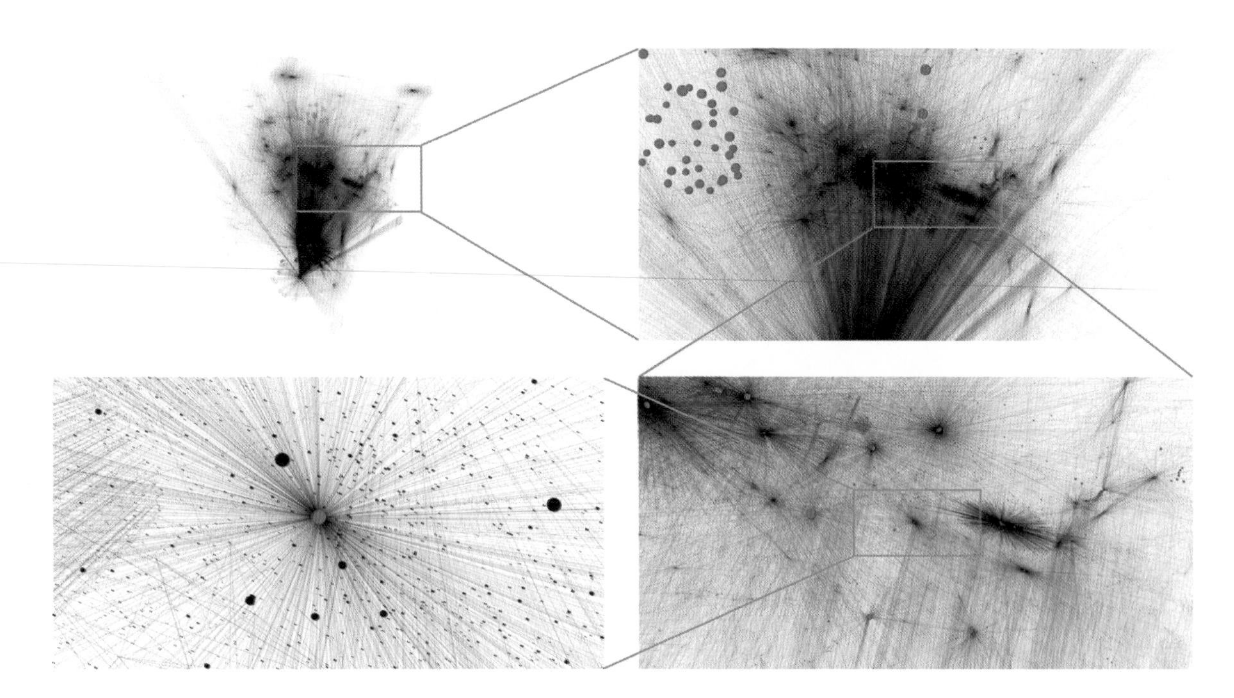

Figure 4.1 **The Topology of the World Wide Web**
Snapshots of the WWW sample mapped out by Hawoong Jeong in 1998 [9]. The sequence of images shows an increasingly magnified local region of the network. The first panel displays all 325,729 nodes, offering a global view of the full dataset. Nodes with more than 50 links are shown in red and nodes with more than 500 links in purple. The closeups reveal the presence of a few highly connected nodes, called *hubs*, which accompany scale-free networks.
Courtesy of M. Martino.

identifying the URLs on it. Next it downloads the documents these links point to and identifies the links on these documents, and so on. This process iteratively returns a local map of the Web. Search engines like Google or Bing operate crawlers to find and index new documents and to maintain a detailed map of the WWW.

The first map of the WWW obtained with the explicit goal of understanding the structure of the network behind it was generated by Hawoong Jeong at the University of Notre Dame. He mapped out the nd.edu domain [9], consisting of about 300,000 documents and 1.5 million links (**Online resource 4.1**). The purpose of the map was to compare the properties of the Web graph to the random network model. Indeed, in 1998 there were reasons to believe that the WWW could be well approximated by a random network. The content of each document reflects the personal and professional interests of its creator, from individuals to organizations. Given the diversity of these interests, the links on these documents might appear to point to randomly chosen documents.

A quick look at the map in **Figure 4.1** supports this view: there appears to be considerable randomness behind the Web's wiring diagram. Yet, a closer inspection reveals some puzzling

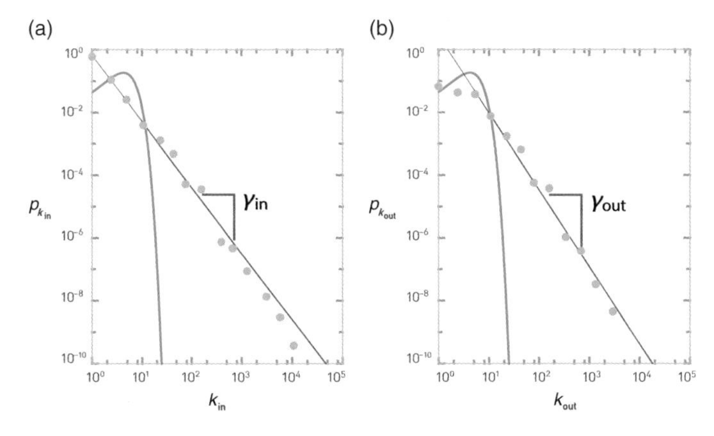

Figure 4.2 **The Degree Distribution of the WWW**
The incoming (**a**) and outgoing (**b**) degree distribution of the WWW sample mapped in the 1999 study of Albert *et al.* [9]. The degree distribution is shown on double logarithmic axes (log–log plot), in which a power law (**Online resource 4.2**) follows a straight line. The symbols correspond to the empirical data and the line corresponds to the power-law fit, with degree exponents $\gamma_{in} = 2.1$ and $\gamma_{out} = 2.45$. We also show as a green line the degree distribution predicted by a Poisson function with the average degree $\langle k_{in} \rangle = \langle k_{out} \rangle = 4.60$ of the WWW sample.

differences between this map and a random network. Indeed, in a random network highly connected nodes, or hubs, are effectively forbidden. In contrast, in **Figure 4.1**, numerous small-degree nodes coexist with a few hubs, nodes with an exceptionally large number of links.

In this chapter we show that hubs are not unique to the Web, but we encounter them in most real networks. They represent a signature of a deeper organizing principle that we call the scale-free property. We therefore explore the degree distribution of real networks, which allows us to uncover and characterize scale-free networks. The analytical and empirical results discussed here represent the foundations of the modeling efforts the rest of this book is based on. Indeed, we will come to see that no matter what network property we are interested in, from communities to spreading processes, it must be inspected in the light of the network's degree distribution.

4.2 Power Laws and Scale-Free Networks

If the WWW were to be a random network, the degrees of the Web documents should follow a Poisson distribution. Yet, as **Figure 4.2** indicates, the Poisson form offers a poor fit for the WWW's degree distribution. Instead, on a log–log scale the data points form an approximate straight line, suggesting that the degree distribution of the WWW is well approximated by

Online Resource 4.2

Fitting a Power Law

The algorithmic tools to perform the fitting procedure described in this section are available at http://tuvalu.santafe.edu/~aaronc/powerlaws/.

$$p_k \sim k^{-\gamma}. \tag{4.1}$$

Equation (**4.1**) is called a *power law distribution* and the exponent γ is its *degree exponent* (**Box 4.1**). If we take a logarithm of (4.1), we obtain

$$\ln p_k \sim \gamma \ln k. \tag{4.2}$$

If (**4.1**) holds, lm p_k is expected to depend linearly on lm k, the slope of this line being the degree exponent γ (**Figure 4.2**).

The WWW is a directed network, hence each document is characterized by an *out-degree* k_{out}, representing the number of links that point from the document to other documents, and an *in-degree* k_{in}, representing the number of other documents that

Box 4.1 The 80/20 Rule and the Top 1%

Figure 4.3 **Vilfredo Federico Damaso Pareto (1848–1923)**
Italian economist, political scientist and philosopher, who had important contributions to our understanding of income distribution and to the analysis of individual choices. A number of fundamental principles are named after him, like *Pareto efficiency*, *Pareto distribution* (another name for a power-law distribution) and the *Pareto principle* (or 80/20 law).

Vilfredo Pareto (**Figure 4.3**), a nineteenth-century economist, noticed that in Italy a few wealthy individuals earned most of the money, while the majority of the population earned rather small amounts. He connected this disparity to the observation that incomes follow a power law, representing the first known report of a power-law distribution [98]. His finding entered the popular literature as the *80/20 rule*: roughly 80% of all money is earned by only 20% of the population.

The 80/20 rule emerges in many areas. For example, in management it is often stated that 80% of the profits are produced by only 20% of the employees. Similarly, 80% of decisions are made during 20% of meeting times.

The 80/20 rule is present in networks as well: 80% of links on the Web point to only 15% of webpages; 80% of citations go to only 38% of scientists; 80% of links in Hollywood are connected to 30% of actors [50]. Most quantities following a power-law distribution obey the 80/20 rule.

During the 2009 economic crisis, power laws gained a new meaning: the Occupy Wall Street movement drew attention to the fact that in the USA, 1% of the population earns a disproportionate 15% of the total US income. This 1% phenomenon, a signature of a profound income disparity, is again a consequence of the power-law nature of the income distribution.

point to the selected document. We must therefore distinguish two degree distributions: the probability that a randomly chosen document points to k_{out} web documents, or $p_{k_{out}}$, and the probability that a randomly chosen node has k_{in} web documents pointing to it, or $p_{k_{in}}$. In the case of the WWW, both $p_{k_{in}}$ and $p_{k_{out}}$ can be approximated by a power law (**Box 4.1**)

$$p_{k_{in}} \sim k^{-\gamma_{in}}, \tag{4.3}$$

$$p_{k_{out}} \sim k^{-\gamma_{out}}, \tag{4.4}$$

where γ_{in} and γ_{out} are the degree exponents for the in- and out-degrees, respectively (**Figure 4.2**). In general, γ_{in} can differ from γ_{out}. For example, in **Figure 4.1** we have $\gamma_{in} \approx 2.1$ and $\gamma_{out} \approx 2.45$.

The empirical results shown in **Figure 4.2** document the existence of a network whose degree distribution is quite different from the Poisson distribution characterizing random networks. We will call such networks *scale-free*, defined as follows [10]:

A scale-free network is a network whose degree distribution follows a power law.

As **Figure 4.2** indicates, for the WWW the power law persists for almost four orders of magnitude, prompting us to call the Web graph a scale-free network. In this case the scale-free property applies to both in- and out-degrees.

To better understand the scale-free property, we have to define the power-law distribution in more precise terms. Therefore, we next discuss the discrete and the continuum formalisms used throughout this book.

4.2.1 Discrete Formalism

As node degrees are positive integers, $k = 0, 1, 2, \ldots,$ the discrete formalism provides the probability p_k that a node has exactly k links

$$p_k = Ck^{-\gamma}. \tag{4.5}$$

The constant C is determined by the normalization condition

$$\sum_{k=1}^{\infty} p_k = 1. \tag{4.6}$$

Using (4.5) we obtain

$$C = \frac{1}{\sum_{k=1}^{\infty} k^{-\gamma}} = \frac{1}{\zeta(\gamma)}, \tag{4.7}$$

where $\zeta(\gamma)$ is the Riemann-zeta function. Thus, for $k > 0$ the discrete power-law distribution has the form

$$p_k = \frac{k^{-\gamma}}{\zeta(\gamma)}. \tag{4.8}$$

Note that (**4.8**) diverges at $k = 0$. If needed, we can specify p_0 separately, representing the fraction of nodes that have no links to other nodes. In that case the calculation of C in (**4.7**) needs to incorporate p_0.

4.2.2 Continuum Formalism

In analytical calculations it is often convenient to assume that the degrees can have any positive real value. In this case we write the power-law degree distribution as

$$p(k) = Ck^{-\gamma}. \tag{4.9}$$

Using the normalization condition

$$\int_{k_{\min}}^{\infty} p(k)dk = 1, \tag{4.10}$$

we obtain

$$C = \frac{1}{\int_{k_{\min}}^{\infty} k^{-\gamma}dk} = (\gamma - 1)k_{\min}^{\gamma-1}. \tag{4.11}$$

Therefore, in the continuum formalism the degree distribution has the form

$$p(k) = (\gamma - 1)k_{\min}^{\gamma-1}k^{-\gamma}. \tag{4.12}$$

Here $k_{\min}$ is the smallest degree for which the power law (**4.8**) holds.

Note that p_k encountered in the discrete formalism has a precise meaning: it is the probability that a randomly selected node has degree k. In contrast, only the integral of $p(k)$ encountered in the continuum formalism has a physical interpretation:

$$\int_{k_1}^{k_2} p(k)dk \tag{4.13}$$

is the probability that a randomly chosen node has degree between k_1 and k_2.

In summary, networks whose degree distribution follows a power law are called scale-free networks. If a network is directed, the scale-free property applies separately to the in- and the out-degrees. To study the properties of scale-free networks mathematically, we can use either the discrete or the continuum formalism. The scale-free property is independent of the formalism we use.

4.3 Hubs

The main difference between a random and a scale-free network comes in the *tail* of the degree distribution, representing the high-k region of p_k. To illustrate this, in **Figure 4.4** we compare a power law with a Poisson function. We find that:

- For small k the power law is above the Poisson function, indicating that a scale-free network has a large number of small-degree nodes, most of which are absent in a random network.
- For k in the vicinity of $\langle k \rangle$ the Poisson distribution is above the power law, indicating that in a random network there is an excess of nodes with degree $k \approx \langle k \rangle$.
- For large k the power law is again above the Poisson curve. The difference is particularly visible if we show p_k on a log–log plot (**Figure 4.4b**), indicating that the probability of observing a high-degree node, or *hub*, is several orders of magnitude higher in a scale-free than in a random network.

Let us use the WWW to illustrate the magnitude of these differences. The probability of having a node with $k = 100$ is about $p_{100} \approx 10^{-94}$ in a Poisson distribution, while it is about

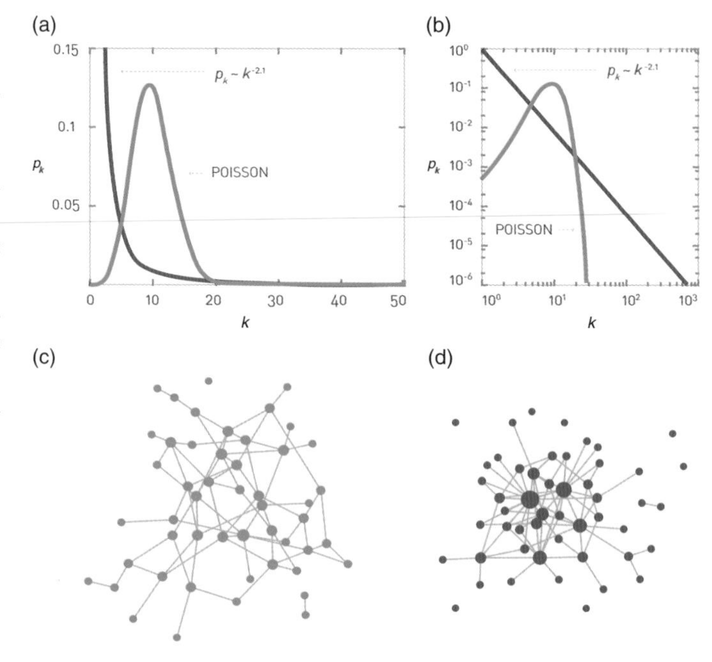

Figure 4.4 **Poisson vs. Power-Law Distributions**
(a) Comparing a Poisson function with a power-law function ($\gamma = 2.1$) on a linear plot. Both distributions have $\langle k \rangle = 11$.
(b) The same curves as in (a), but shown on a log–log plot, allowing us to inspect the difference between the two functions in the high-k regime.
(c) A random network with $\langle k \rangle = 3$ and $N = 50$, illustrating that most nodes have comparable degree $k \approx \langle k \rangle$.
(d) A scale-free network with $\gamma = 2.1$ and $\langle k \rangle = 3$, illustrating that numerous small-degree nodes coexist with a few highly connected hubs. The size of each node is proportional to its degree.

$p_{100} \approx 4 \times 10^{-4}$ if p_k follows a power law. Consequently, if the WWW were to be a random network with $\langle k \rangle = 4.6$ and size $N \approx 10^{12}$, we would expect

$$N_{k \geq 100} = 10^2 \sum_{k=100}^{\infty} \frac{(4.6)^k}{k!} e^{-4.6} \simeq 10^{-82} \tag{4.14}$$

nodes with at least 100 links, or effectively none. In contrast, given the WWW's power-law degree distribution, with $\gamma_{in} = 2.1$ we have $N_{k \geq 100} = 4 \times 10^9$, more than four billion nodes with degree $k \geq 100$.

4.3.1 The Largest Hub

All real networks are finite. The size of the WWW is estimated to be $N \approx 10^{12}$ nodes; the size of the social network is the Earth's population, about $N \approx 7 \times 10^9$. These numbers are huge, but finite. Other networks pale in comparison: the genetic network in a human cell has approximately 20,000 genes, while the metabolic network of the *E. coli* bacteria has only about a thousand metabolites. This prompts us to ask: How does the network size affect the size of its hubs?

To answer this we calculate the maximum degree k_{max}, called the *natural cutoff* of the degree distribution p_k. It represents the expected size of the largest hub in a network.

It is instructive to perform the calculation first for the exponential distribution

$$p(k) = Ce^{-\lambda k}.$$

For a network with minimum degree k_{min}, the normalization condition

$$\int_{k_{min}}^{\infty} p(k)dk = 1 \quad (4.15)$$

provides $C = \lambda e^{\lambda k_{min}}$. To calculate k_{max} we assume that in a network of N nodes we expect at most one node in the (k_{max}, ∞) regime (**Advanced topics 3.B**). In other words, the probability of observing a node whose degree exceeds k_{max} is $1/N$:

$$\int_{k_{max}}^{\infty} p(k)dk = \frac{1}{N}. \quad (4.16)$$

Equation (**4.16**) yields

$$k_{max} = k_{min} + \frac{\ln N}{\lambda}. \quad (4.17)$$

As $\ln N$ is a slow function of the system size, (**4.17**) tells us that the maximum degree will not be significantly different from k_{min}. For a Poisson degree distribution the calculation is a bit more involved, but the obtained dependence of k_{max} on N is even slower than the logarithmic dependence predicted by (**4.17**) (**Advanced topics 3.B**).

For a scale-free network, according to (**4.12**) and (**4.16**), the natural cutoff follows

$$k_{max} = k_{min} N^{\frac{1}{\gamma - 1}}. \quad (4.18)$$

Hence the larger a network, the larger is the degree of its biggest hub. The polynomial dependence of k_{max} on N implies that in a large scale-free network there can be orders of magnitude differences in size between the smallest node, k_{min}, and the biggest hub, k_{max} (**Figure 4.5**).

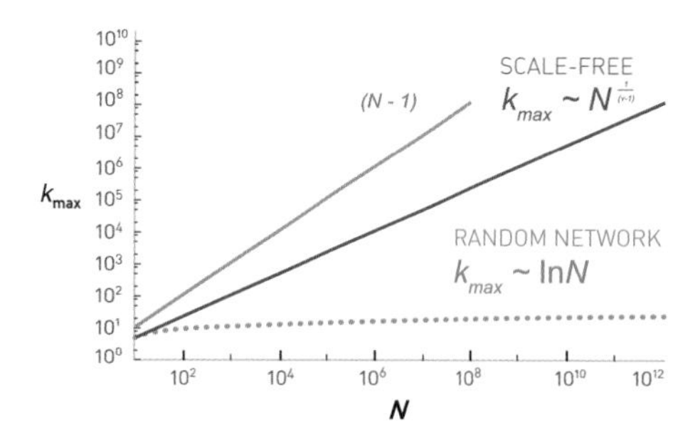

Figure 4.5 **Hubs are Large in Scale-Free Networks**

The estimated degree of the largest node (natural cutoff) in scale-free and random networks with the same average degree $\langle k \rangle = 3$. For the scale-free network we chose $\gamma = 2.5$. For comparison, we also show the linear behavior, $k_{max} \sim N - 1$, expected for a complete network. Overall, hubs in a scale-free network are several orders of magnitude larger than the biggest node in a random network with the same N and $\langle k \rangle$.

To illustrate the difference in the maximum degree of an exponential and a scale-free network, let us return to the WWW sample of **Figure 4.1**, consisting of $N \approx 3 \times 10^5$ nodes. As $k_{\min} = 1$, if the degree distribution were to follow an exponential, (**4.17**) predicts that the maximum degree should be $k_{\max} \approx 14$ for $\lambda = 1$. In a scale-free network of similar size and $\gamma = 2.1$, (**4.18**) predicts $k_{\max} \approx 95{,}000$ – a remarkable difference. Note that the largest in-degree of the WWW map of **Figure 4.1** is 10,721, which is comparable with the $k_{\max}$ predicted by a scale-free network. This reinforces our conclusion that *in a random network hubs are effectively forbidden, while in scale-free networks they are naturally present.*

In summary, the key difference between a random and a scale-free network is rooted in the different shape of the Poisson and of the power-law function. In a random network most nodes have comparable degrees and hence hubs are forbidden. Hubs are not only tolerated, but are expected in scale-free networks (**Figure 4.6**). Furthermore, the more nodes a scale-free network has, the larger are its hubs. Indeed, the size of the hubs grows polynomially with the network size; hence, they can grow quite large in scale-free networks. In contrast, in a random network the size of the largest node grows logarithmically or slower with N, implying that hubs will be tiny even in a very large random network.

4.4 The Meaning of Scale-Free

The term "scale-free" is rooted in a branch of statistical physics called the *theory of phase transitions*, which explored power laws extensively in the 1960s and 1970s (**Advanced topics 3.F**). To best understand the meaning of the scale-free term, we need to familiarize ourselves with the moments of the degree distribution.

The nth moment of the degree distribution is defined as

$$\langle k^n \rangle = \sum_{k_{\min}}^{\infty} k^n p_k \approx \int_{k_{\min}}^{\infty} k^n p(k) dk. \tag{4.19}$$

The lower moments have an important interpretation.

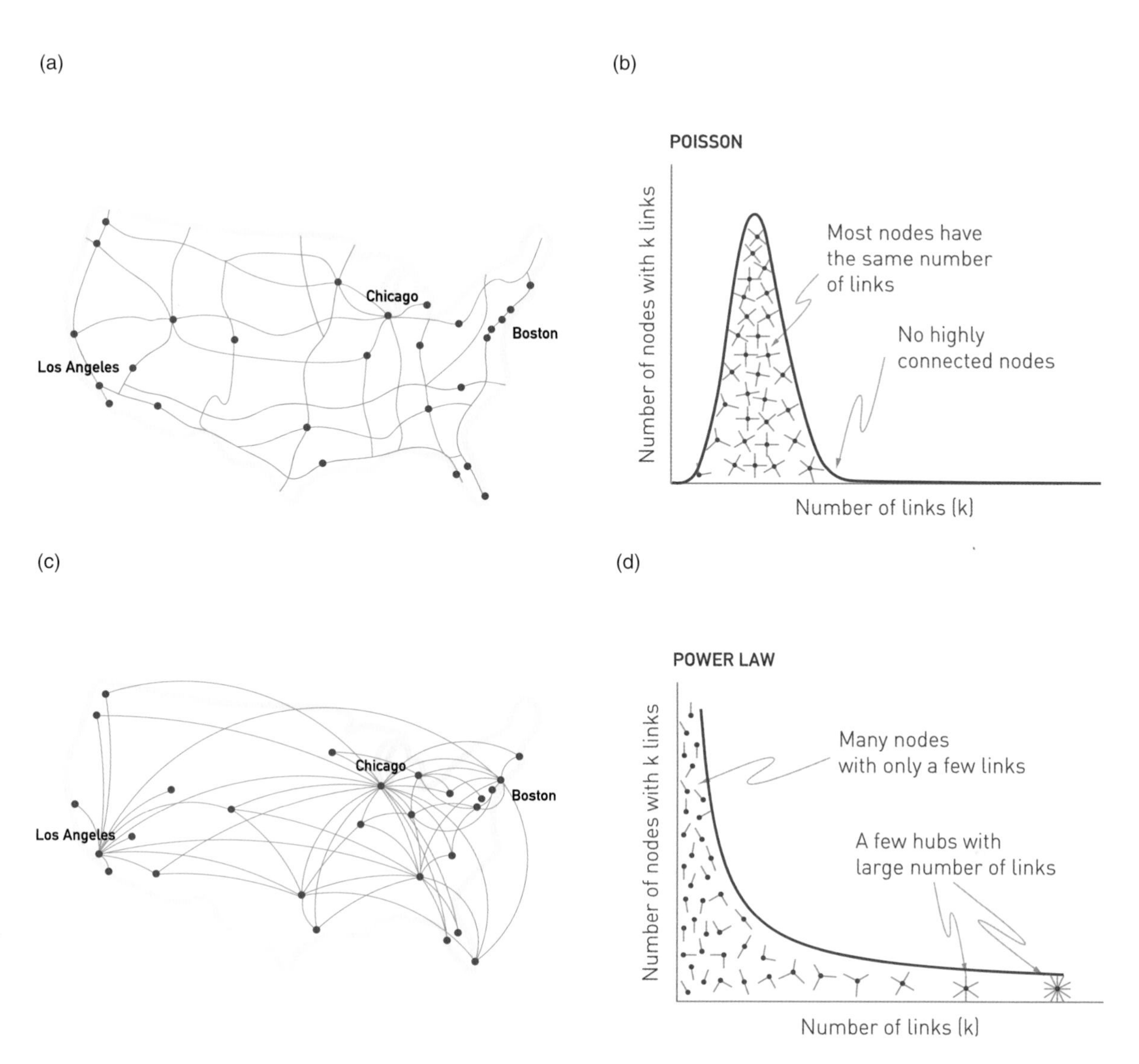

Figure 4.6 **Random vs. Scale-Free Networks**

(a) The degrees of a random network follow a Poisson distribution, rather similar to a bell curve. Therefore, most nodes have comparable degrees and nodes with a large number of links are absent.

(b) A random network looks a bit like the national highway network in which nodes are cities and links are the major highways. There are no cities with hundreds of highways and no city is disconnected from the highway system.

(c) In a network with a power-law degree distribution most nodes have only a few links. These numerous small nodes are held together by a few highly connected hubs.

(d) A scale-free network looks like the air-traffic network, whose nodes are airports and whose links are the direct flights between them. Most airports are tiny, with only a few flights. Yet we have a few very large airports, like Chicago or Los Angeles, that act as major hubs, connecting many smaller airports. Once hubs are present, they change the way we navigate the network. For example, if we travel from Boston to Los Angeles by car, we must drive through many cities. On the airplane network, however, we can reach most destinations via a single hub, like Chicago.

After [50].

- $n = 1$: The first moment is the average degree, $\langle k \rangle$.
- $n = 2$: The second moment, $\langle k^2 \rangle$, helps us calculate the *variance* $\sigma_k^2 = \langle k^2 \rangle - \langle k \rangle^2$ *measuring the spread in the degrees. Its square root,* σ_k, *is the* standard deviation.
- $n = 3$: The third moment, $\langle k^3 \rangle$, determines the *skewness* of a distribution, telling us how symmetric p_k is around the average $\langle k \rangle$.

For a scale-free network, the nth moment of the degree distribution is

$$\langle k^n \rangle = \int_{k_{min}}^{k_{max}} k^n p(k) dk = C \frac{k_{max}^{n-\gamma+1} - k_{min}^{n-\gamma+1}}{n - \gamma + 1}. \quad (4.20)$$

While typically k_{min} is fixed, the degree of the largest hub, k_{max}, increases with the system size, following (**4.18**). Hence, to understand the behavior of $\langle k^n \rangle$ we need to take the asymptotic limit $k_{max} \to \infty$ in (**4.20**), probing the properties of very large networks. In this limit (**4.20**) predicts that the value of $\langle k^n \rangle$ depends on the interplay between n and γ:

- If $n - \gamma + 1 \leq 0$ then the first term on the r.h.s. of (**4.20**), $k_{max}^{n-\gamma+1}$, goes to zero as k_{max} increases. Therefore, all moments that satisfy $n \leq \gamma - 1$ are finite.
- If $n - \gamma + 1 > 0$ then $\langle k^n \rangle$ goes to infinity as $k_{max} \to \infty$. Therefore all maximum moments larger than $\gamma - 1$ diverge.

For many scale-free networks the degree exponent γ is between 2 and 3 (**Table 4.1**). Hence, for these in the $N \to \infty$ limit the first moment $\langle k \rangle$ is finite, but the second and higher moments, $\langle k^2 \rangle$, $\langle k^3 \rangle$, go to infinity. This divergence helps us understand the origin of the "scale-free" term. Indeed, if the degrees follow a normal distribution, then the degree of a randomly chosen node is typically in the range

$$k = \langle k \rangle \pm \sigma_k. \quad (4.21)$$

Yet, the average degree $\langle k \rangle$ and the standard deviation σ_k have rather different magnitude in random and in scale-free networks:

- **Random Networks Have a Scale**
  For a random network with a Poisson degree distribution, $\sigma_k = \langle k \rangle^{1/2}$, which is always smaller than $\langle k \rangle$. Hence, the network's nodes have degrees in the range $k = \langle k \rangle \pm \langle k \rangle^{1/2}$.

Table 4.1 **Degree Fluctuations in Real Networks**

The table shows the first moment $\langle k \rangle$ and the second moment $\langle k^2 \rangle$ ($\langle k_{in}^2 \rangle$ and $\langle k_{out}^2 \rangle$ for directed networks) for ten reference networks. For directed networks we list $\langle k \rangle = \langle k_{in} \rangle = \langle k_{out} \rangle$. We also list the estimated degree exponent, γ, for each network, determined using the procedure discussed in **Advanced topics 4.A**. The stars next to the reported values indicate the confidence of the fit to the degree distribution. That is, * means that the fit shows statistical confidence for a power law ($k^{-\gamma}$) while ** marks statistical confidence for a fit **(4.39)** with an exponential cutoff. Note that the power grid is not scale-free. For this network a degree distribution of the form $e^{-\lambda k}$ offers a statistically significant fit, which is why we placed an "Exp." in the last column.

Network	N	L	$\langle k \rangle$	$\langle k_{in}^2 \rangle$	$\langle k_{out}^2 \rangle$	$\langle k^2 \rangle$	γ_{in}	γ_{out}	γ
Internet	192,244	609,066	6.34	–	–	240.1	–	–	3.42*
WWW	325,729	1,497,134	4.60	1546.0	482.4	–	2.00	2.31	–
Power grid	4,941	6,594	2.67	–	–	10.3	–	–	Exp.
Mobile phone calls	36,595	91,826	2.51	12.0	11.7	–	4.69*	5.01*	–
Email	57,194	91,826	1.81	94.7	1163.9	–	3.43*	2.03*	–
Science collaboration	23,133	93,439	8.08	–	–	178.2	–	–	3.35*
Actor network	702,388	29,397,908	83.71	–	–	47.353.7	–	–	2.12*
Citation network	449,673	4,689,479	10.43	971.5	198.8	–	3.03**	4.00*	–
***E. coli* metabolism**	1,039	5,802	5.58	535.7	396.7	–	2.43*	2.90*	–
Protein interactions	2,018	2,930	2.90	–	–	32.3	–	–	2.89*

In other words, nodes in a random network have comparable degrees and the average degree $\langle k \rangle$ serves as the "scale" of a random network.

- **Scale-Free Networks Lack a Scale**

  For a network with a power-law degree distribution with $\gamma < 3$, the first moment is finite but the second moment is infinite. The divergence of $\langle k^2 \rangle$ (and of σ_k) for large N indicates that the fluctuations around the average can be arbitrarily large. This means that when we randomly choose a node, we do not know what to expect: the selected node's degree could be tiny or arbitrarily large. Hence, networks with $\gamma < 3$ do not have a meaningful internal scale, but are "scale-free" (**Figure 4.7**).

For example, the average degree of the WWW sample is $\langle k \rangle = 4.60$ (**Table 4.1**). Given that $\gamma \approx 2.1$, the second moment diverges, which means that our expectation for the in-degree of a randomly chosen WWW document is $k = 4.60 \pm \infty$ in the $N \to \infty$ limit. That is, a randomly chosen web document could easily yield a document of degree one or two, as 74.02% of nodes

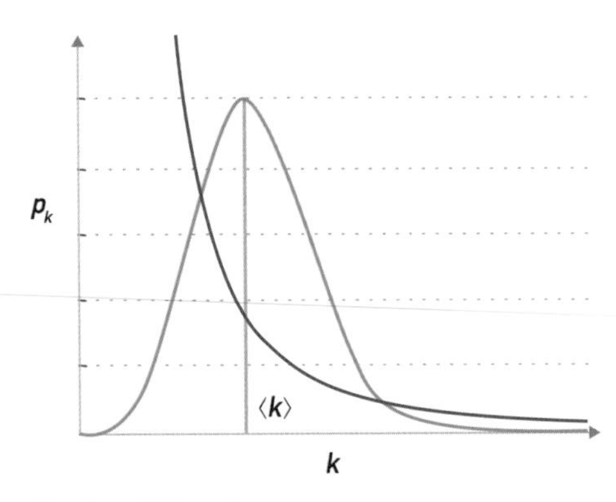

Random Network
Randomly chosen node: $k = \langle k \rangle \pm \langle k \rangle^{1/2}$
Scale: $\langle k \rangle$

Scale-Free Network
Randomly chosen node: $k = \langle k \rangle \pm \infty$
Scale: none

Figure 4.7 **Lack of an Internal Scale**
For any exponentially bounded distribution, like a Poisson or a Gaussian, the degree of a randomly chosen node is in the vicinity of $\langle k \rangle$. Hence, $\langle k \rangle$ serves as the network's *scale*. For a power-law distribution the second moment can diverge, and the degree of a randomly chosen node can be significantly different from $\langle k \rangle$. Hence, $\langle k \rangle$ does not serve as an intrinsic scale. As a network with a power-law degree distribution lacks an intrinsic scale, we call it scale-free.

have in-degree less than $\langle k \rangle$. Yet, it could also yield a node with hundreds of millions of links, like google.com or facebook.com.

Strictly speaking, $\langle k^2 \rangle$ diverges only in the $N \to \infty$ limit. Yet, the divergence is relevant for finite networks as well. To illustrate this, **Table 4.1** lists $\langle k^2 \rangle$ and **Figure 4.8** shows the standard deviation for ten real networks. For most of these networks σ is significantly larger than $\langle k \rangle$, documenting large variations in node degrees. For example, the degree of a randomly chosen node in the WWW sample is $k_{\text{in}} = 4.60 \pm 1{,}546$, indicating once again that the average is not informative.

In summary, the scale-free name captures the lack of an internal scale, a consequence of the fact that nodes with widely different degrees coexist in the same network. This feature distinguishes scale-free networks from lattices, in which all nodes have exactly the same degree ($\sigma_k = 0$), or from random networks, whose degrees vary in a narrow range ($\sigma_k = \langle k \rangle^{1/2}$). As we will see in the coming chapters, this divergence is the origin of some of the most intriguing properties of scale-free networks, from their robustness to random failures to the anomalous spread of viruses.

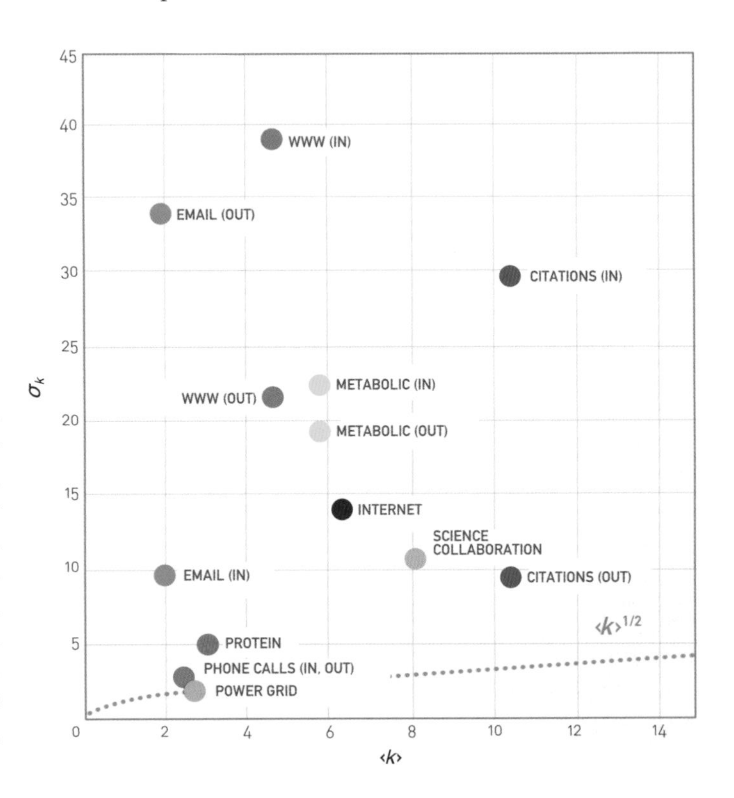

Figure 4.8 **Standard Deviation is Large in Real Networks**
For a random network the standard deviation follows $\sigma_k = \langle k \rangle^{1/2}$, shown as a green dashed line on the figure. The symbols show σ_k for nine of the ten reference networks, calculated using the values shown in **Table 4.1**. The actor network has a very large $\langle k \rangle$ and σ_k, hence it is omitted for clarity. For each network, σ_k is larger than the value expected for a random network with the same $\langle k \rangle$. The only exception is the power grid, which is not scale-free. While the phone-call network is scale-free, it has a large γ, hence it is well approximated by a random network.

4.5 Universality

While the terms WWW and Internet are often used interchangeably in the media, they refer to different systems. The WWW is an information network, whose nodes are documents and whose links are URLs. In contrast, the Internet is an infrastructural network, whose nodes are computers called routers and whose links correspond to physical connections, like copper and optical cables or wireless links.

This difference has important consequences: the cost of linking a Boston-based web page to a document residing on the same computer or to one on a Budapest-based computer is the same. In contrast, establishing a direct Internet link between routers in Boston and Budapest would require us to lay a cable between North America and Europe, which is prohibitively expensive. Despite these differences, the degree distribution of both networks is well approximated by a power law [9, 99, 100]. The signatures of the Internet's scale-free nature are visible in **Figure 4.9**, showing that a few high-degree routers hold together a large number of routers with only a few links.

In the past decade many real networks of major scientific, technological and societal importance were found to display the scale-free property. This is illustrated in **Figure 4.10**, where we show the degree distribution of an infrastructural network (Internet), a biological network (protein interactions), a

Figure 4.9 **The Topology of the Internet**
An iconic representation of the Internet topology at the beginning of the twenty-first century. The image was produced by CAIDA, an organization based at the University of California in San Diego, devoted to collecting, analyzing and visualizing Internet data. The map illustrates the Internet's scale-free nature: a few highly connected hubs hold together numerous small nodes.

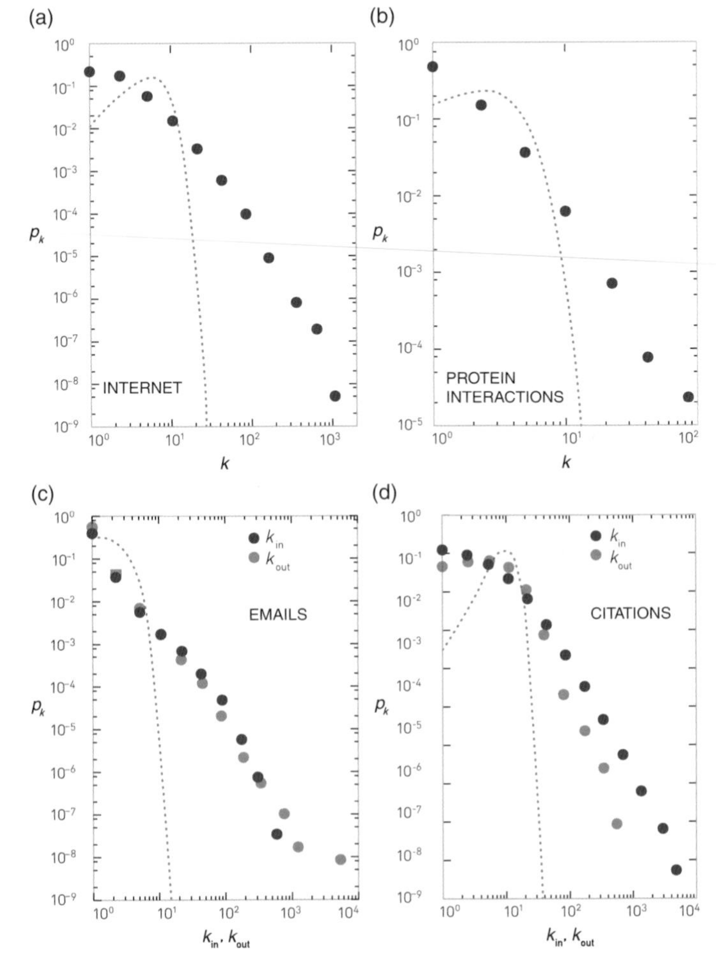

Figure 4.10 **Many Real Networks are Scale-Free**
The degree distribution of four networks listed in **Table 4.1**.
(**a**) Internet at the router level.
(**b**) Protein–protein interaction network.
(**c**) Email network.
(**d**) Citation network.
In each panel the green dotted line shows the Poisson distribution with the same $\langle k \rangle$ as the real network, illustrating that the random network model cannot account for the observed p_k. For directed networks we show separately the incoming and outgoing degree distributions.

communication network (emails) and a network characterizing scientific communications (citations). For each network the degree distribution deviates significantly from a Poisson distribution, being better approximated by a power law.

The diversity of the systems that share the scale-free property is remarkable (**Box 4.2**). Indeed, the WWW is a manmade network with a history of little more than two decades, while the protein interaction network is the product of four billion years of evolution. In some of these networks the nodes are molecules, in others they are computers. It is this diversity that prompts us to call the scale-free property a *universal* network characteristic.

From the perspective of a researcher, a crucial question is the following: How do we know if a network is scale-free? On the

Box 4.2 Timeline: Scale-Free Networks

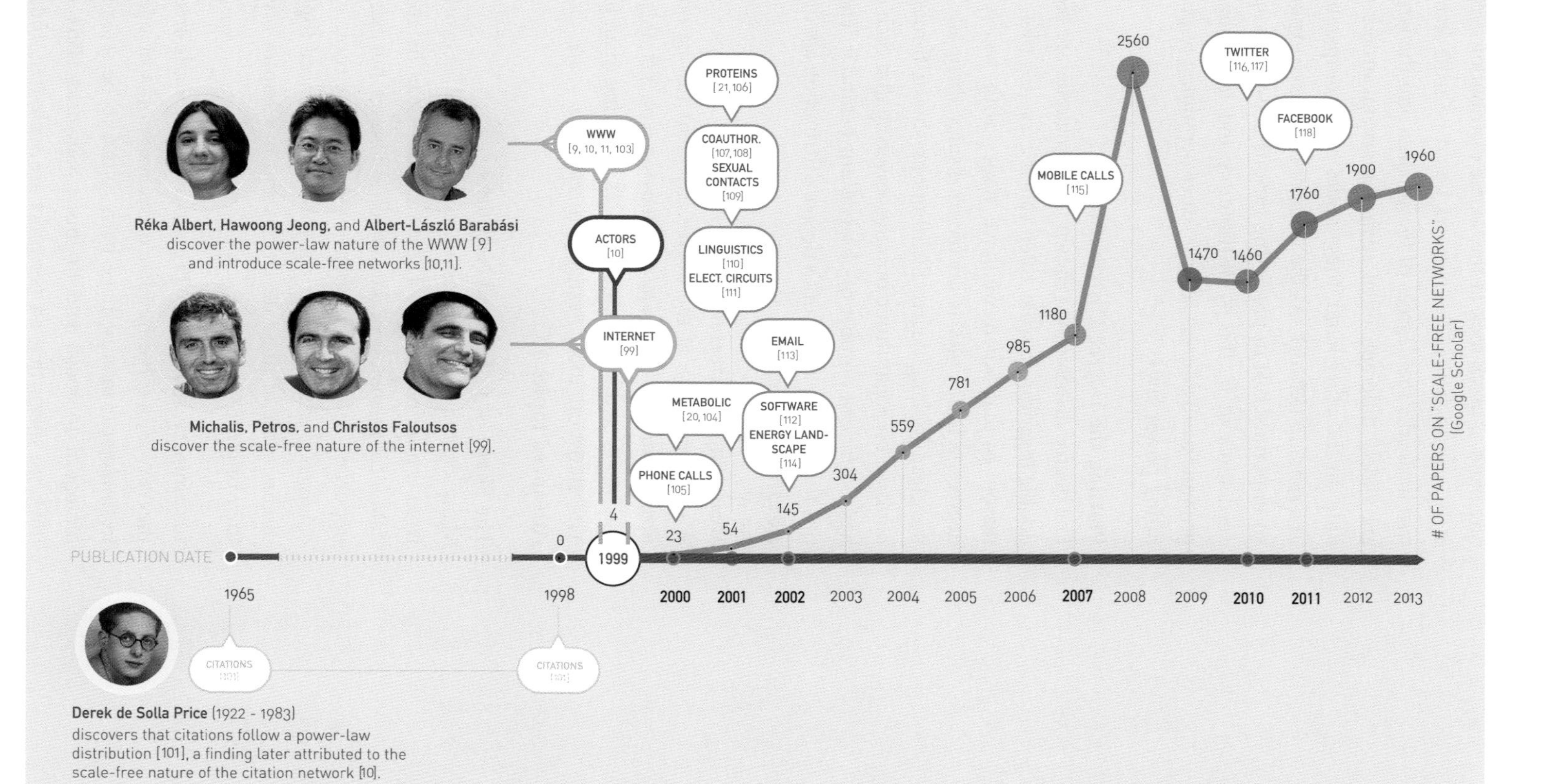

"we expect that the scale-invariant state observed in all systems for which detailed data has been available to us is a generic property of many complex networks, with applicability reaching far beyond the quoted examples."

Barabási and Albert, 1999

one hand, a quick look at the degree distribution will immediately reveal whether the network could be scale-free: in scale-free networks the degrees of the smallest and the largest nodes are widely different, often spanning several orders of magnitude. In the other hand, these nodes have comparable degrees in a random network. As the value of the degree exponent plays an important role in predicting various network properties, we need tools to fit the p_k distribution and to estimate γ. This prompts us to address several issues pertaining to plotting and fitting power laws.

Plotting the Degree Distribution
The degree distributions shown in this chapter are plotted on a double logarithmic scale, often called a log–log plot. The main reason is that when we have nodes with widely different degrees, a linear plot is unable to display them all. To obtain the clean-looking degree distributions shown throughout this book we use logarithmic binning, ensuring that each data point has a sufficient number of observations behind it. The practical tips for plotting a network's degree distribution are discussed in **Advanced topics 4.B**.

Measuring the Degree Exponent
A quick estimate of the degree exponent can be obtained by fitting a straight line to p_k on a log–log plot. Yet, this approach can be affected by systematic biases, resulting in an incorrect γ. The statistical tools available to estimate γ are discussed in **Advanced topics 4.C**.

The Shape of p_k for Real Networks
Many degree distributions observed in real networks deviate from a pure power law. These deviations can be attributed to data incompleteness or data collection biases, but can also carry important information about processes that contribute to the emergence of a particular network. In **Advanced topics 4.B** we discuss some of these deviations and in **Chapter 6** we explore their origins.

In summary, since the 1999 discovery of the scale-free nature of the WWW, a large number of real networks of scientific and technological interest have been found to be scale-free, from biological to social and linguistic networks (**Box 4.2**). This does not mean that all networks are scale-free. Indeed, many

Box 4.3 Not all Networks are Scale-Free

The ubiquity of the scale-free property does not mean that *all* real networks are scale-free. On the contrary, several important networks do not share this property:

- Networks appearing in materials science, describing the bonds between the atoms in crystalline or amorphous materials. In these networks each node has exactly the same degree, determined by chemistry (**Figure 4.11**).
- The neural network of the *C. elegans* worm [119].
- The power grid, consisting of generators and switches connected by transmission lines.

For the scale-free property to emerge, the nodes need to have the capacity to link to an arbitrary number of other nodes. These links do not need to be concurrent: we do not constantly chat with each of our acquaintances and a protein in a cell does not simultaneously bind to each of its potential interaction partners. The scale-free property is absent in systems that limit the number of links a node can have, effectively restricting the maximum size of the hubs. Such limitations are common in materials (**Figure 4.11**), explaining why they cannot develop a scale-free topology.

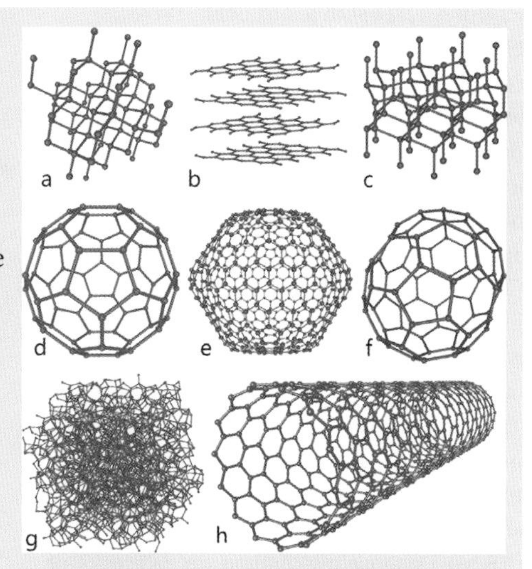

Figure 4.11 **The Material Network**
A carbon atom can share only four electrons with other atoms, hence no matter how we arrange these atoms relative to each other, in the resulting network a node can never have more than four links. Hence, hubs are forbidden and the scale-free property cannot emerge. The figure shows several carbon allotropes, that is materials made of carbon that differ in the structure of the network the carbon atoms arrange themselves in. This different arrangement results in materials with widely different physical and electronic characteristics, like (**a**) diamond; (**b**) graphite; (**c**) lonsdaleite; (**d**) C60 (buckminsterfullerene); (**e**) C540 (a fullerene); (**f**) C70 (another fullerene); (**g**) amorphous carbon; (**h**) single-walled carbon nanotube.

important networks, from the power grid to networks observed in materials science, do not display the scale-free property (**Box 4.3**).

4.6 Ultra-Small-World Property

The presence of hubs in scale-free networks raises an interesting question: Do hubs affect the small-world property? **Figure 4.4** suggests that they do: airlines build hubs precisely to decrease the number of hops between two airports. The

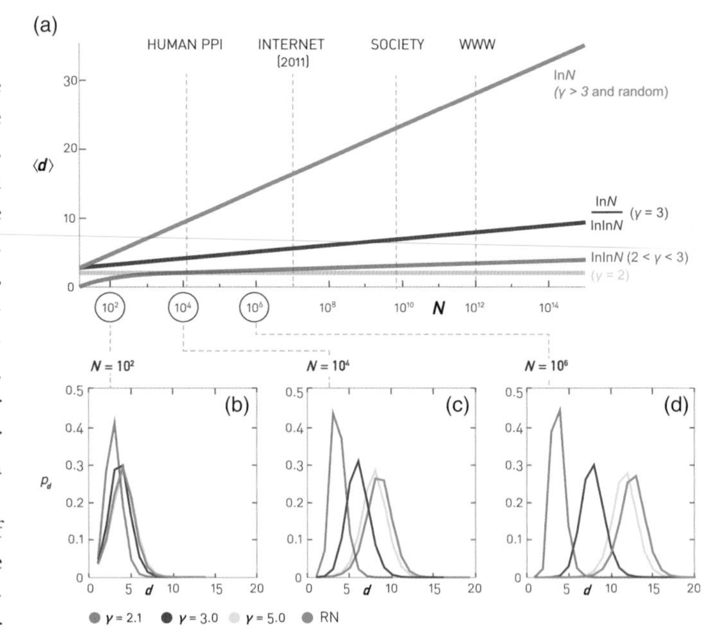

Figure 4.12 **Distances in Scale-Free Networks**
(a) The scaling of the average path length in the four scaling regimes characterizing a scale-free network: constant ($\gamma = 2$), $\ln \ln N$ ($2 < \gamma < 3$), $\ln N / \ln \ln N$ ($\gamma = 3$), $\ln N$ ($\gamma > 3$ and random networks). The dotted lines mark the approximate size of several real networks. Given their modest size, in biological networks, like the human protein–protein interaction network, the differences in the node-to-node distances are relatively small in the four regimes. The difference in $\langle d \rangle$ is quite significant for networks of the size of the social network or the WWW. For these, the small-world formula significantly underestimates the real $\langle d \rangle$.
(b)–(d) Distance distribution for networks of size $N = 10^2$, 10^4, 10^6, illustrating that while for small networks ($N = 10^2$) the distance distributions are not too sensitive to γ, for large networks ($N = 10^6$), p and $\langle d \rangle$ change visibly with γ. The networks were generated using the static model [123] with $\langle k \rangle = 3$.

calculations support this expectation, finding that *the distances in a scale-free network are smaller than the distances observed in an equivalent random network.*

The dependence of the average distance $\langle d \rangle$ on the system size N and the degree exponent γ are captured by the formula [120, 121]

$$\langle d \rangle \sim \begin{cases} \text{const.} & \gamma = 2 \\ \ln \ln N & 2 < \gamma < 3 \\ \dfrac{\ln N}{\ln \ln N} & \gamma = 3 \\ \ln N & \gamma > 3. \end{cases} \tag{4.22}$$

Next we discuss the behavior of $\langle d \rangle$ in the four regimes predicted by **(4.22)**, as summarized in **Figure 4.12**.

Anomalous Regime ($\gamma = 2$)

According to **(4.18)**, for $\gamma = 2$ the degree of the biggest hub grows linearly with the system size, that is $k_{\max} \sim N$. This forces the network into a *hub-and-spoke* configuration in which all nodes are close to each other because they all connect to the same central hub. In this regime the average path length does not depend on N.

Ultra-Small World ($2 < \gamma < 3$)
Equation (**4.22**) predicts that in this regime the average distance increases as $\ln \ln N$, a significantly slower growth than the $\ln N$ derived for random networks. We call networks in this regime *ultra-small*, as the hubs radically reduce the path length [120]. They do so by linking to a large number of small-degree nodes, creating short distances between them.

To see the implication of the ultra-small-world property, consider again the world's social network with $N \approx 7 \times 10^9$. If society is described by a random network, the N-dependent term is $\ln N = 22.66$. In contrast, for a scale-free network the N-dependent term is $\ln \ln N = 3.12$, indicating that the hubs radically shrink the distance between the nodes.

Critical Point ($\gamma = 3$)
This value is of particular theoretical interest, as the second moment of the degree distribution does not diverge any longer. We therefore call $\gamma = 3$ the *critical point*. At this critical point the $\ln N$ dependence encountered for random networks returns. Yet, the calculations indicate the presence of a double logarithmic correction $\ln \ln N$ [120, 122], which shrinks the distances compared with a random network of similar size.

Small World ($\gamma > 3$)
In this regime $\langle k^2 \rangle$ is finite and the average distance follows the small-world result derived for random networks (**Box 4.4**). While hubs continue to be present, for $\gamma > 3$ they are not sufficiently large and numerous to have a significant impact on the distance between the nodes.

Taken together, (**4.22**) indicates that the more pronounced the hubs are, the more effectively they shrink the distances between nodes. This conclusion is supported by **Figure 4.12a**, which shows the scaling of the average path length for scale-free networks with different γ. The figure indicates that while for small N the distances in the four regimes are comparable, for large N we observe remarkable differences.

Further support is provided by the path-length distribution for scale-free networks with different γ and N (**Figure 4.12b–d**). For $N = 10^2$ the path-length distributions overlap, indicating that at this size differences in γ result in undetectable differences in the path length. For $N = 10^6$, however, the $\langle d \rangle$ observed for different γ are well separated. **Figure 4.12d** also

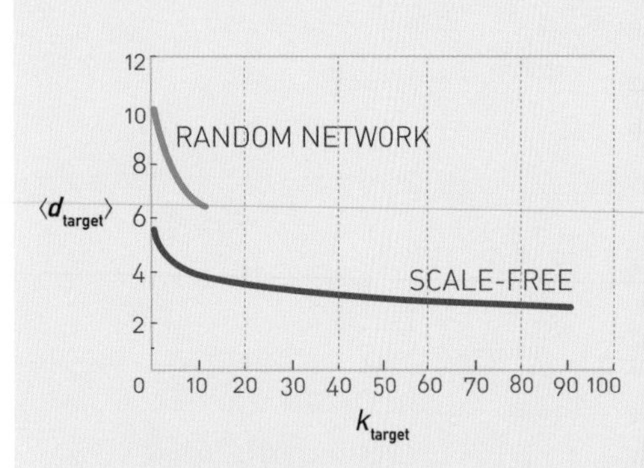

Figure 4.13 **Closing on the Hubs**
The distance $\langle d_{\mathrm{target}} \rangle$ of a node with degree $k \approx \langle k \rangle$ from a *target node* with degree k_{target} in a random and a scale-free network. In scale-free networks we are closer to the hubs than in random networks. The figure also illustrates that in a random network the largest-degree nodes are considerably smaller and hence the path lengths are visibly longer than in a scale-free network. Both networks have $\langle k \rangle = 2$ and $N = 1,000$ and for the scale-free network we choose $\gamma = 2.5$.

Box 4.4 We are Always Close to the Hubs

Frigyes Karinthy, in his 1929 short story [90] that first described the small-world concept, cautions that "it's always easier to find someone who knows a famous or popular figure than some run-of-the-mill, insignificant person." In other words, we are typically closer to hubs than to less-connected nodes. This effect is particularly pronounced in scale-free networks (**Figure 4.13**).

The implications are obvious: there are always short paths linking us to famous individuals, like well-known scientists or the President of the United States, as they are hubs with an exceptional number of acquaintances. It also means that many of the shortest paths go through these hubs.

In contrast to this expectation, measurements aiming to replicate the six-degrees concept in the online world find that individuals involved in chains that reached their target were less likely to send a message to a hub than individuals involved in incomplete chains [124]. The reason may be self-imposed: we perceive hubs as being busy, so we contact them only in real need. We therefore avoid them in online experiments of no perceived value to them.

shows that the larger the degree exponent, the larger are the distances between the nodes.

In summary, the scale-free property has several effects on network distances:

- It shrinks the average path lengths. Therefore, most scale-free networks of practical interest are not only "small" but "ultra small." This is a consequence of the hubs, which act as bridges between many small-degree nodes.
- It changes the dependence of $\langle d \rangle$ on the system size, as predicted by (**4.22**). The smaller is γ, the shorter are the distances between the nodes.
- Only for $\gamma > 3$ do we recover the $\ln N$ dependence, the signature of the small-world property characterizing random networks (**Figure 4.12**).

4.7 The Role of the Degree Exponent

Many properties of a scale-free network depend on the value of the degree exponent γ. A close inspection of **Table 4.1** indicates that:

- γ varies from system to system, prompting us to explore how the properties of a network change with γ;
- for most real systems the degree exponent is above 2, making us wonder why we don't see networks with $\gamma < 2$.

To address these questions, we next discuss how the properties of a scale-free network change with γ (**Box 4.5**).

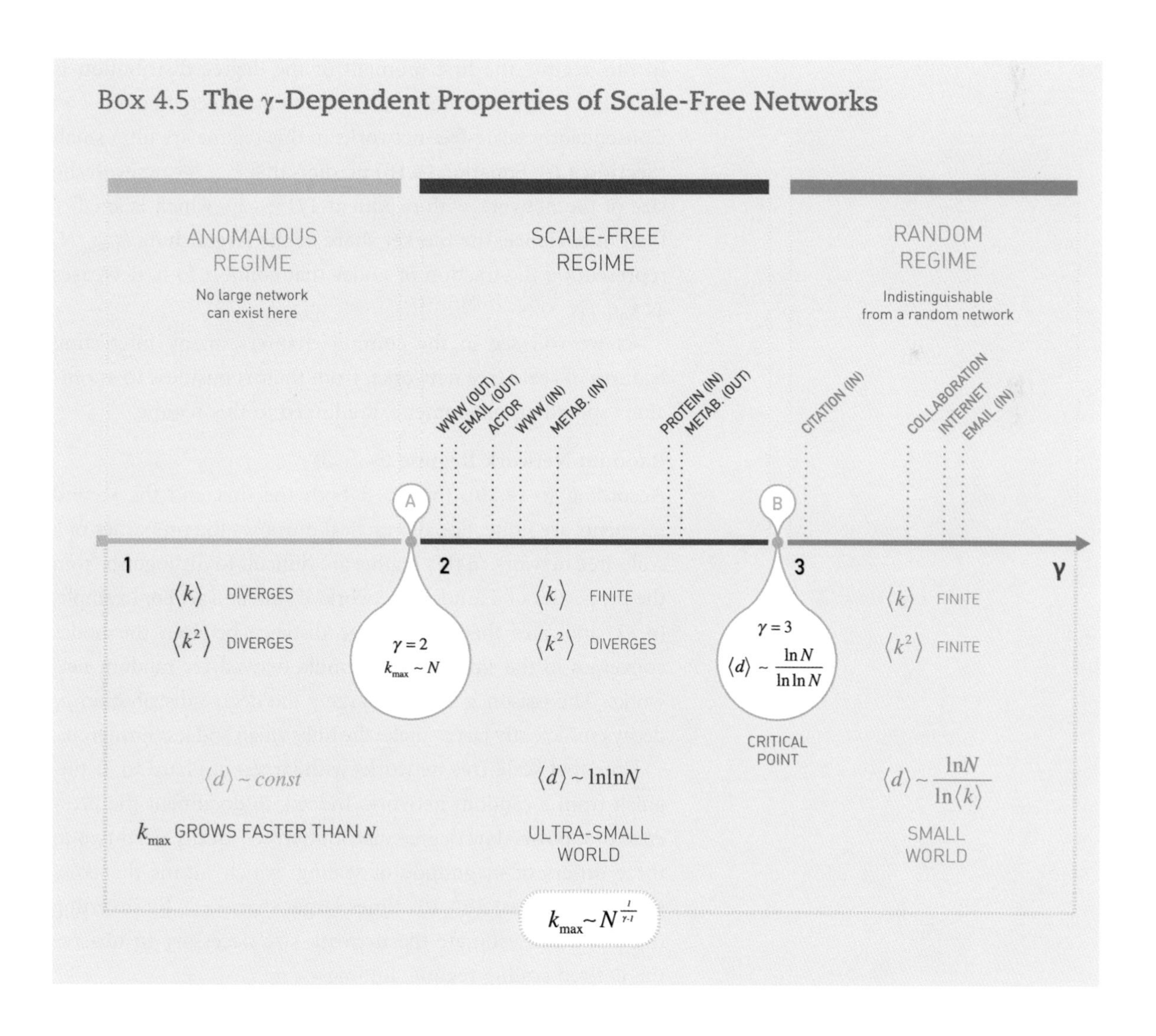

Anomalous Regime ($\gamma \leq 2$)
For $\gamma < 2$ the exponent $1/(\gamma - 1)$ in (**4.18**) is larger than one, hence the number of links connected to the largest hub grows faster than the size of the network. This means that for sufficiently large N, the degree of the largest hub must exceed the total number of nodes in the network, hence it will run out of nodes to connect to. Similarly, for $\gamma < 2$ the average degree $\langle k \rangle$ diverges in the $N \to \infty$ limit. These odd predictions are only two of the many anomalous features of scale-free networks in this regime. They are signatures of a deeper problem: a large scale-free network with $\gamma < 2$, that lack multi-links, cannot exist (**Box 4.6**).

Scale-Free Regime ($2 < \gamma < 3$)
In this regime the first moment of the degree distribution is finite but the second and higher moments diverge as $N \to \infty$. Consequently, scale-free networks in this regime are ultra small (**Section 4.6**). Equation (**4.18**) predicts that k_{max} grows with the size of the network with exponent $1/(\gamma - 1)$, which is smaller than one. Hence, the market share of the largest hub, k_{max}/N, representing the fraction of nodes that connect to it, decreases as $k_{max}/N \sim N^{-(\gamma-2)/(\gamma-1)}$.

As we will see in the coming chapters, many interesting features of scale-free networks, from their robustness to anomalous spreading phenomena, are linked to this regime.

Random Network Regime ($\gamma > 3$)
According to (**4.20**), for $\gamma > 3$ both the first and the second moments are finite. For all practical purposes the properties of a scale-free network in this regime are difficult to distinguish from the properties of a random network of similar size. For example, (**4.22**) indicates that the average distance between the nodes converges to the small-world formula derived for random networks. The reason is that for large γ the degree distribution p_k decays sufficiently fast to make the hubs small and less numerous.

Note that scale-free networks with large γ are hard to distinguish from a random network. Indeed, to document the presence of a power-law degree distribution we ideally need two to three orders of magnitude of scaling, which means that k_{max} should be at least 10^2–10^3 times larger than k_{min}. By inverting (**4.18**) we can estimate the network size necessary to observe the desired scaling regime, finding

Box 4.6 Why Scale-Free Networks with $\gamma < 2$ do not Exist

To see why networks with $\gamma < 2$ are problematic, we need to attempt to build one. A degree sequence that can be turned into a *simple graph* (i.e., a graph lacking multi-links or self-loops) is called *graphical* [125]. Yet, not all degree sequences are graphical: for example, if the number of stubs is odd, then we will always have an unmatched stub (**Figure 4.14b**).

The graphicality of a degree sequence can be tested with an algorithm proposed by Erdős and Gallai [125–129]. If we apply the algorithm to scale-free networks, we find that the number of graphical degree sequences drops to zero for $\gamma < 2$ (**Figure 4.14c**). Hence, degree distributions with $\gamma < 2$ cannot be turned into simple networks. Indeed, for networks in this regime the largest hub grows faster than N. If we do not allow self-loops and multi-links, then the largest hub will run out of nodes to connect to once its degree exceeds $N - 1$.

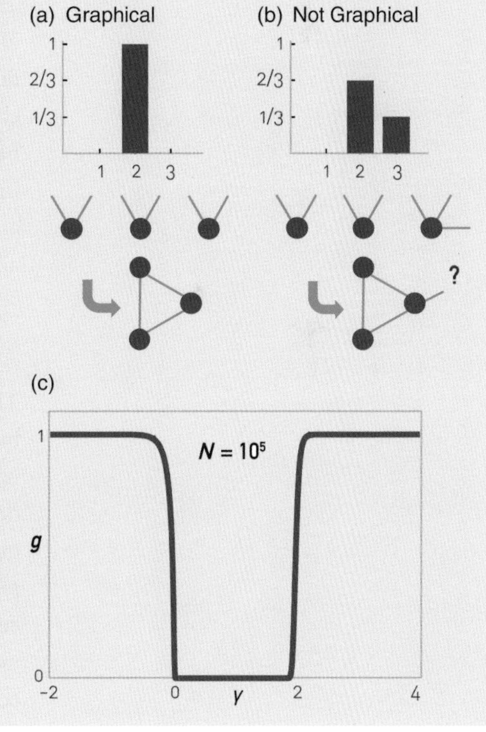

Figure 4.14 Networks with $\gamma < 2$ are Not Graphical

(a), (b) Degree distributions and the corresponding degree sequences for two small networks. The difference between them lies in the degree of a single node. While we can build a simple network using the degree distribution (a), it is impossible to build one using (b), as one stub always remains unmatched. Hence, (a) is *graphical* while (b) is not.

(c) Fraction of networks, g, for a given γ that are graphical. A large number of degree sequences with degree exponent γ and $N = 10^5$ were generated, testing the graphicality of each network. The figure indicates that while virtually all networks with $\gamma > 2$ are graphical, it is impossible to find graphical networks in the $0 < \gamma < 2$ range.

After [129].

$$N = \left(\frac{k_{\max}}{k_{\min}}\right)^{\gamma-1}. \tag{4.23}$$

For example, if we wish to document the scale-free nature of a network with $\gamma = 5$ and require scaling that spans at least two orders of magnitude (e.g., $k_{\min} \sim 1$ and $k_{\max} \simeq 10$), according to (**4.23**) the size of the network must exceed $N > 10^8$. There are very few network maps of this size. Therefore, there may be many networks with large degree exponent. Given, however, their limited size, it is difficult to obtain convincing evidence of their scale-free nature.

In summary, we find that the behavior of scale-free networks is sensitive to the value of the degree exponent γ. Theoretically the most interesting regime is $2 < \gamma < 3$, where $\langle k^2 \rangle$ diverges, making scale-free networks ultra small. Interestingly, many networks of practical interest, from the WWW to protein interaction networks, are in this regime.

4.8 Generating Networks with Arbitrary Degree Distribution

Networks generated by the Erdős–Rényi model have a Poisson degree distribution. The empirical results discussed in this chapter indicate, however, that the degree distribution of real networks deviates significantly from a Poisson form, raising an important question: How do we generate networks with an arbitrary p_k? In this section we discuss three frequently used algorithms designed for this purpose.

4.8.1 Configuration Model

The configuration model, described in **Figure 4.15**, helps us build a network with a pre-defined degree sequence. In the network generated by the model, each node has a pre-defined degree k_i but otherwise the network is wired randomly. Consequently, the network is often called a *random network with a pre-defined degree sequence.* By repeatedly applying this procedure to the same degree sequence we can generate different networks with the same p_k (**Figure 4.15b–d**). There are a couple of caveats to consider:

- The probability of having a link between nodes of degree k_i and k_j is

$$p_{ij} = \frac{k_i k_j}{2L - 1}. \tag{4.24}$$

  Indeed, a stub starting from node i can connect to $2L - 1$ other stubs. Of these, k_j are attached to node j. So, the probability that a particular stub is connected to a stub of node j is $k_j/(2L-1)$. As node i has k_i stubs, it has k_i attempts to link to j, resulting in (**4.24**).

- The obtained network contains self-loops and multi-links, as there is nothing in the algorithm to forbid a node connecting to itself, or generating multiple links between two nodes. We can choose to reject stub pairs that lead to these, but if we do so, we may not be able to complete the network. Rejecting self-loops or multi-links also means that not all possible matchings appear with equal probability. Hence (**4.24**) will not be valid, making analytical calculations difficult. Yet, the number of self-loops and multi-links remains negligible, as the number of connection choices increases with N, so typically we do not need to exclude them [94].
- The configuration model is frequently used in calculations, as (**4.24**) and its inherently random character helps us analytically calculate numerous network measures.

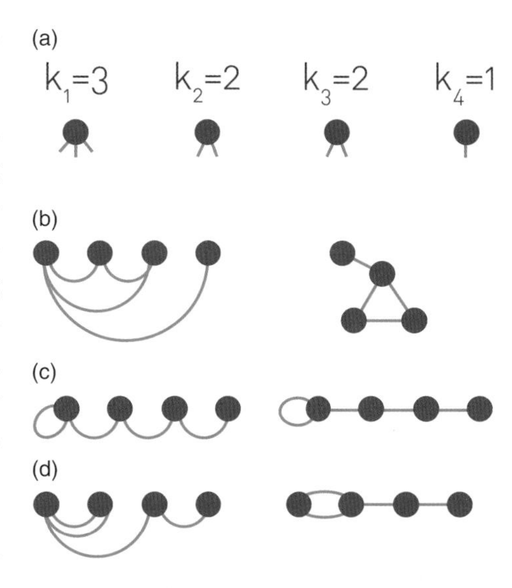

Figure 4.15 **The Configuration Model**
The configuration model builds a network whose nodes have pre-defined degrees [130, 131]. The algorithm consists of the following steps:

(a) Degree Sequence
Assign a degree to each node, represented as stubs or half-links. The degree sequence is either generated analytically from a pre-selected p_k distribution (**Box 4.7**), or it is extracted from the adjacency matrix of a real network. We must start from an even number of stubs, otherwise we are left with unpaired stubs.

(b)–(d) Network Assembly
Randomly select a stub pair and connect them. Then randomly choose another pair from the remaining $2L-2$ stubs and connect them. This procedure is repeated until all stubs are paired up. Depending on the order in which the stubs were chosen, we obtain different networks. Some networks include cycles (b), others self-loops (c) or multi-links (d). Yet the expected number of self-loops and multi-links goes to zero in the $N \to \infty$ limit.

4.8.2 Degree-Preserving Randomization

As we explore the properties of a real network, we often need to ask if a certain network property is predicted by its degree distribution alone, or if it represents some additional property not contained in p_k. To answer this question we need to generate networks that are wired randomly, but whose p_k is *identical* to the original network. This can be achieved through *degree-preserving randomization* [132], described in **Figure 4.17b**. The idea behind the algorithm is simple: we randomly select two links and swap them, if the swap does not lead to multi-links. Hence, the degree of each of the four involved nodes in the swap remains unchanged. Consequently, hubs stay hubs and small-degree nodes retain their small degree, but the wiring diagram of the generated network is randomized. Note that degree-preserving randomization is different from *full randomization*, where we swap links without preserving the node degrees (**Figure 4.17a**). Full randomization turns any network into an Erdős–Rényi network with a Poisson degree distribution that is independent of the original p_k.

4.8.3 Hidden-Parameter Model

The configuration model generates self-loops and multi-links, features that are absent in many real networks. We can use the *hidden-parameter model* (**Figure 4.18**) to generate networks

Box 4.7 Generating a Degree Sequence with Power-Law Distribution

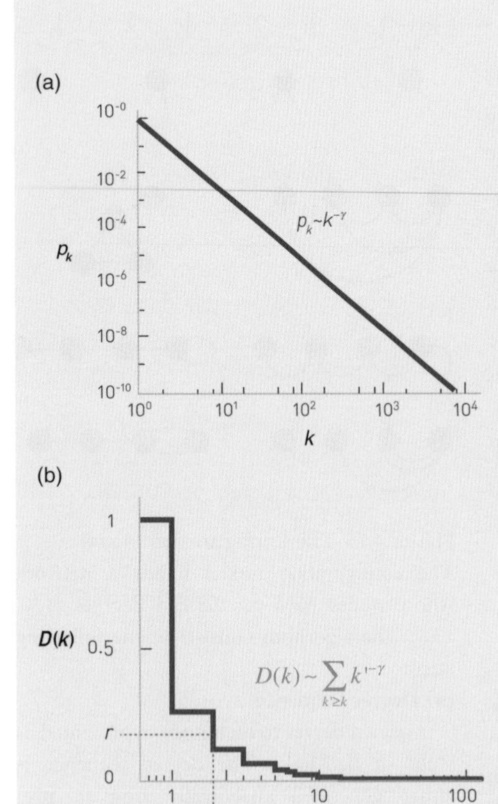

Figure 4.16 **Generating a Degree Sequence**
(a) The power-law degree distribution of the degree sequence we wish to generate.
(b) The function (4.25) that allows us to assign degrees k to uniformly distributed random numbers r.

The *degree sequence* of an undirected network is a sequence of node degrees. For example, the degree sequence of each of the networks shown in **Figure 4.15a** is {3, 2, 2, 1}. As **Figure 4.15a** illustrates, the degree sequence does not uniquely identify a graph, as there are multiple ways we can pair up the stubs.

To generate a degree sequence from a pre-defined degree distribution we start from an analytically pre-defined degree distribution, like $p_k \sim k^{-\gamma}$, shown in **Figure 4.16a**. Our goal is to generate a degree sequence $\{k_1, k_2, \ldots, k_N\}$ that follows the distribution p_k. We start by calculating the function

$$D(k) = \sum_{k' \geq k} p_{k'} \tag{4.25}$$

shown in **Figure 4.16b**. $D(k)$ is between 0 and 1, and the step size at any k equals p_k. To generate a sequence of N degrees following p_k, we generate N random numbers r_i, $i = 1, \ldots, N$, chosen uniformly from the (0, 1) interval. For each r_i we use the plot in (b) to assign a degree k_i. The obtained $k_i = D^{-1}(r_i)$ set of numbers follows the desired p_k distribution. Note that the degree sequence assigned to a p_k is not unique – we can generate multiple sets of $\{k_1, \ldots, k_N\}$ sequences compatible with the same p_k.

with a pre-defined p_k but without multi-links and self-loops [133, 134, 135].

We start from N isolated nodes and assign each node i a hidden parameter η_i, chosen from a distribution $\rho(\eta)$. The nature of the generated network depends on the selection of the $\{\eta_i\}$ hidden parameter sequence. There are two ways to generate the appropriate hidden parameters:

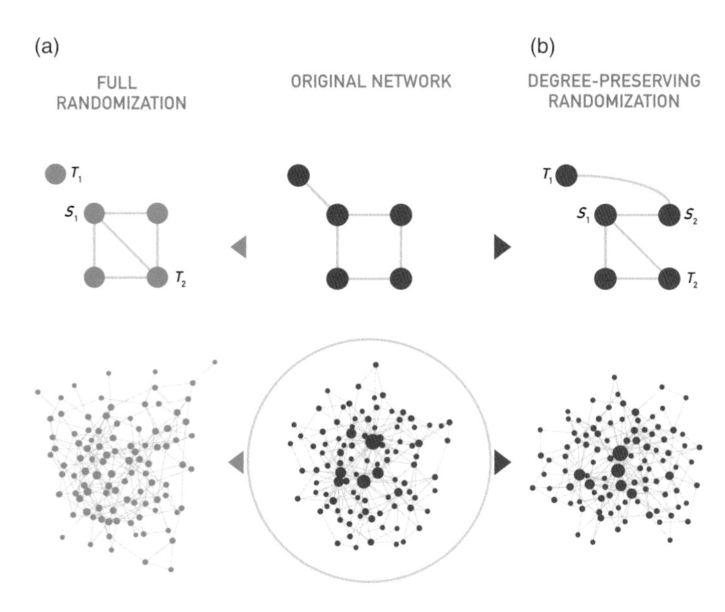

Figure 4.17 **Degree-Preserving Randomization**
Two algorithms can generate a randomized version of a given network [132], with different outcomes.
(a) Full Randomization
This algorithm generates a random (Erdős–Rényi) network with the same N and L as the original network. We select randomly a source node (S_1) and two target nodes, where the first target (T_1) is linked directly to the source node and the second target (T_2) is not. We rewire the S_1–T_1 link, turning it into an S_1–T_2 link. As a result, the degree of the target nodes T_1 and T_2 changes. We perform this procedure once for each link in the network.
(b) Degree-Preserving Randomization
This algorithm generates a network in which each node has exactly the same degree as in the original network, but the network's wiring diagram has been randomized. We select two sources (S_1, S_2) and two target nodes (T_1, T_2), such that initially there is a link between S_1 and T_1 and a link between S_2 and T_2. We then swap the two links, creating an S_1–T_2 and an S_2–T_1 link. The swap leaves the degree of each node unchanged. We repeat this procedure until we rewire each link at least once.
Bottom panels. Starting from a scale-free network (middle), full randomization eliminates the hubs and turns the network into a random network (left). In contrast, degree-preserving randomization leaves the hubs in place and the network remains scale-free (right).

- η_i can be a sequence of N random numbers chosen from a pre-defined $\rho(\eta)$ distribution. The degree distribution of the obtained network is

$$p_k = \int \frac{e^{-\eta}\eta^k}{k!}\rho(\eta)d\eta. \tag{4.26}$$

- η_i can come from a deterministic sequence $\{\eta_1, \eta_2, \ldots, \eta_N\}$. The degree i distribution of the obtained network is

$$p_k = \frac{1}{N}\sum_j \frac{e^{-\eta_j}\eta_j^k}{k!}. \tag{4.27}$$

The hidden-parameter model offers a particularly simple method to generate a scale-free network. Indeed, using

$$\eta_i = \frac{c}{i^\alpha},\ i = 1, \ldots, N \tag{4.28}$$

as the sequence of hidden parameters, according to (**4.27**) the obtained network will have the degree distribution

$$p_k \sim k^{-\left(1+\frac{1}{\alpha}\right)} \tag{4.29}$$

for large k. Hence, by choosing the appropriate α we can tune $\gamma = 1 + 1/\alpha$. We can also use $\langle\eta\rangle$ to tune $\langle k\rangle$ as (**4.26**) and (**4.27**) imply that $\langle k\rangle = \langle\eta\rangle$.

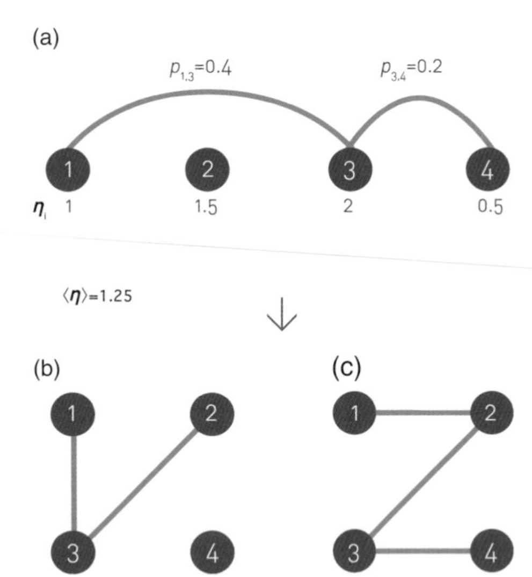

Figure 4.18 **Hidden-Parameter Model**
(**a**) We start with N isolated nodes and assign to each node a *hidden parameter* η_i, which is either selected from a $\rho(\eta)$ distribution or is provided by a sequence $\{\eta_i\}$. We connect each node pair with probability

$$p(\eta_i, \eta_j) = \frac{\eta_i \eta_j}{\langle \eta \rangle N}.$$

The figure shows the probability of connecting nodes (1,3) and (3,4).
(**b**), (**c**) After connecting the nodes, we obtain the networks shown in (b) or (c), representing two independent realizations generated by the same hidden parameter sequence (a).
The expected number of links in the network generated by the model is

$$L = \frac{1}{2} \sum_{i,j}^{N} \frac{\eta_i \eta_j}{\langle \eta \rangle N} = \frac{1}{2} \langle \eta \rangle N.$$

Similar to the random network model, L will vary from network to network, following an exponentially bounded distribution. If we wish to control the average degree $\langle k \rangle$, we can add L links to the network one by one. The end points i and j of each link are then chosen randomly with a probability proportional to η_i and η_j. In this case we connect i and j only if they were not connected previously.

In summary, the configuration model, degree-preserving randomization and hidden-parameter model can generate networks with a pre-defined degree distribution and help us analytically calculate key network characteristics. We will turn to these algorithms each time we explore whether a certain network property is a consequence of the network's degree distribution, or if it represents some emergent property (**Box 4.8**). As we use these algorithms, we must be aware of their limitations:

- The algorithms do not tell us *why* a network has a certain degree distribution. Understanding the origin of the observed p_k will be the subject of **Chapters 6 and 7**.
- Several important network characteristics, from communities (**Chapter 9**) to degree correlations (**Chapter** 7), are lost during randomization.

Hence, the networks generated by these algorithms are a bit like a photograph of a painting: at first they appear to be the same as the original, but on closer inspection we realize that many details, from the texture of the canvas to the brush strokes, have been lost.

The three algorithms discussed above raise the following question: How do we decide which one to use? Our choice depends on whether we start from a degree sequence $\{k_i\}$ or a degree distribution p_k and whether we can tolerate self-loops and multi-links between two nodes. The decision tree involved in this choice is provided in **Figure 4.20**.

4.9 Summary

The scale-free property has played an important role in the development of network science for two main reasons:

- Many networks of scientific and practical interest, from the WWW to subcellular networks, are scale-free. This universality made the scale-free property an unavoidable issue in many disciplines.
- Once the hubs are present, they fundamentally change the system's behavior. The ultra-small property offers a first hint of their impact on a network's properties; we will encounter many more examples in the coming chapters.

Box 4.8 Testing the Small-World Property

In the literature, the distances observed in a real network are often compared with the small-world formula (**3.19**). Yet (**3.19**) was derived for random networks, while real networks do not have a Poisson degree distribution. If the network is scale-free, then (**4.22**) offers the appropriate formula. Yet (**4.22**) provides only the scaling of the distance with N, and not its absolute value. Instead of fitting the average distance, we often ask: Are the distances observed in a real network comparable with the distances observed in a randomized network with the same degree distribution? Degree-preserving randomization helps answer this question. We illustrate the procedure on the protein interaction network.

(i) **Original Network**
We start by measuring the distance distribution p_d of the original network, obtaining $\langle d \rangle = 5.61$ (**Figure 4.19**).

(ii) **Full Randomization**
We generate a random network with the same N and L as the original network. The obtained p_d shifts visibly to the right, providing $\langle d \rangle = 7.13$, much larger than the original $\langle d \rangle = 5.61$. It is tempting to conclude that the protein interaction network is affected by some unknown organizing principle that keeps the distances shorter. This would be a flawed conclusion, however, as the bulk of the difference is due to the fact that full randomization changed the degree distribution.

(iii) **Degree-Preserving Randomization**
As the original network is scale free, the proper random reference should maintain the original degree distribution. Hence we determine p_d after degree-preserving randomization, finding that it is comparable with the original p_d.

In summary, *a random network overestimates the distances between the nodes, as it is missing the hubs.* The network obtained by degree-preserving randomization retains the hubs, so the distances of the randomized network are comparable with the original network. This example illustrates the importance of choosing the proper randomization procedure when exploring networks.

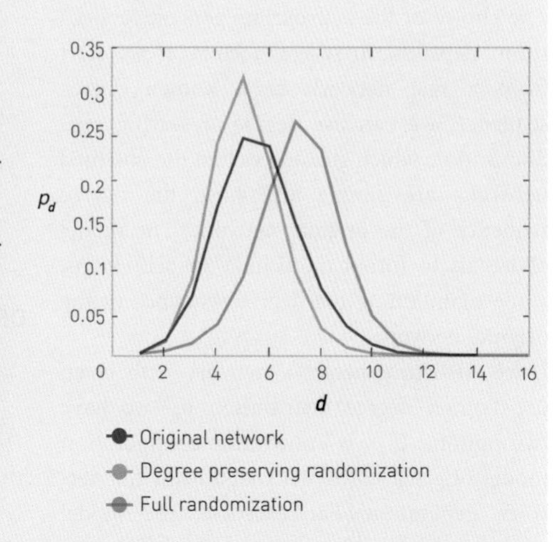

Figure 4.19 **Randomizing Real Networks**
The distance distribution p_d between each node pair in the protein–protein interaction network (**Table 4.1**). The green line provides the path-length distribution obtained under *full randomization*, which turns the network into an Erdős–Rényi network, while keeping N and L unchanged (**Figure 4.17**).
The light purple curve corresponds to the p_d of the network obtained after *degree-preserving randomization*, which keeps the degree of each node unchanged.
We have: $\langle d \rangle = 5.61 \pm 1.64$ (original), $\langle d \rangle = 7.13 \pm 1.62$ (full randomization), $\langle d \rangle = 5.08 \pm 1.34$ (degree-preserving randomization).

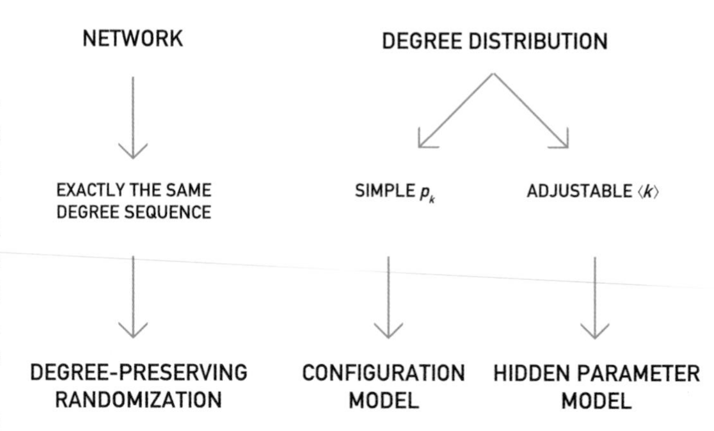

Figure 4.20 **Choosing a Generative Algorithm**
The choice of the appropriate generative algorithm depends on several factors. If we start from a real network or a known degree sequence, we can use degree-preserving randomization, which guarantees that the obtained networks are simple and have the degree sequence of the original network. The model allows us to forbid multi-links or self- loops, while maintaining the degree sequence of the original network.
If we wish to generate a network with given pre-defined degree distribution p_k, we have two options. If p_k is known, the configuration model offers a convenient algorithm for network generation. For example, the model allows us to generate a network with a pure power-law degree distribution $p_k = Ck^{-\gamma}$ for $k \geq k_{\min}$.
However, tuning the average degree $\langle k \rangle$ of a scale-free network within the configuration model is a tedious task, because the only available free parameter is $k_{\min}$. Therefore, if we wish to alter $\langle k \rangle$, it is more convenient to use the hidden-parameter model with parameter sequence (4.28). This way the tail of the degree distribution follows $\sim k^{-\alpha}$ and by changing the number of links L we can control $\langle k \rangle$.

As we continue to explore the consequences of the scale-free property, we must keep in mind that the power-law form (**4.1**) is rarely seen in this pure form in real systems. The reason is that a host of processes affect the topology of each network, which also influence the shape of the degree distribution. We will discuss these processes in the coming chapters. The diversity of these processes and the complexity of the resulting p_k confuses those who approach these networks through the narrow perspective of the quality of fit to a pure power law. Instead, the scale-free property tells us that we must distinguish two rather different classes of networks:

Exponentially Bounded Networks
We call a network *exponentially bounded* if its degree distribution decreases exponentially or faster for high k. As a consequence, $\langle k^2 \rangle$ is smaller than $\langle k \rangle$, implying that we lack significant degree variations. Examples of p_k in this class include the Poisson, Gaussian or simple exponential distribution (**Table 4.2**). Erdős–Rényi and Watts–Strogatz networks are the best known model networks belonging to this class. Exponentially bounded networks lack outliers, consequently most nodes have a comparable degree. Real networks in this class include highway networks and the power grid.

Fat-Tailed Networks
We call a network *fat tailed* if its degree distribution has a power-law tail in the high-k region. As a consequence, $\langle k^2 \rangle$ is much larger than $\langle k \rangle$, resulting in considerable degree variations. Scale-free networks with a power-law degree distribution (**4.1**) offer the best-known example of networks belonging

to this class. Outliers, or exceptionally high-degree nodes, are not only allowed but are expected in these networks. Networks in this class include the WWW, the Internet, protein interaction networks, and most social and online networks.

While it would be desirable to statistically validate the precise form of the degree distribution, often it is sufficient to decide if a given network has an exponentially bounded or a fat-tailed degree distribution (see **Advanced topics 4.A**). If the degree distribution is exponentially bounded, the random network model offers a reasonable starting point to understand its topology. If the degree distribution is fat tailed, a scale-free network offers a better approximation. We will also see in the coming chapters that the key signature of the fat-tailed behavior is the magnitude of $\langle k^2 \rangle$: if $\langle k^2 \rangle$ is large, systems behave like scale-free networks; if $\langle k^2 \rangle$ is small, being comparable with $\langle k \rangle(\langle k \rangle + 1)$, systems are well approximated by random networks.

In summary, to understand the properties of real networks, it is often sufficient to remember that in scale-free networks a few highly connected hubs coexist with a large number of small nodes. The presence of these hubs plays an important role in the system's behavior. In this chapter we explored the basic characteristics of scale-free networks. We are left, therefore, with an important question: Why are so many real networks scale free? The next chapter provides the answer.

4.10 Homework

4.10.1 Hubs

Calculate the expected maximum degree k_{max} for the undirected networks listed in **Table 4.1**.

4.10.2. Friendship Paradox

The degree distribution p_k expresses the probability that a randomly selected node has k neighbors. However, if we randomly select a link, the probability that a node at one of its ends has degree k is $q_k = Akp_k$, where A is a normalization factor.

(a) Find the normalization factor A, assuming that the network has a power-law degree distribution with $2 < \gamma < 3$, with minimum degree k_{min} and maximum degree k_{max}.

(b) In the configuration model q_k is also the probability that a randomly chosen node has a neighbor with degree k. What is the average degree of the neighbors of a randomly chosen node?
(c) Calculate the average degree of the neighbors of a randomly chosen node in a network with $N = 10^4$, $\gamma = 2.3$, $k_{min} = 1$ and $k_{max} = 1{,}000$. Compare the result with the average degree of the network, $\langle k \rangle$.
(d) How can you explain the "paradox" of (c) – that is a node's friends have more friends than the node itself?

4.10.3. Generating Scale-Free Networks

Write a computer code to generate networks of size N with a power-law degree distribution with degree exponent γ. Refer to **Section 4.9** for the procedure. Generate three networks with $\gamma = 2.2$ and with $N = 10^3$, $N = 10^4$ and $N = 10^5$ nodes, respectively. What is the percentage of multi-link and self-loops in each network? Generate more networks to plot this percentage as a function of N. Do the same for networks with $\gamma = 3$.

4.10.4. Mastering Distributions

Use software which includes a statistics package, like Matlab, Mathematica or Numpy in Python, to generate three synthetic datasets, each containing 10,000 integers that follow a power-law distribution with $\gamma = 2.2$, $\gamma = 2.5$ and $\gamma = 3$. Use $k_{min} = 1$. Apply the techniques described in **Advanced topics 4.C** to fit the three distributions.

4.11 Advanced Topics 4.A Power Laws

Power laws have a convoluted history in natural and social sciences, being interchangeably (and occasionally incorrectly) called *fat-tailed*, *heavy-tailed*, *long-tailed*, *Pareto* or *Bradford distributions*. They also have a series of close relatives, like *log-normal*, *Weibull* or *Lévy distributions*. In this section we discuss some of the most frequently encountered distributions in network science and their relationship to power laws.

Box 4.9 At A Glance: Scale-Free Networks

Degree Distribution

Discrete form:

$$p_k = \frac{k^{-\gamma}}{\zeta(\gamma)}$$

Continuous form:

$$p(k) = (\gamma - 1)k_{min}^{\gamma-1}k^{-\gamma}$$

Size of the Largest Hub

$$k_{max} = k_{min}N^{\frac{1}{\gamma-1}}$$

Moments of p_k for $N \to \infty$

$2 < \gamma \leq 3$: $\langle k \rangle$ finite, $\langle k^2 \rangle$ diverges

$\gamma > 3$: $\langle k \rangle$ and $\langle k^2 \rangle$ finite

Distances

$$\langle d \rangle \sim \begin{cases} \text{const.} & \gamma = 2 \\ \ln\ln N & 2 < \gamma < 3 \\ \dfrac{\ln N}{\ln\ln N} & \gamma = 3 \\ \ln N & \gamma > 3 \end{cases}$$

4.11.1 Exponentially Bounded Distributions

Many quantities in nature, from the height of humans to the probability of being in a car accident, follow bounded distributions. A common property of these is that p_x decays either exponentially (e^{-x}) or faster than exponentially (e^{-x^2/σ^2}) for high x. Consequently, the largest expected x is bounded by some upper value x_{max} that is not too different from $\langle x \rangle$. Indeed, the expected largest x obtained after we draw N numbers from a bounded p_x grows as $x_{max} \sim \log N$ or slower. This means that outliers, representing unusually high x values, are rare. They are so rare that they are effectively forbidden, meaning that they do not occur with any meaningful probability. Instead, most events drawn from a bounded distribution are in the vicinity of $\langle x \rangle$.

The high-x regime is called the *tail of a distribution*. Given the absence of numerous events in the tail, these distributions are also called *thin tailed.*

Analytically, the simplest bounded distribution is the exponential distribution $e^{-\lambda x}$. Within network science the most frequently encountered bounded distribution is the Poisson distribution (or its parent, the binomial distribution), which describes the degree distribution of a random network. Outside network science the most frequently encountered member of this class is the normal (Gaussian) distribution (**Table 4.2**).

4.11.2 Fat-Tailed Distributions

The terms *fat tailed*, *heavy tailed* or *long tailed* refer to a p_x whose decay at large x is slower than exponential. In these distributions we often encounter events characterized by very large x values, usually called *outliers* or *rare events*. The power-law distribution (**4.1**) represents the best-known example of a fat-tailed distribution. An instantly recognizable feature of a fat-tailed distribution is that the magnitude of the events x drawn from it can span several orders of magnitude. Indeed, in these distributions the size of the largest event after N trials scales as $x_{max} \sim N^{\zeta}$, where ζ is determined by the exponent γ characterizing the tail of the p_x distribution. As N^{ζ} grows fast, rare events or outliers occur with a noticeable frequency, often dominating the properties of the system.

The relevance of fat-tailed distributions to networks is provided by several factors:

- Many quantities occurring in network science, like degrees, link weights and betweenness centrality, follow a power-law distribution in both real and model networks.
- The power-law form is analytically predicted by appropriate network models (**Chapter 5**).

4.11.3 Crossover Distribution (Log-Normal, Stretched Exponential)

When an empirically observed distribution appears to be between a power law and exponential, *crossover distributions* are often used to fit the data. These distributions may be exponentially bounded (power law with exponential cutoff), or not bounded but decay faster than a power law (log-normal or stretched exponential). Next we discuss the properties of several frequently encountered crossover distributions. A *power law with exponential cutoff* is often used to fit the degree distribution of real networks. Its density function has the form

$$p(x) = Cx^{-\gamma}e^{-\lambda x} \tag{4.30}$$

$$C = \frac{\lambda^{1-\gamma}}{\Gamma(1-\gamma, \lambda x_{min})}, \tag{4.31}$$

where $x > 0$, $\gamma > 0$ and $\Gamma(s, y)$ denotes the upper incomplete gamma function. The analytical form (**4.30**) directly captures its crossover nature: it combines a power-law term, a key component of fat-tailed distributions, with an exponential term, responsible for its exponentially bounded tail. To highlight its crossover characteristics we take the logarithm of (**4.30**):

$$\ln p(x) = \ln C - \gamma \ln x - \lambda x. \tag{4.32}$$

For $x \ll 1/\lambda$ the second term on the r.h.s. dominates, suggesting that the distribution follows a power law with exponent γ. Once $x \gg 1/\lambda$, the λx term overcomes the $\ln x$ term, resulting in an exponential cutoff for high x.

A *stretched exponential (Weibull) distribution* is formally similar to (**4.30**), except that there is a fractional power law in the exponential. Its name comes from the fact that its cumulative distribution function is one minus a stretched exponential function $P(x) = e^{-(\lambda x)^{\beta}}$ (**4.32**), which leads to the density function

$$P'(x) = Cx^{\beta-1}e^{-(\lambda x)^{\beta}} \quad (4.33)$$

$$C = \beta\lambda^{\beta}. \quad (4.34)$$

In most applications x varies between 0 and $+\infty$. In (**4.32**), β is the *stretching exponent*, determining the properties of $p(x)$:

- For $\beta = 1$ we recover a simple exponential function.
- If β is between 0 and 1, the graph of log $p(x)$ versus x is "stretched," meaning that it spans several orders of magnitude in x. This is the regime where a stretched exponential is difficult to distinguish from a pure power law. The closer β is to 0, the more similar is $p(x)$ to the power law x^{-1}.
- If $\beta > 1$ we have a "compressed" exponential function, meaning that x varies in a very narrow range.
- For $\beta = 2$ (**4.33**) reduces to the Rayleigh distribution.

As we will see in **Chapters 5 and 6**, several network models predict a streched exponential degree distribution.

A *log-normal (Galton or Gibrat) distribution* emerges if ln$x$ follows a normal distribution. Typically a variable follows a log-normal distribution if it is the product of many independent positive random numbers. We encounter log-normal distributions in finance, representing the compound return from a sequence of trades.

The probability density function of a log-normal distribution is

$$p(x) = \frac{1}{\sqrt{2\pi}\sigma_x} \exp\left[-\frac{(\ln x - \mu)^2}{2\sigma^2}\right]. \quad (4.35)$$

Hence a log-normal is like a normal distribution except that its variable in the exponential term is not x, but ln$x$.

To understand why a log-normal is occasionally used to fit a power-law distribution, we note that

$$\sigma^2 = \langle(\ln x)^2\rangle - \langle\ln x\rangle^2 \quad (4.36)$$

captures the typical variation of the order of magnitude of x. Therefore $\ln x$ now follows a normal distribution, which means that x can vary rather widely. Depending on the value of σ, the log-normal distribution may resemble a power law for several orders

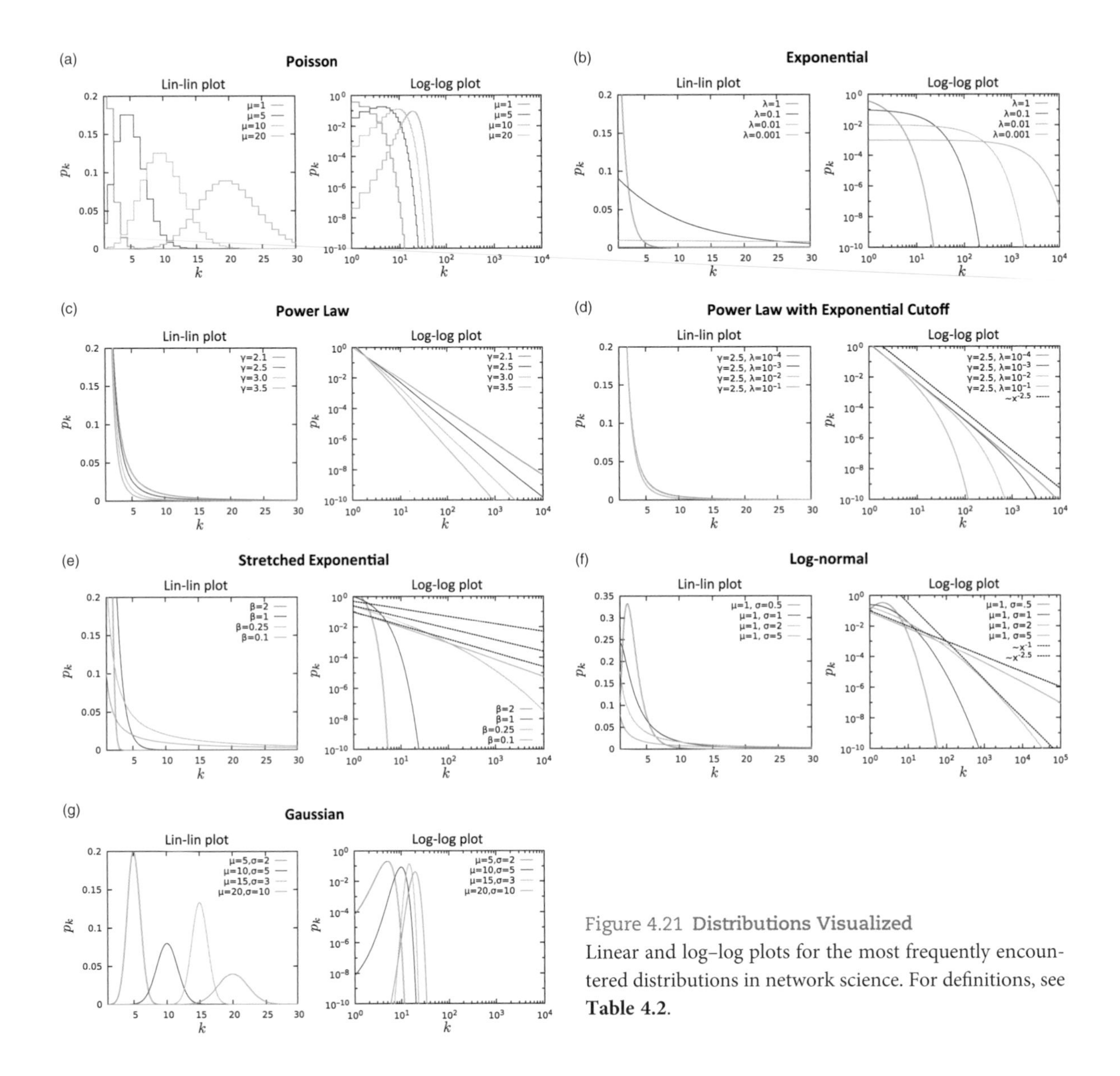

Figure 4.21 **Distributions Visualized**
Linear and log–log plots for the most frequently encountered distributions in network science. For definitions, see **Table 4.2**.

of magnitude. This is also illustrated in **Table 4.2**, which shows that $\langle x^2 \rangle$ grows exponentially with σ and hence can be very large.

In summary, in most areas where we encounter fat-tailed distributions, there is an ongoing debate asking which distribution offers the best fit to the data. Frequently encountered candidates include a power law, a stretched exponential or a log-normal function (**Figure 4.21**). In many systems empirical data is not sufficient to distinguish these

Table 4.2 **Distributions in Network Science**

The table lists frequently encountered distributions in network science. For each distribution we show the density function p_x and the appropriate normalization constant C for the continuous and the discrete case. Given that $\langle x \rangle$ and $\langle x^2 \rangle$ play an important role in network theory, we show the analytical form of these two quantities for each distribution.

NAME	$p_x/p(x)$	$\langle x \rangle$	$\langle x^2 \rangle$
Poisson [discrete]	$e^{-\mu}\mu^x/x!$	μ	$\mu(1+\mu)$
Exponential [discrete]	$(1-e^{-\lambda})e^{-\lambda x}$	$1/(e^{\lambda}-1)$	$(e^{\lambda}+1)/(e^{\lambda}-1)^2$
Exponential [continuous]	$\lambda e^{-\lambda x}$	$1/\lambda$	$2/\lambda^2$
Power law (discrete)	$x^{-\alpha}/\zeta(\alpha)$	$\begin{cases} \zeta(\alpha-2)/\zeta(\alpha) & \text{if } \alpha > 2 \\ \infty, & \text{if } \alpha \leq 2 \end{cases}$	$\begin{cases} \zeta(\alpha-1)/\zeta(\alpha) & \text{if } \alpha > 3 \\ \infty, & \text{if } \alpha \leq 3 \end{cases}$
Power law (continuous)	$(\alpha-1)x^{-\alpha}$	$\begin{cases} (\alpha-1)/(\alpha-2), & \text{if } \alpha > 2 \\ \infty, & \text{if } \alpha \leq 2 \end{cases}$	$\begin{cases} (\alpha-1)/(\alpha-3), & \text{if } \alpha > 3 \\ \infty, & \text{if } \alpha \leq 3 \end{cases}$
Power law with cutoff (continuous)	$\frac{\lambda^{1-\alpha}}{\Gamma(1-\alpha,1)}x^{-\alpha}e^{-\lambda x}$	$\lambda^{-1}\frac{\Gamma(2-\alpha,1)}{\Gamma(1-\alpha,1)}$	$\lambda^{-2}\frac{\Gamma(3-\alpha,1)}{\Gamma(1-\alpha,1)}$
Stretched exponential (continuous)	$\beta\lambda^{\beta}x^{\beta-1}e^{-(\lambda x)^{\beta}}$	$\lambda^{-1}\Gamma(1+\beta^{-1})$	$\lambda^{-2}\Gamma(1+2\beta^{-1})$
Log-normal (continuous)	$\frac{1}{x\sqrt{2\pi\sigma^2}}e^{-(\ln x-\mu)^2/(2\sigma^2)}$	$e^{\mu+\sigma^2/2}$	$e^{2(\mu+\sigma^2)}$
Normal (continuous)	$\frac{1}{\sqrt{2\pi\sigma^2}}e^{-(x-\mu)^2/(2\sigma^2)}$	μ	$\mu^2+\sigma^2$

distributions. Hence, as long as there is empirical data to be fitted, the debate surrounding the best fit will never die out.

The debate is resolved by accurate mechanistic models, which analytically predict the expected degree distribution. We will see in the coming chapters that in the context of networks the models predict Poisson, simple exponential, stretched exponential and power law distributions. The remaining distributions in **Table 4.2** are occasionally used to fit the degrees of some networks, despite the fact that we lack a theoretical basis for their relevance to networks.

4.12 Advanced Topics 4.B Plotting Power Laws

Plotting the degree distribution is an integral part of analyzing the properties of a network. The process starts with obtaining N_k, the number of nodes with degree k. This can be provided by direct measurement or by a model. From N_k we calculate $p_k = N_k/N$. The question is, how to plot p_k to best extract its properties.

4.12.1 Use a Log–Log Plot

In a scale-free network, numerous nodes with one or two links coexist with a few hubs, representing nodes with thousands or even millions of links. Using a linear k-axis compresses the numerous small-degree nodes in the small-k region, rendering them invisible. Similarly, as there can be orders of magnitude difference in p_k for $k = 1$ and for large k, if we plot p_k on a linear vertical axis, its value for large k will appear to be zero (**Figure 4.22a**). The use of a log–log plot avoids these problems. We can either use logarithmic axes, with powers of 10 (used throughout this book, see **Figure 4.22b**) or we can plot $\log p_k$ as a function of $\log k$ (equally correct, but slightly harder to read). Note that points with $p_k = 0$ or $k = 0$ are not shown on a log–log plot as $\log 0 = -\infty$.

4.12.2 Avoid Linear Binning

A flawed method (yet frequently seen in the literature) is to simply plot $p_k = N_k/N$ on a log–log plot (**Figure 4.22b**). This is called *linear binning*, as each bin has the same size $\Delta k = 1$. For a scale-free network, linear binning results in an instantly recognizable plateau at large k, consisting of numerous data points that form a horizontal line (**Figure 4.22b**). This plateau has a simple explanation: typically we have only one copy of each high-degree node, hence in the high-k region we either have $N_k = 0$ (no node with degree k) or $N_k = 1$ (a single node with degree k). Consequently, linear binning will either provide

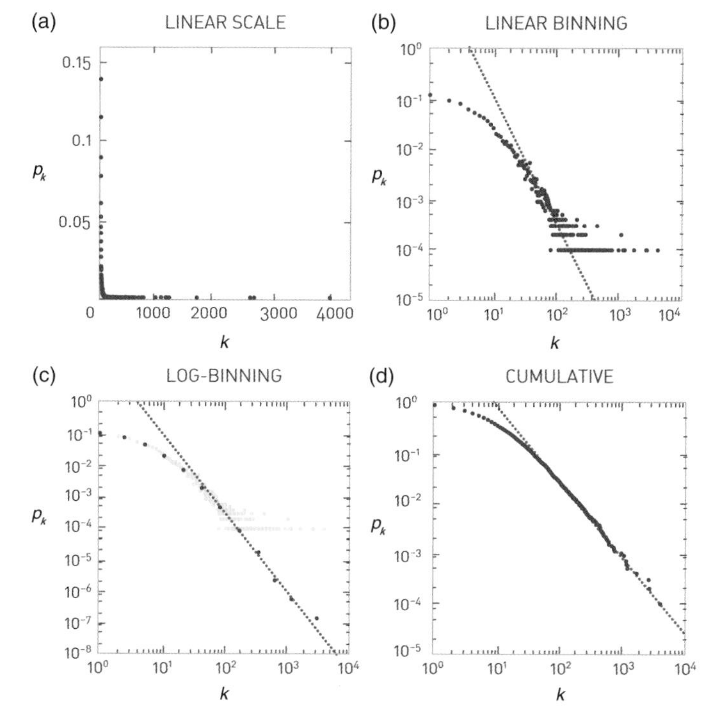

Figure 4.22 **Plotting a Degree Distribution**
A degree distribution of the form $p_k \sim (k + k_0)^{-\gamma}$, with $k_0 = 10$ and $\gamma = 2.5$, plotted using the four procedures described in the text.

(a) Linear Scale, Linear Binning
It is impossible to see the distribution on a lin–lin scale. This is the reason why we always use a log–log plot for scale-free networks.

(b) Log–Log Scale, Linear Binning
Now the tail of the distribution is visible but there is a plateau in the high-k regime, a consequence of linear binning.

(c) Log–Log Scale, Log-Binning
With log-binning the plateau disappears and the scaling extends into the high-k regime. For reference we show in light gray the data of (b) with linear binning.

(d) Log–Log Scale, Cumulative
The cumulative degree distribution shown on a log–log plot.

$p_k = 0$, not shown on a log–log plot, or $p_k = 1/N$, which applies to all hubs, generating a plateau at $p_k = 1/N$.

This plateau affects our ability to estimate the degree exponent γ. For example, if we attempt to fit a power law to the data shown in **Figure 4.22b** using linear binning, the obtained γ is quite different from the real value $\gamma = 2.5$. The reason is that under linear binning we have a large number of nodes in small-k bins, allowing us to confidently fit p_k in this regime. In the large-k bins we have too few nodes for a proper statistical estimate of p_k. Instead, the emerging plateau biases our fit. Yet it is precisely this high-k regime that plays a key role in determining γ. Increasing the bin size will not solve this problem. It is therefore recommended to avoid linear binning for fat-tailed distributions.

4.12.3 Use Logarithmic Binning

Logarithmic binning corrects the non-uniform sampling of linear binning. For log-binning we let the bin sizes increase with the degree, making sure that each bin has a comparable number of nodes. For example, we can choose the bin sizes to be multiples of 2, so that the first bin has size $b_0 = 1$, containing all nodes with $k = 1$; the second has size $b_1 = 2$, containing nodes with degree $k = 2, 3$; the third bin has size $b_2 = 4$, containing nodes with degree $k = 4, 5, 6, 7$. By induction, the nth bin has size 2^n and contains all nodes with degree $k = 2^{n-1}, 2^{n-1} + 1, \ldots, 2^{n-1} - 1$. Note that the bin size can increase with arbitrary increments, $b = c^n$, where $c > 1$. The degree distribution is given by $p_{\langle k_n \rangle} = N_n / b_n$, where N_n is the number of nodes found in bin n of size b_n and $\langle k_n \rangle$ is the average degree of the nodes in bin b_n.

The logarithmically binned p_k is shown in **Figure 4.22c**. Note that now the scaling extends into the high-k plateau, invisible under linear binning. Therefore, logarithmic binning extracts useful information from the rare high-degree nodes as well (**Box 4.10**).

4.12.4 Use Cumulative Distribution

Another way to extract information from the tail of p_k is to plot the complementary cumulative distribution

$$P_k = \sum_{q=k+1}^{\infty} p_q, \tag{4.37}$$

which again enhances the statistical significance the high-degree region. If p_k follows the power law (**4.1**), then the cumulative distribution scales as

$$P_k \sim k^{-\gamma+1}. \tag{4.38}$$

The cumulative distribution again eliminates the plateau observed for linear binning and leads to an extended scaling region (**Figure 4.22d**), allowing for a more accurate estimate of the degree exponent.

In summary, plotting the degree distribution to extract its features requires special attention. Mastering the appropriate tools can help us better explore the properties of real networks (**Box 4.10**).

Box 4.10 **Degree Distribution of Real Networks**

In real systems we rarely observe a degree distribution that follows a pure power law. Instead, for most real systems p_k has the shape shown in **Figure 4.23a**, with some recurring features:

- *Low-degree saturation* is a common deviation from the power-law behavior. Its signature is a flattened p_k for $k < k_{sat}$. This indicates that we have fewer small-degree nodes than expected for a pure power law.
- *High-degree cutoff* appears as a rapid drop in p_k for $k > k_{cut}$, indicating that we have fewer high-degree nodes than expected in a pure power law. This makes the size of the largest hub smaller than predicted by (**4.18**). High-degree cutoffs emerge if there are inherent limitations in the number of links a node can have, like the difficulty of maintaining meaningful relationships with an exceptionally large number of acquaintances.

Given the widespread presence of such cutoffs, the degree distribution is occasionally fitted to

$$p_k = a(k + k_{sat})^{-\gamma} \exp\left(-\frac{k}{k_{cut}}\right), \quad (4.39)$$

where k_{sat} accounts for degree saturation, and the exponential term accounts for high-k cutoff. To extract the full extent of the scaling we plot

$$\tilde{p}_x = p_x \exp\left(\frac{k}{k_{cut}}\right) \quad (4.40)$$

as a function of $\tilde{k} = k + k_{sat}$. According to (**4.40**) $\tilde{p} \sim \tilde{k}^{-\gamma}$, correcting for the two cutoffs, as seen in **Figure 4.23b**.

It is occasionally claimed that the presence of low-degree or high-degree cutoffs implies that the network is not scale-free. This is a misunderstanding of the scale-free property: virtually all properties of scale-free networks are insensitive to the low-degree saturation. Only the high-degree cutoff affects the system's properties by limiting the divergence of the second moment, $\langle k^2 \rangle$.

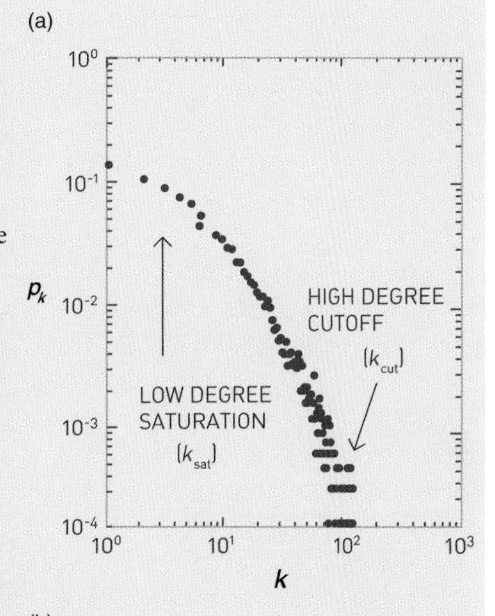

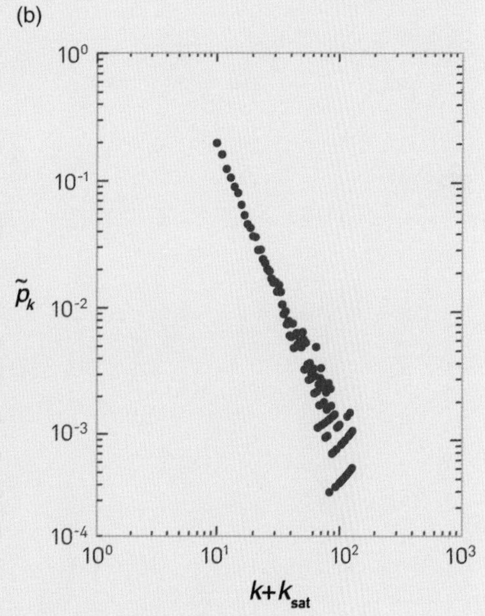

Figure 4.23 **Rescaling the Degree Distribution**

(**a**) In real networks the degree distribution frequently deviates from a pure power law by showing a *low-degree saturation* and a *high-degree cutoff*.

(**b**) By plotting the rescaled function of $(k + k_{sat})$, as suggested by (4.40), the degree distribution follows a power law for all degrees.

4.13 Advanced Topics 4.C Estimating the Degree Exponent

As the properties of scale-free networks depend on the degree exponent (**Section 4.7**), we need to determine the value of γ. We face several difficulties, however, when we try to fit a power law to real data. The most important is the fact that the scaling is rarely valid for the full range of the degree distribution. Rather we observe small-degree and high-degree cutoffs (**Box 4.10**), denoted in this section by K_{min} and K_{max}, within which we have a clear scaling region. Note that K_{min} and K_{max} are different from k_{min} and k_{max}, the latter corresponding to the smallest and largest degrees in a network. They can be the same as k_{sat} and k_{cut} discussed in **Box 4.10**. Here we focus on estimating the small-degree cutoff K_{min}, as the high-degree cutoff can be determined in a similar fashion. The reader is advised to consult the discussion on systematic problems provided at the end of this section before implementing this procedure.

4.13.1 Fitting Procedure

As the degree distribution is typically provided as a list of positive integers $k_{min}, \ldots, k_{max}$, we aim to estimate γ from a discrete set of data points [136]. We use the citation network to illustrate the procedure. The network consists of $N = 384,362$ nodes, each node representing a research paper published between 1890 and 2009 in journals published by the American Physical Society. The network has $L = 2,353,984$ links, each representing a citation from a published research paper to some other publication in the dataset (outside citations are ignored). For no particular reason, this is not the citation dataset listed in **Table 4.1**. See [137] for an overall characterization of this data. The steps of the fitting process are [136]:

1. Choose a value of K_{min} between k_{min} and k_{max}. Estimate the value of the degree exponent corresponding to this K_{min} using

$$\gamma = 1 + N\left[\sum_{i=1}^{N} \ln \frac{k_i}{K_{min} - \frac{1}{2}}\right]^{-1}. \quad (4.41)$$

2. With the obtained $(\gamma, K_{\min})$ parameter pair assume that the degree distribution has the form

$$p_k = \frac{1}{\zeta(\gamma, K_{\min})} k^{-\gamma}, \tag{4.42}$$

hence the associated cumulative distribution function (CDF) is

$$P_k = 1 - \frac{\zeta(\gamma, k)}{\zeta(\gamma, K_{\min})}. \tag{4.43}$$

3. Use the Kolmogorov–Smirnov test to determine the maximum distance D between the CDF of the data $S(k)$ and the fitted model provided by (**4.43**) with the selected $(\gamma, k_{\min})$ parameter pair,

$$D = \max_{k \geq K\min} |S(k) - P_k|. \tag{4.44}$$

Equation (**4.44**) identifies the degree for which the difference D between the empirical distribution $S(k)$ and the fitted distribution (**4.43**) is the largest.

4. Repeat steps 1–3 by scanning the whole $K_{\min}$ range from $k_{\min}$ to $k_{\max}$. We aim to identify the $K_{\min}$ value for which D provided by (**4.44**) is minimal. To illustrate the procedure, we plot D as a function of $K_{\min}$ for the citation network (**Figure 4.24b**). The plot indicates that D is minimal for $K_{\min} = 49$, and the corresponding γ estimated by (**4.41**), representing the optimal fit, is $\gamma = 2.79$. The standard error for the obtained degree exponent is

$$\sigma_\gamma = \frac{1}{\sqrt{N\left[\frac{\zeta''(\gamma, K_{\min})}{\zeta(\gamma, K_{\min})} - \left(\frac{\zeta'(\gamma, K_{\min})}{\zeta(\gamma, K_{\min})}\right)^2\right]}}, \tag{4.45}$$

which implies that the best fit is $\gamma \pm \sigma_\gamma$. For the citation network we obtain $\sigma_\gamma = 0.003$, hence $\gamma = 2.79(3)$.

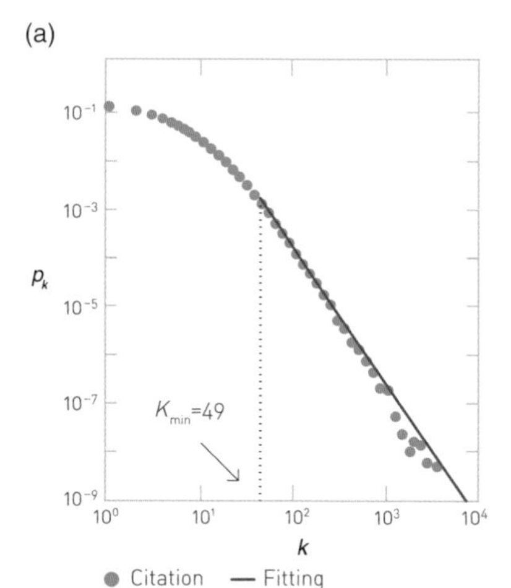

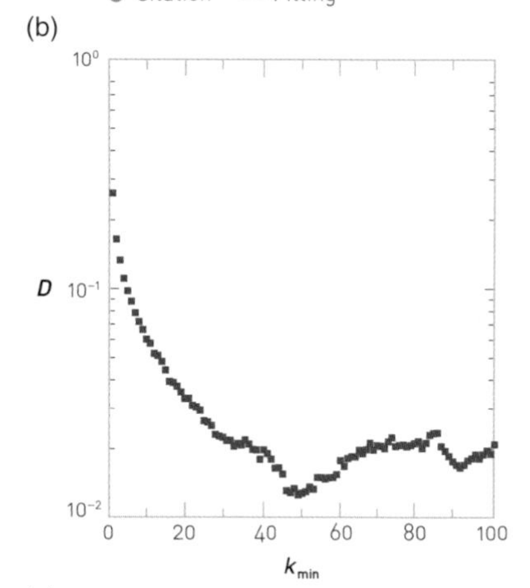

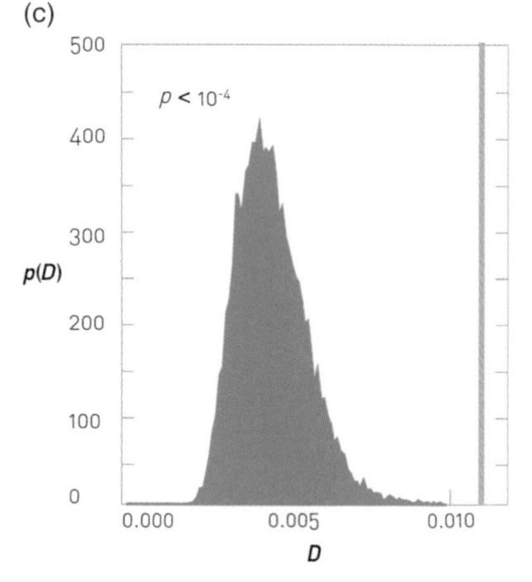

Figure 4.24 **Maximum Likelihood Estimation**
(**a**) The degree distribution p_k of the citation network, where the straight purple line represents the best fit based on the model (4.39).
(**b**) The values of the Kolmogorov–Smirnov test vs. $K_{\min}$ for the citation network.
(**c**) $p(D^{\text{synthetic}})$ for $M = 10,000$ synthetic datasets, where the gray line corresponds to the D^{real} value extracted for the citation network.

Note that in order to estimate γ, networks with fewer than $N = 50$ nodes should be treated with caution.

4.13.2 Goodness-of-Fit

Just because we obtained a (γ, K_{min}) pair that represents an optimal fit to our dataset does not mean that the power law itself is a good model for the studied distribution. We therefore need to use a goodness-of-fit test, which generates a p value that quantifies the plausibility of the power-law hypothesis. The most often used procedure consists of the following steps:

1. Use the cumulative distribution (**4.43**) to estimate the *KS* distance between the real data and the best fit, which we denote by D^{real}. This is step 3 above, taking the value of D for K_{min} that offered the best fit to the data. For the citation data we obtain $D^{real} = 0.01158$ for $K_{min} = 49$ (**Figure 4.24c**).
2. Use (**4.42**) to generate a degree sequence of N degrees (i.e., the same number of random numbers as the number of nodes in the original dataset) and substitute the obtained degree sequence for the empirical data, determining $D^{synthetic}$ for this hypothetical degree sequence. Hence $D^{synthetic}$ represents the distance between a synthetically generated degree sequence, consistent with our degree distribution, and the real data.
3. The goal is to see if the obtained $D^{synthetic}$ is comparable with D^{real}. For this we repeat step 2 M times ($M \gg 1$), and each time we generate a new degree sequence and determine the corresponding $D^{synthetic}$, eventually obtaining the $p(D^{synthetic})$ distribution. Plot $p(D^{synthetic})$ and show D^{real} as a vertical bar (**Figure 4.24c**). If D^{real} is within the $p(D^{synthetic})$ distribution, it means that the distance between the model providing the best fit and the empirical data is comparable with the distance expected from random-degree samples chosen from the best-fit distribution. Hence the power law is a reasonable model for the data. If, however, D^{real} falls outside the $p(D^{synthetic})$ distribution, then the power law is not a good model – some other function is expected to describe the original p_k better.

While the distribution shown in **Figure 4.24c** may in some cases be useful to illustrate the statistical significance of the fit, in general it is better to assign a p number to the fit, given by

$$p = \int_{D}^{\infty} P(D^{\text{synthetic}})dD^{\text{synthetic}}. \tag{4.46}$$

The closer p is to 1, the more likely that the difference between the empirical data and the model can be attributed to statistical fluctuations alone. If p is very small, the model is not a plausible fit to the data.

Typically, the model is accepted if $p > 1\%$. For the citation network we obtain $p < 10^{-4}$, indicating that a pure power law is not a suitable model for the original degree distribution. This outcome is somewhat surprising, as the power-law nature of citation data has been documented repeatedly since the 1960s [101, 102]. This failure indicates the limitation of the blind fitting to a power law, without an analytical understanding of the underlying distribution.

4.13.3 Fitting Real Distributions

To correct the problem, we note that the fitting model (**4.44**) eliminates all the data points with $k < K_{\text{min}}$. As the citation network is fat tailed, choosing $K_{\text{min}} = 49$ forces us to discard over 96% of the data points. Yet there is statistically useful information in the $k < K_{\text{min}}$ regime that is ignored by the previous fit. We must introduce an alternative model that resolves this problem.

As we discussed in **Box 4.10**, the degree distribution of many real networks, like the citation network, does not follow a pure power law. It often has low-degree saturations and high-degree cutoffs, described by the form

$$p_k = \frac{1}{\sum\limits_{k=1} (k' + k_{\text{sat}})^{-\gamma} e^{-k'/k_{\text{cut}}}} (k + k_{\text{sat}})^{-\gamma} e^{-k/k_{\text{cut}}} \tag{4.47}$$

and the associated CDF is

$$P_k = \frac{1}{\sum\limits_{k'=1} (k' + k_{\text{sat}})^{-\gamma} e^{-k'/k_{\text{cut}}}} \sum_{k'=1}^{k} (k' + k_{\text{sat}})^{-\gamma} e^{-k'/k_{\text{cut}}}, \tag{4.48}$$

where k_{sat} and k_{cut} correspond to low-k saturation and large-k cutoff, respectively. The difference between our earlier procedure and (**4.47**) is that we now do not discard the points that

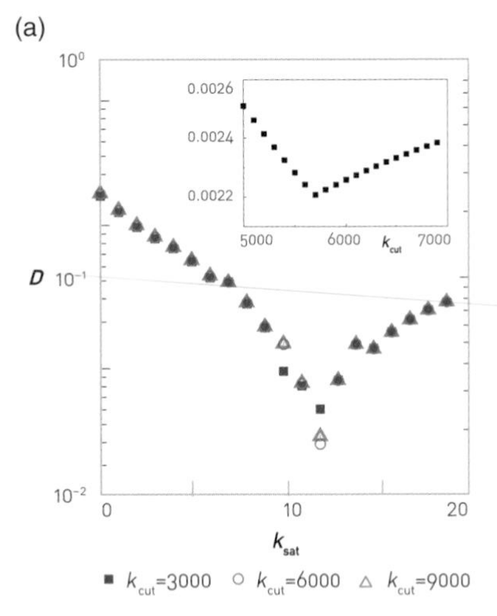

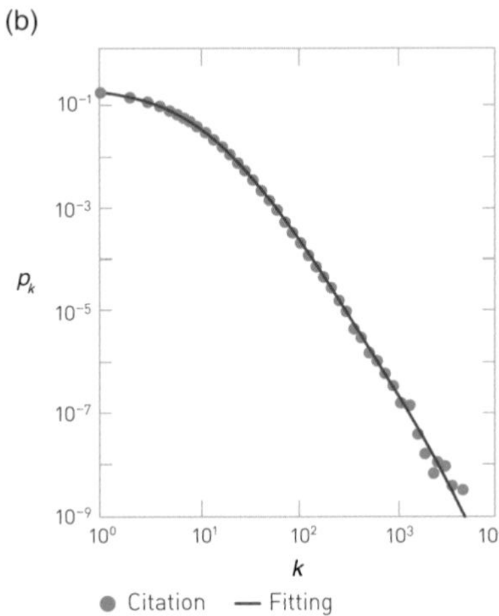

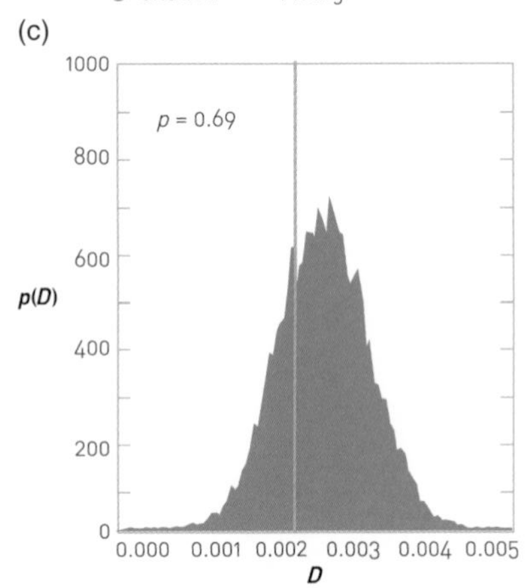

Figure 4.25 **Estimating the Scaling Parameters for Citation Networks**

(a) The Kolmogorov–Smirnov parameter D vs. k_{sat} for $k_{cut} = 3000$, 6000, 9000, respectively. The curve indicates that $k_{sat} = 12$ corresponds to the minimal D. Inset: D vs. k_{cut} for $k_{sat} = 12$, indicating that $k_{cut} = 5691$ minimizes D.

(b) Degree distribution p_k, where the straight line represents the best estimate from (a). Now the fit accurately captures the whole curve, not only its tail (or it did in **Figure 4.24a**).

(c) $p(D^{synthetic})$ for $M = 10,000$ synthetic datasets. The gray line corresponds to the D^{real} value for the citation network.

deviate from a pure power law, but instead use a function that offers a better fit to the whole degree distribution, from k_{min} to k_{max}.

Our goal is to find the fitting parameters k_{sat}, k_{cut} and γ of the model **(4.47)**, which we achieve through the following steps (**Figure 4.25**):

1. Pick a value for k_{sat} and k_{cut} between K_{min} and K_{max}. Estimate the value of the degree exponent γ using the steepest descent method that maximizes the log-likelihood function

$$\log \mathcal{L}(\gamma|k_{sat}, k_{cut}) = \sum_{i=1}^{N} \log p(k_i|\gamma, k_{sat}, k_{cut}). \qquad (4.49)$$

That is, for fixed ($k_{sat,}$ k_{cut}) we vary γ until we find the maximum of **(4.49)**.

2. With the obtained $\gamma(k_{sat}, k_{cut})$, assume that the degree distribution has the form **(4.47)**. Calculate the Kolmogorov–Smirnov parameter D between the CDF of the original data and the fitted model provided by **(4.47)**.
3. Change k_{sat} and k_{cut}, and repeat steps 1–3, scanning with k_{sat} from $k_{min} = 0$ to k_{max} and scanning with k_{cut} from $k_{min} = k_0$ to k_{max}. The goal is to identify k_{sat} and k_{cut} values for which D is minimal. We illustrate this by plotting D as a function of k_{sat} for several k_{cut} values in **Figure 4.25a** for our citation network. The (k_{sat}, k_{cut}) for which D is minimal, and the corresponding γ is provided by **(4.41)**, represent the optimal parameters of the fit. For our dataset the optimal fit is obtained for $k_{sat} = 12$ and $k_{cut} = 5,691$, providing the degree exponent $\gamma = 3.028$.

We find now that D for the real data is within the generated $p(D)$ distribution (**Figure 4.25c**), and the associated p value is 69%.

4.13.4 Systematic Fitting Issues

The procedure described above may offer the impression that determining the degree exponent is a cumbersome but straightforward process. In reality, these fitting methods have some well-known limitations:

1. A pure power law is an idealized distribution that emerges in its form (4.1) only in simple models (**Chapter 5**). In reality, a whole range of processes contribute to the topology of real networks, affecting the precise shape of the degree distribution. These processes will be discussed in **Chapter 6**. If p_k does not follow a pure power law, the methods described above, designed to fit a power law to the data, will inevitably fail to detect statistical significance. While this finding can mean that the network is not scale-free, it most often means that we have not yet gained a proper understanding of the precise form of the degree distribution. Hence we are fitting the wrong functional form of p_k to the dataset.
2. The statistical tools used above to test the goodness-of-fit rely on the Kolmogorov–Smirnov criterion, which measures the maximum distance between the fitted model and the dataset. If almost all data points follow a perfect power law, but a *single* point for some reason deviates from the curve, we will lose the fit's statistical significance. In real systems there are numerous reasons for such local deviations that have little impact on the system's overall behavior. Yet, removing these "outliers" could be seen as data manipulation; if kept, however, one cannot detect the statistical significance of the power-law fit.

   A good example is provided by the actor network, whose degree distribution follows a power law for most degrees. There is, however, a prominent outlier at $k = 1,287$, thanks to the 1956 movie *Around the World in Eighty Days*. This is the only movie where imdb.com (the source of the actor network) lists all the normally uncredited

Table 4.3 **Exponential Fitting**

For the power grid, a power-law degree distribution does not offer a statistically significant fit. Indeed, we will encounter numerous evidence that the underlying network is not scale-free. We used the fitting procedure described in this section to fit the exponential function $e^{-\lambda k}$ to the degree distribution of the power grid, obtaining a statistically significant fit. The table shows the obtained λ parameters, the k_{min} over which the fit is valid, the obtained p value and the percentage of data points included in the fit.

	λ	k_{min}	p-Value	Percentage
Power grid	0.517	4	0.91	12%

Table 4.4 **Fitting Parameters for Real Networks**

The estimated degree exponents and the appropriate fit parameters for the reference networks studied in this book. We implement two fitting strategies, the first aiming to fit a pure power law in the region (K_{min}, ∞) and the second fitting a power law with saturation and exponential cutoff to the whole dataset. In the table we show the obtained γ exponent and K_{min} for the fit with the best statistical significance, the p value for the best fit and the percentage of the data included in the fit. In the second case we again show the exponent γ, the two fit parameters, k_{sat} and k_{cut}, and the p value of the obtained fit. Note that $p > 0.01$ is considered to be statistically significant.

	γ	K_{min}	p-Value	Percentage	γ	k_{sat}	k_{cut}	p-Value
Internet	3.42	72	0.13	0.6%	3.55	8	8500	0.00
WWW (in)	2.00	1	0.00	100%	1.97	0	660	0.00
WWW (out)	2.31	7	0.00	15%	2.82	8	8500	0.00
Power grid	4.00	5	0.00	12%	8.56	19	14	0.00
Mobile phone calls (in)	4.69	9	0.34	2.6%	6.95	15	10	0.00
Mobile phone calls (out)	5.01	11	0.77	1.7%	7.23	15	10	0.00
Email-pre (in)	3.43	88	0.11	0.2%	2.27	0	8500	0.00
Email-pre (out)	2.03	3	0.00	1.2%	2.55	0	8500	0.00
Science collaboration	3.35	25	0.0001	5.4%	1.50	17		
Actor network	2.12	54	0.00	33%	–	–	–	0.00
Citation network (in)	2.79	51	0.00	3.0%	3.03	12	5691	0.69
Citation network (out)	4.00	19	0.00	14%	−0.16	5	10	0.00
***E. coli* metabolism (in)**	2.43	3	0.00	57%	3.85	19	12	0.00
***E. coli* metabolism (out)**	2.90	5	0.00	34%	2.56	15	10	0.00
Yeast–protein interactions	2.89	7	0.67	8.3%	2.95	2	90	0.52

extras in the cast. Hence the movie appears to have 1,288 actors. The second-largest movie in the dataset has only 340 actors. Since each extra has links only to the 1,287 extras that played in the same movie, we have a local peak in p_k at $k = 1{,}287$. Thanks to this peak, the degree distribution, fitted to a power law, fails to pass the Kolmogorov–Smirnov criterion. Indeed, as indicated in **Table 4.3**, neither the pure power-law fit, nor a power law with high-degree cutoff, offers a statistically significant fit. Yet, ultimately this single point does not alter the power-law nature of the degree distribution.

3. As a result of the issues discussed above, the methodology described to fit a power-law distribution often predicts a small scaling regime (99%, see **Table 4.4**) to obtain a statistically significant fit. Once plotted next to the original dataset, the obtained fit can at times be ridiculous, even if the method predits statistical significance.

In summary, estimating the degree exponent is still not yet an exact science. We continue to lack methods to estimate the statistical significance in a manner that would be acceptable to a practitioner. The blind application of the tools described above often leads to either fits that obviously do not capture the trends in the data, or to a false rejection of the power-law hypothesis. An important improvement is our ability to derive the expected form of the degree distribution, a problem discussed in **Chapter 6**.

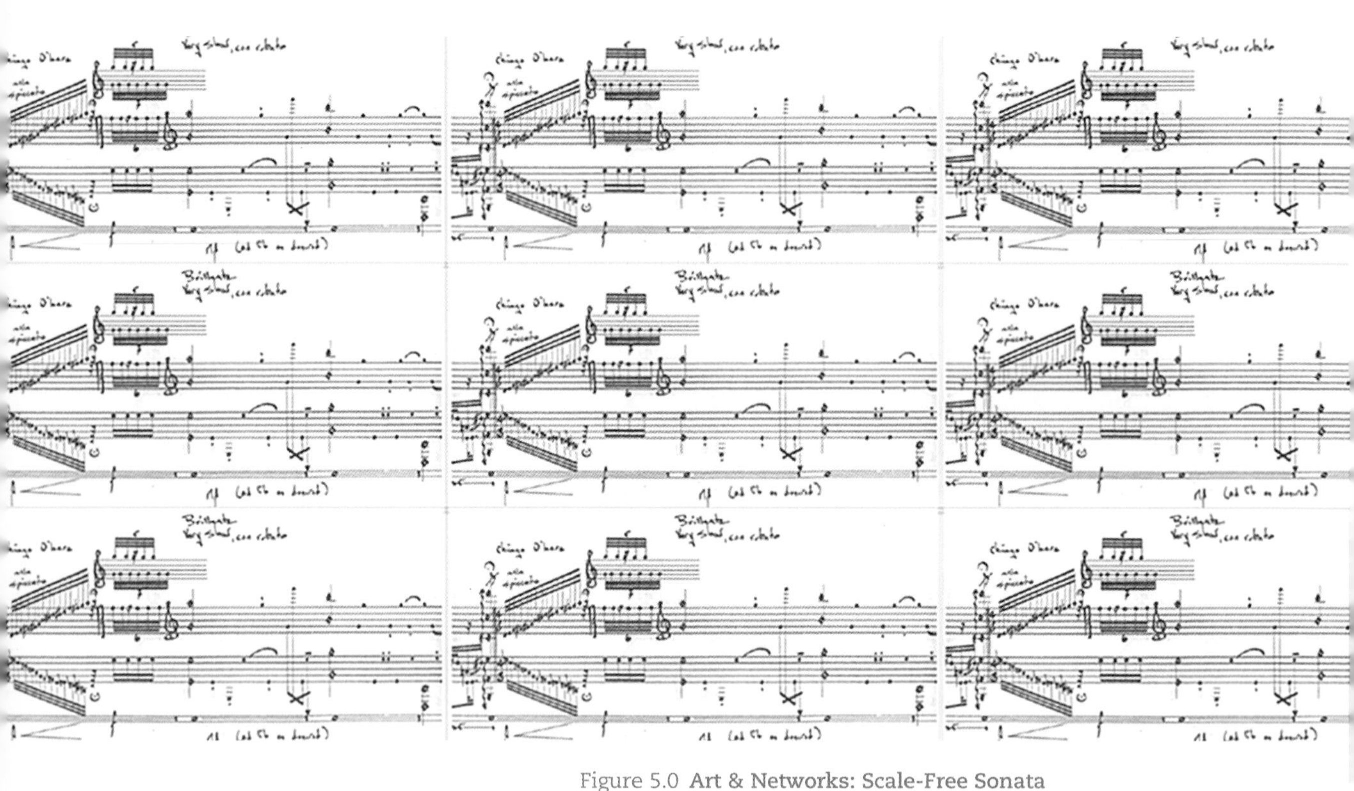

Figure 5.0 **Art & Networks: Scale-Free Sonata**
Composed by Michael Edward Edgerton in 2003, *1 sonata for piano* incorporates growth and preferential attachment to mimic the emergence of a scale-free network. The image shows the beginning of what Edgerton calls Hub #5. The relationship between the music and networks is explained by the composer:

"6 hubs of different length and procedure were distributed over the 2nd and 3rd movements. Musically, the notion of an airport was utilized by diverting all traffic into a limited landing space, while the density of procedure and duration were varied considerably between the 6 differing occurrences" (Online Resource 5.1).

CHAPTER 5

The Barabási–Albert Model

5.1 Introduction

Hubs represent the most striking difference between a random and a scale-free network. On the WWW, they are websites with an exceptional number of links, like google.com or facebook.com; in the metabolic network they are molecules like ATP or ADP, energy carriers involved in an exceptional number of chemical reactions. The very existence of these hubs and the related scale-free topology raises two fundamental questions:

- Why do such different systems as the WWW and the cell converge to a similar scale-free architecture?
- Why does the random network model of Erdős and Rényi fail to reproduce the hubs and the power laws observed in real networks?

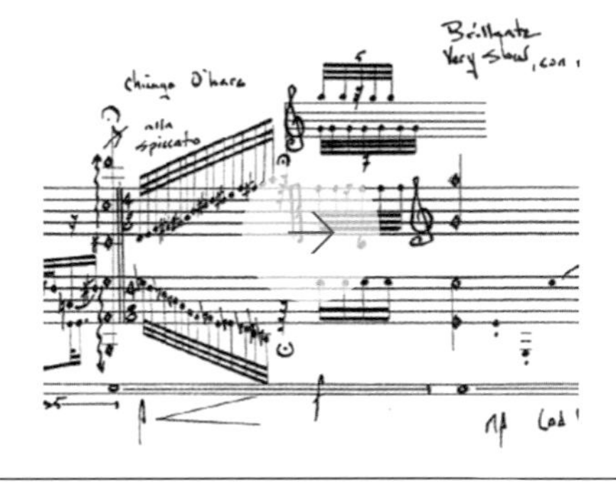

Online Resource 5.1
Scale-Free Sonata
Listen to a recording of Michael Edward Edgerton's *1 sonata for piano*, music inspired by scale-free networks.

The first question is particularly puzzling given the fundamental differences in the nature, origin and scope of the systems that display the scale-free property:

- The *nodes* of the cellular network are metabolites or proteins, while the nodes of the WWW are documents, representing information without a physical manifestation.
- The *links* within a cell are chemical reactions and binding interactions, while the links of the WWW are URLs, or small segments of computer code.
- The *history* of these two systems could not be more different: the cellular network is shaped by four billion years of evolution, while the WWW is less than three decades old.
- The *purpose* of the metabolic network is to produce the chemical components the cell needs to stay alive, while the purpose of the WWW is information access and delivery.

To understand why such *different* systems converge to a *similar* architecture we need to first understand the mechanism responsible for the emergence of the scale-free property. This is the main topic of this chapter. Given the diversity of the systems that display the scale-free property, the explanation must be simple and fundamental. The answers will change the way we model networks, forcing us to move from describing a network's topology to modeling the evolution of a complex system.

5.2 Growth and Preferential Attachment

We start our journey by asking: Why are hubs and power laws absent in random networks? The answer emerged in 1999, highlighting two hidden assumptions of the Erdős–Rényi model, that are violated in real networks [10]. Next we discuss these assumptions separately.

5.2.1 Networks Expand Through the Addition of New Nodes

The random network model assumes that we have a *fixed* number of nodes, N. Yet, *in real networks the number of nodes grows continually thanks to the addition of new nodes.*

Consider a few examples:

- In 1991 the WWW had a single node, the first webpage built by Tim Berners-Lee, the creator of the Web. Today the Web has over a trillion (10^{12}) documents, an extraordinary number that has been reached through the continuous addition of new documents by millions of individuals and institutions (**Figure 5.1a**).
- The collaboration and the citation networks expand continually through the publication of new research papers (**Figure 5.1b**).
- The actor network continues to expand through the release of new movies (**Figure 5.1c**).
- The protein interaction network may appear to be static, as we inherit our genes (and hence our proteins) from our parents, yet it is not: the number of genes in a human cell has grown from a few to over 20,000 in four billion years.

Consequently, if we wish to model these networks, we cannot resort to a static model. Our modeling approach must instead acknowledge that networks are the product of a steady growth process.

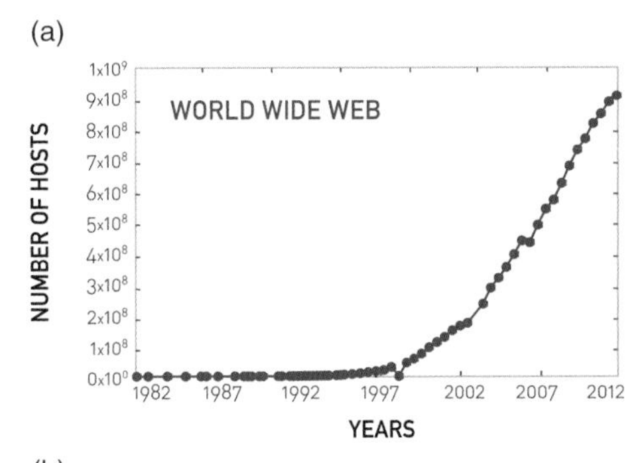

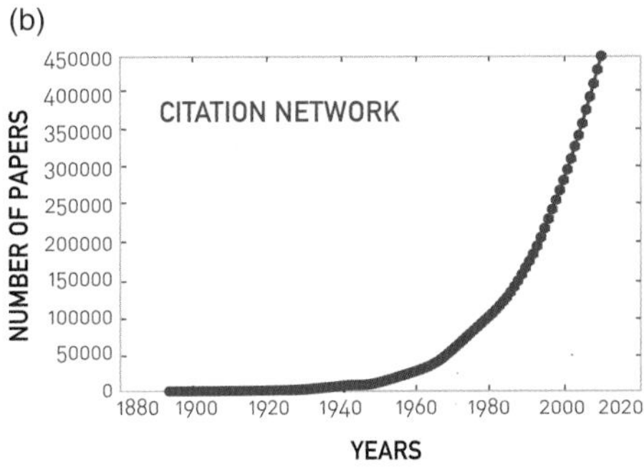

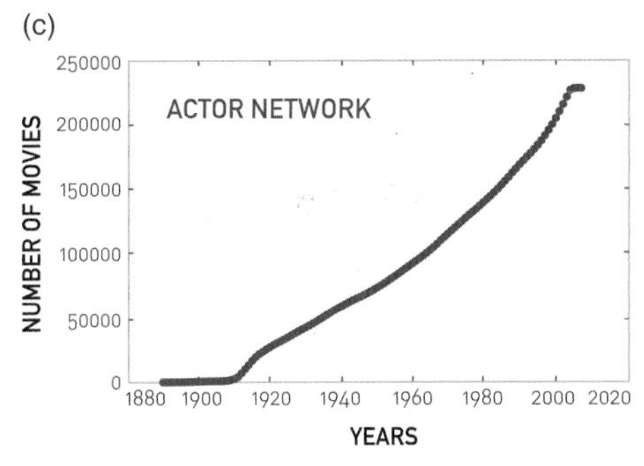

Figure 5.1 **The Growth of Networks**
Networks are not static, but grow via the addition of new nodes:
(a) The evolution of the number of WWW hosts, documenting the Web's rapid growth. After http://www.isc.org/solutions/survey/history.
(b) The number of scientific papers published in *Physical Review* since the journal's founding. The increasing number of papers drives the growth of both the science collaboration network and the citation network shown in the figure.
(c) Number of movies listed on imdb.com, driving the growth of the actor network.

5.2.2 Nodes Prefer to Link with More Connected Nodes

The random network model assumes that we randomly choose the interaction partners of a node. Yet *most real networks' new nodes prefer to link to the more connected nodes*, a process called *preferential attachment* (**Figure 5.2**).

Consider a few examples:

- We are familiar with only a tiny fraction of the trillion or more documents available on the WWW. The nodes we know are not entirely random: we have all heard about Google and Facebook, but we rarely encounter the billions of less-prominent nodes that populate the Web. As our knowledge is biased toward the more popular Web documents, we are more likely to link to a high-degree node than to a node with only a few links.
- No scientist can attempt to read the more than a million scientific papers published each year. Yet the more cited a paper is, the more likely we are to hear about it and eventually read it. As we cite what we read, our citations are biased

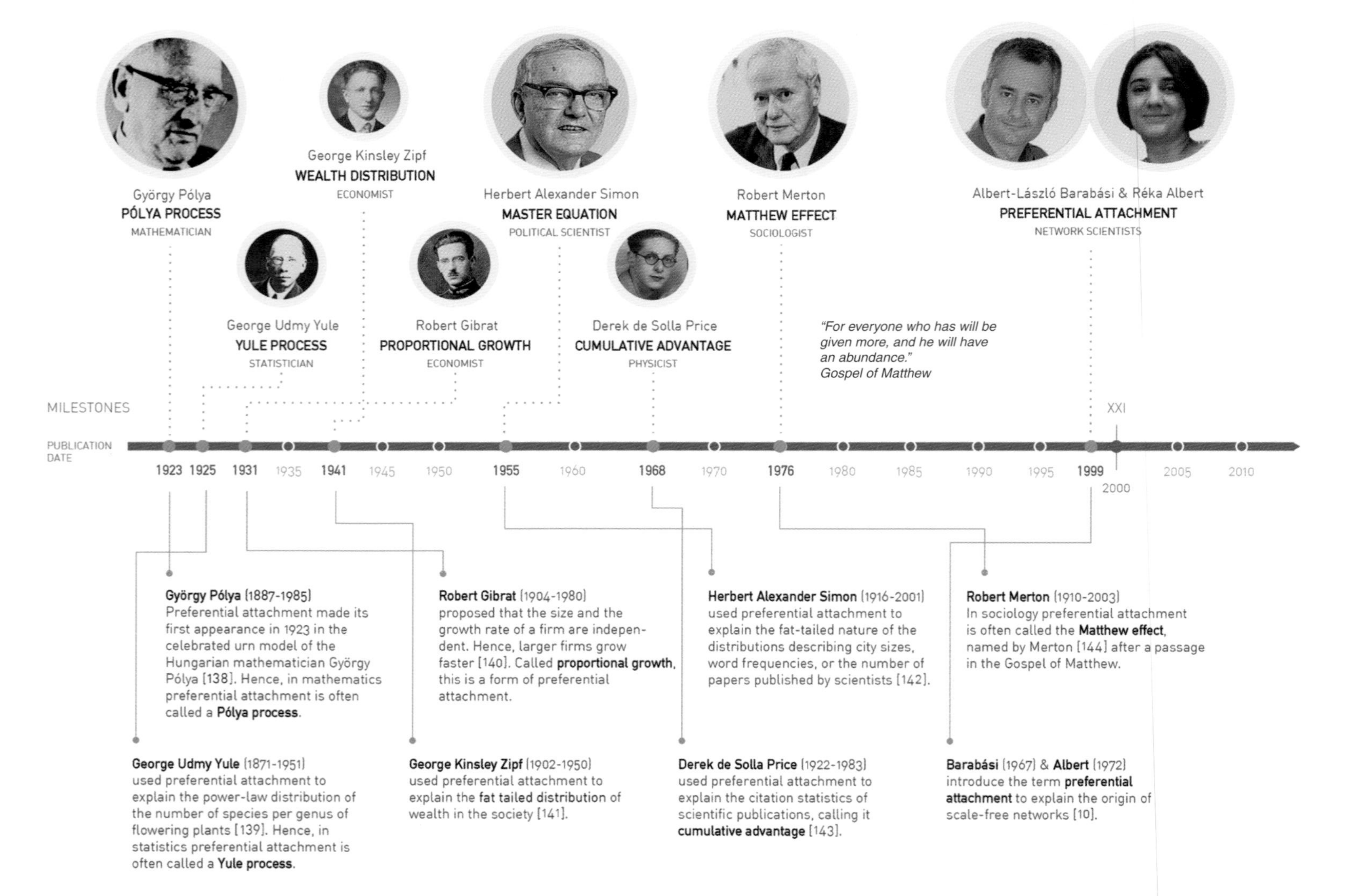

Figure 5.2 **Preferential Attachment: A Brief History**
Preferential attachment has emerged independently in many disciplines, helping explain the presence of power laws characterizing various systems. In the context of *networks*, preferential attachment was introduced in 1999 to explain the empirically observed scale-free property of real networks.

toward the more cited publications, representing the high-degree nodes of the citation network.

- The more movies an actor has played in, the more familiar a casting director will be with his or her skills. Hence, the higher the degree of an actor in the actor network, the higher the chances that they will be considered for a new role.

In summary, the random network model differs from real networks in two important characteristics:

(A) **Growth**
Real networks are the result of a growth process that increases continuously with N. In contrast, the random network model assumes that the number of nodes, N, is fixed.

(B) **Preferential Attachment**
In real networks, new nodes tend to link to more connected nodes. In contrast, nodes in random networks choose their interaction partners randomly.

There are many other differences between real and random networks, some of which will be discussed in the coming chapters. Yet as we show next, these two – *growth* and *preferential attachment* – play a particularly important role in shaping a network's degree distribution.

5.3 The Barabási–Albert Model

The recognition that growth and preferential attachment coexist in real networks has inspired a minimal model called the *Barabási–Albert model*, which can generate scale-free networks [10]. Also known as the *BA model* or the *scale-free model*, it is defined as follows.

We start with m_0 nodes, the links between which are chosen arbitrarily, as long as each node has at least one link. The network develops following two steps (**Figure 5.3**):

(A) **Growth**
At each time step we add a new node with $m(\leq m_0)$ links that connect the new node to m nodes already in the network.

(B) **Preferential Attachment**
The probability $\Pi(k_i)$ that a link of the new node connects to node i depends on the degree k_i as

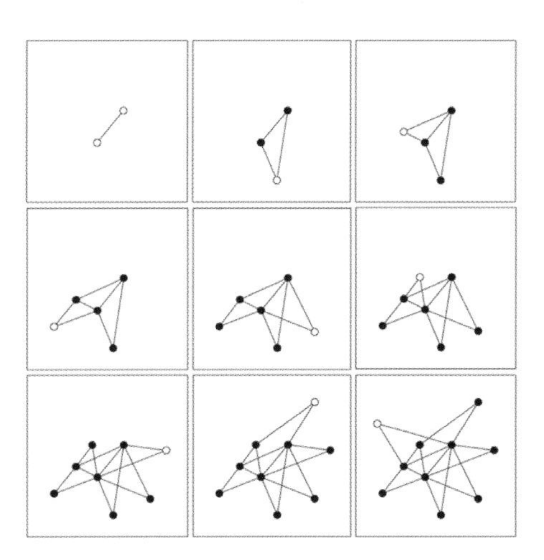

Figure 5.3 **Evolution of the Barabási–Albert Model**
The sequence of images shows nine subsequent steps of the Barabási–Albert model. Empty circles mark the newly added nodes to the network, deciding where to connect their two links ($m = 2$) using preferential attachment (**5.1**). After [50].

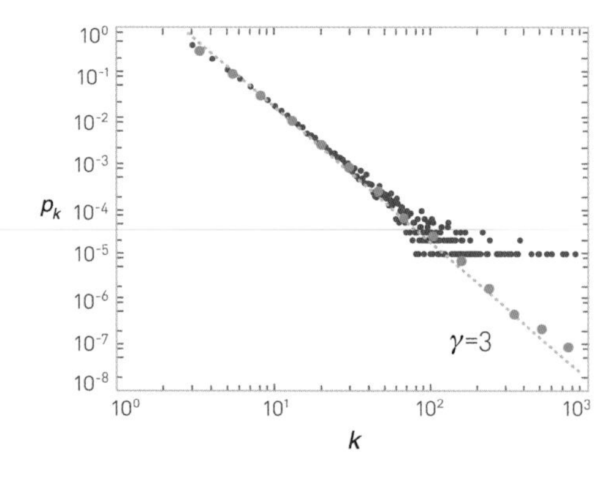

Figure 5.4 **Degree Distribution**
The degree distribution of a network generated by the Barabási–Albert model. The figure shows p_k for a single network of size $N = 100,000$ and $m = 3$. It shows both the linearly binned (purple) and the log-binned version (green) of p_k. The straight line is added to guide the eye and has slope $\gamma = 3$, corresponding to the network's predicted degree exponent.

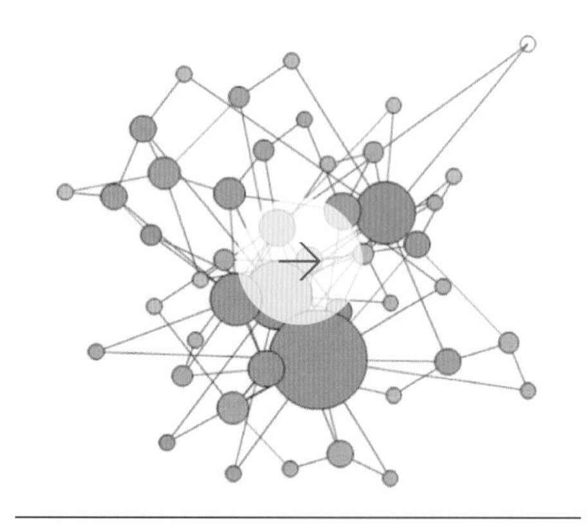

Online Resource 5.2
Emergence of a Scale-Free Network
Watch a video that shows the growth of a scale-free network and the emergence of the hubs in the Barabási–Albert model. Courtesy of Dashun Wang.

$$\Pi(k_i) = \frac{k_i}{\sum_j k_j}. \tag{5.1}$$

Preferential attachment is a probabilistic mechanism: a new node is free to connect to *any* node in the network, whether it is a hub or has a single link. Equation **(5.1)** implies, however, that if a new node has a choice between a degree-two and a degree-four node, it is twice as likely to connect to the degree-four node.

After t time steps the Barabási–Albert model generates a network with $N = t + m_0$ nodes and $m_0 + mt$ links. As **Figure 5.4** shows, the obtained network has a power-law degree distribution with degree exponent $\gamma = 3$. A mathematically self-consistent definition of the model is provided in **Box 5.1**.

As **Figure 5.3** and **Online resource 5.2** indicate, while most nodes in the network have only a few links, a few gradually turn into hubs. These hubs are the result of a *rich-gets-richer phenomenon*: owing to preferential attachment, new nodes are more likely to connect to the more connected nodes than to the smaller nodes. Hence, the larger nodes will acquire links at the expense of the smaller nodes, eventually becoming hubs.

In summary, the Barabási–Albert model indicates that two simple mechanisms, *growth* and *preferential attachment*, are responsible for the emergence of scale-free networks. The origin of the power law and the associated hubs is a *rich-gets-richer phenomenon* induced by the coexistence of these two ingredients. To understand the model's behavior and to quantify the emergence of the scale-free property, we need to become familiar with the model's mathematical properties, which is the subject of the next section.

5.4 Degree Dynamics

To understand the emergence of the scale-free property, we need to focus on the time evolution of the Barabási–Albert model. We begin by exploring the time-dependent degree of a single node [11].

In the model an existing node can increase its degree each time a *new* node enters the network. This new node will link

Box 5.1 The Mathematical Definition of the Barabási–Albert Model

The definition of the Barabási–Albert model leaves many mathematical details open:

- It does not specify the precise initial configuration of the first m_0 nodes.
- It does not specify whether the m links assigned to a new node are added one by one, or simultaneously. Therefore, if the links are truly independent, they could connect to the same node i, resulting in multi-links.

Bollobás and collaborators [15] proposed the *linearized chord diagram* (LCD) to resolve these problems. According to the LCD, for $m = 1$ we build a graph $G_1^{(t)}$ as follows (**Figure 5.5**):

(1) Start with $G_1^{(0)}$, an empty graph with no nodes.
(2) Given $G_1^{(t-1)}$, generate $G_1^{(t)}$ by adding the node v_t and a single link between v_t and v_i, where v_i is chosen with probability

$$p = \begin{cases} \dfrac{k_i}{2t-1} & \text{if } 1 \leq i \leq t-1 \\ \dfrac{1}{2t-1}, & \text{if } i = t. \end{cases} \tag{5.2}$$

That is, we place a link from the new node v_t to node v_i with probability $k_i/(2t-1)$, where the new link already contributes to the degree of v_t. Consequently, node v_t can also link to itself with probability $1/(2t-1)$, the second term in **(5.2)**. Note also that the model permits self-loops and multi-links, yet their number becomes negligible in the $t \to \infty$ limit.

For $m > 1$ we build $G_m^{(t)}$ by adding m links from the new node v_t one by one, in each step allowing the outward half of the newly added link to contribute to the degrees.

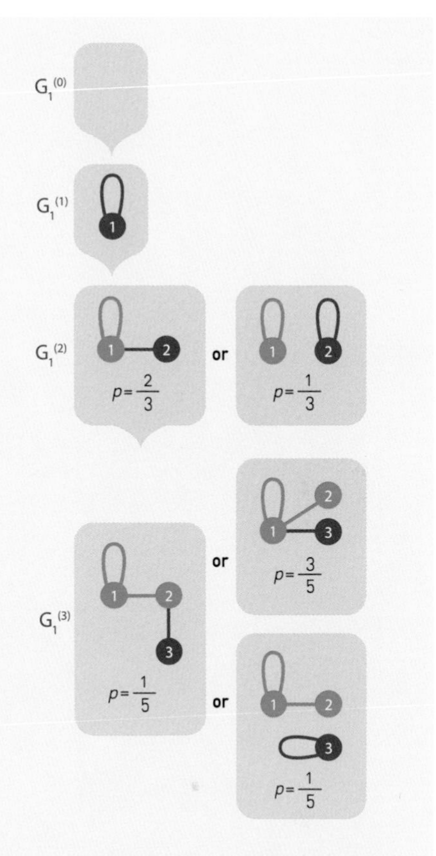

Figure 5.5 **The Linearized Chord Diagram**
The construction of the LCD [15]. The figure shows the first four steps of the network's evolution for $m = 1$.

$G_1^{(0)}$: We start with an empty network.

$G_1^{(1)}$: The first node can only link to itself, forming a self-loop. Self-loops are allowed, and so are multi-links for $m > 1$.

$G_1^{(2)}$: Node 2 can either connect to node 1 with probability 2/3, or to itself with probability 1/3. According to **(5.2)**, half of the links that the new node 2 brings along are already counted as present. Consequently, node 1 has degree $k_1 = 2$ and node 2 has degree $k_2 = 1$, the normalization constant being 3.

$G_1^{(3)}$: Let us assume that the first of the two $G_1^{(2)}$ network possibilities have materialized. When node 3 comes along, it again has three choices: it can connect to node 2 with probability 1/5, to node 1 with probability 3/5 or to itself with probability 1/5.

to m of the $N(t)$ nodes already present in the system. The probability that one of these links connects to node i is given by **(5.1)**.

Let us approximate the degree k_i with a continuous real variable, representing its expectation value over many realizations of the growth process. The rate at which an existing node i acquires links as a result of new nodes connecting to it is

$$\frac{dk_i}{dt} = m\Pi(k_i) = m\frac{k_i}{\sum_{j=1}^{N-1} k_j}. \tag{5.3}$$

The coefficient m describes that each new node arrives with m links. Hence, node i has m chances of being chosen. The sum in the denominator of **(5.3)** goes over all nodes in the network except the newly added node, thus

$$\sum_{j=1}^{N-1} k_j = 2mt - m. \tag{5.4}$$

Therefore, **(5.3)** becomes

$$\frac{dk_i}{dt} = \frac{k_i}{2t - 1}. \tag{5.5}$$

For large t the (-1) term can be neglected in the denominator, obtaining

$$\frac{dk_i}{k_i} = \frac{1}{2}\frac{dt}{t}. \tag{5.6}$$

By integrating **(5.6)** and using the fact that $k_i(t_i) = m$, meaning that node i joins the network at time t_i with m links, we obtain

$$k_i(t) = m\left(\frac{t}{t_i}\right)^{\beta}. \tag{5.7}$$

We call β the *dynamical exponent* and it has the value

$$\beta = \frac{1}{2}.$$

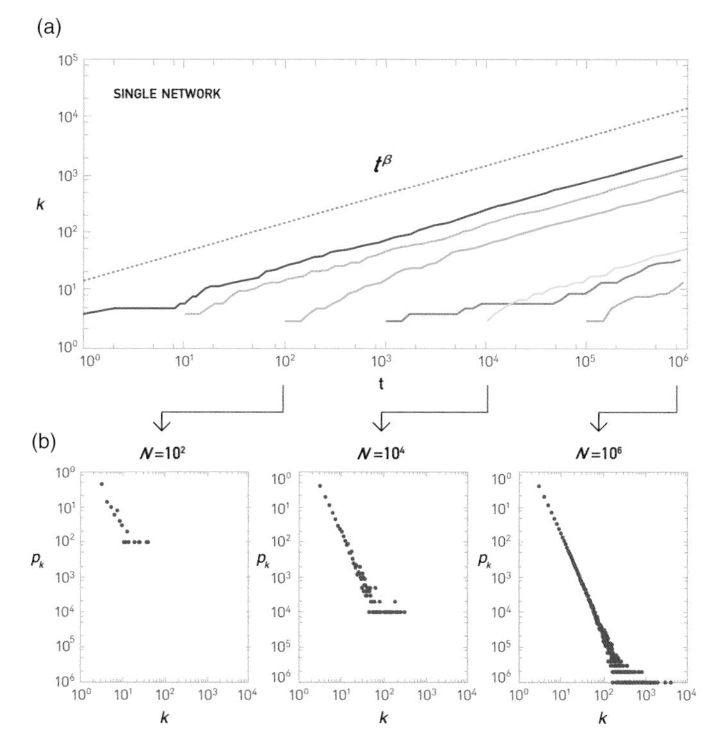

Figure 5.6 **Degree Dynamics**
(a) The growth of the degrees of nodes added at time $t = 1, 10, 10^2, 10^3, 10^4, 10^5$ (continuous lines from left to right) in the Barabási–Albert model. Each node increases its degree following **(5.7)**. Consequently, at any moment the older nodes have higher degrees. The dotted line corresponds to the analytical prediction **(5.7)** with $\beta = 1/2$.
(b) Degree distribution of the network after adding $N = 10^2$, 10^4 and 10^6 nodes, i.e. at time $t = 10^2$, 10^4 and 10^6 (illustrated by arrows in (a)). The larger the network, the more obvious is the power-law nature of the degree distribution. Note that we used linear binning for p_k to better observe the gradual emergence of the scale-free state.

Equation **(5.7)** offers a number of predictions:

- The degree of each node increases following a power law with the same dynamical exponent $\beta = 1/2$ (**Figure 5.6a**). Hence all nodes follow the same dynamical law.
- The growth in the degrees is sublinear (i.e., $\beta < 1$). This is a consequence of the growing nature of the Barabási–Albert model: each new node has more nodes to link to than the previous node. Hence, with time the existing nodes compete for links with an increasing pool of other nodes.
- The earlier node i was added, the higher is its degree $k_i(t)$. Hence, hubs are large because they arrived earlier, a phenomenon called *first-mover advantage* in marketing and business.
- The rate at which node i acquires new links is given by the derivative of **(5.7)**

$$\frac{dk_i(t)}{dt} = \frac{m}{2}\frac{1}{\sqrt{t_i t}}, \tag{5.8}$$

Box 5.2 Time in Networks

As we compare the predictions of the network models with real data, we have to decide how to measure *time* in networks. Real networks evolve over rather different time scales.

World Wide Web
The first webpage was created in 1991. Given its current trillion documents, the WWW has added a node each millisecond (10^{-3} s).

Cell
The cell is the result of four billion years of evolution. With roughly 20,000 genes in a human cell, on average the cellular network has added a node every 200,000 years ($\sim 10^{13}$ s).

Given these enormous time-scale differences, it is impossible to use real time to compare the dynamics of different networks. Therefore, in network theory we use *event time*, advancing our time step by one each time there is a change in the network topology.

For example, in the Barabási-Albert model the addition of each new node corresponds to a new time step, hence $t = N$. In other models time is also advanced by the arrival of a new link or the deletion of a node. If needed, we can establish a direct mapping between event time and physical time.

indicating that in each time step older nodes acquire more links (as they have smaller t_i). Furthermore, the rate at which a node acquires links decreases with time as $t^{-1/2}$. Hence, fewer and fewer links go to a node.

In summary, the Barabási–Albert model captures the fact that in real networks nodes arrive one after the other, offering a dynamical description of a network's evolution. This generates competition for links, during which the older nodes have an advantage over the younger ones, eventually turning into hubs (**Box 5.2**).

5.5 Degree Distribution

The distinguishing feature of the networks generated by the Barabási–Albert model is their power-law degree distribution (**Figure 5.4**). In this section we calculate the functional form of p_k, helping us understand its origin.

A number of analytical tools are available to calculate the degree distribution of the Barabási–Albert network. The simplest is the *continuum theory* that we started to develop in the previous section [10, 11]. It predicts the degree distribution (**Box 5.3**)

$$p(k) \approx 2m^{1/\beta}k^{-\gamma} \tag{5.9}$$

with

$$\gamma = \frac{1}{\beta} + 1 = 3. \tag{5.10}$$

Therefore the degree distribution follows a power law with degree exponent $\gamma = 3$, in agreement with the numerical results (**Figures 5.4 and 5.7**). Moreover, **(5.10)** links the degree exponent, γ, a quantity characterizing the network topology, with the dynamical exponent, β, that characterizes a node's temporal evolution, revealing a deep relationship between the network's topology and dynamics.

While the continuum theory predicts the correct degree exponent, it fails to accurately predict the pre-factors of **(5.9)**. The correct pre-factors can be obtained using a master [13] or a rate equation [145] approach, or calculated exactly using the LCD model [15] (**Box 5.2**). Consequently, the *exact degree distribution* of the Barabási–Albert model is (**Advanced topics 5.A**)

$$p_k = \frac{2m(m+1)}{k(k+1)(k+2)}. \quad (5.11)$$

Equation (**5.11**) has several implications:

- For large k, (**5.11**) reduces to $p_k \sim k^{-3}$, or $\gamma = 3$, in line with (**5.9**) and (**5.10**).
- The degree exponent γ is independent of m, a prediction that agrees with the numerical results (**Figure 5.7a**).
- The power-law degree distribution observed in real networks describes systems of rather different age and size. Hence, an approporiate model should lead to a time-independent degree distribution. Indeed, according to (**5.11**) the degree distribution of the Barabási–Albert model is independent of both t and N. Hence the model predicts the emergence of a *stationary scale-free state*. Numerical simulations support this prediction, indicating that the p_k observed for different t (or N) fully overlap (**Figure 5.7b**).
- Equation (**5.11**) predicts that the coefficient of the power-law distribution is proportional to $m(m+1)$ (or m^2 for large m), again confirmed by numerical simulations (**Figure 5.7a**, inset).

In summary, the analytical calculations predict that the Barabási–Albert model generates a scale-free network with degree exponent $\gamma = 3$. The degree exponent is independent of the m and m_0 parameters. Furthermore, the degree distribution is stationary (i.e., time invariant), explaining why networks with different history, size and age develop a similar degree distribution.

Box 5.3 Continuum Theory

To calculate the degree distribution of the Barabási–Albert model in the continuum approximation we first calculate the number of nodes with degree smaller than k, that is $k_i(t) < k$. Using (**5.7**), we write

$$t_i < t\left(\frac{m}{k}\right)^{1/\beta}. \quad (5.12)$$

In the model we add a node at equal time steps (**Box 5.2**). Therefore, the number of nodes with degree smaller than k is

$$t\left(\frac{m}{k}\right)^{1/\beta}. \quad (5.13)$$

Altogether there are $N = m_0 + t$ nodes, which becomes $N \approx t$ in the large-t limit. Therefore, the probability that a randomly chosen node has degree k or smaller, which is the cumulative degree distribution, follows

$$P(k) = 1 - \left(\frac{m}{k}\right)^{1/\beta}. \quad (5.14)$$

By taking the derivative of (**5.14**) we obtain the degree distribution

$$p_k = \frac{\partial P(k)}{\partial k} = \frac{1}{\beta}\frac{m^{1/\beta}}{k^{1/\beta+1}} = 2m^2k^{-3}, \quad (5.15)$$

which is (**5.9**).

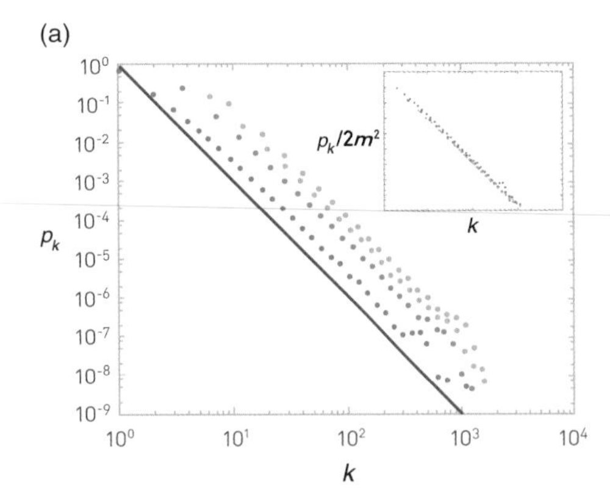

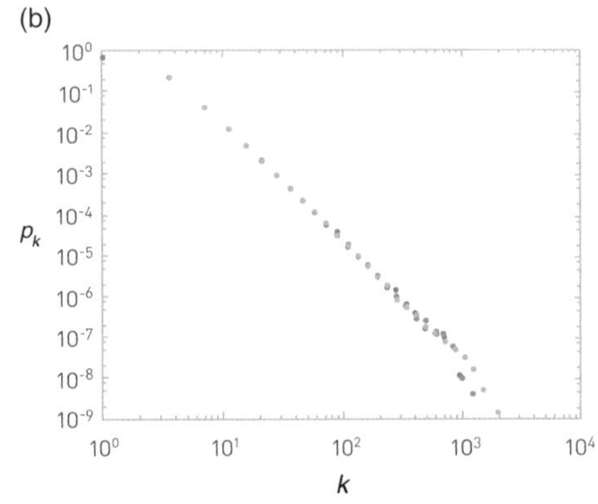

Figure 5.7 **Probing the Analytical Predictions**

(a) We generated networks with $N = 100,000$ and $m_0 = m = 1$ (blue), 3 (green), 5 (gray) and 7 (orange). The fact that the curves are parallel to each other indicates that γ is independent of m and m_0. The slope of the purple line is -3, corresponding to the predicted degree exponent $\gamma = 3$. Inset: **(5.11)** predicts $p \sim 2m^2$, hence $p/2m^2$ should be independent of m. Indeed, by plotting $p_k/2m^2$ vs. k, the data points shown in the main plot collapse into a single curve.

(b) The Barabási–Albert model predicts that p_k is independent of N. To test this we plot p_k for $N = 50,000$ (blue), 100,000 (green) and 200,000 (gray), with $m_0 = m = 3$. The obtained p_k are practically indistinguishable, indicating that the degree distribution is *stationary*, i.e. independent of time and system size.

5.6 The Absence of Growth or Preferential Attachment

The coexistence of growth and preferential attachment in the Barabási–Albert model raises an important question: Are they both necessary for the emergence of the scale-free property? In other words, could we generate a scale-free network with only one of the two ingredients? To address these questions, next we discuss two limiting cases of the model, each containing only one of the two ingredients [10, 11].

5.6.1 Model A

To test the role of preferential attachment, we keep the growing character of the network (ingredient A) and eliminate preferential attachment (ingredient B). Hence, *Model A* starts with m_0 nodes and evolves via the following steps.

(A) **Growth**
At each time step we add a new node with $m(\leq m_0)$ links that connect to m nodes added earlier.

(B) **Preferential Attachment**
The probability that a new node links to a node with degree k_i is

$$\Pi(k_i) = \frac{1}{(m_0 + t - 1)}. \tag{5.16}$$

That is, $\Pi(k_i)$ is independent of k_i, indicating that new nodes choose randomly the nodes they link to.

The continuum theory predicts that for Model A, $k_i(t)$ increases logarithmically with time:

$$k_i(t) = m \ln\left(e\frac{m_0 + t - 1}{m_0 + t_i - 1}\right), \tag{5.17}$$

a much slower growth than the power-law increase **(5.7)**. Consequently, the degree distribution follows an exponential (**Figure 5.8a**)

$$p(k) = \frac{e}{m}\exp\left(-\frac{k}{m}\right). \tag{5.18}$$

An exponential function decays much faster than a power law, hence it does not support hubs. Therefore, the lack of preferential attachment eliminates the network's scale-free character and the hubs. Indeed, as all nodes acquire links with equal probability, we lack a rich-get-richer process and no clear winner can emerge.

5.6.2 Model B

To test the role of growth, next we keep preferential attachment (ingredient B) and eliminate growth (ingredient A). Hence, *Model B* starts with N nodes and evolves via the following step.

(B) Preferential Attachment
At each time step a node is selected randomly and connected to node i with degree k_i already present in the network, where i is chosen with probability $\Pi(k)$. As $\Pi(0) = 0$, nodes with $k = 0$ are assumed to have $k = 1$, otherwise they cannot acquire links.

In Model B the number of nodes remains constant during the network's evolution, while the number of links increases linearly with time. As a result, for large t the degree of each node also increases linearly with time (**Figure 5.8b**, inset)

$$k_i(t) \approx \frac{2}{N}t. \tag{5.19}$$

Indeed, at each time step we add a new link, without changing the number of nodes.

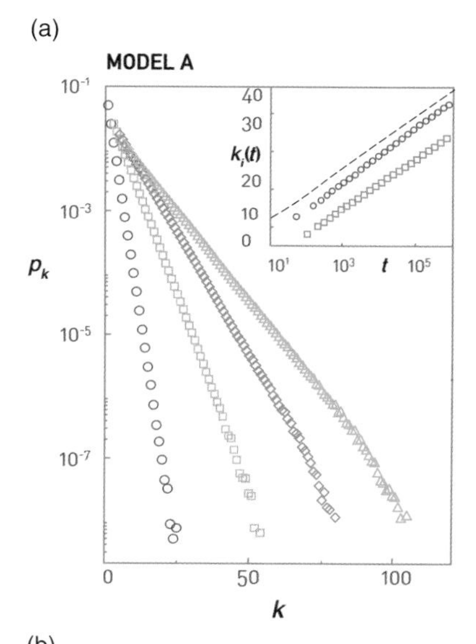

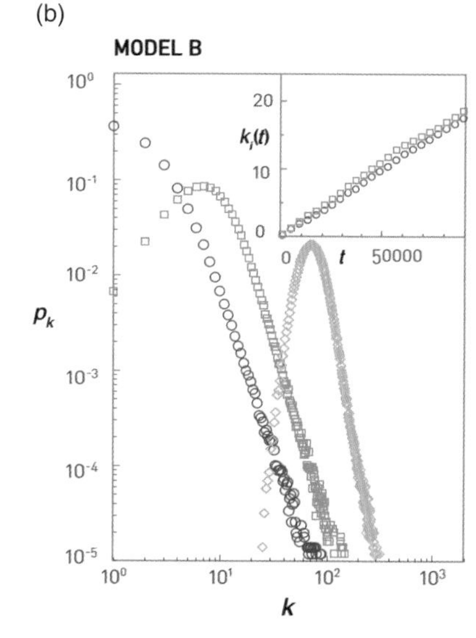

Figure 5.8 **Model A and Model B**
Numerical simulations probing the role of growth and preferential attachment.
(a) **Model A**
Degree distribution for Model A that incorporates growth but lacks preferential attachment. The symbols correspond to $m_0 = m = 1$ (circles), 3 (squares), 5 (diamonds), 7 (triangles) and $N = 800{,}000$. The linear–log plot indicates that the resulting network has an exponential p_k, as predicted by **(5.18)**.
Inset: Time evolution of the degree of two nodes added at $t_1 = 7$ and $t_2 = 97$ for $m_0 = m = 3$. The dashed line follows **(5.17)**.
(b) **Model B**
Degree distribution for Model B that lacks growth but incorporates preferential attachment, shown for $N = 10{,}000$ and $t = N$ (circles), $t = 5N$ (squares) and $t = 40N$ (diamonds). The changing shape of p_k indicates that the degree distribution is not stationary.
Inset: Time-dependent degrees of two nodes ($N = 10{,}000$), indicating that $k_i(t)$ grows linearly, as predicted by **(5.19)**. After [11].

At early times, when there are only a few links in the network (i.e., $L \ll N$), each new link connects previously unconnected nodes. In this stage the model's evolution is indistinguishable from the Barabási–Albert model with $m = 1$. Numerical simulations show that in this regime the model develops a degree distribution with a power-law tail (**Figure 5.8b**).

Yet p_k is not stationary. Indeed, after a transient period the node degrees converge to the average degree **(5.19)** and the degree develops a peak (**Figure 5.8b**). For $t \to N(N-1)/2$ the network becomes a complete graph in which all nodes have degree $k_{\max} = N - 1$, hence $p_k = \delta(N-1)$.

In summary, the absence of preferential attachment leads to a growing network with a stationary but exponential degree distribution. In contrast, the absence of growth leads to the loss of stationarity, forcing the network to converge to a complete graph. This failure of Models A and B to reproduce the empirically observed scale-free distribution indicates that growth and preferential attachment are simultaneously needed for the emergence of the scale-free property.

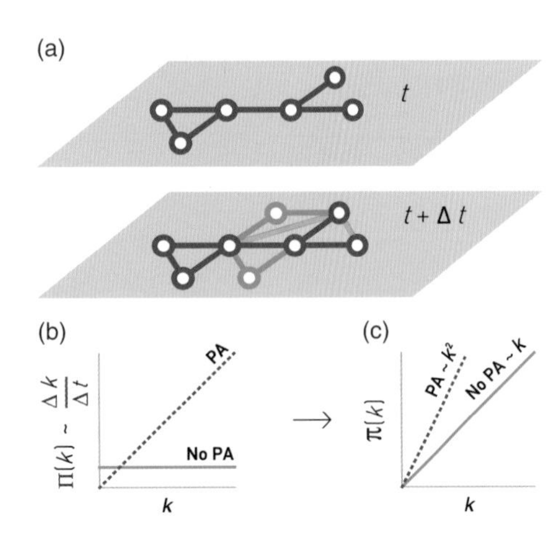

Figure 5.9 **Detecting Preferential Attachment**

(**a**) If we have access to two maps of the same network taken at time t and $t + \Delta t$, comparing them allows us to measure the $\Pi(k)$ function. Specifically, we look at nodes that have gained new links thanks to the arrival of the two new green nodes at $t + \Delta t$. The orange lines correspond to links that connect previously disconnected nodes, called *internal links*. Their role is discussed in **Chapter 6**.

(**b**) In the presence of preferential attachment, $\Delta k/\Delta t$ will depend linearly on a node's degree at time t.

(**c**) The scaling of the cumulative preferential attachment function $\pi(k)$ helps us detect the presence or absence of preferential attachment (**Figure 5.10**).

5.7 Measuring Preferential Attachment

In the previous section we showed that growth and preferential attachment are jointly responsible for the scale-free property. The presence of growth in real systems is obvious: all large networks have reached their current size by adding new nodes. But to convince ourselves that preferential attachment is also present in real networks, we need to detect it experimentally. In this section we show how to detect preferential attachment by measuring the $\Pi(k)$ function in real networks.

Preferential attachment relies on two distinct hypotheses:

Hypothesis 1
The likelihood of connecting to a node depends on that node's degree k. This is in contrast with the random network model, for which $\Pi(k)$ is independent of k.

Hypothesis 2
The functional form of $\Pi(k)$ is linear in k.

Both hypotheses can be tested by measuring $\Pi(k)$. We can determine $\Pi(k)$ for systems for which we know the time at which each node joined the network, or we have at least two network maps collected at not too distant moments in time [146, 147].

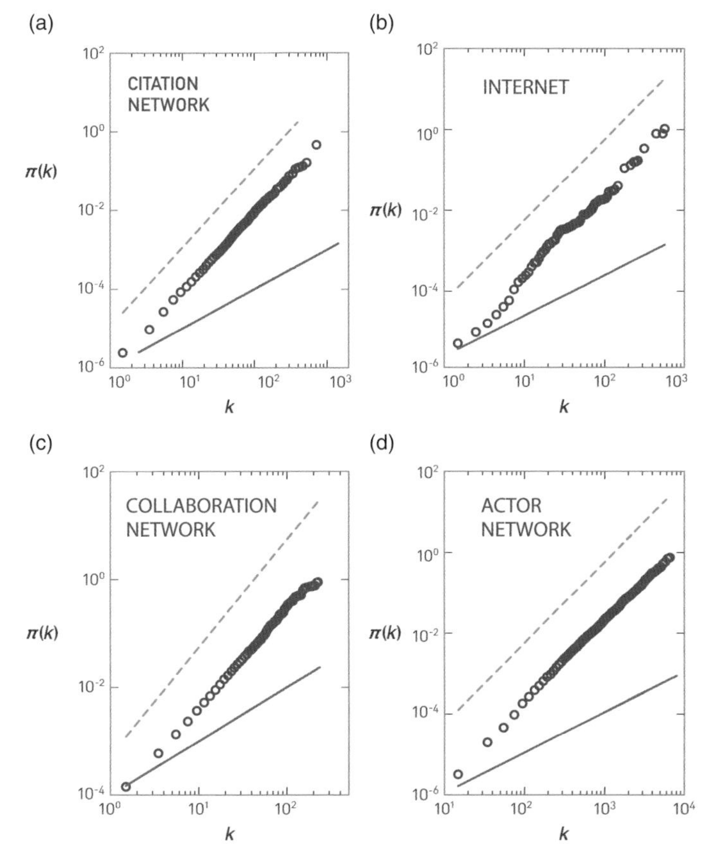

Figure 5.10 **Evidence of Preferential Attachment**

The figure shows the cumulative preferential attachment function $\pi(k)$, defined in **(5.21)**, for several real systems:

(a) Citation network.

(b) Internet.

(c) Scientific collaboration network (neuroscience).

(d) Actor network.

In each panel we have two lines to guide the eye: the dashed line corresponds to linear preferential attachment ($\pi(k) \sim k^2$) and the continuous line indicates the absence of preferential attachment ($\pi(k) \sim k$). In line with Hypothesis 1 we detect a k-dependence in each dataset. Yet, in (c) and (d), $\pi(k)$ grows slower than k^2, indicating that for these systems preferential attachment is sublinear, violating Hypothesis 2. Note that these measurements only consider links added through the arrival of new nodes, ignoring the addition of internal links.

After [146].

Consider a network for which we have two different maps, the first taken at time t and the second at time $t + \Delta t$ (**Figure 5.9a**). For nodes that changed their degree during the Δt time frame, we measure $\Delta k_i = k_i(t + \Delta t) - k_i(t)$. According to **(5.1)**, the relative change $\Delta k_i/\Delta t$ should follow

$$\frac{\Delta k_i}{\Delta t} \sim \Pi(k_i), \tag{5.20}$$

providing the functional form of preferential attachment. For **(5.20)** to be valid we must keep Δt small, so that the changes in Δk are modest. But Δt must not be too small so that there are still detectable differences between the two networks.

In practice, the obtained $\Delta k_i/\Delta t$ curve can be noisy. To reduce this noise we measure the *cumulative preferential attachment function*

$$\pi(k) = \sum_{k_i=0}^{k} \Pi(k_i). \tag{5.21}$$

In the absence of preferential attachment we have $\Pi(k_i) = \text{constant}$, hence, $\pi(k) \sim k$ according to **(5.21)**. If linear preferential attachment is present, that is if $\Pi(k_i) = k_i$, we expect $\pi(k) \sim k^2$.

Figure 5.10 shows the measured $\pi(k)$ for four real networks. For each system we observe a faster than linear increase in $\pi(k)$, indicating the presence of preferential attachment. **Figure 5.10** also suggests that $\Pi(k)$ can be approximated by

$$\Pi(k) \sim k^{\alpha}. \qquad (5.22)$$

For the Internet and citation networks we have $\alpha \approx 1$, indicating that $\Pi(k)$ depends linearly on k, following **(5.1)**. This is in line with Hypotheses 1 and 2. For the co-authorship and the actor networks, the best fit provides $\alpha = 0.9 \pm 0.1$, indicating the presence of a *sublinear preferential attachment.*

In summary, **(5.20)** allows us to detect the presence (or absence) of preferential attachment in real networks. The measurements show that the attachment probability depends on the node degree. We also find that while in some systems preferential attachment is linear, in others it can be sublinear. The implications of this nonlinearity are discussed in the next section.

5.8 Nonlinear Preferential Attachment

The observation of sublinear preferential attachment in **Figure 5.10** raises an important question: What is the impact of this nonlinearity on the network topology? To answer this we replace the linear preferential attachment **(5.1)** with **(5.22)** and calculate the degree distribution of the obtained *nonlinear Barabási–Albert model.*

The behavior for $\alpha = 0$ is clear: in the absence of preferential attachment we are back to Model A discussed in **Section 5.4**. Consequently, the degree distribution follows the exponential **(5.17)**.

For $\alpha = 1$ we recover the Barabási–Albert model, obtaining a scale-free network with degree distribution **(5.14)**.

Next we focus on the case $\alpha \neq 0$ and $\alpha \neq 1$. The calculation of p_k for an arbitrary α predicts several scaling regimes [145] (**Advanced topics 5.B**):

Sublinear Preferential Attachment $(0 < \alpha < 1)$
For any $\alpha > 0$ new nodes favor the more connected nodes over the less connected nodes. Yet, for $\alpha < 1$ the bias is weak, not sufficient to generate a scale-free degree distribution. Instead, in this regime the degrees follow the stretched exponential distribution (**Section 4.10**)

$$p_k \sim k^{-\alpha} \exp\left(\frac{-2\mu(\alpha)}{\langle k \rangle (1-\alpha)} k^{1-\alpha} \right), \qquad (5.23)$$

where $\mu(\alpha)$ depends only weakly on α. The exponential cutoff in **(5.23)** implies that sublinear preferential attachment limits the size and the number of hubs.

Sublinear preferential attachment also alters the size of the largest degree, k_{max}. For a scale-free network k_{max} scales polynomially with time, following **(4.18)**. For sublinear preferential attachment we have

$$k_{\text{max}} \sim (\ln t)^{1/(t-\alpha)}, \qquad (5.24)$$

a logarithmic dependence that predicts a much slower growth of the maximum degree than the polynomial. This slower growth is the reason why the hubs are smaller for $\alpha < 1$ (**Figure 5.11**).

Superlinear Preferential Attachment $(\alpha > 1)$
For $\alpha > 1$ the tendency to link to highly connected nodes is enhanced, accelerating the *rich-gets-richer process*. The consequence of this is most obvious for $\alpha > 2$, when the model predicts a *winner-takes-all* phenomenon: almost all nodes connect to a few super-hubs. Hence we observe the emergence of a hub-and-spoke network, in which most nodes link directly to a few central nodes. The situation for $1 < \alpha < 2$ is less extreme, but similar.

This winner-takes-all process alters the size of the largest hub as well, finding that (**Figure 5.11**)

$$k_{\text{max}} \sim t. \qquad (5.25)$$

Hence, for $\alpha > 1$ the largest hub links to a finite fraction of nodes in the system.

In summary, nonlinear preferential attachment changes the degree distribution, either limiting the size of the hubs ($\alpha < 1$), or leading to super-hubs ($\alpha > 1$, **Figure 5.12**). Consequently, $\Pi(k)$ needs to depend strictly linearly on the

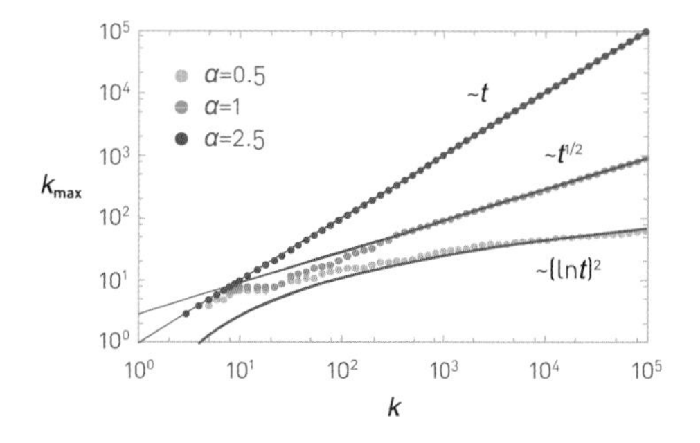

Figure 5.11 **The Growth of the Hubs**
The nature of preferential attachment affects the degree of the largest node. While in a scale-free network ($\alpha = 1$) the biggest hub grows as $t^{1/2}$ (green curve, **(4.18)**), for *sublinear preferential attachment* ($\alpha < 1$) this dependence becomes logarithmic, following **(5.24)**. For *superlinear preferential attachment* ($\alpha > 1$) the biggest hub grows linearly with time, always grabbing a finite fraction of all links, following **(5.25)**. The symbols are provided by numerical simulations; the dotted lines represent the analytical predictions.

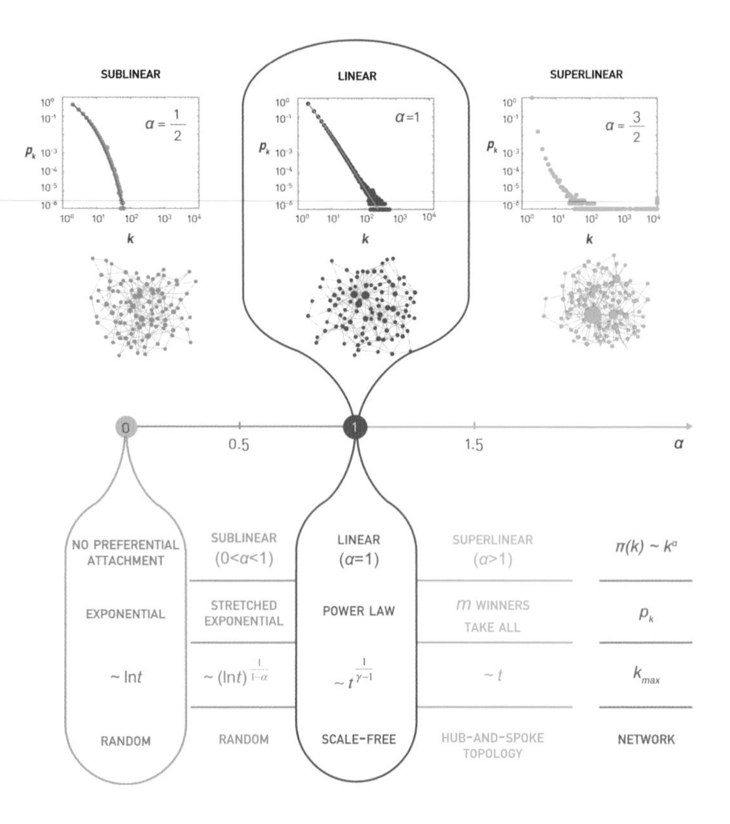

Figure 5.12 **Nonlinear Preferential Attachment**

The scaling regimes characterizing the nonlinear Barabási–Albert model. The three top panels show p_k for different α ($N = 10^4$). The network maps show the corresponding topologies ($N = 100$). The theoretical results predict the existence of four scaling regimes.

No Preferential Attachment ($\alpha = 0$)
The network has a simple exponential degree distribution, following **(5.18)**. Hubs are absent and the resulting network is similar to a random network.

Sublinear Regime ($0 < \alpha < 1$)
The degree distribution follows the stretched exponential **(5.23)**, resulting in fewer and smaller hubs than in a scale-free network. As $\alpha \to 1$ the cutoff length increases and p_k follows a power law over an increasing range of degrees.

Linear Regime ($\alpha = 1$)
This corresponds to the Barabási–Albert model, hence the degree distribution follows a power law.

Superlinear Regime ($\alpha > 1$)
The high-degree nodes are disproportionately attractive. A winner-takes-all dynamic leads to a hub-and-spoke topology. In this configuration the earliest nodes become super-hubs and all subsequent nodes link to them. The degree distribution, shown for $\alpha = 1.5$, indicates the coexistence of many small nodes with a few *super-hubs* in the vicinity of $k = 10^4$.

degrees for the resulting network to have a pure power law p_k. While in many systems we do observe such a linear dependence, in others, like the scientific collaboration network and the actor network, preferential attachment is sublinear. This nonlinear $\Pi(k)$ is one reason why the degree distribution of real networks deviates from a pure power law. Hence, for systems with sublinear $\Pi(k)$ the stretched exponential **(5.23)** should offer a better fit to the degree distribution.

5.9 The Origins of Preferential Attachment

Given the key role preferential attachment plays in the evolution of real networks, we must ask, where does it come from? The question can be broken down into two narrower issues:

> Why does $\Pi(k)$ depend on k?
> Why is the dependence of $\Pi(k)$ linear in k?

In the past decade we have witnessed the emergence of two philosophically different answers to these questions. The first views preferential attachment as the interplay between random events and some structural property of a network. These mechanisms do not require global knowledge of the network but rely on random events, hence we will call them *local* or *random* mechanisms. The second assumes that each new node or link balances conflicting needs, hence they are preceded by a cost–benefit analysis. These models assume familiarity with the whole network and rely on optimization principles, prompting us to call them *global* or *optimized* mechanisms. In this section we discuss both approaches.

5.9.1 Local Mechanisms

The Barabási–Albert model postulates the presence of preferential attachment. Yet, as we show below, we can build models that generate scale-free networks apparently without preferential attachment. They work by *generating* preferential attachment. Next we discuss two such models and derive $\Pi(k)$ for them, allowing us to understand the origins of preferential attachment.

5.9.2 Link-Selection Model

The *link-selection model* offers perhaps the simplest example of a local mechanism that generates a scale-free network without preferential attachment [148]. It is defined as follows (**Figure 5.13**):

- *Growth.* At each time step we add a new node to the network.
- *Link selection.* We select a link at random and connect the new node to one of the two nodes at the two ends of the selected link.

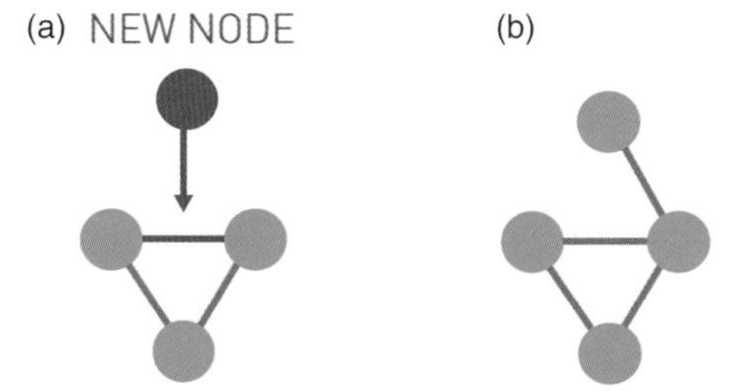

Figure 5.13 **Link-Selection Model**
(**a**) The network grows by adding a new node, which selects randomly a link from the network (shown in purple).
(**b**) The new node connects with equal probability to one of the two nodes at the ends of the selected link. In this case the new node connected to the node at the right end of the selected link.

The model requires no knowledge about the overall network topology, hence it is inherently local and random. Unlike the Barabási–Albert model, it lacks a built-in $\Pi(k)$ function. However, we show next that it generates preferential attachment.

We start by writing the probability q_k that the node at the end of a randomly chosen link has degree k as

$$q_k = Ckp_k. \tag{5.26}$$

Equation **(5.26)** captures two effects:

- The higher is the degree of a node, the higher is the chance that it is located at the end of the chosen link.

- The more degree-k nodes are in the network (i.e., the higher is p_k), the more likely that a degree-k node is at the end of the link.

In **(5.26)** C can be calculated using the normalization condition $\sum q_k = 1$, obtaining $C = 1/\langle k \rangle$. Hence the probability of finding a degree-k node at the end of a randomly chosen link is

$$q_k = \frac{kp_k}{\langle k \rangle}. \quad (5.27)$$

Equation **(5.27)** is the probability that a new node connects to a node with degree k. The fact that the bias in **(5.27)** is linear in k indicates that the link-selection model builds a scale-free network by generating linear preferential attachment.

5.9.3 Copying Model

While the link-selection model offers the simplest mechanism for preferential attachment, it is neither the first nor the most popular in the class of models that rely on local mechanisms. That distinction goes to the *copying model* (**Figure 5.14**). This model mimics a simple phenomenon: the authors of a new webpage tend to borrow links from other webpages on related topics [149, 150]. It is defined as follows:

At each time step a new node is added to the network. To decide where it connects, we randomly select a node u corresponding, for example, to a web document whose content is related to the content of the new node. Then we follow a two-step procedure (**Figure 5.14**):

(i) *Random connection.* With probability p the new node links to u, which means that we link to the randomly selected web document.

(ii) *Copying.* With probability $1 - p$ we randomly choose an *outgoing link* of node u and link the new node to the link's target. In other words, the new webpage *copies* a link of node u and connects to its target, rather than connecting to node u directly.

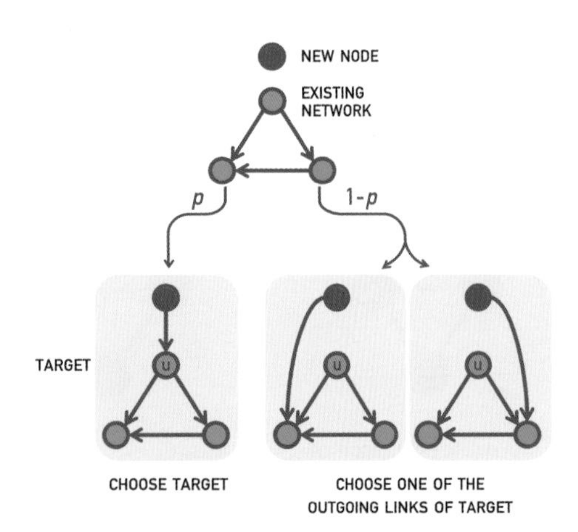

Figure 5.14 **Copying Model**
The main steps of the copying model. A new node connects with probability p to a randomly chosen target node u, or with probability $1 - p$ to one of the nodes the target u points to. In other words, with probabilty $1 - p$ the new node *copies* a link of its target u.

The probability of selecting a particular node in step (i) is $1/N$. Step (ii) is equivalent to selecting a node linked to a randomly selected link. The probability of selecting a degree-k node through this copying step (ii) is $k/2L$ for undirected networks. Combining (i) and (ii), the likelihood that a new node connects to a degree-k node follows

$$\Pi(k) = \frac{p}{N} + \frac{1-p}{2L}k$$

which, being linear in k, predicts a linear preferential attachment.

The popularity of the copying model lies in its relevance to real systems:

- *Social networks.* The more acquaintances an individual has, the higher is the chance that she will be introduced to new individuals by her existing acquaintances. In other words, we "copy" the friends of our friends. Consequently, without friends, it is difficult to make new friends.
- *Citation networks.* No scientist can be familiar with all the papers published on a certain topic. Authors decide what to read and cite by "copying" references from the papers they have read. Consequently, papers with more citations are more likely to be studied and cited again.
- *Protein interactions.* Gene duplication, responsible for the emergence of new genes in a cell, can be mapped into the copying model, explaining the scale-free nature of protein interaction networks [151, 152].

Taken together, we find that both the link-selection model and the copying model generate a linear preferential attachment through random linking.

5.9.4 Optimization

A long-standing assumption of economics is that humans make rational decisions, balancing cost against benefits. In other words, each individual aims to maximize their personal advantage. This is the starting point of rational choice theory in economics [153] and it is a hypothesis central to modern political science, sociology and philosophy. As we show below, such rational decisions can lead to preferential attachment [154, 155, 156].

Consider the Internet, whose nodes are routers connected via cables. Establishing a new Internet connection between two routers requires us to lay down a new cable between them. As this is costly, each new link is preceded by a careful cost–benefit analysis. Each new router (node) will choose its link to balance access to good network performance (i.e., proper bandwith) with the cost of laying down a new cable (i.e., physical distance). This can be a conflicting desire, as the closest node may not offer the best network performance.

For simplicity, let us assume that all nodes are located on a continent with the shape of a unit square. At each time step we add a new node and randomly choose a point within the square as its physical location. When deciding where to connect the new node i, we calculate the cost function [154]

$$C_i = \min_j[\delta d_{ij} + h_j] \tag{5.28}$$

which compares the cost of connecting to each node j already in the network. Here, d_{ij} is the Euclidean distance between the new node i and the potential target j, and h_j is the network-based distance of node j from the *first node* of the network, which we designate as the desireable "center" of the network (**Figure 5.15**), offering the best network performance. Hence, h_j captures the "resources" offered by node j, measured by its distance from the network's center.

The calculations indicate the emergence of three distinct network topologies, depending on the value of the parameter δ in **(5.28)** and N (**Figure 5.15**):

Star Network $\delta < (1/2)^{1/2}$
For $\delta = 0$ the Euclidean distances are irrelevant, hence each node links to the central node, turning the network into a star. We have a star configuration each time the h_j term dominates δd_{ij} in **(5.28)**.

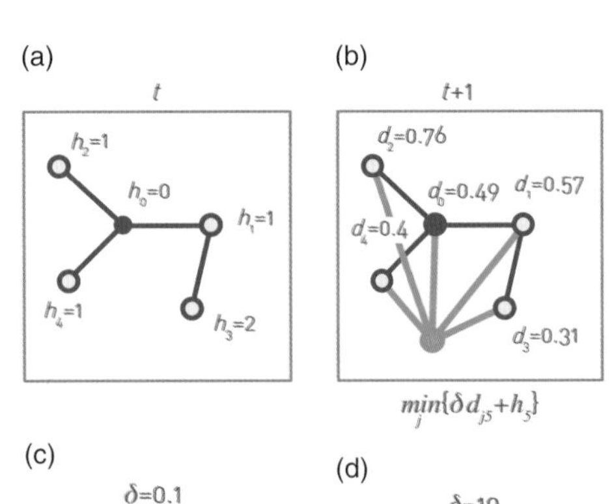

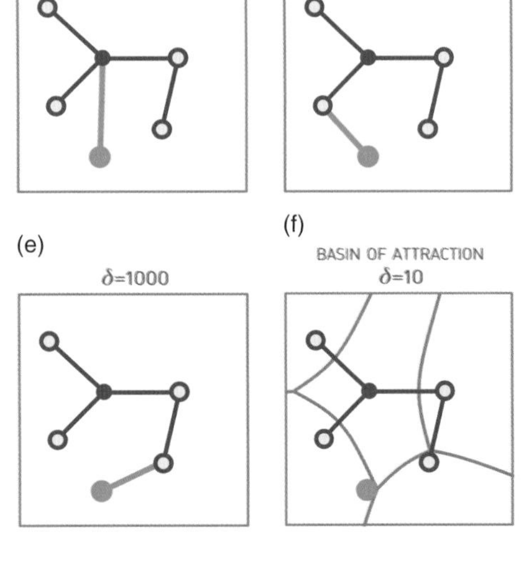

Figure 5.15 **Optimization Model**
(a) A small network, where the h_j term in the cost function **(5.28)** is shown for each node. Here, h_j represents the network-based distance of node j from node $i = 0$, designated as the "center" of the network, offering the best network performance. Therefore, $h_0 = 0$ and $h_3 = 2$.
(b) A new node (green) will choose the node j to which it connects by minimizing C_j of **(5.28)**.
(c)–(e) If δ is small, the new node will connect to the central node with $h_j = 0$. As we increase δ, the balance in **(5.28)** shifts, forcing the new node to connect to more distant nodes. Panels (c)–(e) show the choice of the new green node for different values of δ.
(f) The basin of attraction for each node for $\delta = 10$. A new node arriving inside a basin will always link to the node at the center of the basin. The size of each basin depends on the degree of the node at its center. Indeed, the smaller is h_j, the larger can be the distance from the new node while still minimizing **(5.28)**. Yet, the higher is the degree of node j, the smaller is its expected distance from the central node h_j.

Random Network $\delta \geq N^{1/2}$
For very large δ the contribution provided by the distance term δd_{ij} overwhelms h_j in **(5.28)**. In this case each new node connects to the node closest to it. The resulting network will have a bounded degree distribution, like a random network (**Figure 5.16b**).

Scale-Free Network $4 \leq \delta \leq N^{1/2}$
Numerical simulations and analytical calculations indicate that for intermediate δ values the network develops a scale-free topology [154]. The origin of the power-law distribution in this regime is rooted in two competing mechanisms:

(i) *Optimization.* Each node has a basin of attraction, so that nodes landing in this basin will always link to it. The size of each basin correlates with h_j of node j at its center, which in turn correlates with the node's degree k_j (**Figure 5.15f**).
(ii) *Randomness.* We choose the location of the new node randomly, ending in one of the N basins of attraction. The node with the largest degree has the largest basin of attraction, and hence gains the most new nodes and links. This leads to preferential attachment, as documented in **Figure 5.16d**.

In summary, we can build models that do not have an explicit $\Pi(k)$ function built into their definition, yet they generate a scale-free network. As we have shown in this section, these work by inducing preferential attachment. The mechanism responsible for preferential attachment can have two fundamentally different origins (**Figure 5.17**): it can be rooted in random processes, like link selection or copying, or in optimization, when new nodes balance conflicting criteria as they decide where to connect. Note that each of the mechanisms discussed above lead to linear preferential attachment, as assumed in the Barabási–Albert model. We are not aware of mechanisms capable of generating nonlinear preferential attachment, like those discussed in **Section 5.8**.

The diversity of the mechanisms discussed in this section suggest that linear preferential attachment is present in so many and such different systems precisely because it can come from both rational choice and random actions [157]. Most complex systems are driven by processes that have a bit of both. Hence, luck or reason, preferential attachment wins either way.

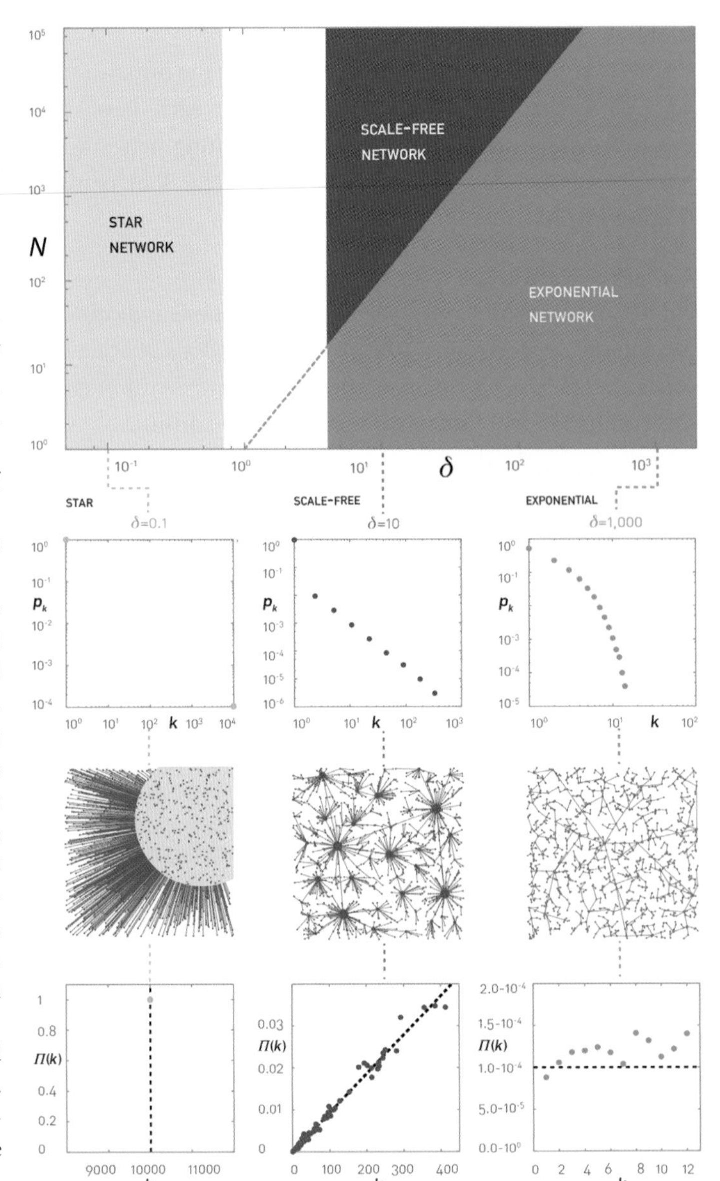

Figure 5.16 **Scaling in the Optimization Model**

(**a**) The three network classes generated by the optimization model: star, scale-free and exponential networks. The topology of the network in the unmarked area is unknown.

The vertical boundary of the star configuration is at $\delta = (1/2)^{1/2}$. This is the inverse of the maximum distance between two nodes on a square lattice with unit length, over which the model is defined. Therefore, if $\delta < (1/2)^{1/2}$, for any new node $\delta d_{ij} < 1$ and the cost **(5.28)** of connecting to the central node is $C_i = \delta d_{ij} + 0$, always lower than connecting to any other node at a cost of $f(i,j) = \delta d_{ij} + 1$. Therefore, for $\delta < (1/2)^{1/2}$ all nodes connect to node 0 (star-and-spoke network (c)).

The oblique boundary of the scale-free regime is $\delta = N^{1/2}$. Indeed, if nodes are placed randomly on the unit square, then the typical distance between neighbors decreases as $N^{-1/2}$. Hence, if $d_{ij} \sim N^{-1/2}$ then $\delta d_{ij} \geq h_{ij}$ for most node pairs. Typically the path length to the central node h_j grows slower than N (in small-world networks $h \sim \log N$). Therefore, C_i is dominated by the δd_{ij} term and the smallest C_i is achieved by minimizing the distance-dependent term. Note that, strictly speaking, the transition only occurs in the $N \longrightarrow \infty$ limit. In the white regime we lack an analytical form for the degree distribution.

(**b**) Degree distribution of networks generated in the three phases marked in (a) for $N = 10^4$.

(**c**) Typical topologies generated by the optimization model for selected δ values. Node size is proportional to its degree.

(**d**) We used the method described in **Section 5.6** to measure the preferential attachment function. Starting from a network with $N = 10{,}000$ nodes we added a new node and measured the degree of the node that it connected to. We repeated this procedure 10,000 times, obtaining $\Pi(k)$. The plots document the presence of linear preferential attachment in the scale-free phase, but its absence in the star and the exponential phases.

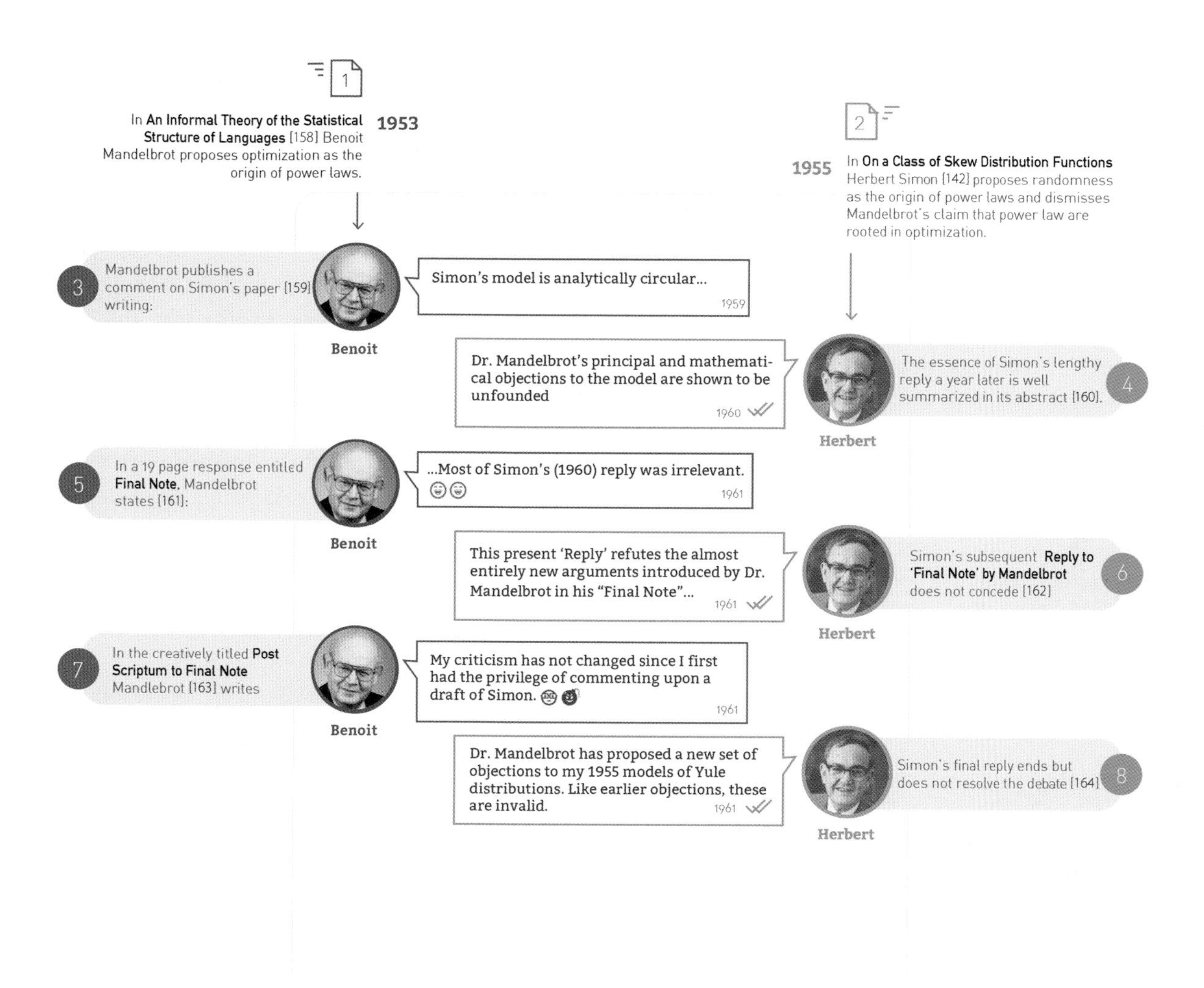

Figure 5.17 **Luck or Reason: An Ancient Fight**

The tension between randomness and optimization, two apparently antagonistic explanations for power laws, is by no means new. In the 1960s Herbert Simon and Benoit Mandelbrot engaged in a fierce public dispute over this very topic. Simon proposed that preferential attachment is responsible for the power-law nature of word frequencies. Mandelbrot fiercely defended an optimization-based framework. The debate spanned seven papers and two years and is one of the most vicious scientific disagreements on record.

In the context of networks today, the argument tilted in Simon's favor: the power laws observed in complex networks appear to be driven by randomness and preferential attachment. Yet the optimization-based ideas proposed by Mandelbrot play an important role in explaining the origins of preferential attachment. So, in the end they were both right.

5.10 Diameter and Clustering Coefficient

To complete the characterization of the Barabási–Albert model we discuss the behavior of the network diameter and the clustering coefficient.

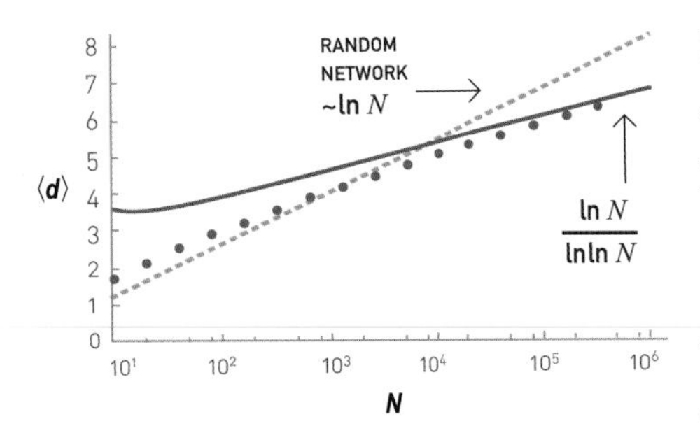

Figure 5.18 **Average Distance**
The dependence of the average distance on the system size in the Barabási–Albert model. The continuous line corresponds to the exact result **(5.29)**, while the dotted line corresponds to the prediction **(3.19)** for a random network. The analytical predictions do not provide the exact pre-factors, hence the lines are not fits but indicate only the predicted N-dependent trends. The results were averaged for ten independent runs for $m = 2$.

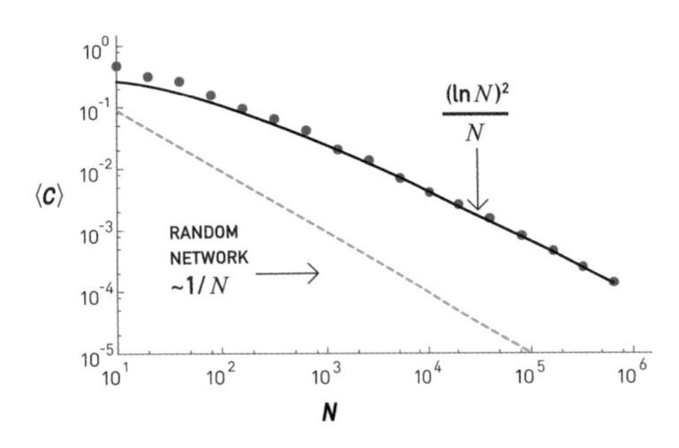

Figure 5.19 **Clustering Coefficient**
The dependence of the average clustering coefficient on the system size N for the Barabási–Albert model. The continuous line corresponds to the analytical prediction **(5.30)**, while the dotted line corresponds to the prediction for a random network, for which $\langle C \rangle \sim 1/N$. The results are averaged for ten independent runs for $m = 2$. The dashed and continuous curves are not fits, but are drawn to indicate the predicted N-dependent trends.

5.10.1 Diameter

The network diameter, representing the maximum distance in the Barabási–Albert network, follows for $m > 1$ and large N [120, 121]

$$\langle d \rangle \sim \frac{\ln N}{\ln \ln N}. \tag{5.29}$$

Therefore the diameter grows slower than $\ln N$, making the distances in the Barabási–Albert model smaller than the distances observed in a random graph of similar size. The difference is particularly relevant for large N.

Note that while **(5.29)** is derived for the diameter, the average distance $\langle d \rangle$ scales in a similar fashion. Indeed, as we show in **Figure 5.18**, for small N the $\ln N$ term captures the scaling of $\langle d \rangle$ with N, but for large N ($\geq 10^4$) the impact of the logarithmic correction $\ln \ln N$ becomes noticeable.

5.10.2 Clustering Coefficient

The clustering coefficient of the Barabási–Albert model follows (**Advanced topics 5.C**) [165]

$$\langle C \rangle \sim \frac{(\ln N)^2}{N}. \tag{5.30}$$

The prediction **(5.30)** is quite different from the $1/N$ dependence obtained for the random network model (**Figure 5.19**). The difference comes in the $(\ln N)^2$ term, which increases the clustering coefficient for large N. Consequently, the Barabási–Albert network is locally more clustered than a random network.

5.11 Summary

The most important message of the Barabási–Albert model is that network structure and evolution are inseparable. Indeed, in the Erdős–Rényi, Watts–Strogatz, configuration and hidden-parameter models the role of the modeler is to cleverly place the links between a *fixed number of nodes*. Returning to our earlier analogy, the networks generated by these models relate to real networks like a photo of a painting relates to the painting itself: it may look like the real one, but the process of generating a photo is drastically different from the process of painting the original painting. The aim of the Barabási–Albert model is to capture the processes that assemble a network in the first place. Hence, it aims to paint the painting again, coming as close as possible to the original brush strokes. Consequently, the modeling philosophy behind the model is simple: *to understand the topology of a complex system, we need to describe how it came into being.*

Random networks, the configuration and hidden-parameter models will continue to play an important role as we explore how certain network characteristics deviate from our expectations. Yet, if we want to explain the origin of a particular network property, we will have to use models that capture the system's genesis.

The Barabási–Albert model raises a fundamental question: Is the combination of growth and preferential attachment the real reason why networks are scale-free? We offered a *necessary* and *sufficient* argument to address this question. First, we showed that growth and preferential attachment are jointly needed to generate scale-free networks, hence if one of them is absent, either the scale-free property or stationarity is lost. Second, we showed that if they are both present, they do lead to scale-free networks. This argument leaves one possibility open, however: Do these two mechanisms explain the scale-free nature of *all* networks? Could there be some real networks that are scale-free thanks to some completely different mechanism? The answer is provided in **Section 5.9**, where we encountered the link-selection, copying and optimization models that do not have a preferential attachment function built into them, yet do lead to a scale-free network. We showed that they do so by generating a linear $\Pi(k)$. This finding underscores a more general pattern: to date, all known models and real systems that are scale-free have been found to have preferential attachment. Hence, the basic mechanisms of the Barabási–Albert model appear to capture the origin of their scale-free topology.

The Barabási–Albert model is unable to describe many characteristics of real systems:

- The model predicts $\gamma = 3$ while the degree exponent of real networks varies between 2 and 5 (**Table 4.2**).
- Many networks, like the WWW or citation networks, are directed, while the model generates undirected networks.
- Many processes observed in networks, from linking to already existing nodes to the disappearance of links and nodes, are absent from the model.

Box 5.4 At a Glance: Barabási–Albert Model

Number of Nodes

$$N = t$$

Number of Links

$$N = mt$$

Average Degree

$$\langle k \rangle = 2m$$

Degree Dynamics

$$k_i(t) = m(t/t_i)^\beta$$

Dynamical Exponent

$$\beta = 1/2$$

Degree Distribution

$$p_k \sim k^{-\gamma}$$

Degree Exponent

$$\gamma = 3$$

Average Distance

$$\langle d \rangle \sim \frac{\ln N}{\ln \ln N}$$

Clustering Coefficient

$$\langle C \rangle \sim (\ln N)^2/N$$

- The model does not allow us to distinguish between nodes based on some intrinsic characteristics, like the novelty of a research paper or the utility of a webpage.
- While the Barabási–Albert model is occasionally used as a model of the Internet or the cell, in reality it is not designed to capture the details of any particular real network. It is a minimal, proof-of-principle model whose main purpose is to capture the basic mechanisms responsible for the emergence of the scale-free property. Therefore, if we want to understand the evolution of systems like the Internet, the cell or the WWW, we need to incorporate the important details that contribute to the time evolution of these systems, like the directed nature of the WWW, the possibility of internal links or node and link removal.

As we show in **Chapter 6**, these limitations can be resolved systematically.

5.12 Homework

5.12.1 Generating Barabási–Albert Networks

With the help of a computer, generate a network with $N = 10^4$ nodes using the Barabási–Albert model with $m = 4$. Use as initial condition a fully connected network with $m = 4$ nodes.

(a) Measure the degree distribution at intermediate steps, namely when the network has 10^2, 10^3 and 10^4 nodes.
(b) Compare the distributions at these intermediate steps by plotting them together and fitting each to a power law with degree exponent γ. Do the distributions "converge"?
(c) Plot together the cumulative degree distributions at intermediate steps.
(d) Measure the average clustering coefficient as a function of N.
(e) Following **Figure 5.6a**, measure the degree dynamics of one of the initial nodes and of the nodes added to the network at time $t = 100$, $t = 1{,}000$ and $t = 5{,}000$.

5.12.2 Directed Barabási–Albert Model

Consider a variation of the Barabási–Albert model, where at each time step a new node arrives and connects with a directed link to a node chosen with probability

$$\Pi(k_i^{\text{in}}) = \frac{k_i^{\text{in}} + A}{\sum_j (k_j^{\text{in}} + A)}.$$

Here, k_i^{in} indicates the in-degree of node i and A is the same constant for all nodes. Each new node has m directed links.

(a) Calculate, using the rate equation approach, the in- and out-degree distribution of the resulting network.
(b) By using the properties of the Gamma and Beta functions, can you find a power-law scaling for the in-degree distribution?
(c) For $A = 0$ the scaling exponent of the in-degree distribution is different from $\gamma = 3$, the exponent of the Barabási–Albert model. Why?

5.12.3 Copying Model

Use the rate equation approach to show that the directed copying model leads to a scale-free network with incoming degree exponent $\gamma_{\text{in}} = \frac{2-p}{1-p}$.

5.12.4 Growth without Preferential Attachment

Derive the degree distribution **(5.18)** of Model A when a network grows by new nodes connecting randomly to m previously existing nodes. With the help of a computer, generate a network of 10^4 nodes using Model A. Measure the degree distribution and check that it is consistent with the prediction **(5.18)**.

5.13 Advanced Topics 5.A Deriving the Degree Distribution

A number of analytical techniques are available to calculate the exact form of the degree exponent **(5.11)**. Next we derive it using the rate-equation approach [145]. The method is sufficiently general to help explore the properties of a wide range of growing networks. Consequently, the calculations described here are of direct relevance for many systems, from models pertaining to the WWW [148, 149, 150] to describing the evolution of the protein interaction network via gene duplication [151, 152].

Let us denote by $N(k,t)$ the number of nodes with degree k at time t. The degree distribution $p_k(t)$ relates to this quantity via $p_k(t) = N(k,t)/N(t)$. Since at each time step we add a new node to the network, we have $N = t$. That is, at any moment the total number of nodes equals the number of time steps (**Box 5.2**).

We write preferential attachment as

$$\Pi(k) = \frac{k}{\sum_j k_j} = \frac{k}{2mt}, \tag{5.31}$$

where the $2m$ term captures the fact that in an undirected network each link contributes to the degree of two nodes. Our goal is to calculate the changes in the number of nodes with degree k after a new node is added to the network. For this we inspect the two events that alter $N(k,t)$ and $p_k(t)$ following the arrival of a new node:

(i) A new node can link to a degree-k node, turning it into a degree-(k+1) node, hence *decreasing* $N(k,t)$.
(ii) A new node can link to a degree-($k-1$) node, turning it into a degree-k node, hence *increasing* $N(k,t)$.

The number of links that are expected to connect to degree-k nodes after the arrival of a new node is

$$\frac{k}{2mt} \times Np_k(t) \times m = \frac{k}{2}p_k(t). \tag{5.32}$$

In **(5.32)** the first term on the l.h.s. captures the probability that the new node will link to a degree-k node (preferential attachment); the second term provides the total number of nodes with degree k, as the more nodes are in this category, the higher the chance that a new node will attach to one of them; the third term is the degree of the incoming node, as the higher is m, the higher is the chance that the new node will link to a degree-k node. We next apply **(5.32)** to cases (i) and (ii) above:

(i′) The number of degree-k nodes that acquire a new link and turn into degree-(k+1) nodes is

$$\frac{k}{2}p_k(t). \tag{5.33}$$

(ii′) The number of degree-($k-1$) nodes that acquire a new link, increasing their degree to k, is

$$\frac{k-1}{2}p_{k-1}(t). \tag{5.34}$$

Combining **(5.33)** and **(5.34)**, we obtain the expected number of degree-k nodes after the addition of a new node:

$$(N+1)p_k(t+1) = Np_k(t) + \frac{k-1}{2}p_{k-1}(t) - \frac{k}{2}p_k(t). \tag{5.35}$$

This equation applies to all nodes with degree $k > m$. As we lack nodes with degree $k = 0, 1, \ldots, m-1$ in the network (each new node arrives with degree m), we need a separate equation for degree-m modes. Following the same arguments we used to derive **(5.35)**, we obtain

$$(N+1)p_m(t+1) = Np_m(t) + \frac{m}{2}p_m(t). \tag{5.36}$$

Equations **(5.35)** and **(5.36)** are the starting point of the recursive process that provides p_k. Let us use the fact that we are looking for a stationary degree distribution, an expectation supported by numerical simulations (**Figure 5.6**). This means that in the $N = t \rightarrow \infty$ limit, $p_k(\infty) = p_k$. Using this we can write the l.h.s. of **(5.35)** and **(5.36)** as

$$\begin{aligned}
&(N+1)p_k(t+1) - Np_k(t) \rightarrow Np_k(\infty) - p_k(\infty) - Np_k(\infty) \\
&\quad = p_k(\infty) = p_k, \\
&(N+1)p_m(t+1) - Np_m(t) \rightarrow p_m.
\end{aligned}$$

Therefore the rate equations **(5.35)** and **(5.36)** take the form

$$p_k = \frac{k-1}{k+2}p_{k-1} \quad k > m, \tag{5.37}$$

$$p_m = \frac{2}{m+2}. \tag{5.38}$$

Note that **(5.37)** can be rewritten as

$$p_{k+1} = \frac{k}{k+3}p_k \tag{5.39}$$

via a $k \rightarrow k+1$ variable change.

We use a recursive approach to obtain the degree distribution. That is, we write the degree distribution for the smallest degree, $k = m$, using **(5.38)** and then use **(5.39)** to calculate p_k for the higher degrees:

$$p_{m+1} = \frac{m}{m+3} p_m = \frac{2m}{(m+2)(m+3)},$$

$$p_{m+2} = \frac{m+1}{m+4} p_{m+1} = \frac{2m(m+1)}{(m+2)(m+3)(m+4)}, \qquad (5.40)$$

$$p_{m+3} = \frac{m+2}{m+5} p_{m+2} = \frac{2m(m+1)}{(m+3)(m+4)(m+5)}.$$

At this point we notice a simple recursive pattern: by replacing $(m+3)$ with k in the denominator we obtain the probability of observing a node with degree k:

$$p_k = \frac{2m(m+1)}{k(k+1)(k+2)}, \qquad (5.41)$$

which represents the exact form of the degree distribution of the Barabási–Albert model.

Note that:

- For large k **(5.41)** becomes $p_k \sim k^{-3}$, in agreement with the numerical result.
- The pre-factor of **(5.11)** or **(5.41)** is different from the pre-factor of **(5.9)**.
- This form was derived independently in [13] and [145], and the exact mathematical proof of its validity is provided in [15].

Finally, the rate equation formalism offers an elegant continuum equation satisfied by the degree distribution [148]. Starting from the equation

$$p_k = \frac{k-1}{2} p_{k-1} - \frac{k}{2} p_k, \qquad (5.42)$$

we can write

$$2p_k = (k-1)p_{k-1} - kp(k) = -p_{k-1} - k[p_k - p_{k-1}], \qquad (5.43)$$

$$2p_k = -p_{k-1} - k\frac{p_k - p_{k-1}}{k-(k-1)} \approx -p_{k-1} - k\frac{\partial p_k}{\partial k}, \qquad (5.44)$$

obtaining

$$p_k = -\frac{1}{2}\frac{\partial [kp_k]}{\partial k}. \qquad (5.45)$$

One can check that the solution of **(5.45)** is

$$p_k \sim k^{-3}. \qquad (5.46)$$

5.14 Advanced Topics 5.B Nonlinear Preferential Attachment

In this section we derive the degree distribution of the nonlinear Barabási–Albert model, governed by the preferential attachment **(5.22)**. We follow [145], but adjust the calculation to cover $m > 1$.

Strictly speaking, a stationary degree distribution only exists if $\alpha \leq 1$ in **(5.22)**. For $\alpha > 1$ a few nodes attract a finite fraction of links, as explained in **Section 5.7**, and we do not have a time-independent p_k. Therefore, we limit ourselves to the $\alpha \leq 1$ case.

We start with the nonlinear Barabási–Albert model, in which at each time step a new node is added with m new links. We connect each new link to an existing node with probability

$$\Pi(k_i) = \frac{k_i^\alpha}{M(\alpha, t)}, \tag{5.47}$$

where k_i is the degree of node i, $0 < \alpha \leq 1$,

$$M(\alpha, t) = t\sum k^\alpha p_k(t) = t\mu(\alpha, t) \tag{5.48}$$

is the normalization factor and $t = N(t)$ represents the number of nodes. Note that $\mu(0,t) = \sum p_k(t) = 1$ and $\mu(1,t) = \sum_k kp_k(t) = \langle k \rangle = 2mt/N$ is the average degree. Since $0 < \alpha \leq 1$,

$$\mu(0, t) \leq \mu(\alpha, t) \leq \mu(1, t). \tag{5.49}$$

Therefore, in the long time limit

$$\mu(\alpha, t \rightarrow \infty) = \text{constant}, \tag{5.50}$$

whose precise value will be calculated later. For simplicity, we adopt the notation $\mu \equiv \mu(\alpha, t \rightarrow \infty)$.

Following the rate-equation approach introduced in **Advanced topics 5.A**, we write the rate equation for the network's degree distribution as

$$(t+1)p_k(t+1) = tp_k(t) + \frac{m}{\mu(\alpha,t)}[(k-1)^\alpha p_{k-1}(t) - k^\alpha p_k(t)] + \delta_{k,m}. \tag{5.51}$$

The first term on the r.h.s. describes the rate at which nodes with degree $(k-1)$ gain new links; the second term describes the loss of degree-k nodes when they gain new links, turning into degree-(k+1) nodes; the last term represents the newly added nodes with degree m.

Asymptotically, in the $t \to \infty$ limit we can write $p_k = p_k(t+1) = p_k(t)$. Substituting $k = m$ in **(5.51)**, we obtain

$$p_m = -\frac{m}{\mu} - m^{\alpha} p_m + 1,$$
$$p_m = -\frac{\mu/m}{\mu/m + m^{\alpha}}. \tag{5.52}$$

For $k > m$,

$$p_k = \frac{m}{\mu}[(k-1)^{\alpha} p_{k-1} - k^{\alpha} p_k], \tag{5.53}$$

$$p_k = \frac{(k-1)^{\alpha}}{\mu/m + k^{\alpha}} p_{k-1}. \tag{5.54}$$

Solving **(5.53)** recursively, we obtain

$$p_m = \frac{\mu/m}{\mu/m + m^{\alpha}}, \tag{5.55}$$

$$p_{m+1} = \frac{m^{\alpha}}{\mu/m + (m+1)} \frac{\mu/m}{\mu/m + m^{\alpha}}, \tag{5.56}$$

$$p_k = \frac{\mu/m}{k^{\alpha}} \prod_{j=m}^{k} \left(1 + \frac{\mu/m}{j^{\alpha}}\right)^{-1}. \tag{5.57}$$

To determine the large-k behavior of p_k, we take the logarithm of **(5.57)**:

$$\ln p_k = \ln(\mu/m) - \alpha \ln k - \sum_{j=m}^{k} (1 + \frac{\mu/m}{j^{\alpha}}). \tag{5.58}$$

Using the series expansion $\ln(1+x) = \sum^{\infty}(1-)^{n+1}/n \cdot x^n$, we obtain

$$\ln p_k = \ln(\mu/m) - \alpha \ln k - \sum_{i=m}^{k} \sum_{n=1}^{\infty} \frac{(-1)^{n+1}}{n} (\mu/m)^n j^{-n\alpha}. \tag{5.59}$$

We approximate the sum over j with the integral

$$\sum_{j=m}^{k} j_X^{-n\alpha} \approx \int_m^k x^{-n\alpha} dx = \frac{1}{1 - n\alpha} (k^{1-n\alpha} - m^{1-n\alpha}), \tag{5.60}$$

which in the special case of $n\alpha = 1$ becomes

$$\sum_{j=m}^{k} j^{-1} \approx \int_{m}^{k} x^{-1} dx = \ln k - \ln m. \tag{5.61}$$

Hence we obtain

$$\begin{aligned}\ln p_k &= \ln(\mu/m) - \alpha \ln k \\ &- \sum_{n=1}^{\infty} \frac{(-1)^{n+1}}{n} \frac{(\mu/m)^n}{1-n\alpha} (k^{1-n\alpha} - m^{1-n\alpha}).\end{aligned} \tag{5.62}$$

Consequently, the degree distribution has the form

$$p_k = C_\alpha k^{-\alpha} e^{-\sum_{\infty}^{n=1} \frac{(-1)^{n+1}}{n} \frac{(\mu+m)^n}{1-n\alpha} k^{1-n\alpha}}, \tag{5.63}$$

where

$$C_\alpha = \frac{\mu}{m} e^{\sum_{\infty}^{n=1} \frac{(-1)^{n+1}}{n} \frac{(\mu/m)^n}{1-n\alpha} m^{1-n\alpha}}. \tag{5.64}$$

The vanishing terms in the exponential do not influence the $k \to \infty$ asymptotic behavior, being relevant only if $1 - n\alpha \geq 1$. Consequently, p_k depends on α as

$$p_k \sim \begin{cases} k^{-\alpha} e^{\frac{-\mu/m}{1-\alpha} k^{1-\alpha}} & 1/2 < \alpha < 1 \\ k^{-\frac{1}{2}+\frac{1}{2}\left(\frac{\mu}{m}\right)^2} e^{-\frac{1\mu}{2m} k^{-2}} & \alpha = 1/2 \\ k^{-\alpha} e^{-\frac{\mu/m}{1-\alpha} k^{1-\alpha} + \frac{1}{2} \frac{(\mu/m)^2}{1-2\alpha} k^{1-2\alpha}} & 1/3 < \alpha < 1/2. \\ \vdots & \end{cases} \tag{5.65}$$

That is, for $1/2 < \alpha < 1$ the degree distribution follows a stretched exponential. As we lower α, new corrections start contributing each time α becomes smaller than $1/n$, where n is an integer.

For $\alpha \to 1$ the degree distribution scales as k^{-3}, as expected for the Barabási–Albert model. Indeed, for $\alpha = 1$ we have $\mu = 2$, and

$$\lim_{\alpha \to 1} \frac{k^{1-\alpha}}{1-\alpha} = \ln k. \tag{5.66}$$

Therefore, $p_k \sim k^{-1} \exp(-2 \ln k) = k^{-3}$.

Finally, we calculate $\mu(\alpha) = \sum_j j^\alpha p_j$. For this we write the sum **(5.58)** as

$$\sum_{k=m}^{\infty} k^{\alpha} p_k = \sum_{k=m}^{\infty} \frac{\mu(\alpha)}{m} \prod_{j=m}^{k} \left(1 + \frac{\mu(\alpha)/m}{j^{\alpha}}\right)^{-1}, \tag{5.67}$$

$$1 = \frac{1}{m} \sum_{k=m}^{\infty} \prod_{i=m}^{k} \left(1 + \frac{\mu(\alpha)/m}{j^{\alpha}}\right)^{-1}. \tag{5.68}$$

We obtain $\mu(\alpha)$ by solving **(5.68)** numerically.

5.15 Advanced Topics 5.C The Clustering Coefficient

In this section we derive the average clustering coefficient **(5.30)** for the Barabási–Albert model. The derivation follows an argument proposed by Klemm and Eguiluz [165], supported by the exact calculation of Bollobás [71].

We aim to calculate the number of triangles expected in the model, which can be linked to the clustering coefficient (**Section 2.10**). We denote by $P(i,j)$ the probability of having a link between node i and node j. Therefore, the probability that three nodes i, j, l form a triangle is $P(i,j)P(i,l)P(j,l)$. The expected number of triangles in which node l with degree k_l participates is thus given by the sum of the probabilities that node l participates in triangles with arbitrarily chosen nodes i and j in the network. We can use the continuous-degree approximation to write

$$Nr_l(\triangleleft) = \int_{i=1}^{N} di \int_{j=1}^{N} dj P(i,j)P(i,l)P(j,l). \tag{5.69}$$

To proceed we need to calculate $P(i,j)$, which requires us to consider how the Barabási–Albert model evolves. Let us denote by $t_j = j$ the time when node j arrived, which we can do as in each time step we added only one new node (event time, **Box 5.2**). Hence the probability that at its arrival node j links to node i with degree k_i is given by preferential attachment:

$$P(i,j) = m\Pi(k_i(j)) = m \frac{k_i(j)}{\sum_{l=1}^{j} k_l(j)} = m \frac{k_i(j)}{2mj}. \tag{5.70}$$

Using **(5.7)**, we can write

$$k_i(t) = m\left(\frac{t}{t_i}\right)^{\frac{1}{2}} = m\left(\frac{j}{i}\right)^{\frac{1}{2}}, \tag{5.71}$$

where we have used the fact that the arrival time of node j is $t_j = j$ and the arrival time of node i is $t_i = i$. Hence, **(5.70)** now becomes

$$P(i,j) = \frac{m}{2}(ij)^{-\frac{1}{2}}. \tag{5.72}$$

Using this result we calculate the number of triangles in **(5.69)**, writing

$$\begin{aligned} Nr_l(\triangleleft) &= \int_{i=1}^{N} di \int_{j=1}^{N} dj P(i,j)P(i,l)P(j,l) \\ &= \frac{m^3}{8}\int_{i=1}^{N} di \int_{j=1}^{N} dj (ij)^{-\frac{1}{2}}(il)^{-\frac{1}{2}}(jl)^{-\frac{1}{2}} \\ &= \frac{m^3}{8l}\int_{i=1}^{N} \frac{di}{i}\int_{j=1}^{N} \frac{dj}{j} = \frac{m^3}{8l}(\ln N)^2. \end{aligned} \tag{5.73}$$

The clustering coefficient can be written as $C_l = \frac{2Nr_l(\triangleleft)}{k_l(k_l-1)}$, hence we obtain

$$C_l = \frac{\frac{m^3}{4l}(\ln N)^2}{k_l(N)\,(k_l(N)-1)}. \tag{5.74}$$

To simplify **(5.74)**, we note that according to **(5.7)** we have

$$k_l(N) = m\left(\frac{N}{l}\right)^{\frac{1}{2}}, \tag{5.75}$$

which is the degree of node l at time $t = N$. Hence, for large k_l we have

$$k_l(N)\,(k_l(N)-1) \approx k_l^2(N) = m^2\frac{N}{l}, \tag{5.76}$$

allowing us to write the clustering coefficient of the Barabási–Albert model as

$$C_l = \frac{m}{4}\frac{(\ln N)^2}{N}, \tag{5.77}$$

which is independent of l, therefore we obtain the result **(5.30)**.

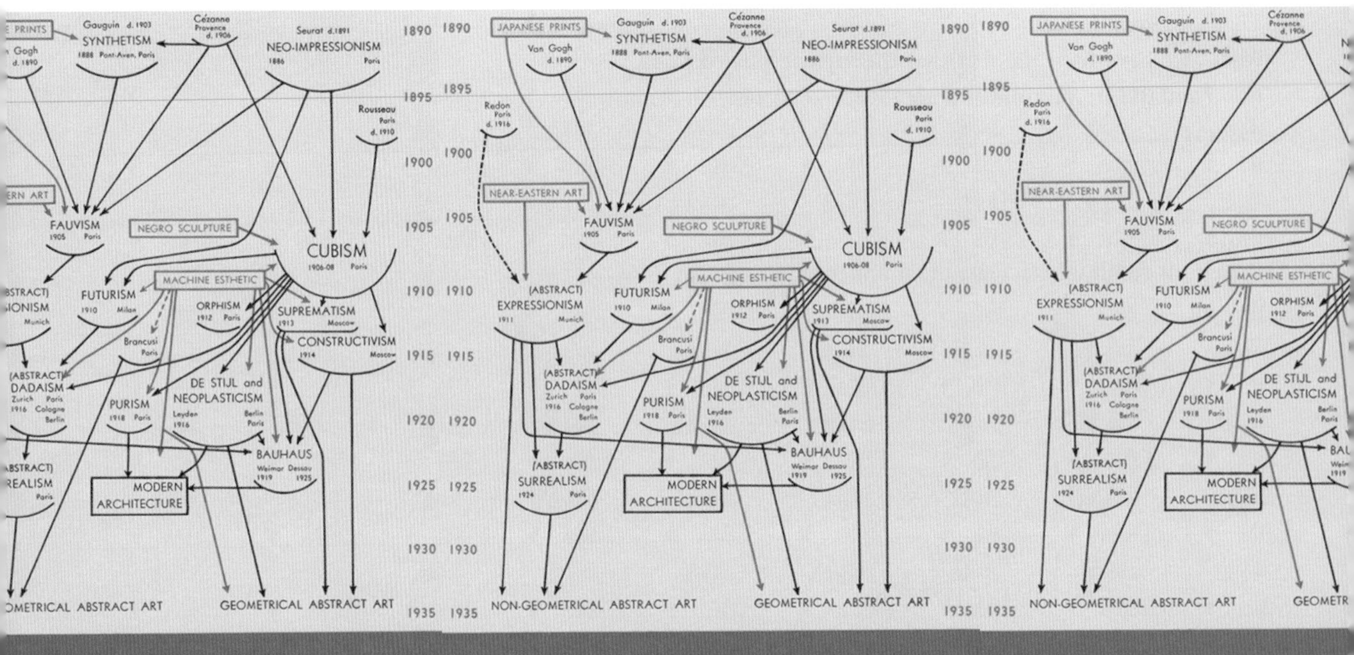

Figure 6.0 **Art & Networks: Alfred Barr**

Alfred H. Barr Jr. (1902–1981), an American art historian and the first director of the Museum of Modern Art (MoMA) in New York City, was one of the most influential forces in the development of popular attitudes toward modern art. This 1936 chart by Barr uses a network framework to illustrate the development and cross-currents of modern art. It appeared on the dust jacket of the catalogue for *Cubism and Abstract Art*, the movement's first major exhibition at MoMA.

CHAPTER 6

Evolving Networks

6.1 Introduction

Founded six years after the birth of the WWW, Google was a latecomer to search. By the late 1990s Alta Vista and Inktomi, two search engines with an early start, dominated the search market. Yet the third mover Google soon not only became the leading search engine, but also acquired links at such an incredible rate that by 2000 it became the biggest hub of the Web as well [50]. But it didn't last: in 2011 Facebook, a youngster by Google's standards, took over as the Web's biggest node.

The Web's competitive landscape highlights an important limitation of our modeling framework: none of the network models we have encountered so far are able to account for it. Indeed, in the Erdős–Rényi model the identity of the biggest node is driven entirely by chance. The Barabási–Albert model offers a more realistic picture, predicting that each node increases its degree following $k(t) \sim t^{1/2}$. This means that the oldest node always has the most links, a phenomenon called the *first-mover's advantage* in the business literature. It also means that a late node can never become the largest hub.

Figure 6.1 **Garment District**
The Garment District is a Manhattan neighborhood located between Fifth and Ninth Avenues, from 34th to 42nd Street. Since the early twentieth century it has been the center for fashion manufacturing and design in the United States. The *Needle threading a button* and the *Jewish tailor*, two sculptures located in the heart of the district, pay tribute to the neighborhood's past.

The garment industry of New York City offers a prominent example of a declining network, helping us understand how the loss of nodes shapes a network's topology (**Box 6.5**). Uncovering the impact of processes like node and link loss on the network topology is one of the goals of this chapter.

In reality, the growth rate of a node does not depend on its age alone. Instead webpages, companies or actors have intrinsic qualities that influence the rate at which they acquire links. Some show up late and nevertheless grab an extraordinary number of links within a short time frame. Others rise early yet never quite make it. The goal of this chapter is to understand how the differences in the node's ability to acquire links affect the network topology. Going beyond this competitive landscape, we also explore how other processes, like node and link deletion (**Figure 6.1**) or the aging of nodes, phenomena frequently observed in real networks, change the way networks evolve and alter their topology. Our goal is to develop a self-consistent theory of evolving networks that can be adjusted at will to predict the dynamics and the topology of a wide range of real networks.

6.2 The Bianconi–Barabási Model

Some people have a knack for turning each random encounter into a lasting social link; some companies turn each consumer into a loyal partner; some webpages turn visitors into addicts. A common feature of these successful nodes is some intrinsic property that propels them ahead of the pack. We will call this property *fitness*.

Fitness is an individual's gift for turning a random encounter into a lasting friendship; it is a company's knack of acquiring consumers relative to its competition; it is a webpage's ability to bring us back on a daily basis despite the many other pages that compete for our attention. Fitness may have genetic roots in people, it may be related to innovativeness and management quality in companies and may depend on the content offered by a website (**Online resource 6.1**).

In the Barabási–Albert model we assumed that a node's growth rate is determined solely by its degree. To incorporate the role of fitness we assume that preferential attachment is driven by the product of a node's fitness, η, and its degree, k. The resulting model, called the *Bianconi–Barabási* or the *fitness model,* consists of the following two steps [166, 167]:

- **Growth**
  In each time step a new node j with m links and fitness η_j is added to the network, where η_j is a random number chosen

from a *fitness distribution* $\rho(\eta)$. Once assigned, a node's fitness does not change.

- **Preferential Attachment**
  The probability that a link of a new node connects to node i is proportional to the product of node i's degree k_i and its fitness η_i,

$$\Pi_i = \frac{\eta_i k_i}{\sum_j \eta_j k_j}. \tag{6.1}$$

In **(6.1)** the dependence of Π_i on k_i captures the fact that higher-degree nodes have more visibility, hence we are more likely to link to them. The dependence of Π_i on η_i implies that between two nodes with the same degree, the one with higher fitness is selected with a higher probability. Hence, **(6.1)** ensures that even a relatively young node, with initially only a few links, can acquire links rapidly if it has larger fitness than the rest of the nodes.

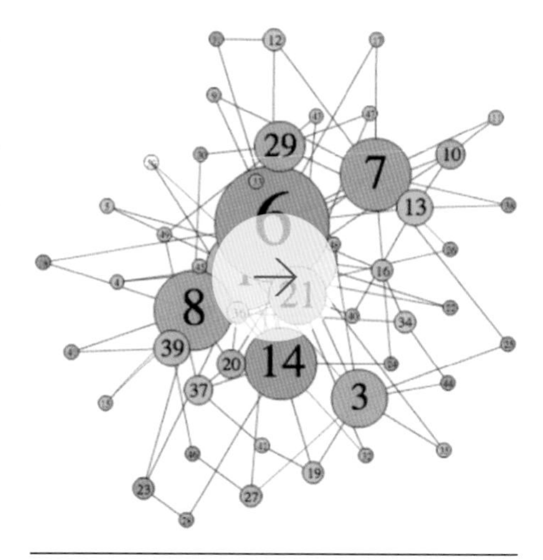

Online Resource 6.1

The Bianconi–Barabási Model

The movie shows a growing network in which each new node acquires a randomly chosen fitness parameter at birth, indicated by the color of the node. Each new node chooses the nodes it links to following generalized preferential attachment **(6.1)**, making a node's growth rate proportional to its fitness. The node size is proportional to its degree, illustrating that with time the nodes with the highest fitness turn into the largest hubs. Video courtesy of Dashun Wang.

6.2.1 Degree Dynamics

We can use the continuum theory to predict each node's temporal evolution. According to **(6.1)**, the degree of node i changes at the rate

$$\frac{\partial k_i}{\partial t} = m \frac{\eta_i k_i}{\sum_j \eta_j k_j}. \tag{6.2}$$

Let us assume that the time evolution of k_i follows a power law with a fitness-dependent exponent $\beta(\eta_i)$ (**Figure 6.2**),

$$k(t, t_i, \eta_i) = m\left(\frac{t}{t_i}\right)^{\beta(\eta_i)}. \tag{6.3}$$

Inserting **(6.3)** into **(6.2)** we find that the *dynamic exponent* satisfies (**Advanced topics 6.A**)

$$\beta(\eta) = \frac{\eta}{C}, \tag{6.4}$$

with

$$C = \int \rho(\eta) \frac{\eta}{1-\beta(\eta)} d\eta. \tag{6.5}$$

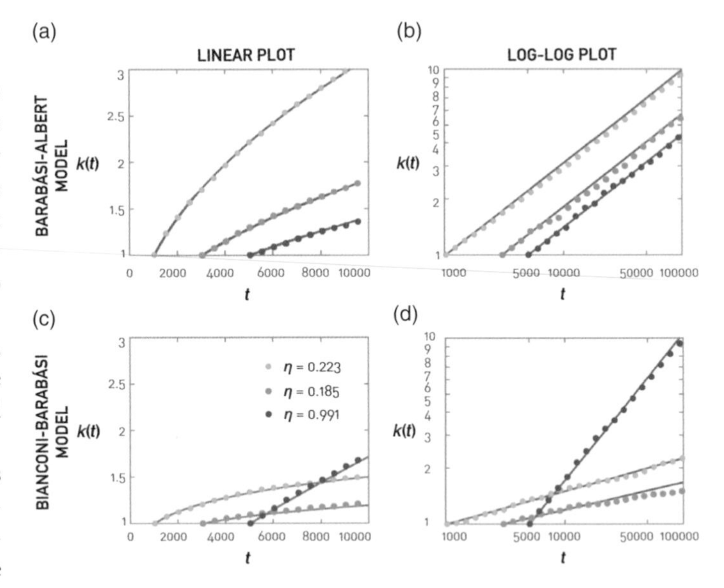

Figure 6.2 **Competition in the Bianconi–Barabási Model**

(**a**) In the Barabási–Albert model all nodes increase their degree at the same rate, hence the earlier a node joins the network, the larger is its degree at any time. The figure shows the time-dependent degree of nodes that arrived at different times ($t_i = 1{,}000, 3{,}000, 5{,}000$), demonstrating that the later nodes are unable to pass the earlier nodes [168].

(**b**) The same as in (a) but on a log–log plot, demonstrating that each node follows the same growth law **(5.7)** with identical dynamical exponents $\beta = 1/2$.

(**c**) In the Bianconi–Barabási model nodes increase their degree at a rate that is determined by their individual fitness. Hence a latecomer node with a higher fitness (purple symbols) can overcome the earlier nodes.

(**d**) The same as in (c) but on a log–log plot, demonstrating that each node increases its degree following a power law with its own fitness-dependent dynamical exponent β, as predicted by **(6.3)** and **(6.4)**.

In (a)–(d) each curve corresponds to the average over 100 independent runs using the same fitness sequence.

In the Barabási–Albert model we have $\beta = 1/2$, hence the degree of each node increases as the square root of time. According to **(6.4)**, in the Bianconi–Barabási model the dynamic exponent is proportional to the node's fitness, η, hence each node has its own dynamic exponent. Consequently, a node with higher fitness will increase its degree faster. Given sufficient time, the fitter node will leave behind nodes with a smaller fitness (**Figure 6.2**). Facebook is a poster child of this phenomenon: a latecomer with an addictive product, it acquired links faster than its competitors, eventually becoming the Web's biggest hub.

6.2.2 Degree Distribution

The degree distribution of the network generated by the Bianconi–Barabási model can be calculated using the continuum theory (**Advanced topics 6.A**), obtaining

$$p_k \approx C \int \frac{\rho(\eta)}{\eta} \left(\frac{m}{k}\right)^{\frac{C}{\eta}+1} d\eta. \tag{6.6}$$

Equation **(6.6)** is a weighted sum of multiple power laws, indicating that p_k depends on the precise form of the fitness distribution, $\rho(\eta)$. To illustrate the properties of the model we use **(6.4)** and **(6.6)** to calculate $\beta(\eta)$ and p_k for two fitness distributions:

- **Equal Fitnesses**
  When all fitnesses are equal, the Bianconi–Barabási model reduces to the Barabási–Albert model. Indeed, let us use $\rho(\eta) = \delta(\eta - 1)$, capturing the fact that each node has the same fitness $\eta = 1$. In this case **(6.5)** yields $C = 2$. Using **(6.4)** we obtain $\beta = 1/2$ and **(6.6)** predicts $p_k \sim k^{-3}$, the known scaling of the degree distribution in the Barabási–Albert model.
- **Uniform Fitness Distribution**
  The model's behavior is more interesting when nodes have different fitnesses. Let us choose η to be uniformly distributed in the $[0, 1]$ interval. In this case C is the solution of the transcendental equation **(6.5),**

$$\exp(-2/C) = 1 - 1/C, \tag{6.7}$$

whose numerical solution is $C^* = 1.255$. Consequently, **(6.4)** predicts that each node i has a different dynamic exponent, $\beta(\eta_i) = \eta_i/C^*$.

Using **(6.6)** we obtain

$$p_k \sim \int_0^1 \frac{C^*}{\eta} \frac{1}{k^{1+C*/\eta}} d\eta \sim \frac{k^{-(1+C^*)}}{\ln k}, \tag{6.8}$$

predicting that the degree distribution follows a power law with degree exponent $\gamma = 2.255$. We do not expect a perfect power law, and the scaling is affected by an inverse logarithmic correction $1/\ln k$.

Numerical support for the above predictions is provided in **Figures 6.2** and **6.3**. The simulations confirm that $k_i(t)$ follows a power law for each η and that the dynamical exponent $\beta(\eta)$ increases with the fitness η. As **Figure 6.3a** indicates, the measured dynamical exponents are in excellent agreement with the prediction **(6.4)**. **Figure 6.3b** also documents an agreement between **(6.8)** and the numerically obtained degree distribution.

In summary, the Bianconi–Barabási model can account for the fact that nodes with different internal characteristics acquire links at different rates. It predicts that a node's growth rate is determined by its fitness η and allows us to calculate the dependence of the degree distribution on the fitness distribution $\rho(\eta)$.

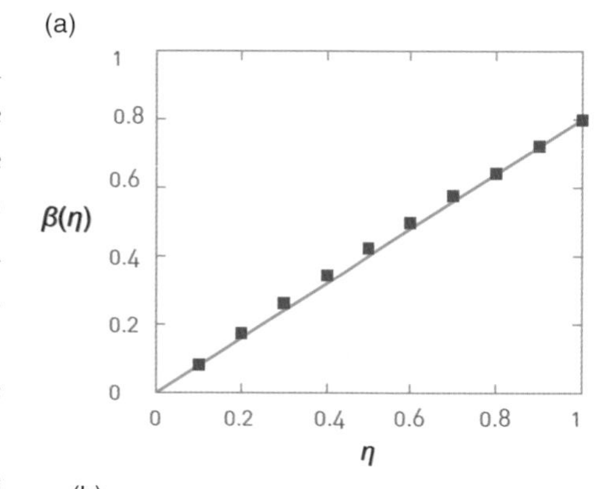

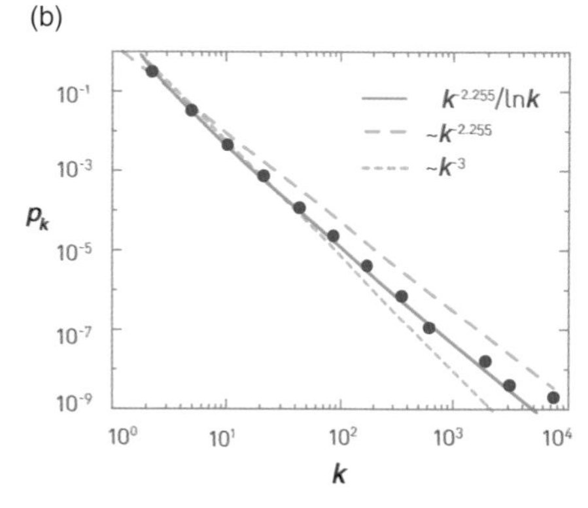

Figure 6.3 **Characterizing the Bianconi–Barabási Model**

(a) The measured dynamic exponent $\beta(\eta)$ shown as a function of η for a uniform $\rho(\eta)$ distribution. The squares were obtained from numerical simulations, while the solid line corresponds to the analytical prediction $\beta(\eta) = \eta/1.255$.

(b) Degree distribution of the model obtained numerically for a network with $m = 2$ and $N = 10^6$ and fitnesses chosen uniformly from the $\eta \in [0, 1]$ interval. The green solid line corresponds to the prediction **(6.8)** with $\gamma = 2.255$. The long-dashed line shows $p_k \sim k^{-2.255}$ without the logarithmic correction, while the short-dashed curve corresponds to $p_k \sim k^{-3}$, expected if all fitnesses are equal. Note that the best fit is provided by **(6.8)**.

6.3 Measuring Fitness

Measuring a node's fitness could help us identify websites that are poised to grow in visibility, research papers that will become influential or actors on their way to stardom (**BOX 6.1**). Yet, our ability to determine the fitness is prone to errors. Consider the challenge of assigning fitness to a webpage on sumo wrestling: while a small segment of the population might find sumo wrestling fascinating, most individuals are indifferent to it and some might even find it odd. Hence, different individuals will inevitably assign different fitnesses to the same node.

According to **(6.1)**, fitness is not assigned by any individual but reflects the network's *collective perception of a node's importance relative to other nodes.* We can, therefore, determine a node's fitness by comparing its time evolution with the time evolution of other nodes in the network. In this section we show that if we have dynamical information about the evolution of the individual nodes, the quantitative framework of the Bianconi–Barabási model allows us to determine the fitness of each node.

To relate a node's growth rate to its fitness we take the logarithm of **(6.3)**,

$$\ln k(t, t_i, \eta_i) = \beta(\eta_i)\ln t + B_i \tag{6.9}$$

where $B_i = \ln(m/t_i^{\beta(\eta_i)})$ is a time-independent parameter. Hence, the slope of $\ln k(t, t_i, \eta_i)$ is a linear function of the dynamical exponent $\beta(\eta_i)$. In turn, $\beta(\eta_i)$ depends linearly on η_i according to **(6.4)**. Therefore, if we can track the time evolution of the degree for a large number of nodes, the distribution of the dynamical exponent $\beta(\eta_i)$ will be identical with the fitness distribution $\rho(\eta)$.

Box 6.1 The Genetic Origins of Fitness

Could fitness, an ability to acquire friends in a social network, have genetic origins? To answer this researchers examined the social network of 1,110 school-age twins [169, 170], using a technique developed to identify the heritability of traits and behaviors. They found that:

- Genetic factors account for 46% of the variation in a student's in-degree (i.e., the number of students that name a given student as a friend).
- Generic factors are not significant for out-degrees (i.e., the number of students a particular student names as friends).

This suggests that an individual's ability to acquire links, or its fitness, is heritable. In other words, in social networks fitness has genetic origins. This conclusion is also supported by research associating a particular genetic trait with variations in popularity [171].

6.3.1 The Fitness of a Web Document

Node fitnesses were systematically measured in the context of the WWW, relying on a dataset that crawled monthly the links of about 22 million web documents for 13 months [172]. While most nodes (documents) did not change their degree during this time frame, 6.5% of nodes showed sufficient changes to determine their dynamical exponent via **(6.9)**. The obtained fitness distribution $\rho(\eta)$ has an exponential form (**Figure 6.4**), indicating that high-fitness nodes are rare.

The shape of the obtained fitness distribution is somewhat unexpected, as one would be tempted to assume that on the Web, fitness varies widely. For example, Google is far more attractive to Web surfers than my personal webpage. Yet the exponential form of $\rho(\eta)$ indicates that the fitness of web documents is bounded, varying in a relatively narrow range. Consequently, the observed large differences in the degree of two web documents is generated by the system's dynamics: growth and preferential attachment amplify small fitness differences, turning nodes with slightly higher fitness into much bigger nodes.

To illustrate this amplification, consider two nodes that arrived at the same time, but have different fitnesses $\eta_2 > \eta_1$. According to **(6.3)** and **(6.4)**, the relative difference between their degrees grows for large t as

$$\frac{k_2 - k_1}{k_1} \sim t^{\frac{\eta_2 - \eta_1}{C}}. \quad (6.10)$$

While the difference between η_2 and η_1 may be small, far into the future (large t) the relative difference between their degrees can become quite significant.

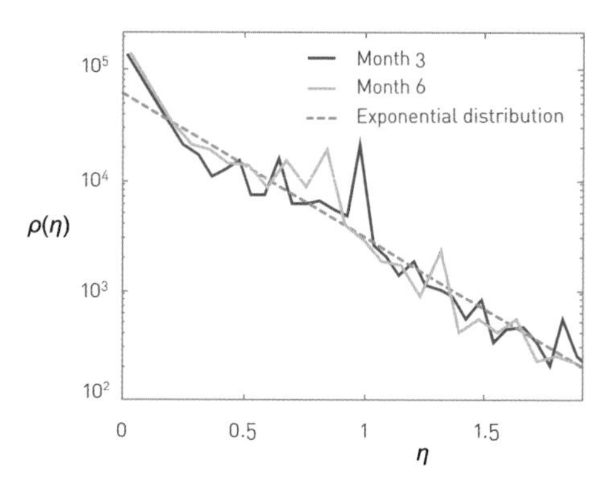

Figure 6.4 **The Fitness Distribution of the WWW**

The fitness distribution obtained by measuring the time evolution of a large number of web documents. The measurements indicate that each node's degree has a power-law time dependence, as predicted by **(6.3)**. The slope of each curve is $\beta(\eta_j)$, which corresponds to the node's fitness η_i up to a multiplicative constant according to **(6.4)**. The plot shows the result of two measurements based on datasets recorded three months apart, demonstrating that the fitness distribution is time independent. The dashed line suggests that the fitness distribution is well approximated by an exponential. After [172].

6.3.2 The Fitness of a Scientific Publication

In some networks the nodes follow more complex dynamics than that predicted by **(6.3)**. To measure their fitness we must first account for their precise growth law. We illustrate this procedure by determining the fitness of a research publication, allowing us to predict its future impact.

While most research papers acquire only a few citations, a small number of publications collect thousands and even tens of thousands of citations [173]. These impact differences mirror differences in the novelty and the relevance of various publications. In general, the probability that a research paper i is cited at time t after publication is [174]

$$\Pi_i \sim \eta_i c_i^t P(t), \quad (6.11)$$

where the paper's fitness η_i accounts for the perceived novelty and importance of the reported discovery; c_i^t is the cumulative number of citations acquired by paper i at time t after publication, accounting for the fact that well-cited papers are more likely to be cited than less-cited contributions (preferential attachment). The last term in **(6.11)** captures the fact that new ideas are integrated into the subsequent work, hence the novelty of each paper fades with time [174, 175]. Measurements indicate that this decay has the log-normal form

$$P_i(t) = \frac{1}{\sqrt{2\pi}t\sigma_i} e^{-\frac{(\ln t - \mu_i)^2}{2\sigma_i^2}}. \tag{6.12}$$

By solving the master equation behind **(6.11)**, we obtain the time-dependent growth of a paper's citations

$$c_i^t = m\left(e^{\frac{\beta\eta_i}{A}\Phi\left(\frac{\ln t - \mu_i}{\sigma_i}\right)} - 1\right), \tag{6.13}$$

where

$$\Phi(x) = \frac{1}{\sqrt{2\pi}} \int_{-\infty}^{x} e^{-y^2/2} dy \tag{6.14}$$

is the cumulative normal distribution and m, β and A are global parameters.

Equation **(6.13)** predicts that the citation history of paper i is characterized by three parameters: the immediacy μ_i governing the time for a paper to reach its citation peak; the longevity σ_i capturing the decay rate; and most importantly the relative fitness $\eta_{i'} \equiv \eta_i\beta/A$ measuring a paper's importance relative to other papers and determining its ultimate impact (**Box 6.2**).

We fit **(6.13)** to the citation history of individual papers published by a journal to obtain the journal's fitness distribution (**Figure 6.5**). The measurements indicate that the fitness distribution of the top cell biology journal, *Cell*, is shifted to the right, indicating that *Cell* papers tend to have high fitness. Not surprisingly, the journal has one of the highest impact factors of all journals. By comparison, the fitness of papers published in *Physical Review* is shifted to the left, indicating that the journal publishes fewer high-fitness papers.

In summary, the framework offered by the Bianconi–Barabási model allows us to determine the fitness of individual nodes and the shape of the fitness distribution $\rho(\eta)$. The measurements show that the fitness distribution is typically exponentially bounded, meaning that fitness differences between different nodes are small. With time these differences are magnified, resulting in a

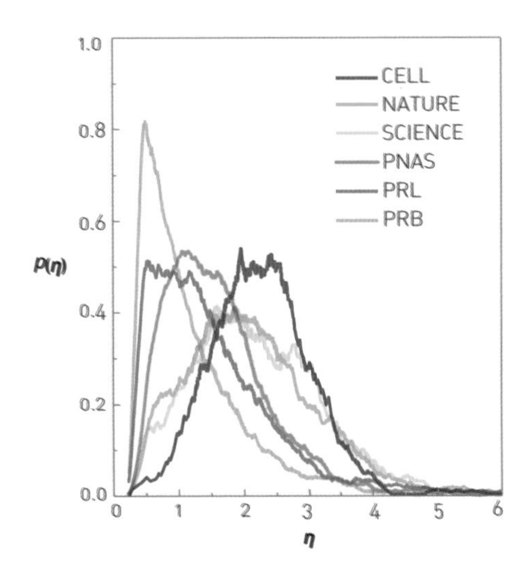

Figure 6.5 **Fitness Distribution of Research Papers**
The fitness distribution of papers published in six journals in 1990. Each paper's fitness was obtained by fitting **(6.13)** to the paper's citation history for a decade-long time interval. Two are physics journals (*Physical Review B* and *Physical Review Letters*), one biology (*Cell*) and three interdisciplinary (*Nature*, *Science* and *PNAS*). The obtained fitness distributions are shifted relative to each other, indicating that *Cell* publishes the highest-fitness papers, followed by *Nature*, *Science*, *PNAS*, *Physical Reviews Letters* and *Physical Review B*.
After [174].

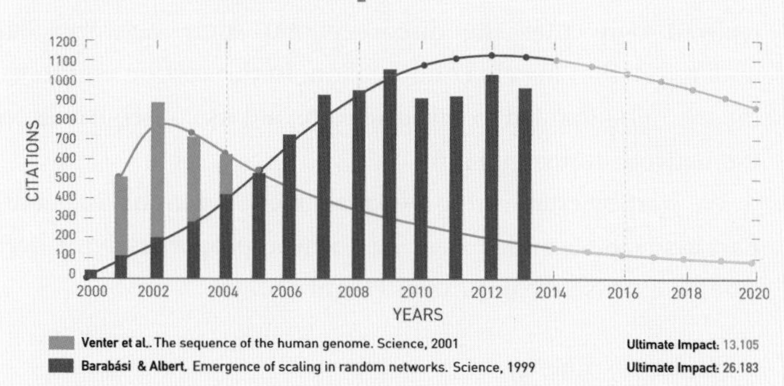

Figure 6.6 **Predicting Ultimate Impact**
The yearly citation history of the paper reporting the first draft of the human genome (Venter *et al.* [27]) and the one reporting the discovery of scale-free networks (Barabási and Albert [10]). The early impact of the two papers cannot be more different: according to the Web of Science, two years after publication the much-anticipated human genome paper collected over 1,400 citations; in contrast, the scale-free network paper was cited only 120 times. Their long-term citation dynamics is also remarkably different: the citations of the human genome paper peaked after year two, a pattern shared by more than 85% of all research papers. In contrast, the yearly citations of the scale-free paper continued to increase for about a decade.
The continuous curves corresponding to the fit **(6.13)** to the respective citation histories, allowing us to determine the papers' future citations and their *ultimate impact.* The ultimate impact corresponds to the total area under each curve for $t \to \infty$. According to **(6.15)**, the ultimate impact of the human genome paper is 13,105, while that of the scale-free paper is 26,183. Therefore the early citation count of a paper is not a strong indicator of its ultimate impact.

Citation counts offer only the historical impact of a research paper. They do not tell us, however, if the paper has already had its run, or if its impact will continue to grow. To gauge a paper's true impact we need to determine how many citations a paper will acquire *during its lifetime*. The citation model **(6.11)** and **(6.14)** allows us to predict this *ultimate impact*. Indeed, by taking the $t \to \infty$ limit of **(6.13)** we obtain [174]

$$c_i^{\infty} = m(e^{\eta_i} - 1). \tag{6.15}$$

Consequently, despite the myriad of factors that contribute to the citation history of a research paper, its ultimate impact is determined only by its fitness η_i. As fitness can be determined by fitting **(6.13)** to a paper's previous citation history, we can use **(6.15)** to predict the ultimate impact of a publication (**Figure 6.6**).

power-law degree distribution of incoming links in the case of the WWW or a broad citation distribution in citation networks.

6.4 Bose–Einstein Condensation

In the previous section we found that the Web's fitness distribution follows a simple exponential (**Figure 6.4**), while the fitness of research papers follows a peaked distribution (**Figure 6.5**). The diversity of the observed fitness distributions raises an important question: How does the network topology depend on the shape of $\rho(\eta)$?

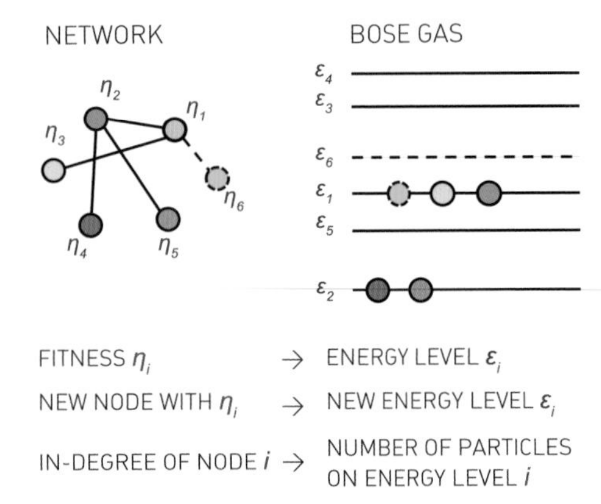

Figure 6.7 **Mapping Networks to a Bose Gas**

Network

A network of six nodes, where each node is characterized by a unique fitness η_i indicated by the node color. The fitnesses are chosen from the fitness distribution $\rho(\eta)$.

Bose Gas

The mapping assigns an energy level ε to each fitness η, resulting in a Bose gas with random energy levels. A link going from a new node i to node j corresponds to one particle at level ε_j.

Growth

The network grows by adding a new node, like the orange node with fitness η_6. For $m = 1$ the new node connects to the gray node (dashed link), chosen following **(6.1)**. In the Bose gas this corresponds to the addition of a new energy level ε_6 (dashed line) and the deposition of a particle at ε_1, the energy level of node 1 to which node η_6 links.

Technically, the answer is provided by **(6.6)**, which links p_k to $\rho(\eta)$. Yet the true impact of the fitness distribution was realized only after the discovery that some networks can undergo Bose–Einstein condensation. In this section we discuss the mapping that led to this discovery and its consequences for the network topology [176].

We start by establishing a formal link between the Bianconi–Barabási model and a Bose gas, whose properties have been studied extensively in physics (**Figure 6.7**):

- **Fitness → Energy**

  We assign to each node with fitness η_i an energy ε_i via

$$\varepsilon_i = -\frac{1}{\beta_T}\log \eta_i. \tag{6.16}$$

  In physical systems β_T plays the role of the inverse temperature. We use the subscript T to distinguish β_T from the dynamic exponent β. According to **(6.16)**, each node in a network corresponds to an energy level in a Bose gas. The larger the node's fitness, the lower is its energy.

- **Links → Particles**

  For each link from node i to node j we add a particle at the energy level ε_j.

- **Nodes → Energy Levels**

  The arrival of a new node with m links corresponds to adding a new energy level ε_j and m new particles to the Bose gas, placed on the energy levels of the nodes to which the new node links.

If we follow the mathematical consequences of this mapping, we find that in the resulting gas the number of particles on each energy level follows Bose statistics, a formula derived by Satyendra Nath Bose in 1924 (**Box 6.3**). Consequently, the links of the fitness model behave like subatomic particles in a quantum gas.

The mapping to a Bose gas is exact and predicts the existence of two distinct phases [176, 177]:

Scale-Free Phase

For most fitness distributions the network displays a fit-gets-rich dynamics, meaning that the degree of each node is ultimately determined by its fitness. While the fittest node inevitably becomes the largest hub, in this phase at any moment the degree distribution follows a power law, indicating that the generated network has a scale-free topology. Consequently, the largest hub follows **(4.18)**, growing only sublinearly. This hub is closely trailed by a few slightly smaller hubs, with almost as many links as the fittest node (**Figure 6.9a**). The model with uniform fitness distribution discussed in **Section 6.2** is in this scale-free phase.

Bose–Einstein Condensation

The unexpected outcome of the mapping to a Bose gas is the possibility of a Bose–Einstein condensation for some fitness distributions. In a Bose–Einstein condensate all particles crowd to the lowest energy level, leaving the rest of the energy levels unpopulated (**Box 6.4**).

In a network, Bose–Einstein condensation means that the fittest node grabs a finite fraction of the links, turning into a super-hub (**Figure 6.9b**). The resulting network is not scale-free but has a hub-and-spoke topology. In this phase the rich-gets-richer process is so dominant that it becomes a qualitatively different *winner-takes-all phenomenon*. Consequently, the network will lose its scale-free nature.

In physical systems, Bose–Einstein condensation is induced by lowering the temperature of the Bose gas below some critical temperature (**Box 6.4**). In networks, the temperature β_T in **(6.16)** is a dummy variable, disappearing from all topologically relevant quantities, like the degree distribution p_k. Hence, the presence or absence of Bose–Einstein condensation depends only on the form of the fitness distribution $\rho(\eta)$. For a network to undergo Bose–Einstein condensation, the fitness distribution needs to satisfy the condition

$$\int_{\eta_{min}}^{\eta_{max}} \frac{\eta\rho(\eta)}{1-\eta} d\eta < 1$$

A fitness distribution that leads to Bose–Einstein condensation is

$$\rho(\eta) = (1-\zeta)(1-\eta)^{\zeta}, \quad (6.22)$$

whereby varying ζ we can induce Bose–Einstein condensation (**Figure 6.9**). Indeed, whether **(6.20)** has a solution depends on the functional form of the energy distribution, $g(\varepsilon)$, which is

Box 6.3 From Fitness to a Bose Gas

In the context of the Bose gas (**Figure 6.7**) the probability that a particle lands on level i is

$$\Pi_i = \frac{e^{-\beta_T \varepsilon_i} k_i}{\sum_j e^{-\beta_T \varepsilon_j} k_j}. \quad (6.17)$$

Hence, the rate at which the energy level ε_i accumulates particles is [176]

$$\frac{\partial k_i(\varepsilon_i, t, t_i)}{\partial t} = m \frac{e^{-\beta_T \varepsilon_i} k_i(\varepsilon_i, t, t_i)}{Z_t} \quad (6.18)$$

where $k_i(\varepsilon_i, t, t_i)$ is the occupation number of level i and

$$Z_t = \sum_{j=1}^{t} t e^{-\beta_T \varepsilon_j} k_j(\varepsilon_i, t, t_j)$$

is the partition function. The solution of **(6.18)** is

$$k_i(\varepsilon_i, t, t_i) = m\left(\frac{t}{t_i}\right)^{f(\varepsilon_i)}, \quad (6.19)$$

where $f(\varepsilon) = e^{-\beta_T(\varepsilon-\mu)}$ and μ is the *chemical potential* satisfying

$$\int \deg(\varepsilon) \frac{1}{e^{\beta_T(\varepsilon-\mu)} - 1} = 1. \quad (6.20)$$

Here, $\deg(\varepsilon)$ is the degeneracy of the energy level ε. Equation **(6.20)** suggests that in the limit $t \to \infty$ the occupation number, representing the number of particles with energy ε, follows the *Bose statistics*

$$\eta(\varepsilon) = \frac{1}{e^{\beta_T(\varepsilon-\mu)} - 1}. \quad (6.21)$$

This mapping of the fitness model to a Bose gas proves that the node degrees in the Bianconi–Barabási model follow Bose statistics.

Box 6.4 Bose–Einstein Condensation

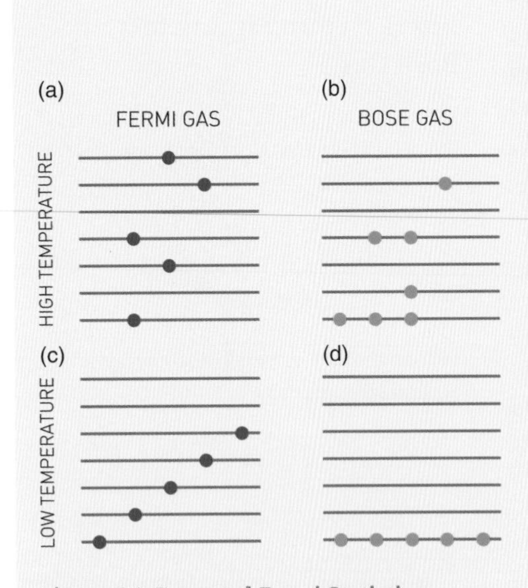

Figure 6.8 **Bose and Fermi Statistics**
In a Fermi gas **(a,c)** only one particle is allowed on each energy level, while in a Bose gas **(b,d)** there is no such restriction. At high temperatures it is hard to notice any difference between the two gases. At low temperatures, however, each particle wants to occupy the lowest possible energy and the difference between the two gases becomes significant.

In classical physics the kinetic energy of a moving particle, $E = mv^2/2$, can have any value between zero (at rest) and an arbitrarily large E (moving very fast). Furthermore, an arbitrary number of particles can have the same energy E if they have the same velocity v. In quantum mechanics energy is quantized, meaning that it can only take up discrete (quantized) values. Furthermore, in quantum mechanics we encounter two different classes of particle. Fermi particles, like electrons, are forbidden to have the same energy within the same system. Hence, only one electron can occupy a given energy level (**Figure 6.8a**). In contrast Bose particles, like photons, are allowed to crowd in arbitrary numbers on the same energy level (**Figure 6.8b**).

At high temperatures, when thermal agitation forces the particles to take up different energies, the difference between a Fermi and a Bose gas is negligible (**Figure 6.8a,b**). The difference becomes significant at low temperatures, when all particles are forced to take up their lowest allowed energy. In a Fermi gas at low temperatures the particles fill the energy levels from the bottom up, just like pouring water fills up a vase (**Figure 6.8c**). However, as any number of Bose particles can share the same energy, they can all crowd at the lowest energy level (**Figure 6.8d**). In other words, no matter how much "Bose liquid" we pour into the vase, it will stay at the bottom of the vessel, never filling it up. This phenomenon is called Bose–Einstein condensation and it was first proposed by Einstein in 1924. Experimental evidence for Bose–Einstein condensation emerged only in 1995 and was recognized with the 2001 Nobel Prize in physics.

determined by the shape of $\rho(\eta)$. Specifically, if **(6.22)** has no non-negative solution for a given $g(\varepsilon)$, then Bose–Einstein condensation emerges and a finite fraction of the particles agglomerate at the lowest energy level (**Online resource 6.2**).

In summary, the precise shape of the fitness distribution determines the topology of a growing network. While fitness distributions like the uniform distribution lead to a scale-free topology, some $\rho(\eta)$ allow for Bose–Einstein condensation. If a network undergoes Bose–Einstein condensation, then one or a few nodes grab most of the links. Hence, the rich-gets-richer process responsible for the scale-free state turns into a winner-takes-all phenomenon. The Bose–Einstein condensation has such an obvious impact on a network's structure that, if present, it is hard to miss: it destroys the hierarchy of hubs characterizing a scale-free network, forcing the network into a star-like hub-and-spoke topology (**Figure 6.9**).

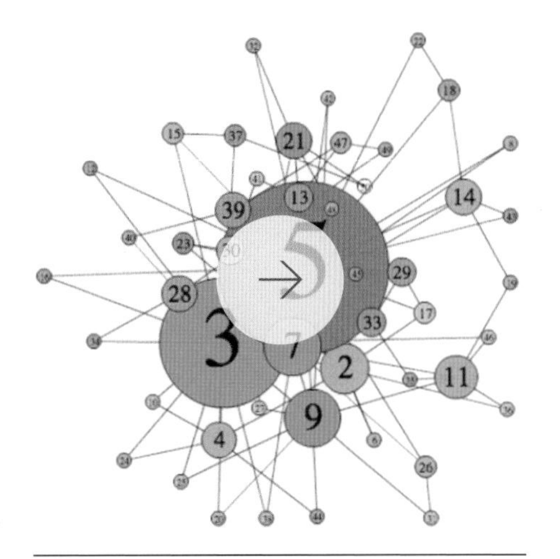

Online Resource 6.2

Bose–Einstein Condensation in Networks

The movie shows the time evolution of a growing network in which one node (purple) has a much higher fitness than the rest of the nodes. This high-fitness node attracts most links, forcing the system to undergo Bose–Einstein condensation. Video courtesy of Dashun Wang.

6.5 Evolving Networks

The Barabási–Albert model is a minimal model, designed to capture the mechanisms responsible for the emergence of the scale-free property. Consequently, it has several well-known limitations (see also **Section 5.10**):

(i) The model predicts $\gamma = 3$, while the experimentally observed degree exponents vary between 2 and 5 (**Table 4.1**).

(ii) The model predicts a power-law degree distribution, while in real systems we observe systematic deviations from a pure power-law function, like small-degree saturation or high-degree cutoff (**Box 4.8**).

(iii) The model ignores a number of elementary processes that are obviously present in many real networks, like the addition of internal links and node or link removal.

These limitations have inspired considerable research, clarifying the role of the numerous elementary processes that influence the network topology. In this section we systematically extend the Barabási–Albert model, arriving at a family of evolving network models that can capture the wide range of phenomena known to shape the topology of real networks.

6.5.1 Initial Attractiveness

In the Barabási–Albert model an isolated node cannot acquire links, as according to preferential attachment **(4.1)** the likelihood that a new node attaches to a $k = 0$ node is strictly zero. In real networks even isolated nodes acquire links. Indeed, each new research paper has a finite probability of being cited for the first time; a person who moves to a new city quickly acquires

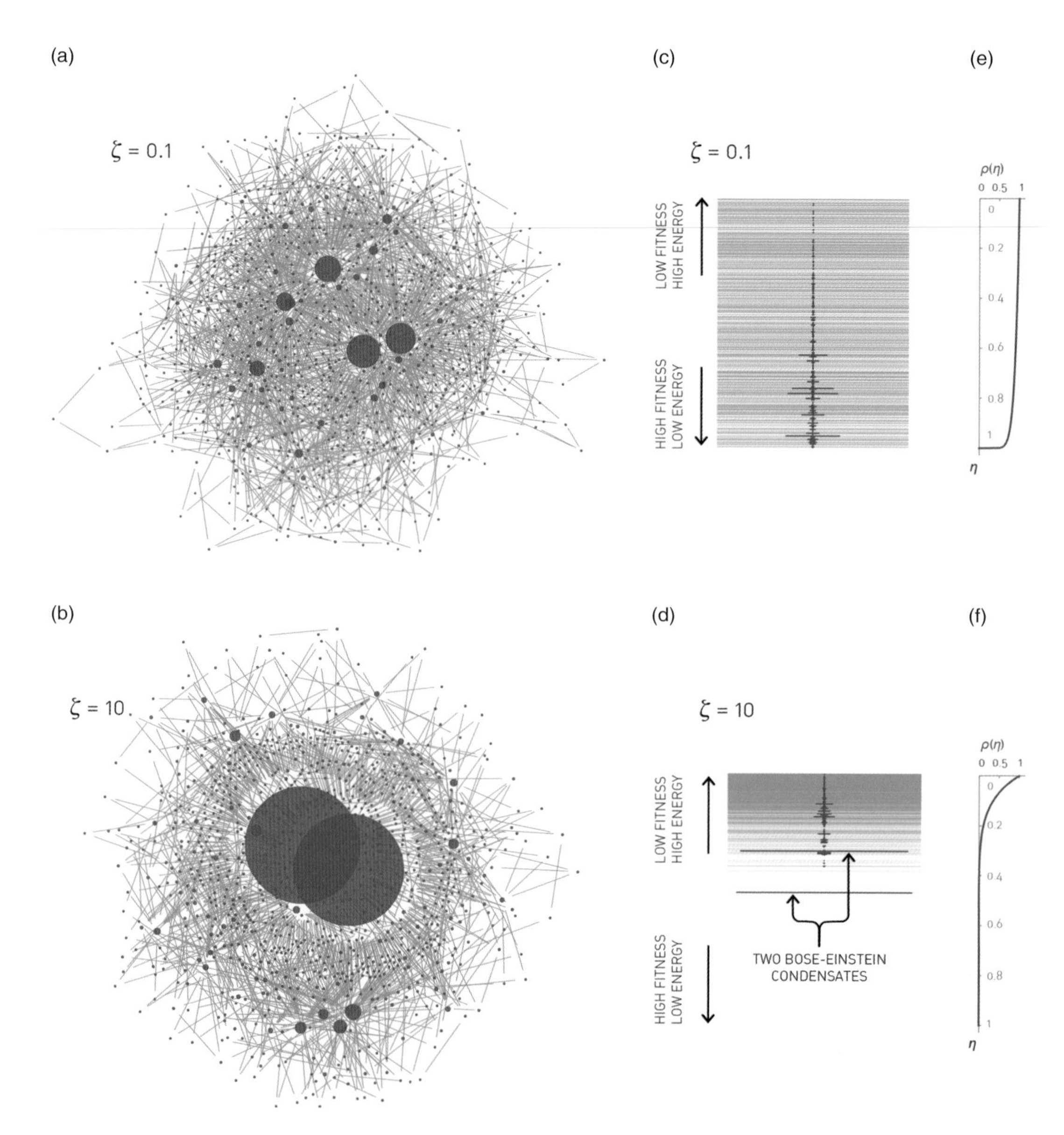

Figure 6.9 **Bose–Einstein Condensation in Networks**

(a,b) A scale-free network (a) and a network that has undergone Bose–Einstein condensation (b). Both networks were generated by the Bianconi–Barabási model with $\rho(\eta)$ following **(6.22)**, but with different exponent ζ. Note that in the condensed phase (b) we have two large hubs with comparable size.

(c,d) The energy levels (green lines) and the deposited particles (purple dots) for a network with $m = 2$ and $N = 1{,}000$. Each energy level corresponds to the fitness of a node of the network shown in (a,b). Each link connected to a node is represented by a particle on the corresponding energy level. As we did not allow multi-links, two highly populated energy levels emerged in (d), indicating that we have two condensates, corresponding to the two hubs seen in (b).

(e,f) The fitness distribution $\rho(\eta)$, given by **(6.22)**, illustrates the difference in shape of the two $\rho(\eta)$ functions. The difference is determined by the parameter ζ, which is (e) $\zeta = 0.1$ and (f) $\zeta = 10$.

acquaintances. To allow unconnected nodes to acquire links we add a constant to the preferential attachment function **(5.1)**,

$$\Pi(k) \sim A + k. \tag{6.23}$$

Here the constant A is called *initial attractiveness*. As $\Pi(0) \sim A$, initial attractiveness is proportional to the probability that a node acquires its first link in the next time step.

Direct measurement of $\Pi(k)$ shows that initial attractiveness is present in real networks (**Figure 6.10**). Once present, it has two consequences:

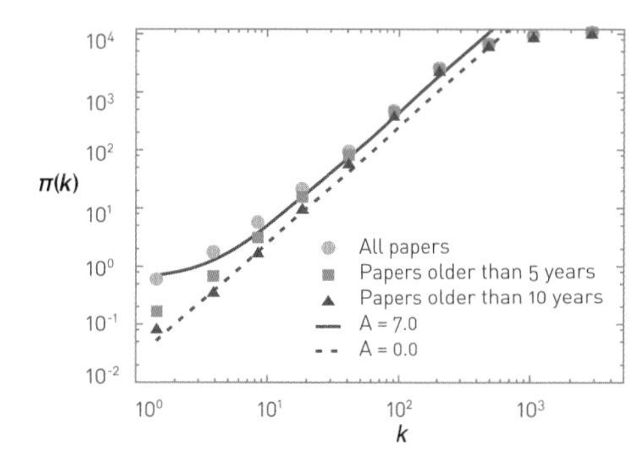

Figure 6.10 **Initial Attractiveness**
The cumulative preferential attachment function **(5.21)** for the citation network, capturing the citation patterns of research papers published from 2007 to 2008. The $\pi(k)$ curve was measured using the methodology described in **Section 5.6**. The continuous line corresponds to initial attractiveness $A \sim 7.0$, while the dashed line corresponds to $A = 0$, i.e. the case without initial attractiveness. $A = 7$ implies that the probability of a new paper being cited for the first time is comparable with the citation probability of a paper with seven citations. After [179].

- **Increases the Degree Exponent**
  If, in the Barabási–Albert model, we replace **(5.1)** with **(6.23)**, the degree exponent becomes [13, 178]

$$\gamma = 3 + \frac{A}{m}. \tag{6.24}$$

  Consequently, initial attractiveness increases γ, making the network more homogenous and reducing the size of the hubs. Indeed, initial attractiveness adds a random component to the probability of attaching to a node. This random component favors the numerous small-degree nodes and weakens the role of preferential attachment. For high-degree nodes the initial attractiveness term A in **(6.23)** is negligible.
- **Generates Small-Degree Saturation**
  The solution of the continuum equation indicates that the degree distribution of a network governed by **(6.23)** does not follow a pure power law, but has the form

$$p_k = C(k + A)^{-\gamma}. \tag{6.25}$$

  Therefore, initial attractiveness induces a small-degree saturation for $k < A$, playing the role of k_{sat} in **(4.39)**. This saturation is rooted in the fact that initial attractiveness enhances the probability of new nodes linking to small-degree nodes, which pushes the small-k nodes toward higher degrees. For high degrees ($k \gg A$) the degree distribution continues to follow a power law, as in this range initial attractiveness does not alter the attachment probability.

6.5.2 Internal Links

In many networks new links not only arrive with new nodes, but are added between pre-existing nodes. For example, the vast majority of new links on the WWW are *internal links*, corresponding to newly added URLs between pre-existing web documents. Similarly, virtually all new social/friendship links form between individuals who already have other friends and acquaintances.

Measurements show that in collaboration networks the internal links follow double preferential attachment, that is the probability of a new internal link connecting nodes with degrees k and k' is [108]

$$\Pi(k,k') \sim (A+Bk)\,(A+Bk'). \tag{6.26}$$

To understand the impact of internal links we explore the limiting cases of **(6.26):**

- **Double Preferential Attachment ($A = 0$)**
Consider an extension of the Barabási–Albert model, where in each time step we add a new node with m links, followed by n internal links, each selected with probability **(6.26)** with $A = 0$. Consequently, the likelihood that a new link emerges is proportional to the degree of the nodes it connects. The degree exponent of the resulting network is [180, 181]

$$\gamma = 2 + \frac{m}{m+2n}, \tag{6.27}$$

indicating that γ is between 2 and 3. This means that double preferential attachment *lowers the degree exponent* from 3 to 2, hence increasing the network's heterogeneity. Indeed, by preferentially connecting the hubs to each other, internal links make both hubs larger at the expense of the smaller nodes.
- **Random Attachment ($B = 0$)**
In this case the internal links are blind to the degree of the nodes they connect. Consequently, the internal links are added between randomly chosen node pairs. Let us again consider the Barabási–Albert model, where after each new node we add n links between randomly selected nodes. The degree exponent of the obtained network is [181]

$$\gamma = 3 + \frac{2n}{m}. \tag{6.28}$$

Hence we have $\gamma \geq 3$ for any n and m combination, indicating that the resulting network will be more homogenous than the network without internal links. Indeed, randomly added internal links mimic the process observed in random networks, making the node degrees more similar to each other.

6.5.3 Node Deletion

In many real systems nodes and links can disappear. For example, nodes are deleted from an organizational network when employees leave the company or from the WWW when web documents are removed. At the same time, in some networks node removal is virtually impossible (**Figure 6.11**).

To explore the impact of node removal, we start from the Barabási–Albert model. At each time step we add a new node with m links and with rate r we remove a node. Depending on r, we observe three distinct scaling regimes [184, 185, 186, 187, 188, 189]:

- **Scale-Free Phase**
  For $r < 1$ the number of removed nodes is smaller than the number of new nodes, hence the network continues to grow. In this case the network is scale-free with degree exponent

$$\gamma = 3 + \frac{2r}{1-r}. \tag{6.29}$$

  Hence, random node removal increases γ, homogenizing the network.
- **Exponential Phase**
  For $r = 1$ nodes arrive and are removed at the same rate, hence the network has a fixed size (N = constant). In this case the network will lose its scale-free nature. Indeed, for $r \to 1$ we have $\gamma \to \infty$ in **(6.29)**.
- **Declining Networks**
  For $r > 1$ the number of removed nodes exceeds the number of new nodes, hence the network declines (**Box 6.5**). Declining networks emerge in several areas. For example, Alzheimer's research focuses on the progressive loss of neurons with age and ecology explores the role of gradual habitat loss [190, 191, 192]. A classical example of a declining network is the telegraph, which dominated long-distance communication in the second part of the nineteenth century and early twentieth century. It was once a growing network: in the United States the length of the telegraph lines grew from 40 miles in 1846 to

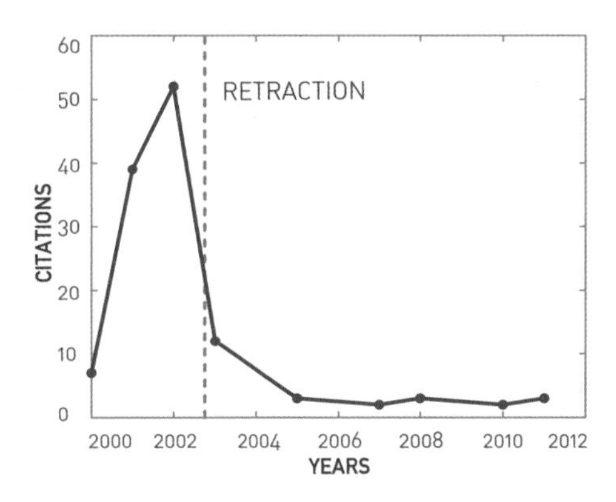

Figure 6.11 **The Impossibility of Node Deletion**
The citation history of a research paper by Jan Hendrik Schön published in *Science* [182] illustrates how difficult it is to remove a node from the citation network. Schön rose to prominence after a series of breakthroughs in the area of semiconductors. His productivity was phenomenal: in 2001 he co-authored one research paper every eight days, published by the most prominent scientific journals, like *Science* and *Nature*.
Soon after Schön published a paper reporting a groundbreaking discovery on single-molecule semiconductors, researchers noticed that he reported identical noise for two experiments carried out at different temperatures [183]. The ensuing questions prompted Lucent Technologies, which ran Bell Labs where Schön worked, to start a formal investigation. Eventually Schön admitted falsifying data. Several dozen of his papers, like the one whose citation pattern is shown in this figure, were retracted. While the papers' formal retraction led to a dramatic drop in citations, the papers continue to be cited after their official "deletion" from the literature, as seen in the figure above. This indicates that it is virtually impossible to remove a node from the citation network.

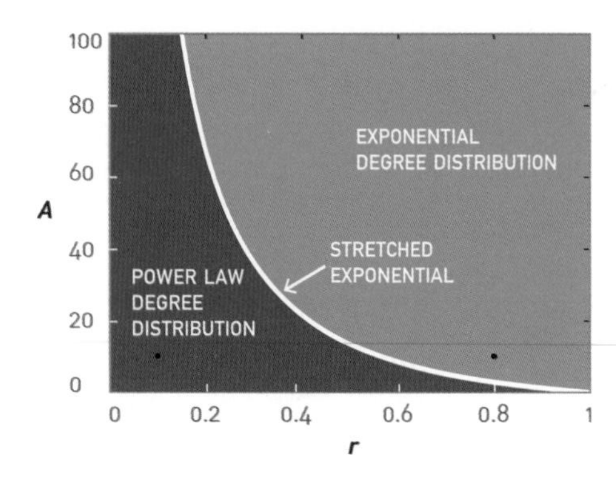

Figure 6.12 **Phase Transitions Induced by Node Removal**

The coexistence of node removal with other elementary processes can lead to interesting topological phase transitions. This is illustrated by a simple model in which the network's growth is governed by **(6.23)**, and we also remove nodes with rate r [189]. The network displays three distinct phases, captured by the phase diagram shown above, whose axes are the node removal rate r and the initial attractiveness A.

Subcritical Node Removal: $r < r^*(A)$

If the rate of node removal is under a critical value $r^*(A)$, shown as the white line on the figure, the network will be scale-free.

Critical Node Removal: $r = r^*(A)$

Once r reaches a critical value $r^*(A)$, the degree distribution turns into a stretched exponential (**Section 4.A**).

Exponential Networks: $r > r^*(A)$

The network loses its scale-free nature, developing an exponential degree distribution. Therefore, the coexistence of multiple elementary processes in a network can lead to sudden changes in the network topology. To be specific, a continuous increase in the node removal rate leads to a phase transition from a scale-free to an exponential network.

23,000 miles in 1852. Yet, following the Second World War, the telegraph gradually disappeared.

The behavior of a network can be rather complex if node removal coexists with other elementary processes. This is illustrated in **Figure 6.12**, indicating that the joint presence of initial attractiveness and node deletion induces phase transitions between scale-free and exponential networks. Finally, note that node removal is not always random, but can depend on the removed node's degree (**Box 6.5**).

In summary, in most networks nodes can disappear. Yet as long as the network continues to grow, its scale-free nature can persist. The degree exponent depends, however, on the details governing the node-removal process.

6.5.4 Accelerated Growth

In the models discussed so far, the number of links increased linearly with the number of nodes. In other words, we assumed that $L=\langle k\rangle N/2$, where $\langle k\rangle$ is independent of time. This is a reasonable assumption for many real networks. Yet, for some real networks the number of links grows faster than N, a phenomena called *accelerated growth*. For example the average degree of the Internet increased from $\langle k\rangle$=3.42 in November 1997 to 3.96 by December 1998 [99]; the WWW increased its average degree from 7.22 to 7.86 during a five month interval [88, 193]; in metabolic networks the average degree of the metabolites grows approximately linearly with the number of metabolites [20]. To explore the consequences of accelerated growth let us assume that in a growing network the number of links arriving with each new node follows [194, 195, 196, 197]

$$m(t) = m_0 t^{\theta}. \tag{6.30}$$

For $\theta = 0$ each new node has the same number of links; for $\theta > 0$, however, the network follows accelerated growth.

The degree exponent of the Barabási–Albert model with accelerated growth **(6.30)** is

$$\gamma = 3 + \frac{2\theta}{1-\theta}. \tag{6.31}$$

Box 6.5 Declining Fashion Networks

The New York City garment industry offers a prominent example of a declining network (**Figure 6.1**). Its nodes are designers and contractors who are connected to each other by the annual co-production of lines of clothing. As the industry has decayed, the network has shrunk persistently: the network's largest connected component collapsed from 3249 nodes in 1985 to 190 nodes in 2003. Interestingly, the network's degree distribution remained unchanged during this period. The analysis of the network's evolution uncovered several properties of declining networks [184]:

- **Preferential Attachment**
  While overall the network was shrinking, new nodes continued to arrive. The measurements indicate that the attachment probability of these new nodes follows $\Pi(k) \sim k^{\alpha}$ with $\alpha = 1.20 \pm 0.06$ (**Figure 6.13a**), offering evidence of superlinear preferential attachment (**Section 5.7**).
- **Link Deletion**
  The probability that a firm lost a link follows $k(t)^{-\eta}$ with $\eta = 0.41 \pm 0.04$, decreases with the firm's degree (**Figure 6.13b**). This documents a *weak-gets-weaker* phenomenon, when the less-connected firms are more likely to lose links.

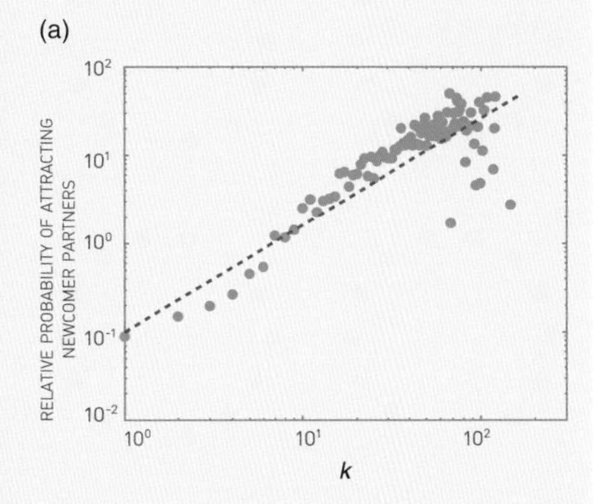

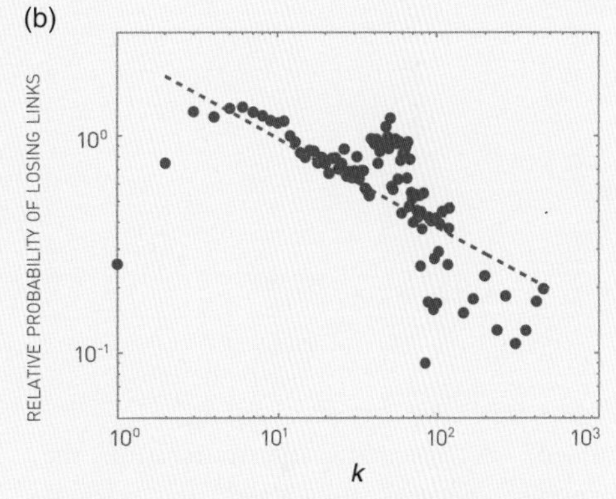

Figure 6.13 The Decline of the Garment Industry
(a) Preferential attachment. The probability that a newcomer firm added at time t connects to an incumbent firm with k links, relative to a random link addition. The dashed line has slope $\alpha = 1.2$. If link addition were to be random, we would expect this quantity to be ≈ 1.
(b) Link deletion. The probability of deleting a link from a degree-k node, relative to random link removal. The dashed line has slope $\eta = 0.41$. If link loss were to be random, the relative probability should be ≈ 1 for any k.

Hence, accelerated growth pushes the degree exponent beyond $\gamma = 3$, making the network more homogenous. For $\theta = 1$ the degree exponent diverges, leading to *hyper-accelerating growth* [195]. In this case $\langle k \rangle$ grows linearly with time and the network loses its scale-free nature.

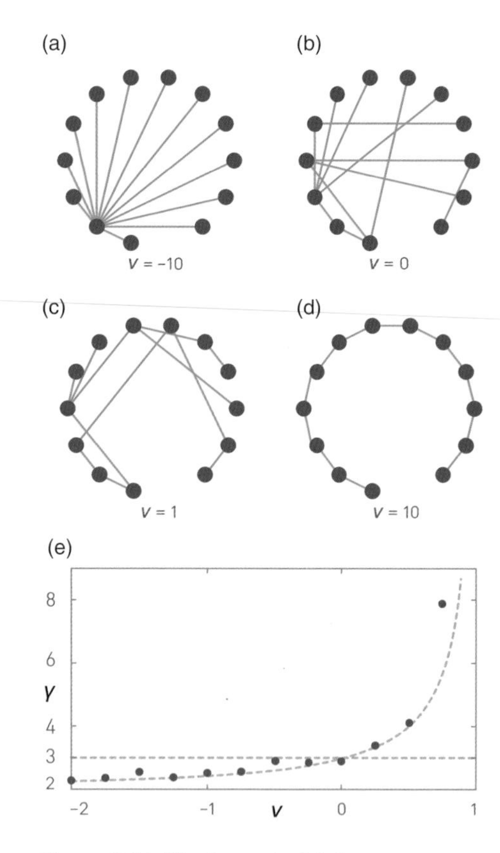

Figure 6.14 **The Impact of Aging**
(a)–(d) A schematic illustration of the expected network topologies for various aging exponents ν in **(6.32)**. In the context of a growing network we assume that the probability of attaching to a node is proportional to $k\tau^{-\nu}$, where τ is the *age* of the node. For negative ν nodes prefer to link to the oldest nodes, turning the network into a hub-and-spoke topology. For positive ν the most recent nodes are the most attractive. For large ν the network turns into a chain, as the last (i.e., the youngest) node is always the most attractive for the new node. The network is shown for $m = 1$ for clarity, but the degree exponent is independent of m.
(e) The degree exponent γ vs. the aging exponent ν predicted by the analytical solution of the aging model. The purple symbols are the result of simulations, each representing a single network with $N = 10{,}000$ and $m = 1$. Redrawn after [198].

6.5.5 Aging

In many real systems nodes have a limited lifetime. For example, actors have a finite professional lifespan, defined as the period when they act in movies. So do scientists, whose professional lifespan typically corresponds to the time frame during which they continue to publish scientific papers. In these networks nodes do not disappear abruptly, but fade away through a slow aging process, gradually reducing the rate at which they acquire new links [119, 198, 199, 200]. Capacity limitations can induce a similar phenomenon: if nodes have finite resources to handle links, once they approach their limit, they will stop accepting new links [119].

To understand the impact of aging we assume that the probability of a new node connecting to node i is $\Pi(k_i, t - t_i)$, where t_i is the time node i was added to the network. Hence, $t - t_i$ is the node's age. Aging is often modeled by choosing [198]

$$\Pi(k_i, t - t_i) \sim k(t - t_i)^{-\nu}, \tag{6.32}$$

where ν is a tunable parameter governing the dependence of the attachment probability on the node's age. Depending on the value of ν, we can distinguish three scaling regimes:

- **Negative ν**
  If $\nu < 0$, new nodes will link to older nodes. Hence, a negative ν *enhances* the role of preferential attachment. In the extreme case $\nu \to -\infty$, each new node connects to the oldest node, resulting in a hub-and-spoke topology (**Figure 6.14a**). The calculations show that the scale-free state persists in this regime, but the degree exponent drops under 3 (**Figure 6.14e**). Hence, $\nu < 0$ makes the network more heterogeneous.
- **Positive ν**
  In this case new nodes are encouraged to attach to younger nodes. In the extreme case $\nu \to \infty$ each node will connect to its immediate predecessor (**Figure 6.14d**). We do not need a very large ν to experience the impact on aging: the degree exponent diverges as we approach $\nu = 1$ (**Figure 6.14e**). Hence, gradual aging homogenizes the network by shadowing the older hubs.
- **$\nu > 1$**
  In this case the aging effect overcomes the role of preferential attachment, leading to the loss of the scale-free property (**Figure 6.14d**).

In summary, the results discussed in this section indicate that a wide range of elementary processes can affect the structure and the dynamics of a growing network (**Figure 6.15**). These results highlight the true power of the evolving network paradigm: it allows us to address, using a mathematically self-consistent and predictive framework, the impact of various processes on the network topology and evolution.

6.6 Summary

As we showed in this chapter, rather diverse processes, from fitness to internal links and aging, can influence the structure of real networks. Through them we learned how to use the theory of evolving networks to predict the impact of various elementary events on a network's topology and evolution. The discussed examples allow us to draw a key conclusion: *if we want to understand the structure of a network, we must first get its dynamics right. The topology is the bonus of this approach.*

The developed tools allow us to reflect on a number of issues that we encountered in the previous chapters, from the correct fit to the degree distribution to the role of the different modeling frameworks. Next we briefly discuss some of these issues.

6.6.1 Topological Diversity

In **Chapter 4** we discussed the difficulties we encounter when we attempt to fit a pure power law to the degree distribution of a real network. The roots of this problem became obvious in this chapter: if we account for the real dynamical processes that contribute to the evolution of a network, we expect systematic deviations from a pure power law. Indeed, we predicted several analytical forms for the degree distribution:

- **Power Law**
  A pure power law emerges if a growing network is governed by linear preferential attachment only, as predicted by the Barabási–Albert model. It is rare to observe such a pure power law in real systems. This idealized model represents the starting point for understanding the degree distribution of real networks.
- **Stretched Exponential**
  If preferential attachment is sublinear, the degree distribution follows a stretched exponential (**Section 5.7**). A similar degree

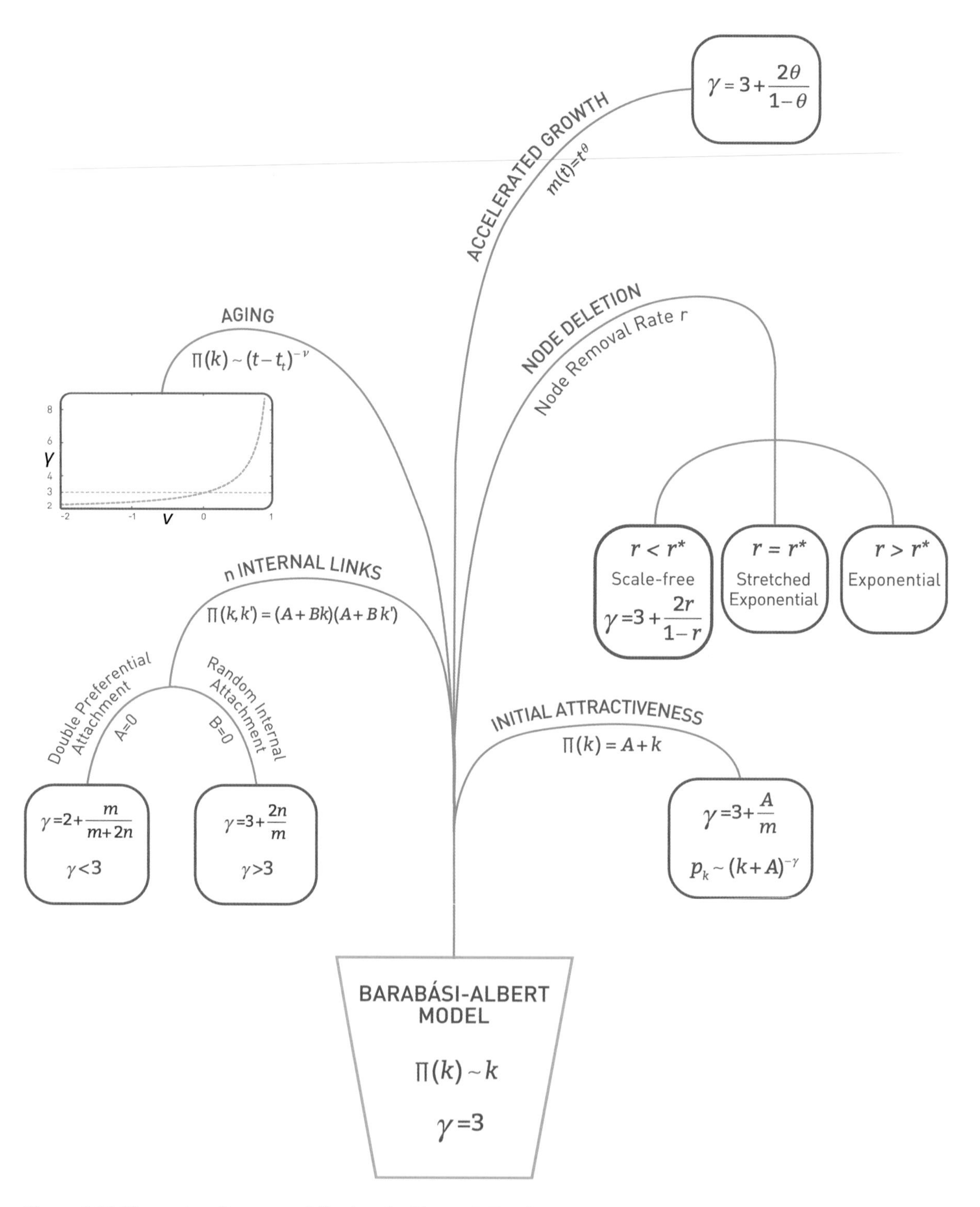

Figure 6.15 **Elementary Processes Affecting the Network Topology**
A summary of the elementary processes discussed in this section and their impact on the degree distribution. Each model is defined as an extension of the Barabási–Albert model.

distribution can also appear under node removal at the critical point (**Figure 6.12**).

- **Fitness-Induced Corrections**
  In the presence of fitness the precise form of p_k depends on the fitness distribution $\rho(\eta)$, which determines p_k via **(6.6)**. For example, a uniform fitness distribution induces a logarithmic correction in p_k as predicted by **(6.8)**. Other forms of $\rho(\eta)$ can lead to rather exotic forms for p_k.
- **Small-Degree Saturation**
  Initial attractiveness adds a random component to preferential attachment. Consequently, the degree distribution develops a small-degree saturation, as seen in **(6.24)**.
- **High-Degree Cutoffs**
  Node and link removal, present in many real systems, can induce exponential high-degree cutoffs in the degree distribution. Furthermore, random node removal can deplete the small-degree nodes, inducing a peak in p_k.

In most real networks several of the elementary processes discussed in this chapter appear together. For example, in the scientific collaboration network we have sublinear preferential attachment with initial attractiveness and the links can be both external and internal. As researchers have different creativity, fitness also plays a role, hence an accurate model requires us to know the appropriate fitness distribution. Therefore, the degree distribution is expected to display small-degree saturation (thanks to initial attractiveness), stretched exponential cutoff at high degrees (thanks to sublinear preferential attachment) and some unknown corrections due to the particular form of the fitness distribution $\rho(\eta)$.

In general, if we wish to obtain an accurate fit to the degree distribution, we first need to build a generative model that analytically predicts the functional form of p_k. Yet, in many systems, developing an accurate theory for p_k may be overkill. It is often sufficient, instead, to establish if we are dealing with an exponentially bounded or a heavy-tailed degree distribution (**Section 4.9**), as the system's properties will be primarily driven by this distinction.

6.6.2 Modeling Diversity

The results of this chapter also allow us to reflect on the role of the network models encountered so far. We can categorize these models into three main classes (**Table 6.1**):

Static Models

The random network model of Erdős and Rényi (**Chapter 3**) and the small-world network model of Watts and Strogatz (**Box 3.8**) have a fixed number of nodes, prompting us to call them *static models*. They both assume that the role of the network modeler is to place the links between the nodes using some random algorithm. To explore their properties we need to rely on combinatorial graph theory, developed by Erdős and Rényi. Both models predict a bounded degree distribution.

Generative Models

The configuration and the hidden-parameter models discussed in **Section 4.8** generate networks with a pre-defined degree distribution. Hence, these models are not mechanistic, in the sense that they do not tell us why a network develops a particular degree distribution. Rather, they help us understand how various network properties, from clustering to path lengths, depend on the degree distribution.

Evolving Network Models

These models capture the mechanisms that govern the time evolution of a network. The most studied example is the Barabási–Albert model, but equally insightful are the extensions discussed in this chapter, from the Bianconi–Barabási model to models involving internal links, aging, node and link deletion, or accelerated

Table 6.1 **Classes of Model in Network Science**

The table summarizes the three main modeling frameworks used in network science, together with their distinguishing features.

Model class	Examples	Characteristics
Static models	Erdős–Rényi Watts–Strogatz	• N fixed • p_k exponentially bounded • Static, time-independent topologies
Generative models	Configuration model Hidden-parameter model	• Arbitrary pre-defined p_k • Static, time-independent topologies
Evolving network models	Barabási–Albert model Bianconi–Barabási model Initial-attractiveness model Internal-links model Node-deletion model Accelerated-growth model Aging model	• p_k is determined by the processes that contribute to the network's evolution • Time-varying network topologies

growth. These models are motivated by the observation that if we correctly capture all microscopic processes that contribute to a network's evolution, then the network's topological characteristics follow from that. To explore the properties of the networks generated by them, we need to use dynamical methods like the continuum theory and the rate equation approach.

Each of these modeling frameworks have an important role in network theory. The Erdős–Rényi model allows us to check if a certain network property could be explained by a purely random connectivity pattern. If our interest is limited to the role of the network environment on some phenomena, like spreading processes or network robustness, the generative models offer an excellent starting point. If, however, we want to understand the origin of a network property, we must resort to evolving network models that capture the processes that built the network in the first place.

6.7 Homework

6.7.1 Accelerated Growth

Calculate the degree exponent of the directed Barabási–Albert model with accelerated growth, assuming that the degree of the newly arriving nodes increases in time as $m(t) = t^{\Theta}$.

6.7.2 The t-Party Evolving Network Model

In the t-party gender plays no role, hence each newcomer is allowed to invite only one other participant to dance. However, attractiveness plays a role: more attractive participants are more likely to be invited to dance by a new participant. The party evolves following these rules:

- Every participant corresponds to a node i and is assigned a time-independent attractiveness coefficient η_i.
- At each time step a new node joins the t-party.
- This new node then invites one already partying node to dance, establishing a new link with it.
- The new node chooses its dance partner with probability proportional to the potential partner's attractiveness. If there are t nodes already in the party, the probability that node i receives a dance invitation is

$$\Pi_i = \frac{\eta_i}{\sum_j \eta_j} = \frac{\eta_i}{t\langle\eta\rangle},$$

where $\langle\eta\rangle$ is the average attractiveness.

(a) Derive the time evolution of the node degrees, telling us how many dances a node had.
(b) Derive the degree distribution of nodes with attractiveness η.
(c) If half of the nodes have $\eta = 2$, and the other half $\eta = 1$, what is the degree distribution of the network after a sufficiently long time?

6.7.3 Bianconi–Barabási Model

Consider the Bianconi–Barabási model with two distinct fitnesses, $\eta = a$ and $\eta = 1$. To be specific, let us assume that the fitness follows the double delta distribution

$$\rho(\eta) = \frac{1}{2}\delta(\eta - a) + \frac{1}{2}\delta(\eta - 1) \text{ with } 0 \leq a \leq 1.$$

(a) Calculate the degree exponent, and its dependence on the parameter a.
(b) Calculate the stationary degree distribution of the network.

6.7.4 Additive Fitness

Assume that the growth of a network is governed by preferential attachment with additive fitness

$$\Pi(k_i) \sim \eta_i + k_i,$$

where a different η_i is assigned to each node, chosen from a $\rho(\eta)$ fitness distribution. Calculate and discuss the degree distribution of the resulting network.

6.8 Advanced Topics 6.A Analytical Solution of the Bianconi–Barabási model

The purpose of this section is to derive the degree distribution of the Bianconi–Barabási model [13, 166, 176, 177]. We start by calculating

$$\left\langle \sum_j \eta_j k_j \right\rangle \tag{6.33}$$

over all possible realizations of the quenched fitnesses η. Since each node is born at a different time t_0, we can write the sum over j as an integral over t_0

$$\left\langle \sum_j \eta_j k_j \right\rangle = \int d\eta \rho(\eta)\eta \int_1^t dt_0 k\eta(t, t_0). \tag{6.34}$$

By replacing $k_\eta(t, t_0)$ with **(6.3)** and performing the integral over t_0, we obtain

$$\left\langle \sum_j \eta_j k_j \right\rangle = \int d\eta \rho(\eta)\eta m \frac{t - t^{\beta(\eta)}}{1 - \beta(\eta)}. \tag{6.35}$$

The dynamic exponent $\beta(\eta)$ is bounded, that is $0 < \beta(\eta) < 1$, because a node can only increase its degree with time $(\beta(\eta) > 0)$ and $k_i(t)$ cannot increase faster than t $(\beta(\eta) < 1)$. Therefore, in the limit $t \to \infty$ in **(6.35)** the term $t^{\beta(n)}$ can be neglected compared with t, obtaining

$$\left\langle \sum_j \eta_j k_j \right\rangle \stackrel{t\to\infty}{=} Cmt(1 - O(t^{-\varepsilon})), \tag{6.36}$$

where $\varepsilon = (1 - \max_\eta \beta(\eta)) > 0$ and

$$C = \int d\eta \rho(\eta) \frac{\eta}{1 - \beta(\eta)}. \tag{6.37}$$

Using **(6.37)** and the notation $k_\eta = k_\eta(t, t_0, \eta)$, we write the dynamic equation **(6.2)** as

$$\frac{\partial k_\eta}{\partial t} = \frac{\eta k_\eta}{Ct}, \tag{6.38}$$

which has a solution of the form **(6.3)**, given that

$$\beta(\eta) = \frac{\eta}{C}, \tag{6.39}$$

confirming the self-consistent nature of the assumption **(6.3)**.

To complete the calculation we need to determine C from **(6.37)**. After substituting $\beta(n)$ with η/C, we obtain

$$1 = \int_0^{\eta_{\max}} d\eta \rho(\eta) \frac{1}{\frac{C}{\eta} - 1}, \tag{6.40}$$

where $\eta_{\max}$ is the maximum possible fitness in the system. The integral **(6.40)** is singular. However, since $\beta(\eta) = \eta/C < 1$ for any η, we have $C > \eta_{\max}$, thus the integration limit never reaches the singularity. Note also that since

$$Cmt = \sum_j \eta_j k_j \leq \eta_{\max} \sum_j k_j = 2mt\eta_{\max} \tag{6.41}$$

we have $C \leq 2\eta_{\max}$.

If there is a single dynamic exponent β, the degree distribution follows the power law $p_k \sim k^{-\gamma}$ with degree exponent $\gamma = 1/\beta + 1$. In the Bianconi–Barabási model we have a spectrum of dynamic exponents $\beta(\eta)$, thus p_k is a weighted sum over different power laws.

To determine the degree distribution in the large-N limit, we first calculate the number of nodes with fitness η and with degree greater than k, that is those satisfying $k_\eta(t) > k$. Using **(6.3)** we find that this condition implies

$$t_0 < t\left(\frac{m}{k}\right)^{C/\eta}. \tag{6.42}$$

Exactly one node is added at each time step and each node has probability $\rho(\eta)d\eta$ of having fitness η. Therefore, $t\left(\frac{m}{k}\right)^{C/\eta}\rho(\eta)d\eta$ nodes satisfy condition **(6.42)**. To obtain the cumulative distribution function (the probability that a random node i has degree smaller than or equal to k), we write

$$\begin{aligned} P(k) = P(k_i \leq k) = 1 - P(k_i > k) &\approx 1 \frac{\int_0^{\eta_{\max}} t\left(\frac{m}{k}\right)^{C/\eta} \rho(\eta)d\eta}{m_0 + t} \\ &\approx 1 - \int_0^{\eta_{\max}} \left(\frac{m}{k}\right)^{C/\eta} \rho(\eta)d\eta, \end{aligned} \tag{6.43}$$

where the last equation is valid asymptotically, for large t. The probability density function for the degree distribution is

$$p(k) = P'(k) = \int_0^{\eta_{\max}} \frac{C}{\eta} m^{C/\eta} k^{-(C/\eta+1)} \rho(\eta)d\eta,$$

recovering **(6.6)**.

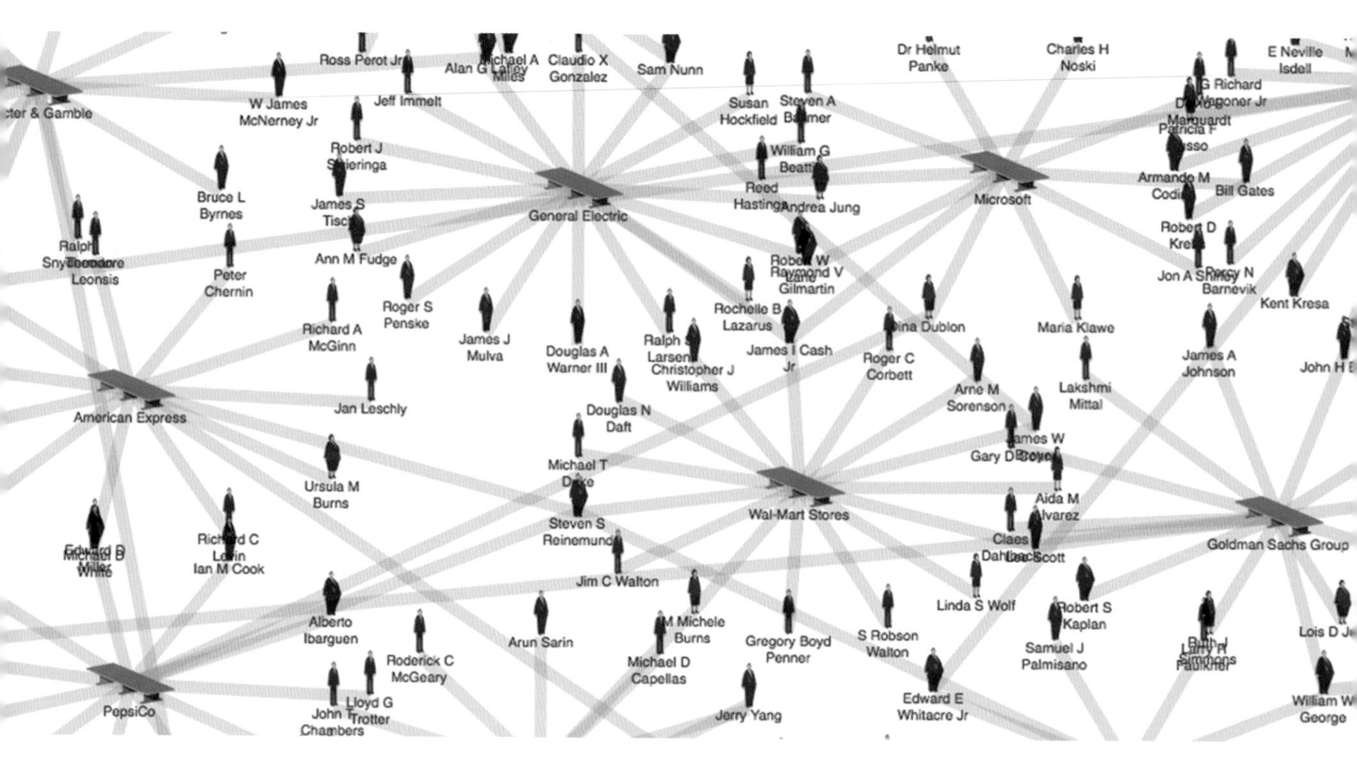

Figure 7.0 **Art & Networks: Josh On**

Created by Josh On, a San Francisco-based designer, the interactive website TheyRule.net uses a network representation to illustrate the interlocking relationship of the US economic class. By mapping out the shared board membership of the most powerful US companies, it reveals the influential role of a small number of individuals who sit on multiple boards. Since its release in 2001, the project has been viewed interchangeably as art and science.

CHAPTER 7

Degree Correlation

7.1 Introduction

Angelina Jolie and Brad Pitt, Ben Affleck and Jennifer Garner, Harrison Ford and Calista Flockhart, Michael Douglas and Catherine Zeta-Jones, Tom Cruise and Katie Holmes, Richard Gere and Cindy Crawford (**Figure 7.1**). An odd list, yet instantly recognizable to those immersed in the headline-driven world of celebrity couples. They are Hollywood stars that are or were married. Their weddings (and breakups) have drawn countless hours of media coverage and sold millions of gossip magazines. Thanks to them we take for granted that celebrities marry each other. We rarely pause to ask: Is this normal? In other words, what is the true chance that a celebrity marries another celebrity?

Assuming that a celebrity could date anyone from a pool of about a hundred million (10^8) eligible individuals worldwide, the chances that their mate would be another celebrity from a generous list of 1,000 other celebrities is only 10^{-5}. Therefore, if dating were driven by random encounters, celebrities would never marry each other.

Figure 7.1 **Hubs Dating Hubs**
Celebrity couples, representing a highly visible proof that in social networks hubs tend to know, date, and marry each other.

Even if we do not care about the dating habits of celebrities, we must pause and explore what this phenomenon tells us about the structure of the social network. Celebrities, political leaders, and CEOs of major corporations tend to know an exceptionally large number of individuals and are known by even more. They are hubs. Hence, celebrity dating (**Figure 7.1**) and joint board memberships (**Figure 7.0**) are manifestations of an interesting property of social networks: hubs tend to have ties to other hubs.

As obvious as this may sound, the property is not present in all networks. Consider, for example, the protein interaction network of yeast, shown in **Figure 7.2**. A quick inspection of the network reveals its scale-free nature: numerous one- and two-degree proteins coexist with a few highly connected hubs. These hubs, however, tend to avoid linking *to each other*. They link instead to many small-degree nodes, generating a hub-and-spoke pattern. This is particularly obvious for the two hubs highlighted in **Figure 7.2**: they almost exclusively interact with small-degree proteins.

A brief calculation illustrates how unusual this pattern is. Let us assume that each node chooses randomly the nodes it connects to. Therefore, the probability that nodes with degrees k and k' link to each other is

$$p_{k,k'} = \frac{kk'}{2L}. \tag{7.1}$$

Equation **(7.1)** tells us that hubs, by virtue of the many links they have, are much more likely to connect to each other than to small-degree nodes. Indeed, if k and k' are large, so is $p_{k,k'}$. Consequently, the likelihood that hubs with degrees $k = 56$ and $k' = 13$ have a direct link between them is $p_{k,k'} = 0.16$,

Figure 7.2 **Hubs Avoiding Hubs**
The protein interaction map of yeast. Each node corresponds to a protein and two proteins are linked if there is experimental evidence that they can bind to each other in the cell. We have highlighted the two largest hubs, with degrees $k = 56$ and $k' = 13$. They both connect to many small-degree nodes and avoid linking to each other.

The network has $N = 1{,}870$ proteins and $L = 2{,}277$ links, representing one of the earliest protein interaction maps [201, 202]. Only the largest component is shown. Note that the protein interaction network of yeast in **Table 4.1** represents a later map, hence it contains more nodes and links than the network shown in this figure. Node color corresponds to the essentiality of each protein: the removal of the red nodes kills the organism, hence they are called lethal or essential proteins. In contrast, the organism can survive without one of its green nodes. After [21].

which is 400 times larger than $p_{1,2} = 0.0004$, the likelihood that a degree-two node links to a degree-one node. Yet, there are no direct links between the hubs in **Figure 7.2**, although we observe numerous direct links between small-degree nodes.

Instead of linking to each other, the hubs highlighted in **Figure 7.2** almost exclusively connect to degree-one nodes. By itself this is not unexpected: we expect that a hub with degree $k = 56$ should link to $N_1 p_{1,56} \approx 12$ nodes with $k = 1$. The problem is that this hub connects to 46 degree-one neighbors (i.e., four times the expected number).

In summary, while in social networks hubs tend to "date" each other, in the protein interaction network the opposite is true. The hubs avoid linking to other hubs, connecting instead to many small-degree nodes. While it is dangerous to derive generic principles from two examples, the purpose of this chapter is to show that these patterns are manifestations of a general property of real networks: they exhibit a phenomenon called *degree correlations*. We discuss how to measure degree correlations and explore their impact on the network topology.

7.2 Assortativity and Disassortativity

Just by virtue of the many links they have, hubs are expected to link to each other. In some networks they do, in others they don't. This is illustrated in **Figure 7.3**, which shows three networks with identical degree sequences but different topologies.

- **Neutral Network**

  Figure 7.3b shows a network whose wiring is random. We call this network *neutral*, meaning that the number of links between the hubs coincides with what we expect by chance, as predicted by **(7.1)**.

- **Assortative Network**

  The network of **Figure 7.3a** has precisely the same degree sequence as that in **Figure 7.3b**, yet the hubs in **Figure 7.3a**

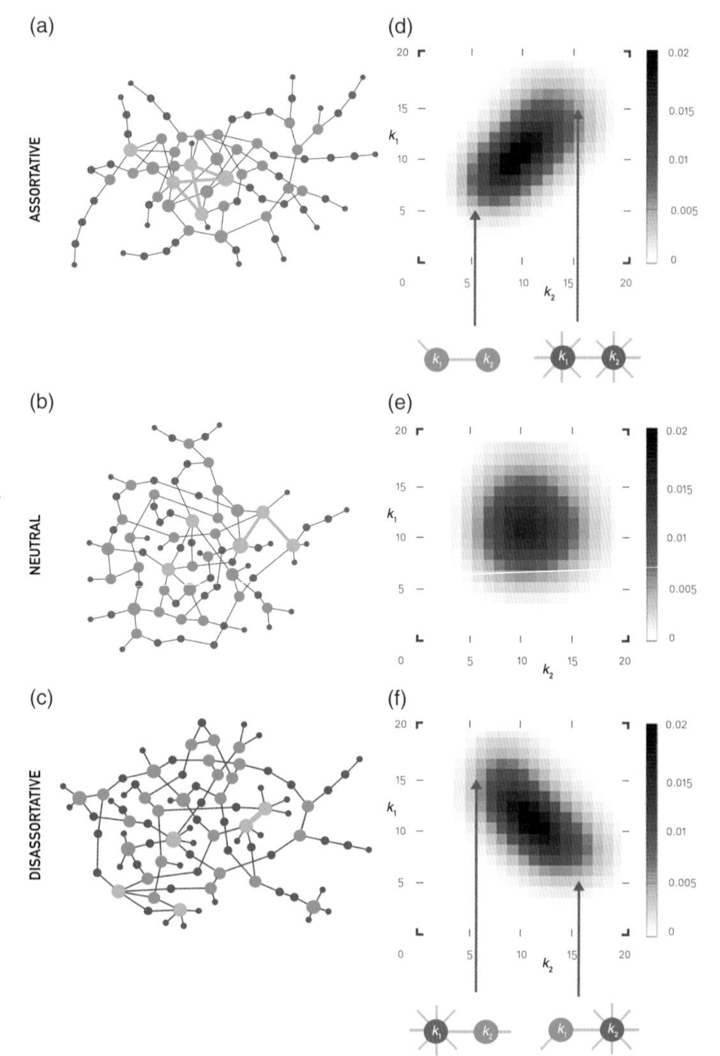

Figure 7.3 **Degree Correlation Matrix**

(a)–(c) Three networks that have precisely the same degree distribution (Poisson p_k), but display different degree correlations. We show only the largest component and we highlight in orange the five highest-degree nodes and the direct links between them.

(d)–(f) The degree correlation matrix e_{ij} for an assortative (d), a neutral (e) and a disassortative network (f) with Poisson degree distribution, $N = 1000$ and $\langle k \rangle = 10$. The colors correspond to the probability that a randomly selected link connects nodes with degrees k_1 and k_2.

(a,d) **Assortative Networks**

For assortative networks e_{ij} is high along the main diagonal. This indicates that nodes of comparable degree tend to link to each other: small-degree nodes to small-degree nodes and hubs to hubs. Indeed, the network in (a) has numerous links between its hubs as well as between its small-degree nodes.

(b,e) **Neutral Networks**

In neutral networks nodes link to each other randomly. Hence the density of links is symmetric around the average degree, indicating the lack of correlations in the linking pattern.

(c,f) **Disassortative Networks**

In disassortative networks e_{ij} is higher along the secondary diagonal, indicating that hubs tend to connect to small-degree nodes and small-degree nodes to hubs. Consequently, these networks have a hub-and-spoke character, as seen in (c).

tend to link to each other and avoid linking to small-degree nodes. At the same time, the small-degree nodes tend to connect to other small-degree nodes. Networks displaying such trends are *assortative*. An extreme manifestation of this pattern is a perfectly assortative network, in which each degree-k node connects only to other degree-k nodes (**Figure 7.4**).

- **Disassortative Network**
  In **Figure 7.3c** the hubs avoid each other, linking instead to small-degree nodes. Consequently, the network displays a hub-and-spoke character, making it *disassortative*.

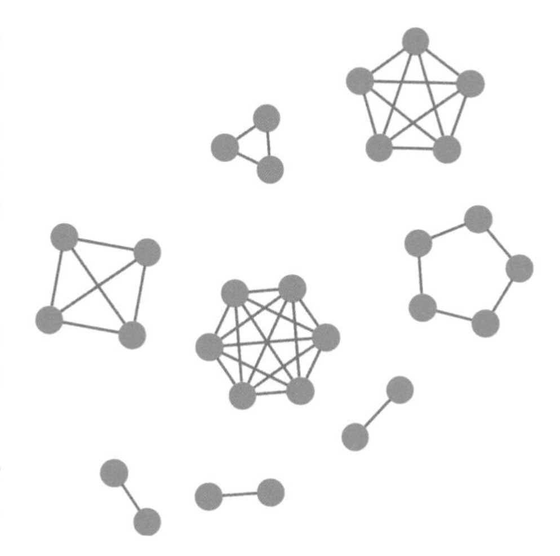

Figure 7.4 **Perfect Assortativity**
In a perfectly assortative network each node links only to nodes with the same degree. Hence $e_{jk} = \delta_{jk} q_k$, where δ_{jk} is the Kronecker delta. In this case all non-diagonal elements of the e_{jk} matrix are zero. The figure shows such a perfectly assortative network, consisting of complete k-cliques.

In general a network displays *degree correlations* if the number of links between the high- and low-degree nodes is systematically different from what is expected by chance. In other words, the number of links between nodes of degree k and k' deviates from (**7.1**).

The information about potential degree correlations is captured by the *degree correlation matrix* e_{ij}, which is the probability of finding a node with degrees i and j at the two ends of a randomly selected link. As e_{ij} is a probability, it is normalized:

$$\sum_{i,j} e_{ij} = 1. \tag{7.2}$$

In (**5.27**) we derived the probability q_k that there is a degree-k node at the end of the randomly selected link, obtaining

$$q_k = \frac{k p_k}{\langle k \rangle}. \tag{7.3}$$

We can connect q_k to e_{ij} via

$$\sum_{j} e_{ij} = q_i. \tag{7.4}$$

In neutral networks, we expect

$$e_{ij} = q_i q_j. \tag{7.5}$$

A network displays degree correlations if e_{ij} deviates from the random expectation (**7.5**). Note that (**7.2**)–(**7.5**) are valid for networks with an arbitrary degree distribution, hence they apply to both random and scale-free networks.

Given that e_{ij} encodes all information about potential degree correlations, we start with its visual inspection. **Figure 7.3d–f** shows e_{ij} for an assortative, a neutral and a disassortative network. In a neutral network small and high-degree nodes connect to each other randomly, hence e_{ij} lacks any trend (**Figure 7.3e**). In contrast, assortative networks show high correlations along the main diagonal, indicating that nodes predominantly connect to nodes with comparable degree. Therefore, low-degree nodes tend to link to other low-degree nodes and hubs to hubs (**Figure 7.3d**). In disassortative networks e_{ij} displays the opposite trend: it has high correlations along the secondary diagonal. Therefore, high-degree nodes tend to connect to low-degree nodes (**Figure 7.3f**).

In summary, information about degree correlations is carried by the degree correlation matrix e_{ij}. Yet the study of degree correlations through the inspection of e_{ij} has numerous disadvantages:

- It is difficult to extract information from the visual inspection of a matrix.
- We are unable to infer the magnitude of the correlations, it is difficult to compare networks with different correlations.
- e_{jk} contains approximately $k_{\max}^2/2$ independent variables, representing a huge amount of information that is difficult to model in analytical calculations and simulations.

We therefore need to develop a more compact way to detect degree correlations. This is the goal of the subsequent sections.

7.3 Measuring Degree Correlations

While e_{ij} contains complete information about the degree correlations characterizing a particular network, it is difficult to interpret its content. In this section we introduce a degree correlation function that offers a simpler way to quantify degree correlations.

Degree correlations capture the relationship between the degrees of nodes that link to each other. One way to quantify their magnitude is to measure for each node i the average degree of its neighbors (**Figure 7.5**):

$$k_{nn}(k_i) = \frac{1}{k_i}\sum_{j=1}^{N} A_{ij}k_j. \tag{7.6}$$

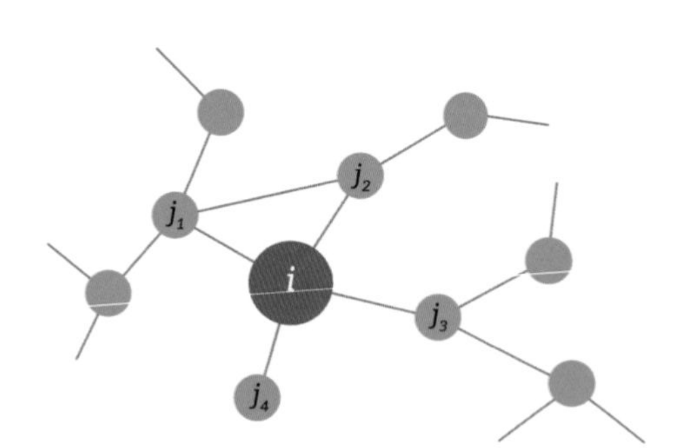

Figure 7.5 Nearest-Neighbor Degree: k
To determine the degree correlation function $k_{nn}(k_i)$ we calculate the average degree of a node's neighbors. The figure illustrates the calculation of $k_{nn}(k_i)$ for node i. As the degree of node i is $k_i = 4$, by averaging the degree of its neighbors j_1, j_2, j_3 and j_4, we obtain $k_{nn}(4) = (4+3+3+1)/4 = 2.75$.

The *degree correlation function* calculates **(7.6)** for all nodes with degree k [203, 204]:

$$k_{nn}(k) = \sum_{k'} k' P(k'|k), \tag{7.7}$$

where $P(k'|k)$ is the conditional probability that by following a link of a degree-k node, we reach a degree-k' node. Therefore, $k_{nn}(k)$ is the average degree of the neighbors of all degree-k

nodes. To quantify degree correlations we inspect the dependence of $k_{nn}(k)$ on k.

- **Neutral Network**
  For a neutral network, **(7.3)–(7.5)** predict

$$P(k'|k) = \frac{e_{kk'}}{\sum_{k'} e_{kk'}} = \frac{e_{kk'}}{q_k} = \frac{q_{k'}q_k}{q_k} = q_{k'}. \tag{7.8}$$

  This allows us to express $k_{nn}(k)$ as

$$k_{nn}(k) = \sum_{k'} k' q_{k'} = \sum_{k'} k' \frac{k'p(k')}{\langle k \rangle} = \frac{\langle k^2 \rangle}{\langle k \rangle}. \tag{7.9}$$

  Therefore, in a neutral network the average degree of a node's neighbors is independent of the node's degree k and depends only on the global network characteristics $\langle k \rangle$ and $\langle k^2 \rangle$. So, plotting $k_{nn}(k)$ as a function of k should result in a horizontal line at $\langle k^2 \rangle / \langle k \rangle$, as observed for the power grid (**Figure 7.6b**). Equation **(7.9)** also captures an intriguing property of real networks: our friends are more popular than we are, a phenomenon called the *friendship paradox* (**Box 7.1**).

- **Assortative Network**
  In assortative networks hubs tend to connect to other hubs, hence the higher is the degree k of a node, the higher is the average degree of its nearest neighbors. Consequently, for assortative networks $k_{nn}(k)$ *increases with* k, as observed for scientific collaboration networks (**Figure 7.6a**).
- **Disassortative Network**
  In disassortative networks hubs prefer to link to low-degree nodes. Consequently, $k_{nn}(k)$ *decreases with* k, as observed for the metabolic network (**Figure 7.6c**).

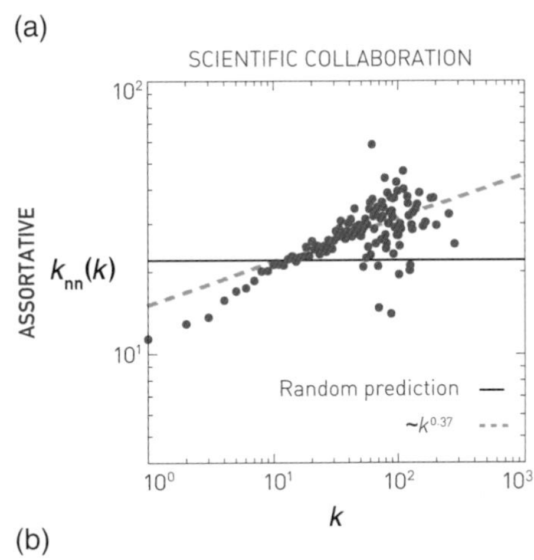

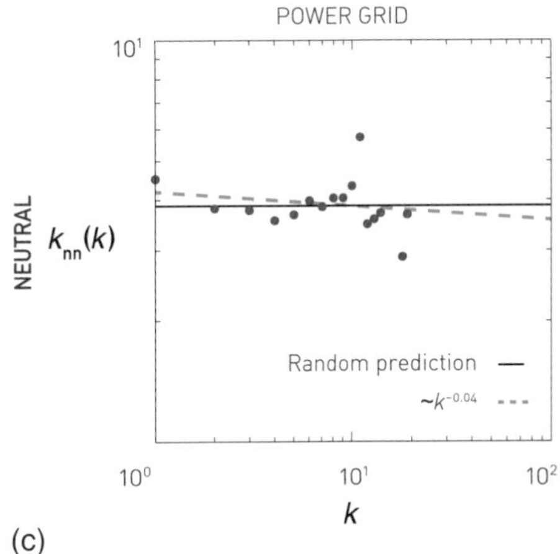

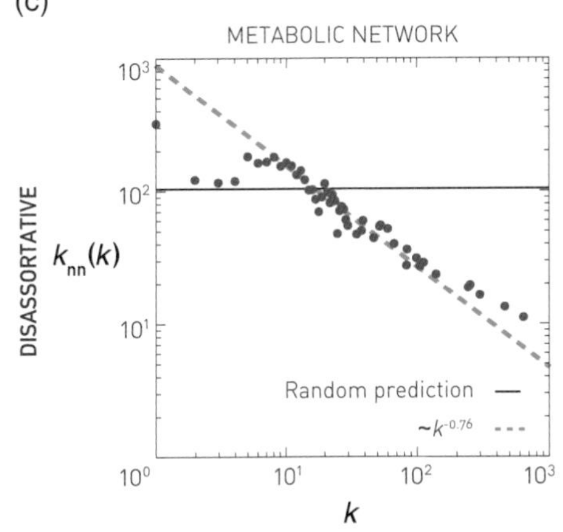

Figure 7.6 **Degree Correlation Function**
The degree correlation function $k_{nn}(k)$ for three real networks. The panels show $k_{nn}(k)$ on a log–log plot to test the validity of the scaling law **(7.10)**.
(a) **Scientific Collaboration Network**
The increasing $k_{nn}(k)$ with k indicates that the network is assortative.
(b) **Power Grid**
The horizontal $k_{nn}(k)$ indicates the lack of degree correlations, in line with **(7.9)** for neutral networks.
(c) **Metabolic Network**
The decreasing $k_{nn}(k)$ documents the network's disassortative nature. On each panel the horizontal line corresponds to the prediction **(7.9)** and the green dashed line is a fit to **(7.10)**.

Box 7.1 Friendship Paradox

The friendship paradox makes a surprising statement: *On average, my friends are more popular than I am* [210, 211]. This claim is rooted in **(7.9)**, telling us that the average degree of a node's neighbors is not simply $\langle k \rangle$, but depends on $\langle k^2 \rangle$ as well.

Consider a random network, for which $\langle k^2 \rangle = \langle k \rangle (1 + \langle k \rangle)$. According to **(7.9)**, $k_{nn}(k) = 1 + \langle k \rangle$. Therefore, the average degree of a node's neighbors is always higher than the average degree of a randomly chosen node, which is $\langle k \rangle$.

The gap between $\langle k \rangle$ and our friends' degree can be particularly large in scale-free networks, for which $\langle k^2 \rangle / \langle k \rangle$ significantly exceeds $\langle k \rangle$ (**Figure 4.8**). Consider, for example, the actor network for which $\langle k^2 \rangle / \langle k \rangle = 565$ (**Table 4.1**). In this network the average degree of a node's friends is hundreds of times the degree of the node itself.

The friendship paradox has a simple origin: we are more likely to be friends with hubs than with small-degree nodes, simply because hubs have more friends than the small nodes.

The behavior observed in **Figure 7.6** prompts us to approximate the degree correlation function with [203]

$$k_{nn}(k) = ak^{\mu}. \tag{7.10}$$

If the scaling **(7.10)** holds, then the nature of degree correlations is determined by the sign of the *correlation exponent* μ.

- **Assortative Networks:** $\mu > 0$
  A fit to $k_{nn}(k)$ for the science collaboration network provides $\mu = 0.37 \pm 0.11$ (**Figure 7.6a**).
- **Neutral Networks:** $\mu = 0$
  According to **(7.9)**, $k_{nn}(k)$ is independent of k. Indeed, for the power grid we obtain $\mu = 0.04 \pm 0.05$, which is indistinguishable from zero (**Figure 7.6b**).
- **Disassortative Networks:** $\mu < 0$
  For the metabolic network we obtain $\mu = -0.76 \pm 0.04$ (**Figure 7.6c**).

In summary, the degree correlation function helps us capture the presence or absence of correlations in real networks. The $k_{nn}(k)$ function also plays an important role in analytical calculations, allowing us to predict the impact of degree correlations on various network characteristics (**Section 7.6**). Yet it is often convenient to use a single number to capture the magnitude of correlations present in a network. This can be achieved either through the correlation exponent μ defined in **(7.10)**, or using the degree correlation coefficient introduced in **Box 7.2**.

7.4 Structural Cutoffs

Throughout this book we have assumed that networks are *simple*, meaning that there is at most one link between two nodes (**Figure 2.17**). For example, in the email network we place a single link between two individuals that are in email contact, despite the fact that they may have exchanged multiple messages. Similarly, in the actor network we connect two actors with a single link if they acted in the same movie, independent of the number of joint movies. All datasets discussed in **Table 4.1** are simple networks.

In simple networks there is a puzzling conflict between the scale-free property and degree correlations [209, 210]. Consider, for example, the scale-free network of **Figure 7.7a**, whose

Box 7.2 Degree Correlation Coefficient

If we wish to characterize degree correlations using a single number, we can use either μ or the *degree correlation coefficient*, defined as [205, 206]

$$r = \sum_{jk} \frac{jk(e_{jk} - q_j q_k)}{\sigma^2} \tag{7.11}$$

with

$$\sigma^2 = \sum_k k^2 q_k - \left[\sum_k k q_k\right]^2. \tag{7.12}$$

Hence, r is the Pearson correlation coefficient between the degrees found at the two ends of the same link. It varies between $-1 \leq r \leq 1$. For $r > 0$ the network is assortative, for $r = 0$ the network is neutral and for $r < 0$ the network is disassortative. For example, for the scientific collaboration network we obtain $r = 0.13$, in line with its assortative nature; for the protein interaction network $r = -0.04$, supporting its disassortative nature and for the power grid we have $r = 0$.

The assumption behind the degree correlation coefficient is that $k_{nn}(k)$ depends linearly on k with slope r. In contrast, the correlation exponent μ assumes that $k_{nn}(k)$ follows the power law **(7.10)**. Naturally, both cannot be valid simultaneously. The analytical models of **Section 7.7** offer some guidance, supporting the validity of **(7.10)**. As we show in **Advanced topics 7.A**, in general r correlates with μ.

two largest hubs have degrees $k = 55$ and $k' = 46$. In a network with degree correlations $e_{kk'}$, the expected number of links between k and k' is

$$E_{kk'} = e_{kk'} \langle k \rangle N. \tag{7.13}$$

For a neutral network, $e_{kk'}$ is given by **(7.5)**, which, using **(7.3)**, predicts

$$E_{kk'} = \frac{k p_k k' p_{k'}}{\langle k \rangle} N = \frac{\frac{55}{300}\frac{46}{300}}{3} 300 = 2.8. \tag{7.14}$$

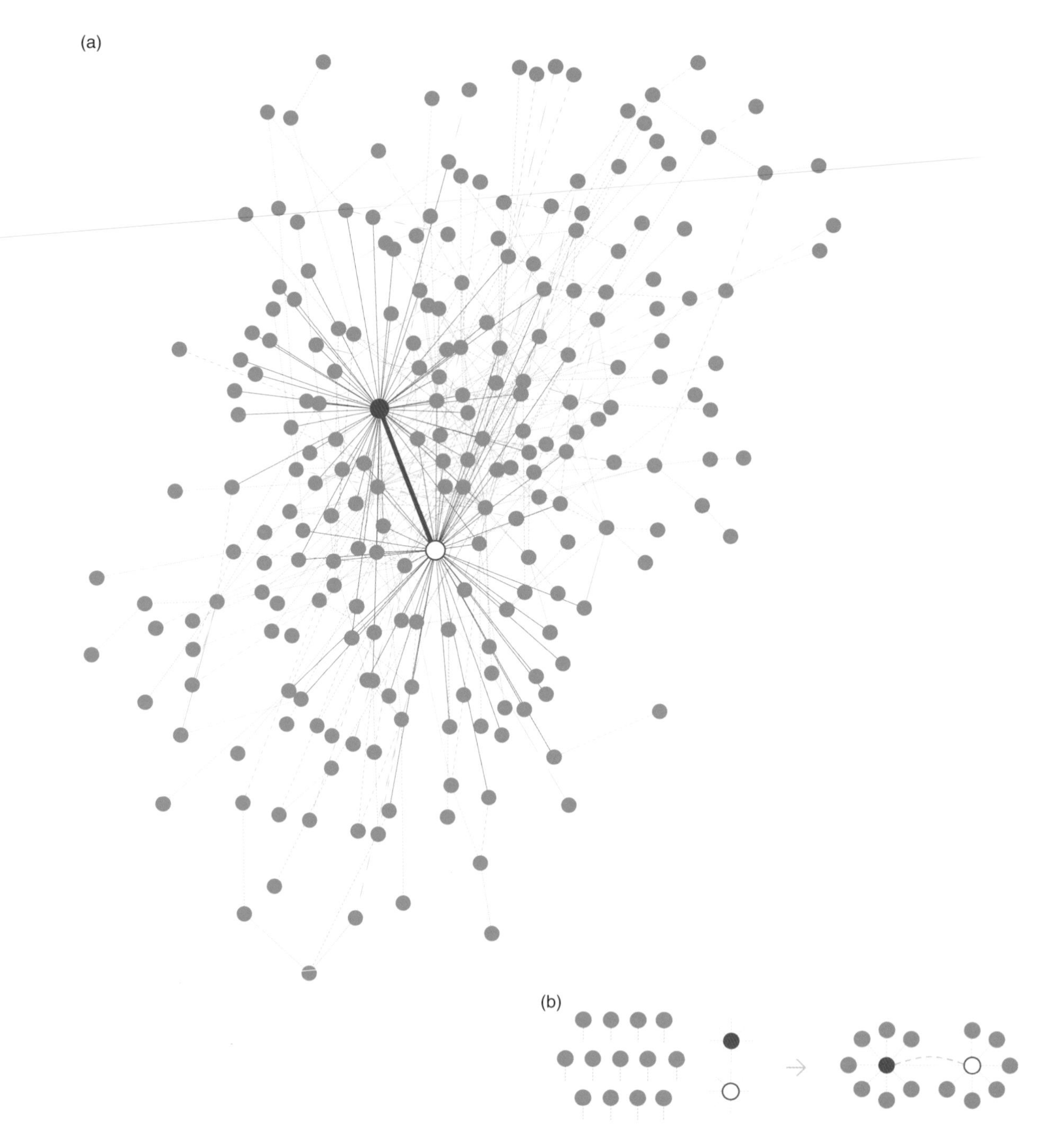

Figure 7.7 **Structural Disassortativity**

(a) A scale-free network with $N = 300$, $L = 450$ and $\gamma = 2.2$, generated by the configuration model (Figure 4.15). By forbidding self-loops and multi-links, we made the network *simple*. We highlight the two largest nodes in the network. As **(7.14)** predicts, to maintain the network's neutral nature, we need approximately three links between these two nodes. The fact that we do not allow multi-links (simple network representation) makes the network disassortative, a phenomenon called *structural disassortativity*.

(b) To illustrate the origins of structural correlations we start from a fixed degree sequence, shown as individual stubs on the left. Next we randomly connect the stubs (configuration model). In this case the expected number of links between the nodes with degree 8 and 7 is $8 \times 7/28 \approx 2$. Yet, if we do not allow multi-links, there can only be one link between these two nodes, making the network structurally disassortative.

Therefore, given the size of these two hubs, they should be connected to each other by *two to three links* to comply with the network's neutral nature. Yet, in a simple network we can have only one link between them, causing a conflict between degree correlations and the scale-free property. The goal of this section is to understand the origin and the consequences of this conflict.

For small k and k', **(7.14)** predicts that $E_{kk'}$ is also small, that is we expect less than one link between the two nodes. Only for nodes whose degree exceeds some threshold k_s does **(7.14)** predict multiple links. As we show in **Advanced topics 7.B**, k_s, called the *structural cutoff*, scales as

$$k_s(N) \sim (\langle k \rangle N)^{1/2}. \tag{7.15}$$

In other words, nodes whose degree exceeds **(7.15)** have $E_{kk'} > 1$, a conflict that, as we show below, gives rise to degree correlations.

To understand the consequences of the structural cutoff we must first ask if a network has nodes whose degrees exceed **(7.15)**. For this we compare the structural cutoff, k_s, with the natural cutoff, k_{max}, which is the expected largest degree in a network. According to (4.18), for a scale-free network $k_{max} \sim N^{\frac{1}{\gamma-1}}$. Comparing k_{max} and k_s allows us to distinguish two regimes:

- **No Structural Cutoff**
  For random networks and scale-free networks with $\gamma \geq 3$ the exponent of k_{max} is smaller than 1/2, hence k_{max} is always smaller than k_s. In other words, the node size at which the structural cutoff turns on exceeds the size of the biggest hub. Consequently we have no nodes for which $E_{kk'} > 1$. For these networks we do not have a conflict between degree correlations and the simple network requirement.
- **Structural Disassortativity**
  For scale-free networks with $\gamma < 3$ we have $1/(\gamma - 1) > 1/2$, that is k_s can be smaller than k_{max}. Consequently, nodes whose degree is between k_s and k_{max} can violate $E_{kk'} > 1$. In other words, the network has fewer links between its hubs than **(7.14)** would predict. These networks will therefore become disassortative, a phenomenon we call *structural disassortativity*. This is illustrated in **Figure 7.8a,b**, showing a simple scale-free network generated by the configuration model. The network shows disassortative scaling, despite the fact that we did not impose degree correlations during its construction.

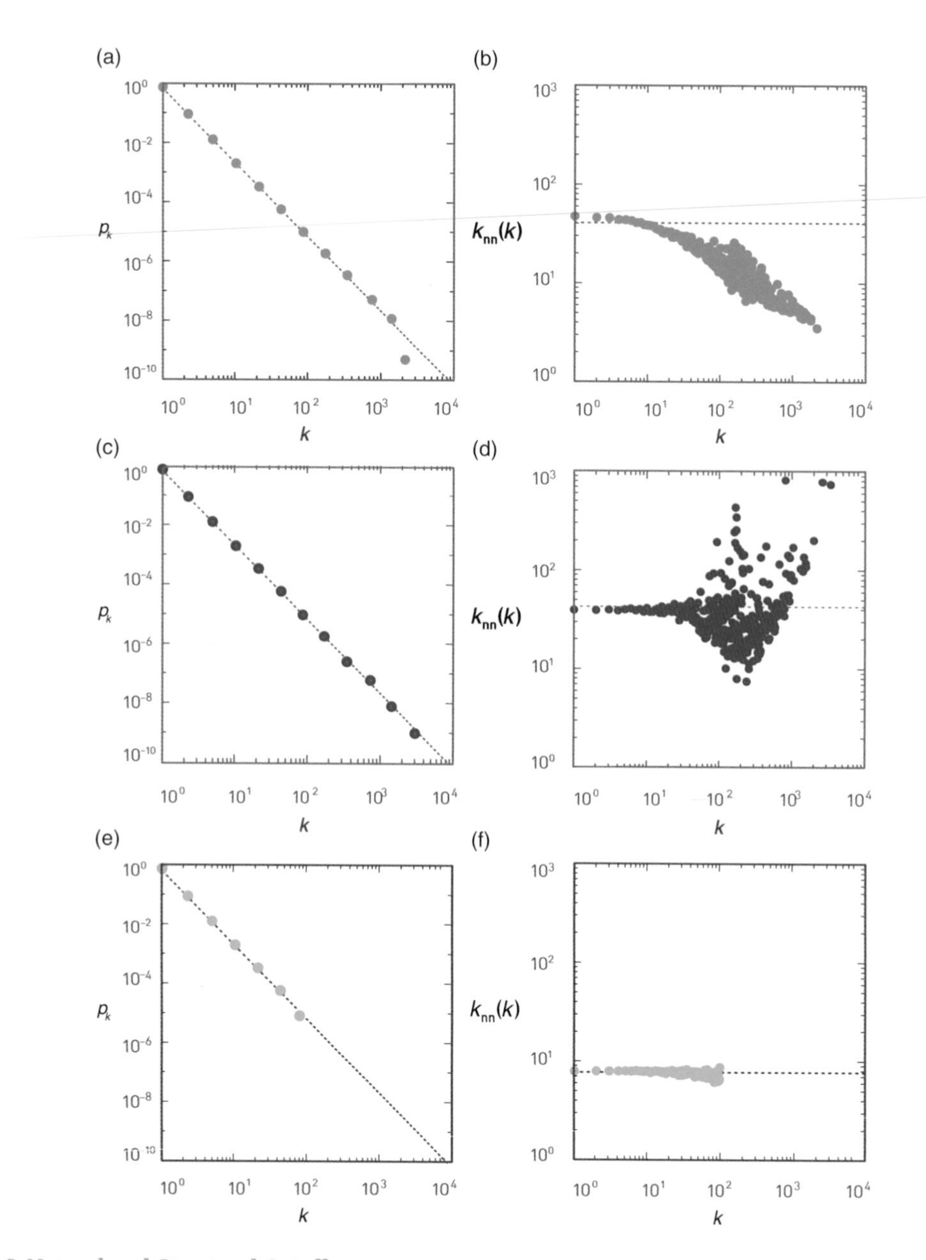

Figure 7.8 **Natural and Structural Cutoffs**
The figure illustrates the tension between the scale-free property and degree correlations. We show the degree distribution (left panels) and the degree correlation function $k_{nn}(k)$ (right panels) of a scale-free network with $N = 10,000$ and $\gamma = 2.5$, generated by the configuration model (**Figure 4.15**).
(a,b) If we generate a scale-free network with the power-law degree distribution shown in (a), and we forbid self-loops and multi-links, the network displays structural disassortativity, as indicated by $k_{nn}(k)$ in (b). In this case, we lack a sufficient number of links between the high-degree nodes to maintain the neutral nature of the network, hence for high k the $k_{nn}(k)$ function must decay.
(c,d) We can eliminate structural disassortativity by relaxing the simple network requirement, that is allowing multiple links between two nodes. As shown in (c,d), in this case we obtain a neutral scale-free network.
(e,f) If we impose an upper cutoff by removing all nodes with $k \geq k_s \simeq 100$, as predicted by **(7.15)**, the network becomes neutral, as seen in (f).

We have two avenues to generate networks that are free of structural disassortativity:

(i) We can relax the simple network requirement, allowing multiple links between the nodes. The conflict disappears and the network will be neutral (**Figure 7.8c,d**).
(ii) If we insist on having a simple scale-free network that is neutral or assortative, we must remove all hubs with degree larger than k_s. This is illustrated in **Figure 7.8e,f**: a network that lacks nodes with $k \geq 100$ is neutral.

Finally, how can we decide whether the correlations observed in a particular network are a consequence of structural disassortativity or are generated by some unknown process that leads to degree correlations? Degree-preserving randomization (**Figure 4.17**) helps us distinguish these two possibilities:

(i) **Degree-Preserving Randomization with Simple Links (R–S)**
We apply degree-preserving randomization to the original network and at each step we make sure that we do *not* permit more than one link between a pair of nodes. On the algorithmic side this means that each rewiring that generates multi-links is discarded. If the real $k_{nn}^{R-S}(k)$ and the randomized $k_{nn}^{R-S}(k)$ are indistinguishable, then the correlations observed in a real system are all structural, fully explained by the degree distribution. If the randomized $k_{nn}^{R-S}(k)$ does not show degree correlations while $k_{nn}(k)$ does, there is some unknown process that generates the observed degree correlations.
(ii) **Degree-Preserving Randomization with Multiple Links (R–M)**
For a self-consistency check it is sometimes useful to perform degree-preserving randomization that allows for multiple links between the nodes. On the algorithmic side this means that we allow each random rewiring, even if it leads to multi-links. This process eliminates all degree correlations.

We performed the randomizations discussed above for three real networks. As **Figure 7.9a** shows, the assortative nature of the scientific collaboration network disappears under both randomizations. This indicates that the assortative correlations

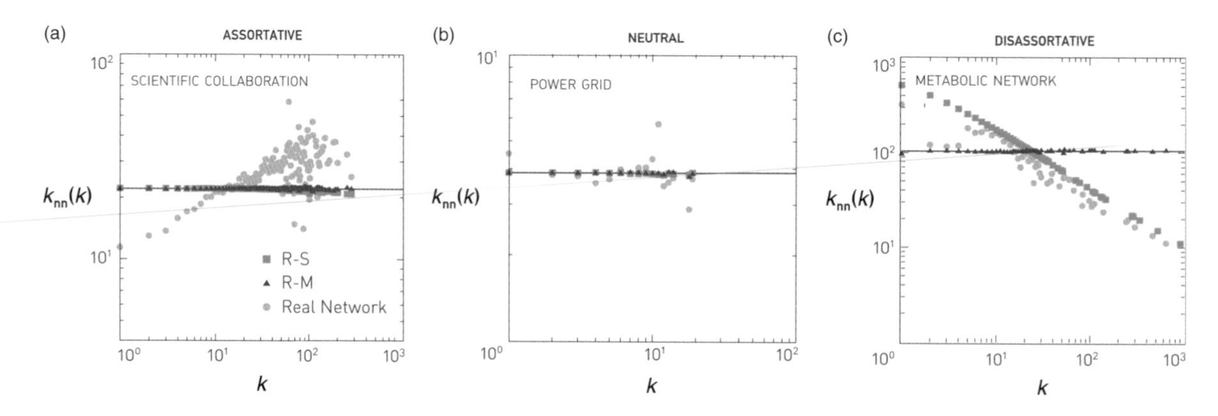

Figure 7.9 **Randomization and Degree Correlations**
To uncover the origin of the observed degree correlations, we must compare $k_{\mathrm{nn}}(k)$ (gray symbols) with $k_{\mathrm{nn}}^{\mathrm{R-S}}(k)$ and $k_{\mathrm{nn}}^{\mathrm{R-M}}(k)$ obtained after degree-preserving randomization. Two degree-preserving randomizations are informative in this context:
Randomization with Simple Links (R–S)
At each step of the randomization process we check that we do not have more than one link between any node pair.
Randomization with Multiple Links (R–M)
We allow multi-links during the randomization processes.
We performed these two randomizations for the networks of **Figure 7.6**. The R–M procedure always generates a neutral network, consequently $k_{\mathrm{nn}}^{\mathrm{R-M}}(k)$ is always horizontal. The true insight is obtained when we compare $k_{\mathrm{nn}}(k)$ with $k_{\mathrm{nn}}^{\mathrm{R-S}}(k)$, helping us to decide if the observed correlations are structural.
(a) **Scientific Collaboration Network**
The increasing $k_{\mathrm{nn}}(k)$ differs from the horizontal $k_{\mathrm{nn}}^{\mathrm{R-S}}(k)$, indicating that the network's assortativity is not structural. Consequently, the assortativity is generated by some process that governs the network's evolution. This is not unexpected: structural effects can generate only disassortativity, not assortativity.
(b) **Power Grid**
The horizontal $k_{\mathrm{nn}}(k)$, $k_{\mathrm{nn}}^{\mathrm{R-S}}(k)$ and $k_{\mathrm{nn}}^{\mathrm{R-M}}(k)$ all support the lack of degree correlations (neutral network).
(c) **Metabolic Network**
As both $k_{\mathrm{nn}}(k)$ and $k_{\mathrm{nn}}^{\mathrm{R-S}}(k)$ decrease, we conclude that the network's disassortativity is induced by its scale-free property. Hence the observed degree correlations are structural.

of the collaboration network are not linked to its scale-free nature. In contrast, for the metabolic network the observed disassortativity remains unchanged under R–S (**Figure 7.9c**). Consequently, the disassortativity of the metabolic network is structural, being induced by its degree distribution.

In summary, the scale-free property can induce disassortativity in simple networks. Indeed, in neutral or assortative networks we expect multiple links between the hubs. If multiple links are forbidden (simple graph), the network will display disassortative tendencies. This conflict vanishes for scale-free networks with $\gamma \geq 3$ and for random networks. It also vanishes if we allow multiple links between the nodes.

7.5 Correlations in Real Networks

To understand the prevalence of degree correlations we need to inspect the correlations characterizing real networks. In **Figure 7.10** we show the $k_{nn}(k)$ function for the ten reference networks, observing several patterns:

- **Power Grid**
  For the power grid, $k_{nn}(k)$ is flat and indistinguishable from its randomized version, indicating a lack of degree correlations **(Figure 7.10a)**. Hence the power grid is neutral.
- **Internet**
  For small degrees ($k \leq 30$), $k_{nn}(k)$ shows a clear assortative trend, an effect that levels off for high degrees **(Figure 7.10b)**. The degree correlations vanish in the randomized version of the Internet map. Hence the Internet is assortative, but structural cutoffs eliminate the effect for high k.
- **Social Networks**
  The three networks capturing social interactions – the mobile phone network, the science collaboration network and the actor network – all have an increasing $k_{nn}(k)$, indicating that they are assortative **(Figure 7.10c–e)**. Hence, in these networks hubs tend to link to other hubs and low-degree nodes tend to link to low-degree nodes. The fact that the observed $k_{nn}(k)$ differs from the $k_{nn}^{R-S}(k)$ indicates that the assortative nature of social networks is not due to their scale-free degree distribution.
- **Email Network**
  While the email network is often seen as a social network, its $k_{nn}(k)$ decreases with k, documenting a clear disassortative behavior **(Figure 7.10f)**. The randomized $k_{nn}^{R-S}(k)$ also decays, indicating that we are observing structural disassortativity, a consequence of the network's scale-free nature.
- **Biological Networks**
  The protein interaction and the metabolic network both have a negative μ, suggesting that these networks are disassortative. Yet, the scaling of $k_{nn}^{R-S}(k)$ is indistinguishable from $k_{nn}(k)$, indicating that we are observing structural disassortativity, rooted in the scale-free nature of these networks **(Figure 7.10g,h)**.
- **WWW**
  The decaying $k_{nn}(k)$ implies disassortative correlations **(Figure 7.10i)**. The randomized $k_{nn}^{R-S}(k)$ also decays, but

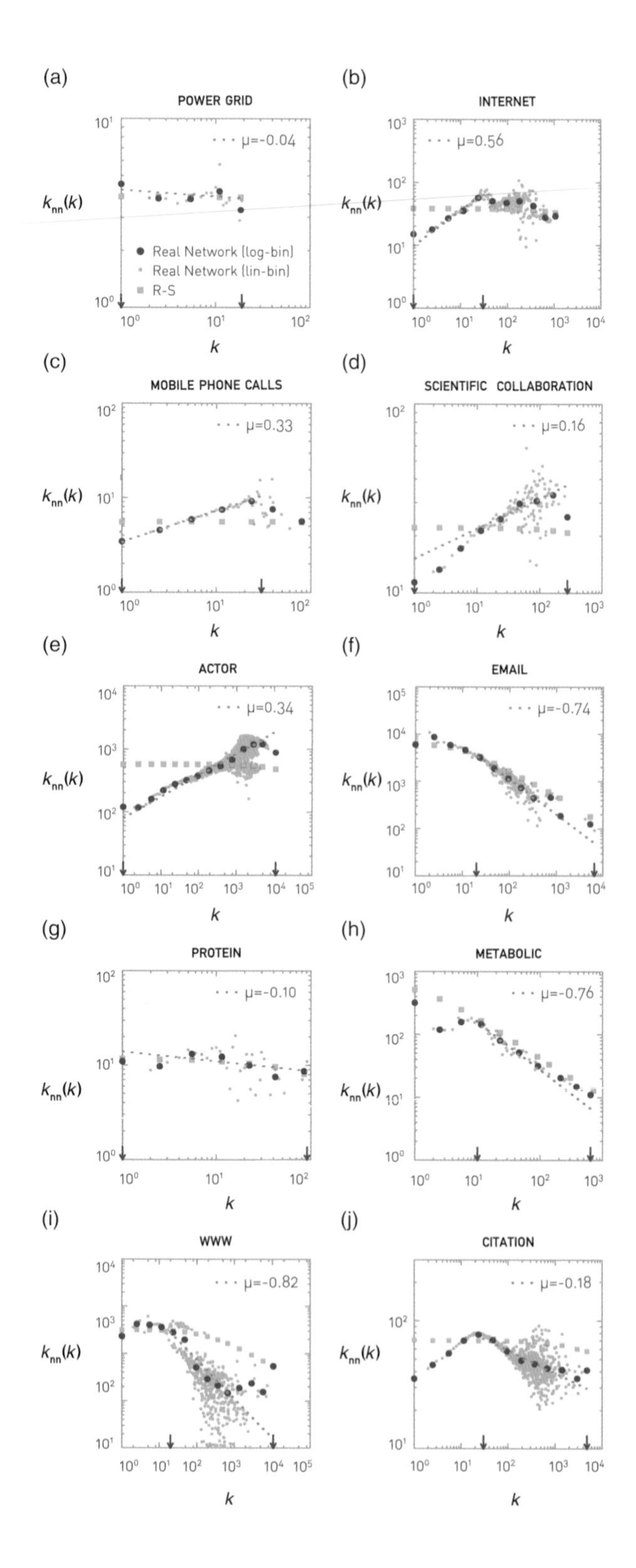

Figure 7.10 **Randomization and Degree Correlations**

The degree correlation function $k_{nn}(k)$ for the ten reference networks **(Table 4.1)**. The gray symbols show the $k_{nn}(k)$ function using linear binning; purple circles represent the same data using log-binning (**Section 4.11**). The green dotted line corresponds to the best fit to **(7.10)** within the fitting interval marked by the arrows at the bottom. Orange squares represent $k_{nn}^{R-S}(k)$ obtained for 100 independent degree-preserving randomizations, while maintaining the *simple* character of these networks. Note that we made directed networks undirected when we measured $k_{nn}(k)$. To fully characterize the correlations emerging in directed networks we must use the directed correlation function (**Box 7.3**).

Box 7.3 Correlations in Directed Networks

The degree correlation function **(7.7)** is defined for undirected networks. To measure correlations in directed networks we must take into account that each node i is characterized by an incoming k_i^{in} and an outgoing k_i^{out} degree [213]. We therefore define four degree correlation functions, $k_{nn}^{\alpha,\beta}(k)$, where α and β refer to the in and out indices, respectively (**Figure 7.11a–d**). In **Figure 7.11e** we show $k_{nn}^{\alpha,\beta}(k)$ for citation networks, indicating a lack of in–out correlations and the presence of assortativity for small k for the other three correlations (in–in, out–in, out–out).

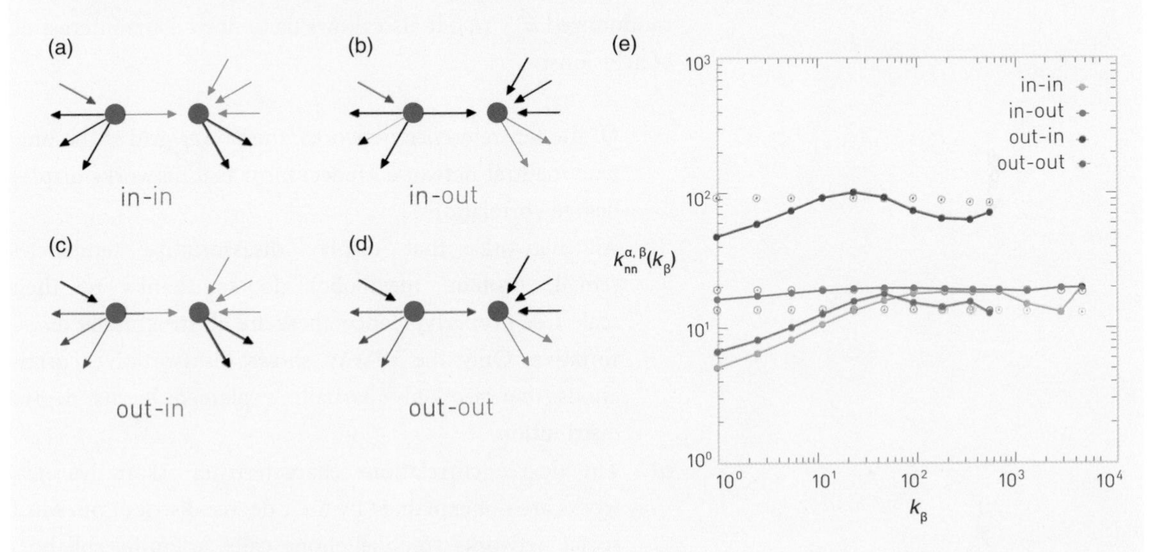

Figure 7.11 **Correlations in Directed Network**
(a)–(d) The four possible correlations characterizing a directed network. We show in purple and green the (α, β) indices that define the appropriate correlation function [213]. For example, $k_{nn}^{in,in}(k)$ describes the correlation between the in-degrees of two nodes connected by a link.
(e) The $k_{nn}^{\alpha,\beta}(k)$ correlation function for (directed) citation networks. For example, $k_{nn}^{in,in}(k)$ is the average in-degree of the in-neighbors of nodes with in-degree k_{in}. These functions show a clear assortative tendency for three of the four functions up to degree $k \simeq 100$. The empty symbols capture the degree-randomized $k_{nn}^{\alpha,\beta}(k)$ for each degree correlation function (R–S randomization).

not as rapidly as $k_{nn}(k)$. Hence the disassortative nature of the WWW is not fully explained by its degree distribution.

- **Citation Network**

  This network displays a puzzling behavior: for $k \leq 20$ the degree correlation function $k_{nn}(k)$ shows a clear assortative trend; for $k > 20$, however, we observe disassortative scaling (**Figure 7.10j**). Such mixed behavior can emerge in networks that display extreme assortativity (**Figure 7.13b**). This suggests that the citation network is strongly assortative, but its scale-free nature induces structural disassortativity, changing the slope of $k_{\mathrm{nn}}(k)$ for $k \gg k_s$.

In summary, **Figure 7.10** indicates that to understand degree correlations, we must always compare $k_{\mathrm{nn}}(k)$ with the degree-randomized $k_{\mathrm{nn}}^{\mathrm{R-S}}(k)$. It also allows us to draw some interesting conclusions:

(i) Of the ten reference networks, the power grid is the only truly neutral network. Hence, most real networks display degree correlations.

(ii) All networks that display disassortative tendencies (email, protein, metabolic) do so thanks to their scale-free property. Hence, these are all structurally disassortative. Only the WWW shows disassortative correlations that are only partially explained by its degree distribution.

(iii) The degree correlations characterizing assortative networks are not explained by their degree distribution. Most social networks (mobile phone calls, scientific collaboration, actor network) are in this class and so is the Internet and the citation network.

A number of mechanisms have been proposed to explain the origin of the observed assortativity. For example, the tendency of individuals to form communities, the topic of **Chapter 9**, can induce assortative correlations [211]. Similarly, society has endless mechanisms, from professional committees to TV shows, to bring hubs together, enhancing the assortative nature of social and professional networks. Finally, homophily, a well-documented social phenomenon [212], indicates that individuals tend to associate with other individuals of similar background and characteristics, hence individuals with comparable degree tend to know each other. This degree-homophily may be responsible for the celebrity marriages as well (**Figure 7.1**).

7.6 Generating Correlated Networks

To explore the impact of degree correlations on various network characteristics, we must first understand the correlations characterizing the network models discussed thus far. It is equally important to develop an algorithm that can generate networks with tunable correlations. As we show in this section, given the conflict between the scale-free property and degree correlations, this is not a trivial task.

7.6.1 Degree Correlations in Static Models

- **Erdős–Rényi Model**
  The random network model is neutral by definition. As it lacks hubs, it does not develop structural correlations either. Hence, for the Erdős–Rényi network $k_{nn}(k)$ is given by **(7.9)**, predicting $\mu = 0$ for any $\langle k \rangle$ and N.
- **Configuration Model**
  The configuration model (**Figure 4.15**) is also neutral, independent of the choice of degree distribution p_k. This is because the model allows for both multi-links and self-loops. Consequently, any conflicts caused by the hubs are resolved by the multiple links between them. If, however, we force the network to be simple, then the generated network will develop structural disassortativity (**Figure 7.8**).
- **Hidden-Parameter Model**
  In the model, e_{jk} is proportional to the product of the randomly chosen hidden variables η_j and η_k (**Figure 4.18**). Consequently, the network is technically uncorrelated. However, if we do not allow multi-links, for scale-free networks we again observe structural disassortativity. Analytical calculations indicate that in this case

$$k_{nn}(k) \sim k^{-1}, \qquad (7.16)$$

  that is the degree correlation function follows **(7.10)** with $\mu = -1$.

Taken together, the static models explored so far generate either neutral networks, or networks characterized by structural disassortativity following **(7.16)**.

7.6.2 Degree Correlations in Evolving Networks

To understand the emergence (or the absence) of degree correlations in growing networks, we start with the initial attractiveness model (**Section 6.5**), which includes as a special case the Barabási–Albert model.

- **Initial Attractiveness Model**
  Consider a growing network in which preferential attachment follows **(6.23)**, that is $\Pi(k) \sim A + k$, where A is the initial attractiveness. Depending on the value of A, we observe three distinct scaling regimes [214]:

  (i) Disassortative Regime:
  $\gamma < 3$ If $-m < A < 0$, we have

$$k_{\mathrm{nn}}(k) \sim m \frac{(m+A)^{1-\frac{A}{m}}}{2m+A} \varsigma\left(\frac{2m}{2m+A}\right) N^{-\frac{A}{2m+A}} k^{\frac{A}{m}}. \tag{7.17}$$

  Hence the resulting network is disassortative, $k_{nn}(k)$ decaying following the power law [214, 215]

$$k_{\mathrm{nn}}(k) \sim k^{-\frac{|A|}{m}}. \tag{7.18}$$

  (ii) Neutral Regime: $\gamma = 3$
  If $A = 0$ the initial attractiveness model reduces to the Barabási–Albert model. In this case

$$k_{\mathrm{nn}}(k) \sim \frac{m}{2} \ln N. \tag{7.19}$$

  Consequently, $k_{\mathrm{nn}}(k)$ is independent of k, hence the network is neutral.

  (iii) Weak Assortativity: $\gamma > 3$
  If $A > 0$ the calculations predict

$$k_{\mathrm{nn}}(k) \approx (m+A) \ln\left(\frac{k}{m+A}\right). \tag{7.20}$$

  As $k_{\mathrm{nn}}(k)$ increases logarithmically with k, the resulting network displays a weak assortative tendency, but does not follow **(7.10)**.

In summary, **(7.17)–(7.20)** indicate that the initial attractiveness model generates rather complex degree correlations, from disassortativity to weak assortativity. Equation **(7.19)** also shows that the network generated by the Barabási–Albert model is neutral. Finally, **(7.17)** predicts a power-law k-dependence for $k_{nn}(k)$, offering analytical support for the empirical scaling **(7.10)**.

- **Bianconi–Barabási Model**
  With a uniform fitness distribution the Bianconi–Barabási model generates a disassortative network [204] (**Figure 7.12**). The fact that the randomized version of the network is also disassortative indicates that the model's disassortativity is structural. Note, however, that the real $k_{nn}(k)$ and the randomized $k_{nn}^{R-S}(k)$ do not overlap, indicating that the disassortativity of the model is not fully explained by its scale-free nature.

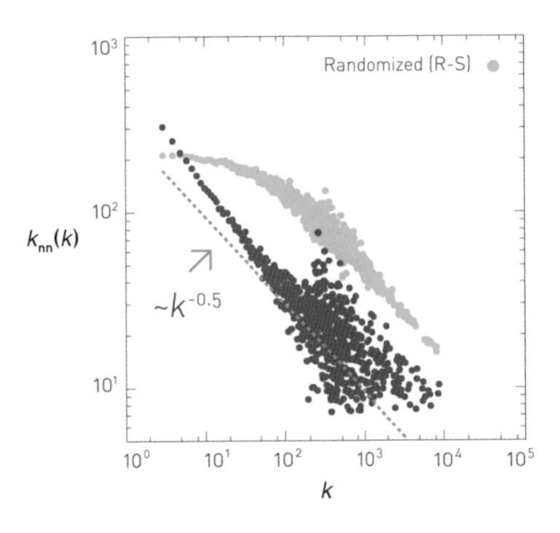

Figure 7.12 **Correlations in the Bianconi–Barabási Model**
The degree correlation function of the Bianconi–Barabási model for $N = 10,000$, $m = 3$ and uniform fitness distribution (**Section 6.2**). As the green dotted line following **(7.10)** indicates, the network is disassortative, consistent with $\mu \simeq -0.5$. The orange symbols correspond to k_{nn}^{R-S}. As $k_{nn}^{R-S}(k)$ also decreases, the bulk of the observed disassortativity is structural. However, the difference between $k_{nn}^{R-S}(k)$ and $k_{nn}(k)$ suggests that structural effects cannot fully account for the observed degree correlation.

7.6.3 Tuning Degree Correlations

Several algorithms can generate networks with desired degree correlations [207, 216]. Next we discuss a simplified version of the algorithm proposed by Xalvi-Brunet and Sokolov that aims to generate maximally correlated networks with a pre-defined degree sequence [217, 218, 219]. It consists of the following steps (**Figure 7.13a**).

- **Step 1: Link Selection**
  Choose at random two links. Label the four nodes at the end of these two links with a, b, c and d such that their degrees are ordered as $k_a \geq k_b \geq k_c \geq k_d$.
- **Step 2: Rewiring**
  Break the selected links and rewire them to form new pairs. Depending on the desired degree correlations, the rewiring is done in two ways.
  - **Step 2A: Assortative**
    By pairing the two highest degree nodes (a with b) and the two lowest degree nodes (c with d), we connect nodes with comparable degrees, enhancing the network's assortative nature.
  - **Step 2B: Disassortative**
    By pairing the highest and the lowest degree nodes (a with d and b with c), we connect nodes with different degrees, enhancing the network's disassortative nature.

By iterating these steps we gradually enhance the network's assortative (step 2A) or disassortative (step 2B) features. If we

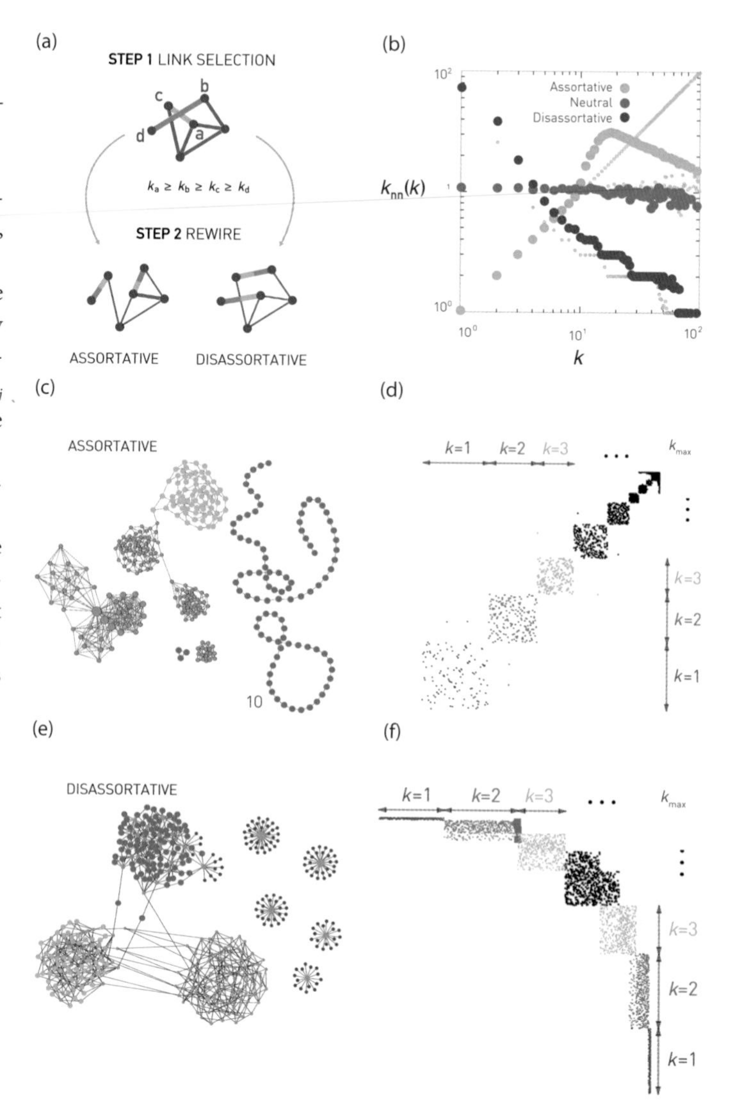

Figure 7.13 **Xulvi-Brunet and Sokolov Algorithm**
The algorithm generates networks with *maximal degree correlations.*
(a) The basic steps of the algorithm.
(b) $k_{nn}(k)$ for networks generated by the algorithm for a scale-free network with $N = 1{,}000$, $L = 2{,}500$, $\gamma = 3.0$.
(c,d) A typical network configuration and the corresponding A_{ij} matrix for the maximally assortative network generated by the algorithm, where the rows and columns of A_{ij} were ordered according to increasing node degrees k.
(e,f) Same as (c,d) for a maximally disassortative network.

The A_{ij} matrices (d) and (f) capture the inner regularity of networks with maximal correlations, consisting of blocks of nodes that connect to nodes with similar degree in (d) and of blocks of nodes that connect to nodes with rather different degrees in (f).

aim to generate a simple network (free of multi-links), after step 2 we check whether the particular rewiring leads to multi-links. If it does, we reject it, returning to step 1.

The correlations characterizing the networks generated by this algorithm converge to the maximal (assortative) or minimal (disassortative) value that we can reach for the given degree sequence (**Figure 7.13b**). The model has no difficulty creating disassortative correlations (**Figure 7.13e, f**). In the assortative limit, simple networks display a mixed $k_{nn}(k)$: assortative for small k and disassortative for high k (**Figure 7.13b**). This is a consequence of structural

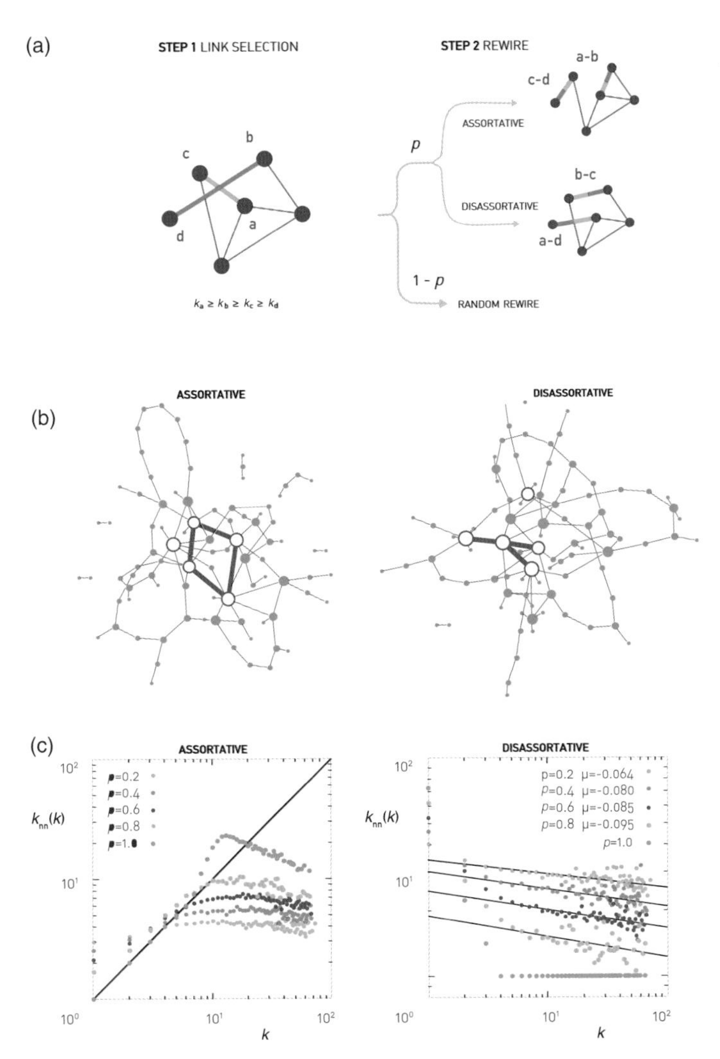

Figure 7.14 **Tuning Degree Correlations**
We can use the Xalvi-Brunet and Sokolov algorithm to tune the magnitude of degree correlations.
(a) We execute the deterministic rewiring step with probability p, and with probability $1 - p$ we randomly pair the a, b, c, d nodes with each other. For $p = 1$ we are back to the algorithm of **Figure 7.13**, generating maximal degree correlations; for $p < 1$ the induced noise tunes the magnitude of the effect.
(b) Typical network configurations generated for $p = 0.5$.
(c) The $k_{nn}(k)$ functions for various p values for a network with $N = 10,000$, $\langle k \rangle = 1$ and $\gamma = 3.0$.
Note that the correlation exponent μ depends on the fitting region, especially in the assortative case.

cutoffs. For scale-free networks the system is unable to sustain assortativity for high k. The observed behavior is reminiscent of the $k_{nn}(k)$ function of citation networks (**Figure 7.10j**).

The version of the Xalvi-Brunet and Sokolov algorithm introduced in **Figure 7.13** generates maximally assortative or disassortative networks. We can tune the magnitude of the generated degree correlations if we use the algorithm discussed in **Figure 7.14**.

In summary, static models, like the configuration or hidden-parameter model, are neutral if we allow multi-links, and develop structural disassortativity if we force them to generate simple networks. To generate networks with tunable correlations, we can use for example the Xalve-Brunet and Sokolov algorithm. An important result of this section is **(7.16)** and **(7.18)**, offering the analytical form of the degree correlation function for the hidden-paramenter model and for a growing network, in both cases predicting a power-law k-dependence. These results offer analytical backing for the scaling hypothesis **(7.10)**, indicating that both structural and dynamical effects can result in a degree correlation function that follows a power law.

7.7 The Impact of Degree Correlations

As we have seen in **Figure 7.10**, most real networks are characterized by some degree correlations. Social networks are assortative; biological networks display structural disassortativity. These correlations raise an important question: Why do we care? In other words, do degree correlations alter the properties of a network? And which network properties do they influence? This section addresses these important questions.

An important property of a random network is the emergence of a phase transition at $\langle k \rangle = 1$, marking the appearance of the giant component (**Section 3.6**). **Figure 7.15** shows the relative size of the giant component for networks with different degree correlations, documenting several patterns [207, 217, 218]:

- **Assortative Networks**
  For assortative networks the phase transition point moves to a lower $\langle k \rangle$, hence a giant component emerges for $\langle k \rangle < 1$. The reason is that it is easier to start a giant component if the high-degree nodes seek each other out.
- **Disassortative Networks**
  The phase transition is delayed in disassortative networks, as in these the hubs tend to connect to small-degree nodes. Consequently, disassortative networks have difficulty forming a giant component.
- **Giant Component**
  For large $\langle k \rangle$ the giant component is smaller in assortative networks than in neutral or disassortative networks. Indeed, assortativity forces the hubs to link to each other, hence they

fail to attract to the giant component the numerous small-degree nodes.

These changes in the size and structure of the giant component have implications for the spread of diseases [220, 221, 222], the topic of **Chapter 10**. Indeed, as we have seen in **Figure 7.10**, social networks tend to be assortative. The high-degree nodes therefore form a giant component that acts as a "reservoir" for the disease, sustaining an epidemic even when on average the network is not sufficiently dense for the virus to persist.

The altered giant component has implications for network robustness as well [223]. As we discuss in **Chapter 8**, the removal of a network's hubs fragments a network. In assortative networks, hub removal causes less damage because the hubs form a core group, hence many of them are redundant. Hub removal is more damaging in disassortative networks, as in these the hubs connect to many small-degree nodes, which fall off the network once a hub is deleted.

Let us mention a few additional consequences of degree correlations:

- **Figure 7.16** shows the path-length distribution of a random network rewired to display different degree correlations. It indicates that in assortative networks the average path length is shorter than in neutral networks. The most dramatic difference is in the network diameter, d_{max}, which is significantly higher for assortative networks. Indeed, assortativity favors links between nodes with similar degree, resulting in long chains of $k = 2$ nodes, enhancing d_{max} (**Figure 7.13c**).
- Degree correlations influence a system's stability against stimuli and perturbations [224], as well as the synchronization of oscillators placed on a network [225, 226].
- Degree correlations have a fundamental impact on the vertex cover problem [227], a much-studied problem in graph theory that requires us to find the minimal set of nodes (cover) such that each link is connected to at least one node in the cover (**Box 7.4**).
- Degree correlations impact our ability to control a network, altering the number of input signals one needs to achieve full control [228].

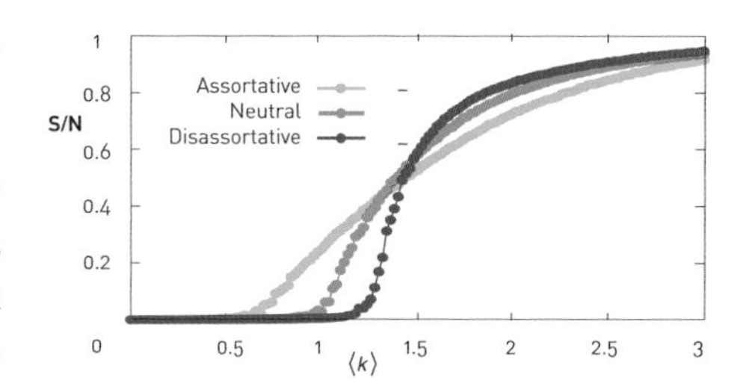

Figure 7.15 **Degree Correlations and the Phase Transition Point**
Relative size of the giant component for an Erdős–Rényi network of size $N = 10,000$ (green curve), which is then rewired using the Xalvi-Brunet and Sokolov algorithm with $p = 0.5$ to induce degree correlations (**Figure 7.14**). The figure indicates that as we move from assortative to disassortative networks, the phase transition point is delayed and the size of the giant component increases for large $\langle k \rangle$. Each point represents an average over 10 independent runs.

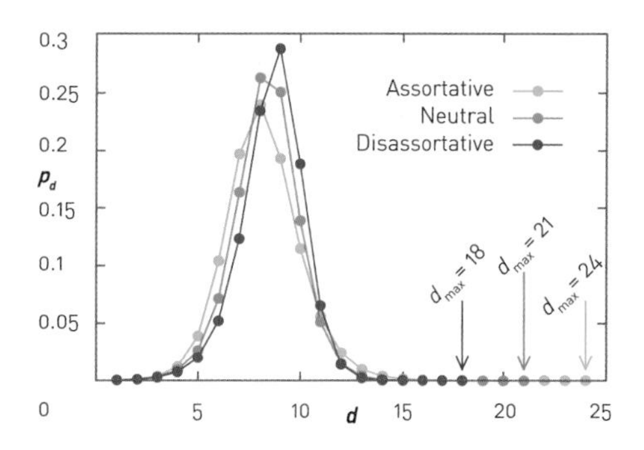

Figure 7.16 **Degree Correlations and Path Lengths**
Distance distribution for a random network with size $N = 10,000$ and $\langle k \rangle = 3$. Correlations are induced using the Xalvi-Brunet and Sokolov algorithm with $p = 0.5$ (**Figure 7.14**). The plots show that as we move from disassortative to assortative networks, the average path length decreases, indicated by the gradual move of the peaks to the left. At the same time the diameter, d_{max}, grows. Each curve represents an average over 10 independent networks.

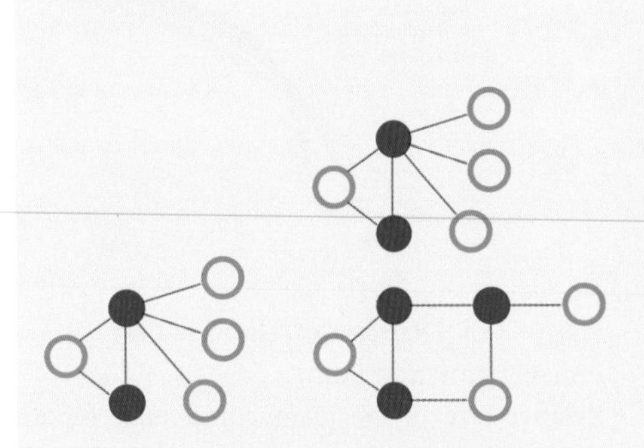

Figure 7.17 **The Minimum Cover**
Formally, a *vertex cover* of a network is a set C of nodes such that each link of the network connects to at least one node in C. A *minimum vertex cover* is a vertex cover of smallest possible size. The figure shows examples of minimum vertex covers in two small networks, where the set C is shown in purple. We can check that if we turn any of the purple nodes into green nodes, at least one link will not connect to a purple node.

Box 7.4 **Vertex Cover and Museum Guards**

Imagine that you are the director of an open-air museum located in a large park. You wish to place guards on the crossroads to observe each path. Yet, to save costs you want to use as few guards as possible. How many guards do you need?

Let N be the number of crossroads and $m < N$ the number of guards you can afford to hire. While there are $\binom{N}{m}$ ways of placing the m guards at N crossroads, most configurations leave some paths unsupervised [229].

The number of trials one needs to place the guards so that they cover all paths grows exponentially with N. Indeed, this is one of the six basic NP-complete problems, called the *vertex cover problem*. The vertex cover of a network is a set of nodes such that each link is connected to at least one node of the set (**Figure 7.17**). NP-completeness means that there is no known algorithm which can identify a minimal vertex cover substantially faster than using an exhaustive search, that is checking each possible configuration individually. The number of nodes in the minimal vertex cover depends on the network topology, being affected by the degree distribution and degree correlations [227].

In summary, degree correlations are not only of academic interest, but they also influence numerous network characteristics and have a discernable impact on many processes that take place on a network.

7.8 Summary

Degree correlations were first discovered in 2001 in the context of the Internet by Romualdo Pastor-Satorras, Alexei Vazquez and Alessandro Vespignani [203, 204], who also introduced the degree correlation function $k_{\mathrm{nn}}(k)$ and the scaling **(7.10)**. A year later Kim Sneppen and Sergey Maslov used the full $p(k_i, k_j)$, related to the e_{ij} matrix, to characterize the degree correlations of protein interaction networks. In 2003 Mark Newman introduced the degree correlation coefficient [207, 208] together with the assortative, neutral and disassortative distinction. These terms have their roots in social sciences [212].

Assortative mating reflects the tendency of individuals to date or marry individuals that are similar to them. For example, low-income individuals marry low-income individuals and college graduates marry college graduates. Network theory uses assortativity in the same spirit, capturing the degree-based similarities between nodes. In assortative networks hubs tend to connect to other hubs and small-degree nodes to other small-degree nodes. In a network environment we can also encounter the traditional assortativity, when nodes of similar properties link to each other (**Figure 7.18**).

Disassortative mixing, when individuals link to individuals who are unlike them, is also common in some social and economic systems. Sexual networks are perhaps the best example, as most sexual relationships are between individuals of different gender. In economic settings trade typically takes place between individuals of different skills: the baker does not sell bread to other bakers, and the shoemaker rarely fixes other shoemakers' shoes.

Taken together, there are several reasons why we care about degree correlations in networks (**Box 7.5**):

- Degree correlations are present in most real networks (**Section 7.5**).
- Once present, degree correlations change a network's behavior (**Section 7.7**).
- Degree correlations force us to move beyond the degree distribution, representing quantifiable patterns that govern the way nodes link to each other that are not captured by p_k alone.

Despite the considerable effort devoted to characterizing degree correlations, our understanding of the phenomena remains incomplete. For example, while in **Section 7.6** we offered an algorithm to tune degree correlations, the problem is far from being fully resolved. Indeed, the most accurate description of a network's degree correlations is contained in the e_{ij} matrix. Generating networks with an arbitrary e_{ij} remains a difficult task.

Box 7.5 At a Glance: Degree Correlations

Degree Correlation Matrix e_{ij}

Neutral networks:

$$e_{ij} = q_i q_i = \frac{k_i p_{k_i} k_j p_{k_j}}{\langle k \rangle^2}$$

Degree Correlation Function

$$k_{nn}(k) = \sum_{k'} k' p(k'|k)$$

Neutral networks:

$$k_{nn}(k) = \frac{\langle k^2 \rangle}{\langle k \rangle}$$

Scaling Hypotheses

$$k_{nn}(k) \sim k^{\mu}$$

$\mu > 0$: *Assortative*
$\mu = 0$: *Neutral*
$\mu < 0$: *Disassortative*

Degree Correlation Coefficient

$$r = \sum_{jk} \frac{jk(e_{jk} - q_j q_k)}{\sigma^2}$$

$r > 0$: *Assortative r*
$r = 0$: *Neutral*
$r < 0$: *Disassortative*

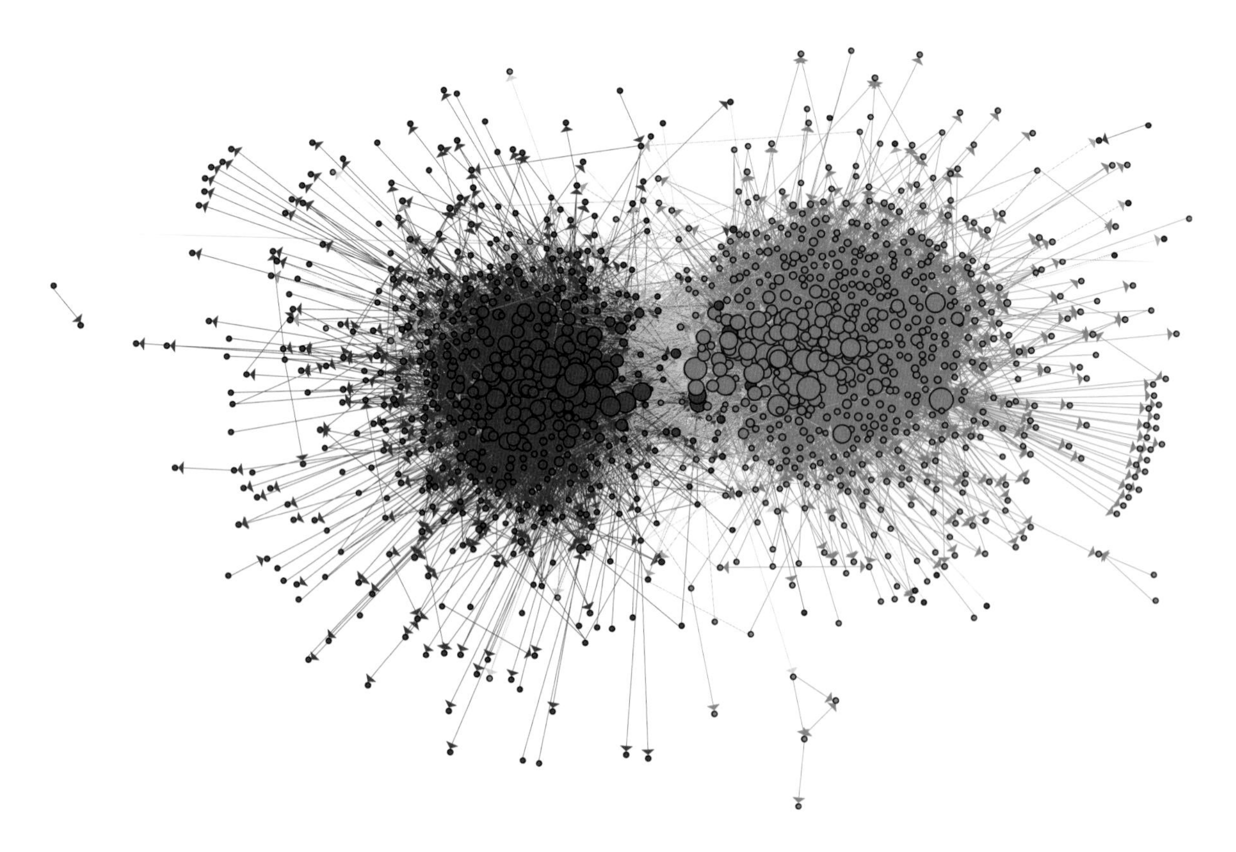

Figure 7.18 **Politics is Never Neutral**
The network behind the US political blogosphere illustrates the presence of assortative mixing, as used in sociology, meaning that nodes of similar characteristics tend to link to each other. In the map, each blue node corresponds to a liberal blog and red nodes are conservative blogs. Blue links connect liberal blogs, red links connect conservative blogs, yellow links go from liberal to conservative and purple from conservative to liberal. As the image indicates, very few blogs link across the political divide, demonstrating the strong assortativity of the political blogosphere. After [230].

Finally, in this chapter we focused on the $k_{nn}(k)$ function, which captures two-point correlations. In principle, higher-order correlations are also present in some networks (**Box 7.6**). The impact of such three- or four-point correlations remains to be understood.

7.9 Homework

7.9.1 Detailed Balance for Degree Correlations

Express the joint probability $e_{kk'}$, the conditional probability $P(k'|k)$ and the probability q_k, discussed in this chapter, in terms of number of nodes N, average degree $\langle k \rangle$, number of nodes with degree k, N_k, and the number of links connecting nodes of degree k and k', $E_{kk'}$ (note that $E_{kk'}$ is twice the number of links when $k = k'$). Based on these expressions, show that for any network we have $e_{kk'} = q_k P(k'|k)$.

Box 7.6 Two-Point, Three-Point Correlations

The complete degree correlations characterizing a network are determined by the conditional probability $P(k^{(1)}, k^{(2)}, \ldots, k^{(k)}|k)$ that a node with degree k connects to nodes with degrees $k^{(1)}, k^{(2)}, \ldots, k^{(k)}$.

Two-Point Correlations

The simplest of these is the two-point correlation discussed in this chapter, being the conditional probability $P(k'|k)$ that a node with degree k is connected to a node with degree k'. For uncorrelated networks this conditional probability is independent of k, that is $P(k'|k) = k'p_{k'}/\langle k \rangle$. As the empirical evaluation of $P(k'|k)$ in real networks is cumbersome, it is more practical to analyze the degree correlation function $k_{\rm nn}(k)$ defined in **(7.7)**.

Three-Point Correlations

Correlations involving three nodes are determined by $P(k^{(1)}, k^{(2)}|k)$. This conditional probability is connected to the clustering coefficient. Indeed, the average clustering coefficient $C(k)$ [220, 221] can be formally written as the probability that a degree-k node is connected to nodes with degrees $k^{(1)}$ and $k^{(2)}$, and that these two are joined by a link, averaged over all the possible values of $k^{(1)}$ and $k^{(2)}$:

$$C(k) = \sum_{k^{(2)}, k^{(2)}} P(k^{(1)}, k^{(2)}|k) p^{k}_{k^{(1)}, k^{(2)}},$$

where $p^{k}_{k^{(1)}, k^{(2)}}$ is the probability that nodes $k^{(1)}$ and $k^{(2)}$ are connected, provided that they have a common neighbor with degree k. For neutral networks $C(k)$ is independent of k, following

$$C = \frac{(\langle k^2 \rangle - \langle k \rangle)^2}{\langle k \rangle^3 N}.$$

7.9.2 Star Network

Consider a star network, where a single node is connected to $N-1$ degree-one nodes. Assume that $N \gg 1$.

(a) What is the degree distribution p_k of this network?

(b) What is the probability q_k that moving along a randomly chosen link we find at its end a node with degree k?

(c) Calculate the degree correlation coefficient r for this network. Use the expressions of $e_{kk'}$ and $P(k'|k)$ calculated in **HOMEWORK 7.1**.

(d) Is this network assortative or disassortative? Explain why.

7.9.3 Structural Cutoffs

Calculate the structural cutoff k_s for the undirected networks listed in **Table 4.1**. Based on the plots in **Figure 7.10**, predict for each network whether k_s is larger or smaller than the maximum expected degree k_{max}. Confirm your prediction by calculating k_{max}.

7.9.4 Degree Correlations in Erdős–Rényi Networks

Consider the Erdős–Rényi $G(N,L)$ model of random networks, introduced in **Chapter 2** (**Box 3.1** and **Section 3.2**), where N labeled nodes are connected with L randomly placed links. In this model, the probability that there is a link connecting nodes i and j depends on the existence of a link between nodes l and s.

(a) Write the probability that there is a link between i and j, e_{ij}, and the probability that there is a link between i and j conditional on the existence of a link between l and s.
(b) What is the ratio of such two probabilities for small networks? And for large networks?
(c) What do you obtain for the quantities discussed in (a) and (b) if you use the Erdős–Rényi $G(N,p)$ model?

Based on the results found for (a)–(c), discuss the implications of using the $G(N,L)$ model instead of the $G(N,p)$ model for generating random networks with small number of nodes.

7.10 Advanced Topics 7.A Degree Correlation Coefficient

In **Box 7.2** we defined the degree correlation coefficient r as an alternative measure of degree correlations [207, 208]. The use of a single number to characterize degree correlations is attractive, as it offers a way to compare the correlations observed in networks of different nature and size. Yet, to effectively use r we must be aware of its origin.

The hypothesis behind the correlation coefficient r implies that the $k_{nn}(k)$ function can be approximated by the linear function

$$k_{nn}(k) = rk. \tag{7.21}$$

This is different from the scaling **(7.10)**, which assumes a power-law dependence on k. Equation **(7.21)** raises several issues:

- The initial attractiveness model predicts a power law **(7.18)** or a logarithmic k-dependence **(7.20)** for the degree correlation function. A similar power law is derived in **(7.16)** for the hidden-parameter model. Consequently, r forces a linear fit to an inherently nonlinear function. This linear dependence is not supported by numerical simulations or analytical calculations. Indeed, as we show in **Figure 7.19**, **(7.21)** offers a poor fit to the data for both assortative and disassortative networks.
- As we have seen in **Figure 7.10**, the dependence of $k_{nn}(k)$ on k is complex, often changing trends for large k thanks to the structural cutoff. A linear fit ignores this inherent complexity.
- The maximally correlated model has a vanishing r for large N, despite the fact that the network maintains its degree correlations (**Box 7.7**). This suggests that the degree correlation coefficient has difficulty detecting correlations characterizing large networks.

Box 7.7 The Problem with Large Networks

The Xalvi-Brunet and Sokolov algorithm helps us calculate the minimal (r_{min}) and the maximal (r_{max}) correlation coefficient for a scale-free network, obtaining [219]

$$r_{min} \sim \begin{cases} -c_1(\gamma, k_0) & \text{for } \gamma > 2 \\ -N^{(2-\gamma)/(\gamma-1)} & \text{for } 2 < \gamma < 3 \\ -N^{(\gamma-4)/(\gamma-1)} & \text{for } 3 < \gamma < 4 \\ -c_2(\gamma, k_0) & \text{for } 4 < \gamma, \end{cases}$$

$$r_{max} \sim \begin{cases} -N^{(-\gamma-2)/(\gamma-1)} & \text{for } 2 < \gamma < \gamma_r \\ -N^{-1(\gamma^2-1)} & \text{for } \gamma_r < \gamma < 3, \end{cases}$$

where $\gamma_r \approx \frac{1}{2} + \sqrt{17/4} \approx {\sim}2.56$.

These expressions indicate that:

(i) For large N both r_{min} and r_{max} vanish, even though the corresponding networks were rewired to have maximal correlations. Consequently, the correlation coefficient r is unable to capture the correlations present in large networks.

(ii) Scale-free networks with $\gamma < 2.6$ always have negative r. This is a consequence of structural correlations (**Section 7.4**).

Given r's limitations, we must inspect $k_{nn}(k)$ to best characterize a large network's degree correlations.

7.10.1 Relationship Between μ and r

On the positive side, r and μ are not independent of each other. To show this we calculated r and μ for the ten reference networks (**Table 7.1**). The results are plotted in **Figure 7.20**, indicating that μ and r correlate for positive r. Note, however, that this correlation breaks down for negative r. To understand the origin of this behavior, next we derive a direct relationship between μ and r. To be specific, we assume the validity of **(7.10)** and determine the value of r for a network with correlation exponent μ.

We start by determining a from **(7.10)**. We can write the second moment of the degree distribution as

$$\langle k^2 \rangle = \langle k_{nn}(k) k \rangle = \sum_k a k^{\mu+1} p_k = a \langle k^{\mu+1} \rangle,$$

Figure 7.19 **Degree Correlation Function**
The degree correlation function $k_{nn}(k)$ for three real networks. The left panels show the cumulative function $k_{nn}(k)$ on a log–log plot to test the validity of **(7.10)**. The right panels show $k_{nn}(k)$ on a lin–lin plot to test the validity of **(7.21)**, that is the assumption that $k_{nn}(k)$ depends linearly on k. This is the hypothesis behind the correlation coefficient r. The slope of the dotted line corresponds to the correlation coefficient r. As the lin–lin plots on the right illustrate, **(7.21)** offers a poor fit for both assortative and disassortative networks.

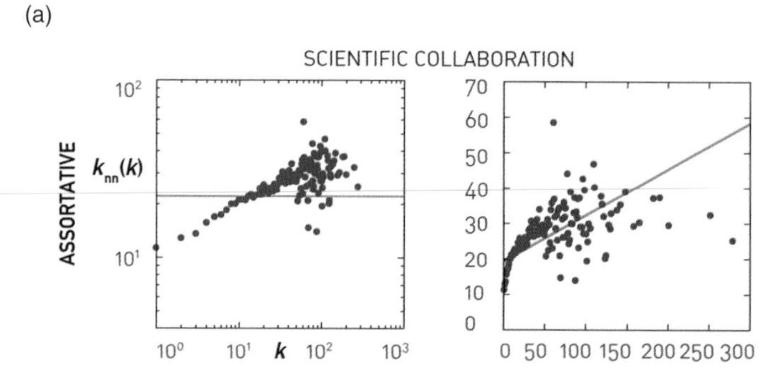

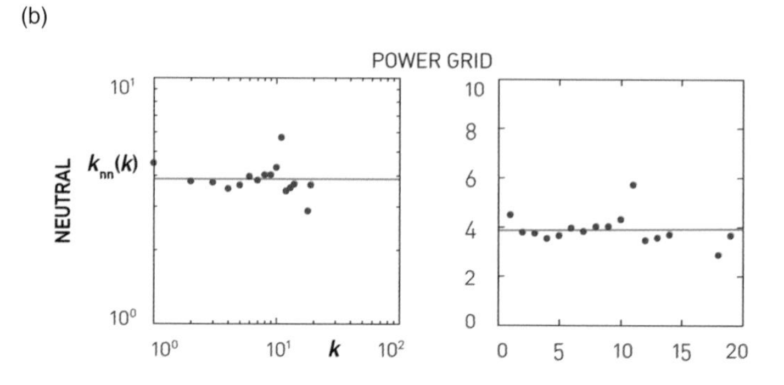

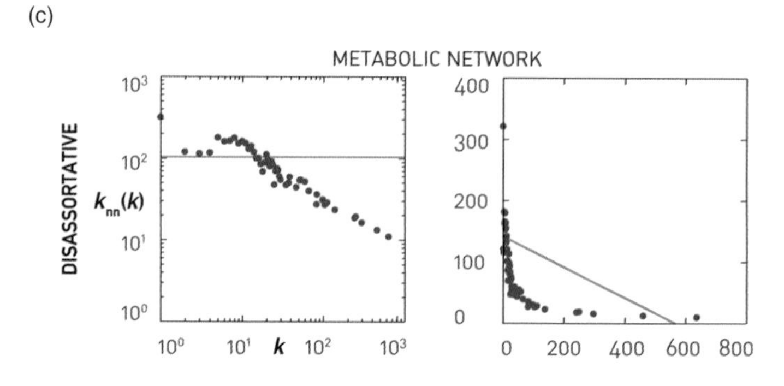

Table 7.1 **Degree Correlations in Reference Networks.**
The table shows the estimated r and μ for the ten reference networks. Directed networks were made undirected to measure r and μ. Alternatively, we can use the directed correlation coefficient to characterize such directed networks (Box 7.7).

Network	N	r	μ
Internet	192,244	0.02	0.56
WWW	325,729	−0.05	−1.11
Power grid	4,941	0.003	0.0
Mobile phone calls	36,595	0.21	0.33
Email	57,194	−0.08	−0.74
Science collaboration	23,133	0.13	0.16
Actor network	702,388	0.31	0.34
Citation network	449,673	−0.02	−0.18
***E. coli* metabolism**	1,039	−0.25	−0.76
Protein interactions	2,018	0.04	−0.1

which leads to

$$a = \frac{\langle k^2 \rangle}{\langle k^{\mu+1} \rangle}.$$

We now calculate r for a network with a given μ:

$$r = \frac{\sum_k kak^{\mu} q_k - \frac{\langle k^2\rangle^2}{\langle k\rangle^2}}{\sigma_r^2} = \frac{\sum_k ak^{\mu+2}\frac{p_k}{\langle k\rangle} - \frac{\langle k^2\rangle^2}{\langle k\rangle^2}}{\sigma_r^2}$$

$$= \frac{\frac{\langle k^2\rangle}{\langle k^{\mu+1}\rangle}\frac{\langle k^{\mu+2}\rangle}{\langle k\rangle} - \frac{\langle k^2\rangle^2}{\langle k^2\rangle}}{\sigma_r^2} == \frac{1}{\sigma_r^2}\frac{\langle k^2\rangle}{\langle k\rangle}\left(\frac{\langle k^{\mu+2}\rangle}{\langle k^{\mu+1}\rangle} - \frac{\langle k^2\rangle}{\langle k\rangle}\right). \tag{7.22}$$

For $\mu = 0$ the term in the last parentheses vanishes, obtaining $r = 0$. Hence if $\mu = 0$ (neutral network), the network will be neutral based on r as well. For $k > 1$ **(7.22)** suggests that for $\mu > 0$ the parentheses are positive, hence $r > 0$, and for $\mu < 0$

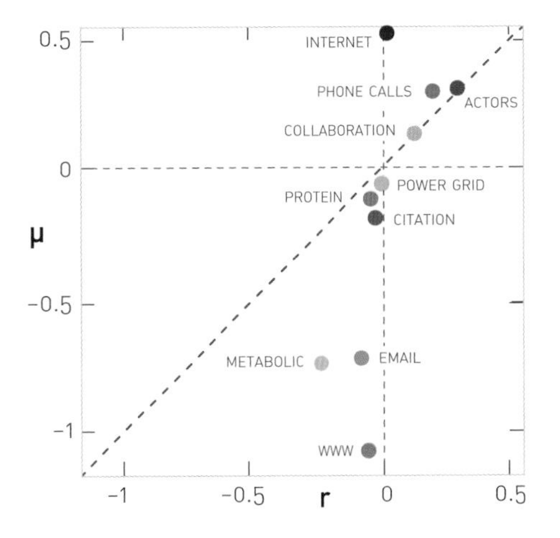

Figure 7.20 Correlation Between r and μ
To illustrate the relationship between r and μ, we estimated μ by fitting the $k_{nn}(k)$ function to **(7.10)**, whether or not the power law scaling was statistically significant.

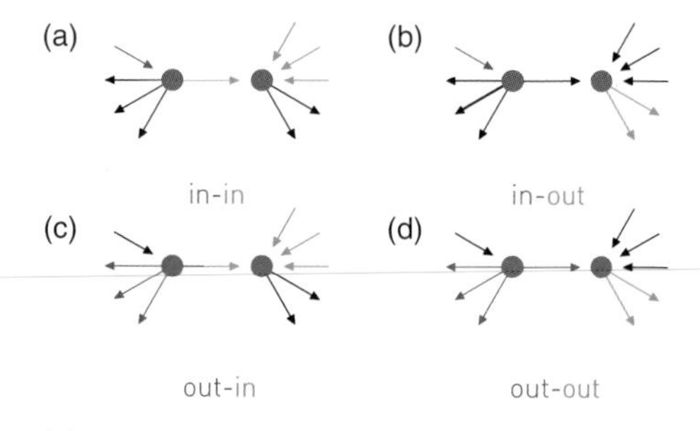

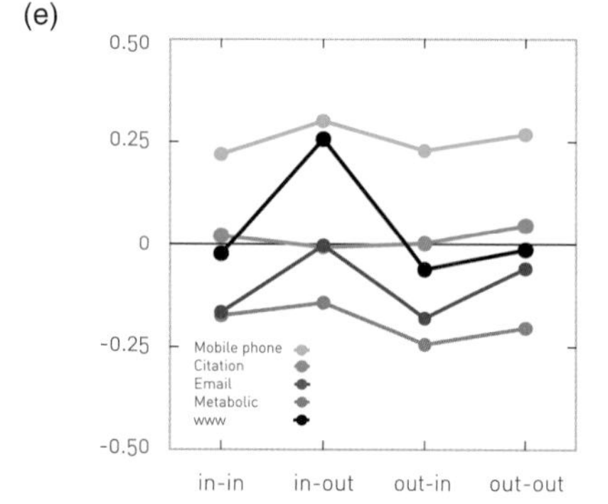

Figure 7.21 **Directed Correlation**
(a)–(d) The purple and green links indicate the α, β indices that define the appropriate correlation coefficient for a directed network. (e) The correlation profile of the five directed networks. While citation networks have negligible correlations, all four correlation coefficients document strong assortative behavior for mobile phone calls and strong disassortative behavior for metabolic networks. The case of the WWW is interesting: while three of its correlation coefficients are close to zero, there is a strong assortative tendency for the in–out degree combination.

the parentheses are negative, hence $r < 0$. Therefore r and μ predict degree correlations of a similar kind.

In summary, if the degree correlation function follows **(7.10)**, then the sign of the degree correlation exponent μ will determine the sign of the coefficient r:

$$\mu < 0 \rightarrow r < 0$$
$$\mu = 0 \rightarrow r = 0$$
$$\mu > 0 \rightarrow r > 0.$$

7.10.2 Directed Networks

To measure correlations in directed networks we must take into account that each node i is characterized by an incoming k_i^{in} and an outgoing k_i^{out} degree. We therefore define four degree correlation coefficients, $r_{\text{in,in}}, r_{\text{in,out}}, r_{\text{out,in}}, r_{\text{out,out}}$, capturing all possible combinations between the incoming and outgoing degrees of two connected nodes (**Figure 7.21a-d**). Formally, we have [213]

$$r_{\alpha,\beta} = \frac{\sum_{jk} jk(e_{jk}^{\alpha,\beta} - q_{\leftarrow j}^{\alpha} q_{\rightarrow k}^{\beta})}{\sigma_{\leftarrow}^{\alpha}\sigma_{\rightarrow}^{\beta}}, \tag{7.23}$$

where α and β refer to the in and out indices and $q_{\leftarrow j}^{\alpha}$ is the probability of finding a node with α-degree j by following a random link backward and $q_{\rightarrow k}^{\beta}$ is the probability of finding a β-link with degree k by following a random link forward. $\sigma_{\leftarrow}^{\alpha}$ and $\sigma_{\rightarrow}^{\beta}$ are the corresponding standard deviations. To illustrate the use of **(7.23)**, in **Figure 7.21e** we show the four correlation coefficients for the five directed reference networks (**Table 7.1**). Note, however, that for a complete characterization of degree correlations it is desirable to measure the four $k_{nn}(k)$ functions as well **(Box 7.3)**.

In summary, the degree correlation coefficient assumes that $k_{\text{nn}}(k)$ scales linearly with k, a hypothesis that lacks numerical and analytical support. Analytical calculations predict the power-law form **(7.10)** or the weaker logarithmic dependence **(7.20)**. Yet, in general the sign of r and μ do agree. Consequently, we can use r to get a quick sense of the nature of the potential correlations present in a network. Yet, the accurate characterization of the underlying degree correlations requires us to measure $k_{\text{nn}}(k)$.

7.11 Advanced Topics 7.B Structural Cutoffs

As discussed in **Section 7.4**, the fundamental conflict between the scale-free property and degree correlations leads to a structural cutoff in simple networks. In this section we derive **(7.15)**, calculating how the structural cutoff depends on the system size N [210].

We start by defining

$$r_{kk'} = \frac{E_{kk'}}{m_{kk'}}, \tag{7.24}$$

where $E_{kk'}$ is the number of links between nodes of degree k and k' for $k \neq k'$ and twice the number of connecting links for $k = k'$, and

$$m_{kk'} = \min\{kN_k, k'N_k, N_kN_{k'}\} \tag{7.25}$$

is the largest possible value of $E_{kk'}$. The origin of **(7.25)** is explained in **Figure 7.22**. Consequently, we can write $r_{kk'}$ as

$$r_{kk'} = \frac{E_{kk'}}{m_{kk'}} = \frac{\langle k \rangle e_{kk'}}{\min\{kP(k), k'P(k'), NP(k)P(k')\}}. \tag{7.26}$$

As $m_{kk'}$ is the maximum of $E_{kk'}$, we must have $r_{kk'} \leq 1$ for any k and k'. Strictly speaking, in simple networks degree pairs for which $r_{kk'} > 1$ cannot exist. Yet, for some networks and for some k, k' pairs, $r_{kk'}$ is larger than one. This is clearly non-physical and signals some conflict in the network configuration. Hence, we define the structural cutoff k_s as the solution of the equation

$$r_{k_s k_s} = 1. \tag{7.27}$$

Note that as soon as $k > Np_{k'}$ and $k' > Np_k$, the effects of the restriction on the multiple links are felt, turning the expression for $r_{kk'}$ into

$$r_{kk'} = \frac{\langle k \rangle e_{kk'}}{Np_k p_{k'}}. \tag{7.28}$$

For scale-free networks these conditions are fulfilled in the region $k, k' > (aN)^{1/(\gamma+1)}$, where a is a constant that depends on p. Note that this value is below the natural cutoff.

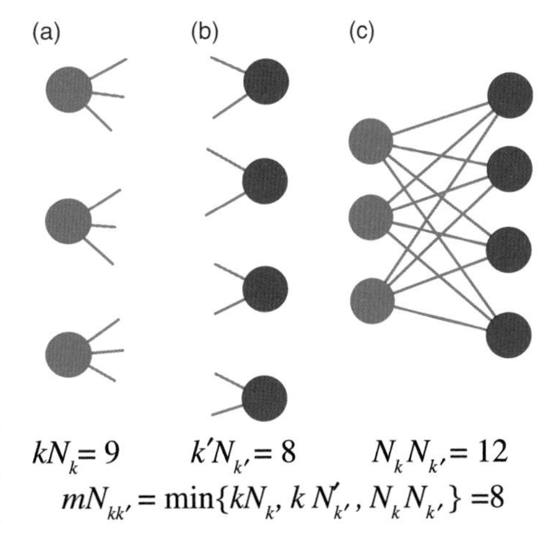

Figure 7.22 Calculating $m_{kk'}$

The maximum number of links one can have between two groups. The figure shows two groups of nodes, with degree $k = 3$ and $k' = 2$. The total number of links between these two groups must not exceed:

(a) The total number of links available in the $k = 3$ group, which is $kN_{k'} = 9$.

(b) The total number of links available in the $k' = 2$ group, which is $k'N_{k'} = 8$.

(c) The total number of links one can potentially place between the two groups, which is $N_k N_{k'}$.

In the example shown above, the smallest of the three is $k'N_{k'} = 8$ of (b).

Consequently this scaling provides a lower bound for the structural cutoff, in the sense that whenever the cutoff of the degree distribution falls below this limit, the condition $r_{kk'} < 1$ is always satisfied.

For neutral networks the joint distribution factorizes as

$$e_{kk'} = \frac{kk' p_k p_{k'}}{\langle k \rangle^2}. \tag{7.29}$$

Hence, the ratio **(7.28)** becomes

$$r_{kk'} = \frac{kk'}{\langle k \rangle N}. \tag{7.30}$$

Therefore, the structural cutoff needed to preserve the condition $r_{kk'} \leq 1$ has the form [210, 231, 232, 233]

$$k_s(N) \sim (\langle k \rangle N)^{1/2}, \tag{7.31}$$

which is **(7.15)**. Note that **(7.31)** is independent of the degree distribution of the underlying network. Consequently, for a scale-free network $k_s(N)$ is independent of the degree exponent γ.

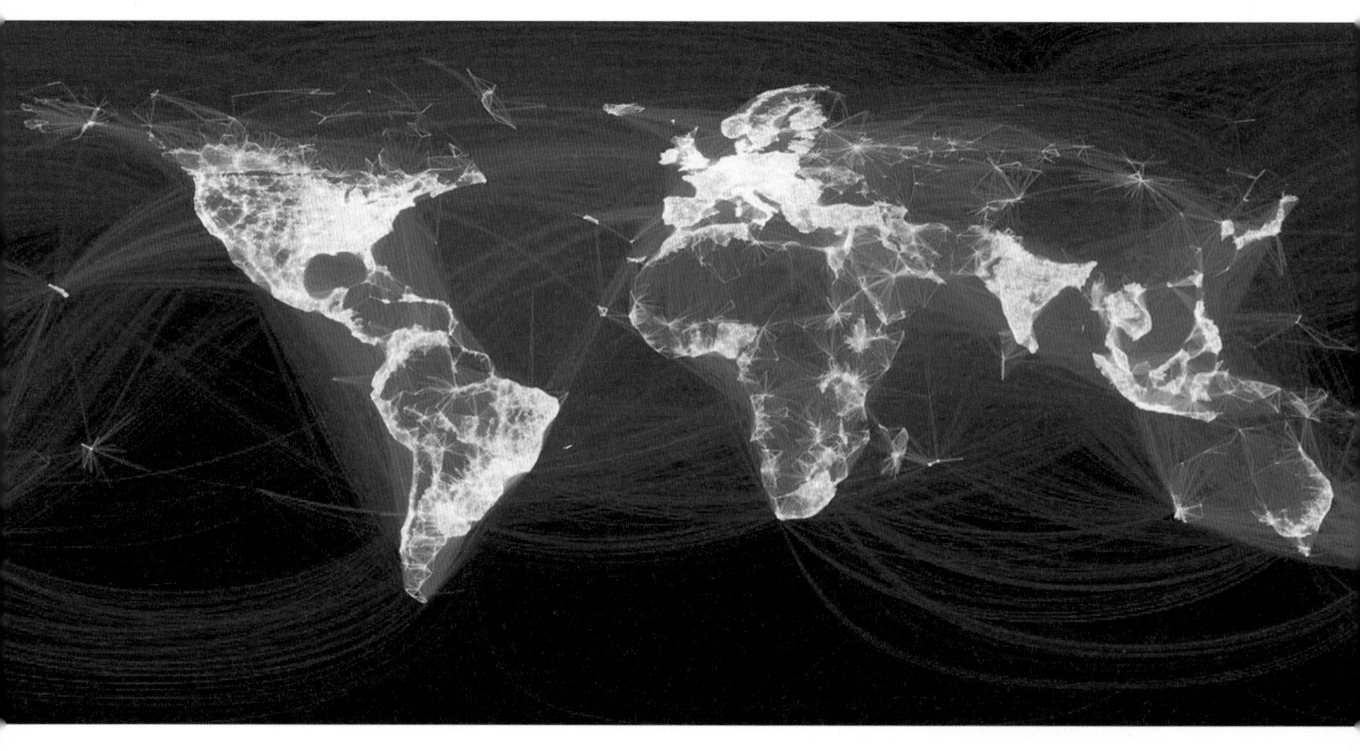

Figure 8.0 **Art & Networks: The Social Graph**
Created by Paul Butler, a Toronto-based data scientist during a Facebook internship in 2010, the image depicts the network connecting the users of the social network company. It highlights the links within and across continents. The presence of dense local links in the USA, Europe and India is just as revealing as the lack of nodes and links in some areas, like China, where the site is banned, and Africa, reflecting a lack of Internet access.

CHAPTER 8

Network Robustness

8.1 Introduction

Errors and failures can corrupt all human designs. The failure of a component in your car's engine may force you to call for a tow truck or a wiring error in your computer chip can make your computer useless. Many natural and social systems have, however, a remarkable ability to sustain their basic functions even when some of their components fail. Indeed, while there are countless protein misfolding errors and missed reactions in our cells, we rarely notice their consequences. Similarly, large organizations can function despite numerous absent employees. Understanding the origins of this robustness is important for many disciplines:

- Robustness is a central question in biology and medicine, helping us understand why some mutations lead to diseases and others do not.
- It is of concern for social scientists and economists, who explore the stability of human societies and institutions in the face of such disrupting forces as famine, war and changes in social and economic order.
- It is a key issue for ecologists and environmental scientists, who seek to predict the failure of an ecosystem when faced with the disruptive effects of human activity.
- It is the ultimate goal in engineering, aiming to design communication systems, cars or airplanes that can carry out their basic functions despite occasional component failures (**Figure 8.1**).

Networks play a key role in the robustness of biological, social and technological systems. Indeed, a cell's robustness is encoded in intricate regulatory, signaling and metabolic networks; society's resilience cannot be divorced from the interwoven social, professional and communication web behind it; an ecosystem's survivability cannot be understood without a careful analysis of the food web that sustains each species. Whenever nature seeks robustness, it resorts to networks.

Figure 8.1 **The Achilles' Heel of Complex Networks**
The cover of the July 27, 2000 issue of *Nature*, highlighting the paper entitled "Attack and error tolerance of complex networks" that began the scientific exploration of network robustness [12].

Figure 8.2 **Robust, Robustness**
"Robust" comes from the Latin *quercus robur*, meaning oak, the symbol of strength and longevity in the ancient world. The tree in the figure stands near the Hungarian village of Diósviszló and is documented at www.dendromania.hu, a site that catalogs Hungary's oldest and largest trees.

Image courtesy of György Pósfai.

The purpose of this chapter is to understand the role networks play in ensuring the robustness of a complex system. We show that the structure of the underlying network plays an essential role in a system's ability to survive random failures or deliberate attacks. We explore the role of networks in the emergence of cascading failures, a damaging phenomenon frequently encountered in real systems. Most important, we show that the laws governing the error and attack tolerance of complex networks and the emergence of cascading failures are universal. Hence, uncovering them helps us understand the robustness of a wide range of complex systems (**Figure 8.2**).

8.2 Percolation Theory

The removal of a single node has only limited impact on a network's integrity (**Figure 8.3a**). The removal of several nodes, however, can break a network into several isolated components (**Figure 8.3d**). Obviously, the more nodes we remove, the higher are the chances that we damage a network, prompting us to ask: How many nodes do we have to delete to fragment a network into isolated components? For example, what fraction of Internet routers must break down so that the Internet turns into clusters of computers that are unable to communicate with each other? To answer these questions, we must first familiarize ourselves with the mathematical underpinnings of network robustness, offered by *percolation theory.*

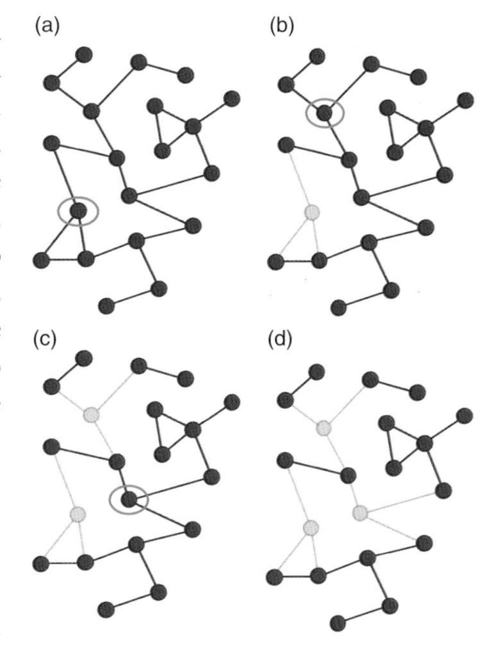

Figure 8.3 **The Impact of Node Removal** The gradual fragmentation of a small network following the breakdown of its nodes. In each panel we remove a different node (highlighted with a green circle), together with its links. While the removal of the first node has only limited impact on the network's integrity, the removal of the second node isolates two small clusters from the rest of the network. Finally, the removal of the third node fragments the network, breaking it into five non-communicating clusters of sizes $s = 2, 2, 2, 5, 6$.

8.2.1 Percolation

Percolation theory is a highly developed subfield of statistical physics and mathematics [235–238]. A typical problem addressed by it is illustrated in **Figure 8.4a,b**, showing a square lattice, where we place pebbles with probability p at each intersection. Neighboring pebbles are considered connected, forming clusters of size two or more. Given that the position of each pebble is decided by chance, we ask:

- What is the expected size of the largest cluster?
- What is the average cluster size?

Obviously, the higher is p, the larger are the clusters. A key prediction of percolation theory is that the cluster size does not

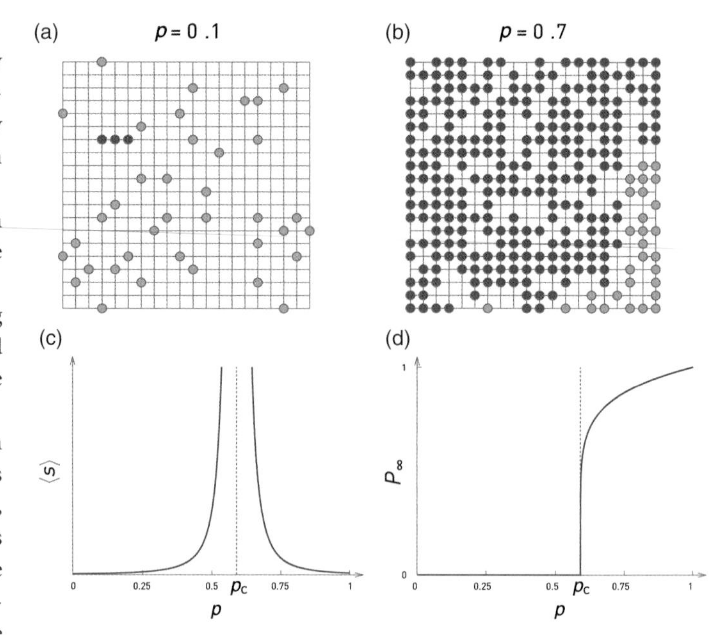

Figure 8.4 **Percolation**
A classical problem in percolation theory explores the random placement with probability p of pebbles on a square lattice. To quantify the nature of this phase transition, we focus on three quantities:
(a) For small p most pebbles are isolated. In this case the largest cluster has only three nodes, highlighted in purple.
(b) For large p most (but not all) pebbles belong to a single cluster, colored purple. This is called the *percolating cluster*, as it spans the whole lattice (see also **Figure 8.6**).
(c) The average cluster size, $\langle s \rangle$, as a function of p. As we approach p_c from below, numerous small clusters coalesce and $\langle s \rangle$ diverges, following (8.1). The same divergence is observed above p_c, where to calculate $\langle s \rangle$ we remove the percolating cluster from the average. The same exponent γ_p characterizes the divergence on both sides of the critical point.
(d) A schematic illustration of the p-dependence of the probability P_∞ that a pebble belongs to the largest connected component. For $p<p_c$ all components arer small, so P_∞ is zero. Once p reaches p_c a giant component emerges. Consequently beyond p_c there is a finite probability that a node belongs to the largest component, as predicted by (**8.2**).

change gradually with p. Rather, for a wide range of p the lattice is populated with numerous tiny clusters (**Figure 8.4a**). If p approaches a critical value p_c, these small clusters grow and coalesce, leading to the emergence of a large cluster at p_c. We call this the *percolating cluster* as it reaches the end of the lattice. In other words, at p_c we observe a phase transition from many small clusters to a percolating cluster that percolates the whole lattice (**Figure 8.4b**).

- **Average Cluster Size: $\langle s \rangle$**
  According to percolation theory, the average size of all finite clusters follows

$$\langle s \rangle \sim |p - p_c|^{-\gamma_p} \qquad (8.1)$$

  In other words, the average cluster size diverges as we approach p_c (**Figure 8.4c**).

- **Order Parameter: P_∞**
  The probability P_∞ that a randomly chosen pebble belongs to the largest cluster follows

$$P_\infty \sim (p - p_c)^{\beta_p}. \qquad (8.2)$$

  Therefore, as p decreases toward p_c the probability that a pebble belongs to the largest cluster drops to zero (**Figure 8.4d**).

- **Correlation Length: ξ**
  The mean distance between two pebbles that belong to the same cluster follows

$$\xi \sim |p - p_c|^{-\nu}. \tag{8.3}$$

  Therefore, while for $p < p_c$ the distance between the pebbles in the same cluster is finite, at p_c this distance diverges. This means that at p_c the size of the largest cluster becomes infinite, allowing it to percolate the whole lattice.

The exponents γ_p, β_p and ν are called *critical exponents*, as they characterize the system's behavior near the critical point p_c. Percolation theory predicts that these exponents are *universal*, meaning that they are independent of the nature of the lattice or the precise value of p_c. Therefore, whether we place the pebbles on a triangular or a hexagonal lattice, the behavior of $\langle s \rangle$, P_∞ and ξ is characterized by the same γ_p, β_p and ν exponents.

Consider the following examples to better understand this universality:

- The value of p_c depends on the lattice type, hence it is not universal. For example, for a two-dimensional square lattice (**Figure 8.4**) we have $p_c \approx 0.593$, while for a two-dimensional triangular lattice $p_c = 1/2$ (site percolation).
- The value of p_c also changes with the lattice dimension: for a square lattice $p_c \approx 0.593$ ($d = 2$); for a simple cubic lattice $p_c \approx 0.3116$ ($d = 3$). Therefore in $d = 3$ we need to cover a smaller fraction of the nodes with pebbles to reach the percolation transition.
- In contrast, with p_c the critical exponents do not depend on the lattice type but only on the lattice dimension. In two dimensions, the case shown in **Figure 8.4**, we have $\gamma_p = 43/18$, $\beta_p = 5/36$ and $\nu = 4/3$, for any lattice. In three dimensions $\gamma = 1.80$, $\beta = 0.41$ and $\nu = 0.88$. For any $d > 6$ we have $\gamma_p = 1$, $\beta_p = 1$, $\nu = 1/2$, hence for large d the exponents are independent of d as well [235].

8.2.2 Inverse Percolation Transition and Robustness

The phenomenon of primary interest in robustness is the impact of node failures on the integrity of a network. We can use percolation theory to describe this process.

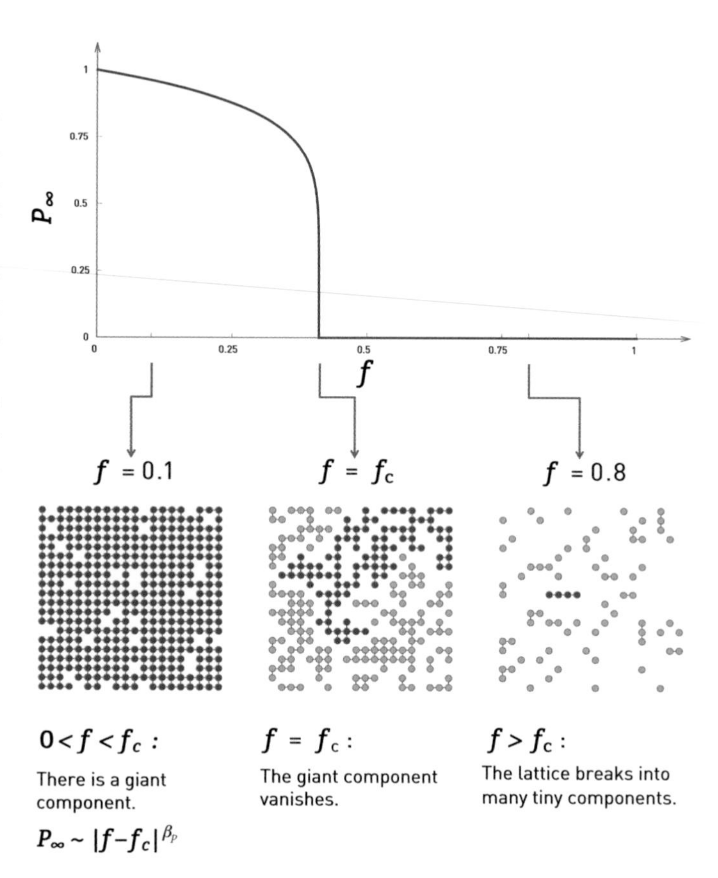

Figure 8.5 **Network Breakdown as Inverse Percolation**
The consequences of node removal are accurately captured by the inverse of the percolation process discussed in **Figure 8.4**. We start from a square lattice, which we view as a network whose nodes are the intersections. We randomly select and remove a fraction f of nodes and measure the size of the largest component formed by the remaining nodes. This size is accurately captured by P_∞, which is the probability that a randomly selected node belongs to the largest component. The observed networks are shown on the bottom panels. Under each panel we list the characteristics of the corresponding phases.

Let us view a square lattice as a network whose nodes are the intersections (**Figure 8.5**). We randomly remove a fraction f of nodes, asking how their absence impacts the integrity of the lattice.

If f is small, the missing nodes do little damage to the network. Increasing f, however, can isolate chunks of nodes from the giant component. Finally, for sufficiently large f the giant component breaks into tiny disconnected components (**Figure 8.5**).

This fragmentation process is not gradual, but it is characterized by a critical threshold f_c. For any $f < f_c$ we continue to have a giant component. Once f exceeds f_c, the giant component vanishes. This is illustrated by the f-dependence of P_∞, representing the probability that a node is part of the giant component (**Figure 8.5**): P_∞ is nonzero under f_c, but it drops to zero as we approach f_c. The critical exponents characterizing this breakdown, γ_p, β_p, ν, are the same as those encountered in

(8.1)–(8.3). Indeed, the two processes can be mapped into each other by choosing $f = 1 - p$.

What, however, if the underlying network is not as regular as a square lattice? As we will see in the coming sections, the answer depends on the precise network topology. Yet, for random networks the answer continues to be provided by percolation theory: random networks under random node failures share the same scaling exponents as infinite-dimensional percolation. Hence the critical exponents for a random network are $\gamma_p = 1$, $\beta_p = 1$ and $\nu = 1/2$, corresponding to the $d > 6$ percolation exponents encountered earlier. The critical exponents for a scale-free network are provided in **Advanced topics 8.A**.

In summary, the breakdown of a network under random node removal is not a gradual process. Rather, removing a small fraction of nodes has only limited impact on a network's integrity. But once the fraction of removed nodes reaches a critical threshold, the network abruptly breaks into disconnected components. In other words, random node failures induce a phase transition from a connected to a fragmented network. We can use the tools of percolation theory to characterize this transition in both regular and random networks (**Box 8.1**). For scale-free networks key aspects of the described phenomena change, however, as we discuss in the next section.

8.3 Robustness of Scale-free Networks

Percolation theory focuses mainly on regular lattices, whose nodes have identical degrees, or on random networks, whose nodes have comparable degrees. What happens, however, if the network is scale-free? How do the hubs affect the percolation transition?

To answer these questions, let us start from the router-level map of the Internet and randomly select and remove nodes one by one. According to percolation theory, once the number of removed nodes reaches a critical value f_c, the Internet should fragment into many isolated subgraphs (**Figure 8.5**). The simulations indicate otherwise: the Internet refuses to break apart even under rather extensive node failures. Instead, the size of the largest component decreases gradually, vanishing only in

Box 8.1 From Forest Fires to Percolation

We can use the spread of a fire in a forest to illustrate the basic concepts of percolation theory. Let us assume that each pebble in **Figure 8.4a,b** is a tree and that the lattice describes a forest. If a tree catches fire, it ignites the neighboring trees; these, in turn, ignite their neighbors. The fire continues to spread until no burning tree has a non-burning neighbor. We must therefore ask: If we randomly ignite a tree, what fraction of the forest burns down? And how long does it take the fire to burn out?

The answer depends on the tree density, controlled by the parameter p. For small p the forest consists of many small islands of trees ($p = 0.55$, **Figure 8.6a**), hence igniting any tree will at most burn down one of these small islands. Consequently, the fire will die out quickly. For large p most trees belong to a single large cluster, hence the fire rapidly sweeps through the dense forest ($p = 0.62$, **Figure 8.6c**).

The simulations indicate that there is a critical p_c at which it takes an extremely long time for the fire to end. This p_c is the critical threshold of the percolation problem. Indeed, at $p = p_c$ the giant component just emerges through the union of many small clusters (**Figure 8.6b**). Hence the fire has to follow a long winding path to reach all trees in the loosely connected clusters, which can be rather time consuming.

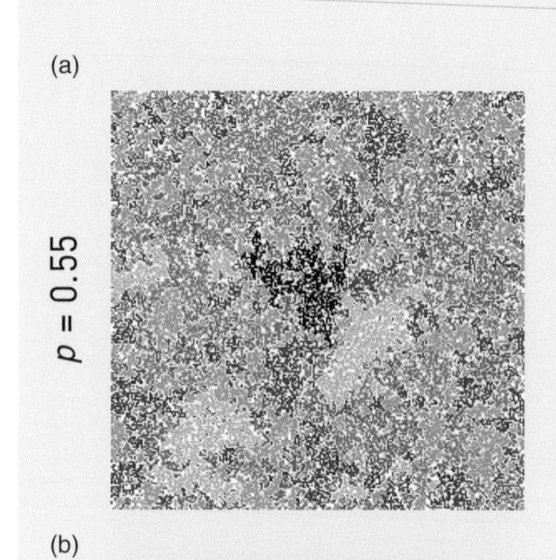

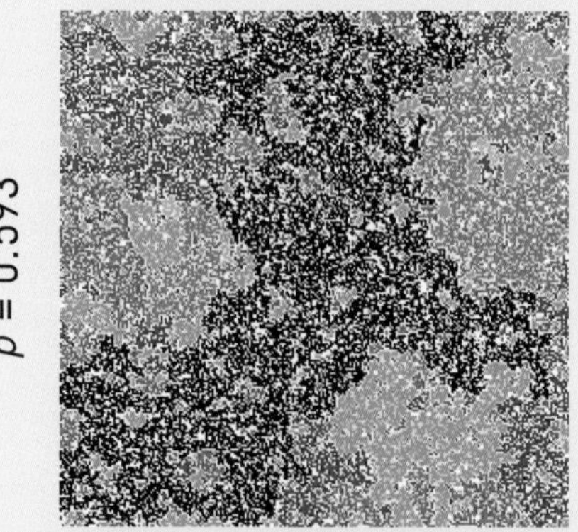

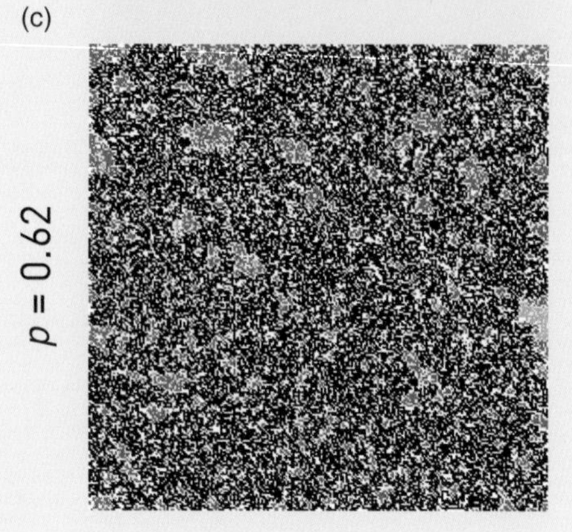

Figure 8.6 **Forest Fire**
The emergence of the giant component as we change the occupation probability. Each panel corresponds to a different p in the vicinity of p_c shown for a lattice of 250×250 sites. The largest cluster is colored black. For $p < p_c$ the largest cluster is tiny, as seen in **(a)**. If this is a forest and the pebbles are trees, any fire can at most consume only a small fraction of the trees, burning out quickly. Once p reaches $p_c \approx 0.593$, shown in **(b)**, the largest cluster percolates the whole lattice and the fire can reach many trees, burning slowly through the forest. Increasing p beyond p_c connects more pebbles (trees) to the largest component, as seen for $p = 0.62$ in **(c)**. Hence, the fire can sweep through the forest, burning out quickly again.

the vicinity of $f = 1$ (**Figure 8.7a**). This means that the network behind the Internet shows an unusual robustness to random node failures: we must remove all of its nodes to destroy its giant component. This conclusion disagrees with percolation on lattices, which predicts that a network must fall apart after the removal of a finite fraction of its nodes.

The behavior observed above is not unique to the Internet. To show this we repeated the above measurement for a scale-free network with degree exponent $\gamma = 2.5$, observing an identical pattern (**Figure 8.7b**): under random node removal the giant component fails to collapse at some finite f_c, but vanishes only gradually near $f = 1$ (**Online resource 8.1**). This hints that the Internet's observed robustness is rooted in its scale-free topology. The goal of this section is to uncover and quantify the origin of this remarkable robustness.

(a)

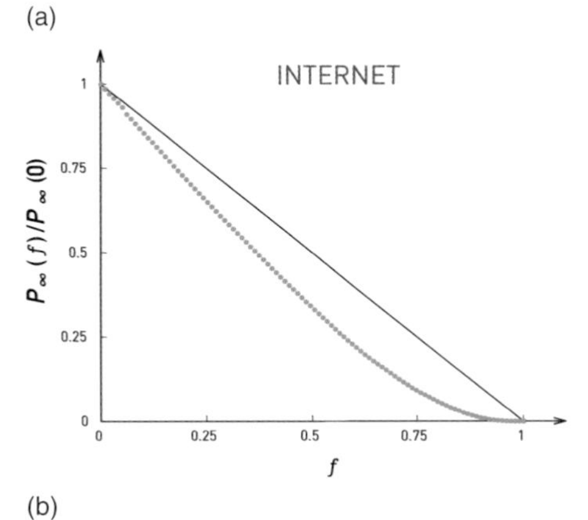

(b)

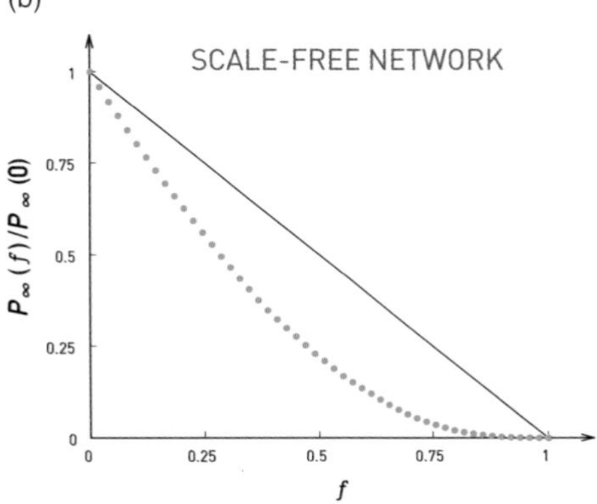

Figure 8.7 **Robustness of Scale-Free Networks**
(a) The fraction of Internet routers that belong to the giant component after a fraction f of routers are randomly removed. The ratio

$$P_\infty(f)/P_\infty(0)$$

provides the relative size of the giant component. The simulations use the router level Internet topology of **Table 4.1**.
(b) The fraction of nodes that belong to the giant component after a fraction f of nodes are removed from a scale-free network with $\gamma = 2.5$, $N = 10,000$ and $k_{\min} = 1$.
The plots indicate that the Internet and in general a scale-free network do not fall apart after the removal of a finite fraction of nodes. We need to remove almost all nodes (i.e., $f_c = 1$) to fragment these networks.

8.3.1 Molloy–Reed Criterion

To understand the origin of the anomalously high f_c characterizing the Internet and scale-free networks, we calculate f_c for a network with an arbitrary degree distribution. To do so we rely on a simple observation: for a network to have a giant component, most nodes that belong to it must be connected to at least two other nodes (**Figure 8.8**). This leads to the *Molloy–Reed criterion* (**Advanced topics 8.B**), stating that a randomly wired network has a giant component if

$$\kappa = \frac{\langle k^2 \rangle}{\langle k \rangle} > 2. \qquad (8.4)$$

Networks with $\kappa < 2$ lack a giant component, being fragmented into many disconnected components. The Molloy–Reed criterion **(8.4)** links the network's integrity, as expressed by the presence or absence of a giant component, to $\langle k \rangle$ and $\langle k^2 \rangle$. It is valid for any degree distribution p_k.

To illustrate the predictive power of **(8.4)**, let us apply it to a random network. As in this case $\langle k^2 \rangle = \langle k \rangle(1 + \langle k \rangle)$, a random network has a giant component if

$$\kappa = \frac{\langle k^2 \rangle}{\langle k \rangle} = \frac{\langle k \rangle(1 + \langle k \rangle)}{\langle k \rangle} = 1 + \langle k \rangle > 2 \qquad (8.5)$$

or

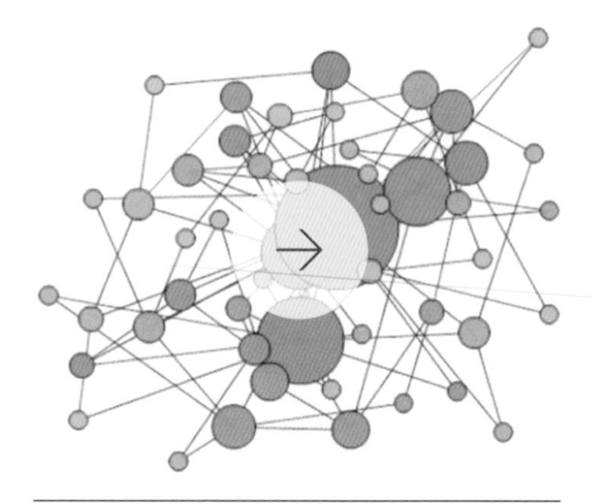

Online Resource 8.1

Scale-Free Network Under Node Failures

To illustrate the robustness of a scale-free network we start from the network we constructed in **Online resource 4.1**, that is a scale-free network generated by the Barabási–Albert model. Next we randomly select and remove nodes one by one. As the movie illustrates, despite the fact that we remove a significant fraction of the nodes, the network refuses to break apart. Visualization by Dashun Wang.

$$\langle k \rangle > 1. \tag{8.6}$$

This prediction coincides with the necessary condition (3.10) for the existence of a giant component.

8.3.2 Critical Threshold

To understand the mathematical origin of the robustness observed in **Figure 8.7**, we ask at what threshold a scale-free network will lose its giant component. By applying the Molloy–Reed criterion to a network with an arbitrary degree distribution, we find that the critical threshold follows (**Advanced topics 8.C**)

$$f_c = 1 - \frac{1}{\frac{\langle k^2 \rangle}{\langle k \rangle} - 1}. \tag{8.7}$$

The most remarkable prediction of **(8.7)** is that the critical threshold f_c depends only on $\langle k \rangle$ and $\langle k^2 \rangle$, quantities that are uniquely determined by the degree distribution p_k.

Let us illustrate the utility of **(8.7)** by calculating the breakdown threshold of a random network. Using $\langle k^2 \rangle = \langle k \rangle (1 + \langle k \rangle)$, we obtain (**Advanced topics 8.D**)

$$f_c^{ER} = 1 - \frac{1}{\langle k \rangle}. \tag{8.8}$$

Hence, the denser is a random network, the higher is its f_c (i.e., the more nodes we need to remove to break it apart). Furthermore, **(8.8)** predicts that f_c is always finite, hence a random network must break apart after the removal of a finite fraction of nodes.

Equation **(8.7)** helps us understand the roots of the enhanced robustness observed in **Figure 8.7**. Indeed, for scale-free networks with $\gamma < 3$ the second moment $\langle k^2 \rangle$ diverges in the $N \to \infty$ limit. If we insert $\langle k^2 \rangle \to \infty$ into **(8.7)**, we find that f_c converges to $f_c = 1$. This means that *to fragment a scale-free network we must remove all of its nodes*. In other words, the random removal of a finite fraction of its nodes does not break apart a large scale-free network.

To better understand this result we express $\langle k \rangle$ and $\langle k^2 \rangle$ in terms of the parameters characterizing a scale-free network: the degree exponent γ and the minimal and maximal degrees k_{min} and k_{max}, obtaining

$$f_c = \begin{cases} 1 - \dfrac{1}{\dfrac{\gamma-2}{3-\gamma} k_{\min}^{\gamma-2} k_{\max}^{3-\gamma} - 1} & 2 < \gamma < 3 \\ 1 - \dfrac{1}{\dfrac{\gamma-2}{\gamma-3} k_{\min} - 1} & \gamma > 3. \end{cases} \quad (8.9)$$

Equation **(8.9)** predicts that (**Figure 8.9**):

- For $\gamma > 3$ the critical threshold f_c depends only on γ and $k_{\min}$, hence f_c is independent of the network size N. In this regime a scale-free network behaves like a random network: it falls apart once a finite fraction of its nodes are removed.
- For $\gamma < 3$, $k_{\max}$ diverges for large N, following **(4.18)**. Therefore, in the $N \to \infty$ limit **(8.9)** predicts $f_c \to 1$. In other words, to fragment an infinite scale-free network we must remove all of its nodes.

Equations **(8.6)–(8.9)** are the key results of this chapter, predicting that scale-free networks can withstand an arbitrary level of random failures without breaking apart. The hubs are responsible for this remarkable robustness. Indeed, random node failures by definition are blind to degree, affecting with the same probability a small- or a large-degree node. Yet, in a scale-free network we have far more small-degree nodes than hubs. Therefore, random node removal will predominantly remove one of the numerous small nodes, as the chances of selecting randomly one of the few large hubs is negligible. These small nodes contribute little to a network's integrity, hence their removal does little damage.

Returning to the airport analogy of **Figure 4.6**, if we close a randomly selected airport, we will most likely shut down one of the numerous small airports. Its absence will hardly be noticed elsewhere in the world: you can still travel from New York to Tokyo, or from Los Angeles to Rio de Janeiro.

8.3.3 Robustness of Finite Networks

Equation **(8.9)** predicts that for a scale-free network f_c converges to one only if $k_{\max} \to \infty$, which corresponds to the $N \to \infty$ limit. While many networks of practical interest are very large, they are still finite, prompting us to ask if the observed anomaly is relevant for finite networks. To address

Figure 8.8 **Molloy–Reed Criterion**
To form a chain each individual must hold the hand of two other individuals. Similarly, to have a giant component in a network, on average each of its nodes should have at least two neighbors. The Molloy–Reed criterion **(8.4)** exploits this property, allowing us to calculate the critical point at which a network breaks apart. See **Advanced topics 8.B** for the derivation.

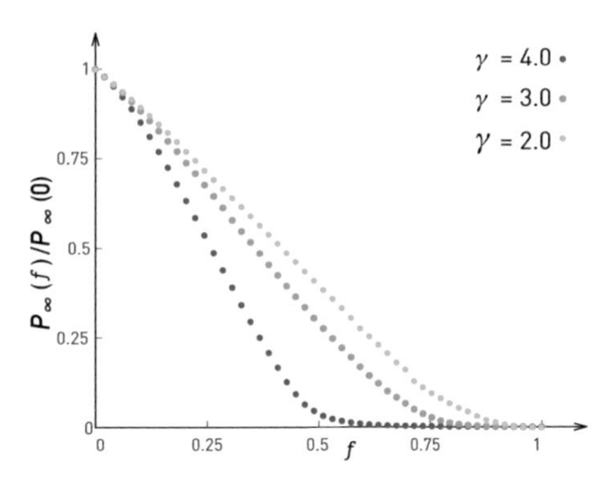

Figure 8.9 **Robustness and Degree Exponent**
The probability that a node belongs to the giant component after the removal of a fraction f of nodes from a scale-free network with degree exponent γ. For $\gamma = 4$ we observe a finite critical point $f_c \simeq 2/3$, as predicted by **(8.9)**. For $\gamma < 3$, however, $f \to 1$. The networks were generated with the configuration model using $k_{\min} = 2$ and $N = 10{,}000$.

this we insert **(4.18)** into **(8.9)**, obtaining that f_c depends on the network size N as (**Advanced topics 8.D**)

$$f_c \approx 1 - \frac{C}{N^{\frac{3-\gamma}{\gamma-1}}}, \tag{8.10}$$

where C collects all terms that do not depend on N. Equation **(8.10)** indicates that the larger a network, the closer is its critical threshold to $f_c = 1$.

To see how close f_c can get to the theoretical limit $f_c = 1$, we calculate f_c for the Internet. The router level map of the Internet has $\langle k^2\rangle / \langle k\rangle = 37.91$ (**Table 4.1**). Inserting this ratio into (8.7) we obtain $f_c = 0.972$. Therefore, we need to remove 97% of the routers to fragment the Internet into disconnected components. The probability that by chance 186,861 routers fail simultaneously, representing 97% of the $N = 192,244$ routers on the Internet, is effectively zero. This is the reason why the topology of the Internet is so robust to random failures.

In general a network displays *enhanced robustness* if its breakdown threshold deviates from the random network prediction (8.8), that is if

$$f_c > f_c^{\mathrm{ER}}. \tag{8.11}$$

Enhanced robustness has several ramifications:

- The inequality **(8.11)** is satisfied for most networks for which $\langle k^2\rangle$ deviates from $\langle k\rangle(\langle k\rangle + 1)$. According to **Figure 4.8**, for virtually all reference networks $\langle k^2\rangle$ exceeds the random expectation. Hence the robustness predicted by **(8.7)** affects most networks of practical interest. This is illustrated in **Table 8.1**, which shows that for most reference networks **(8.11)** holds.
- Equation **(8.7)** predicts that the degree distribution of a network does not need to follow a strict power law to display enhanced robustness. All we need is a larger $\langle k^2\rangle$ than expected for a random network of similar size.
- The scale-free property changes not only f_c, but also the critical exponents γ_p, β_p and ν in the vicinity of f_c. Their dependence on the degree exponent γ is discussed in **Advanced topics 8.A**.

Enhanced robustness is not limited to node removal, but emerges under link removal as well (**Figure 8.10**).

Table 8.1 **Breakdown Thresholds Under Random Failures and Attacks**

The table shows the estimated f_c for random node failures (second column) and attacks (fourth column) for ten reference networks. The procedure for determining f_c is described in **Advanced topics 8.E.** The third column (randomized network) offers f_c for a network whose N and L coincide with the original network, but whose nodes are connected randomly to each other (randomized network, f_c^{ER}, determined by **(8.8)**). For most networks, random failures exceed f_c^{ER} for the corresponding randomized network, indicating that these networks display enhanced robustness, as they satisfy **(8.11)**. Three networks lack this property: the power grid, a consequence of the fact that its degree distribution is exponential (**Figure 8.31a**); the actor and citation networks, which have a very high $\langle k \rangle$, diminishing the role of the high $\langle k^2 \rangle$ in **(8.7)**.

Network	Random failures (real network)	Random failures (randomized network)	Attack (real network)
Internet	0.92	0.84	0.16
WWW	0.88	0.85	0.12
Power grid	0.61	0.63	0.20
Mobile phone calls	0.78	0.68	0.20
Email	0.92	0.69	0.04
Science collaboration	0.92	0.88	0.27
Actor network	0.98	0.99	0.55
Citation network	0.96	0.95	0.76
E. coli metabolism	0.96	0.90	0.49
Yeast protein interactions	0.88	0.60	0.16

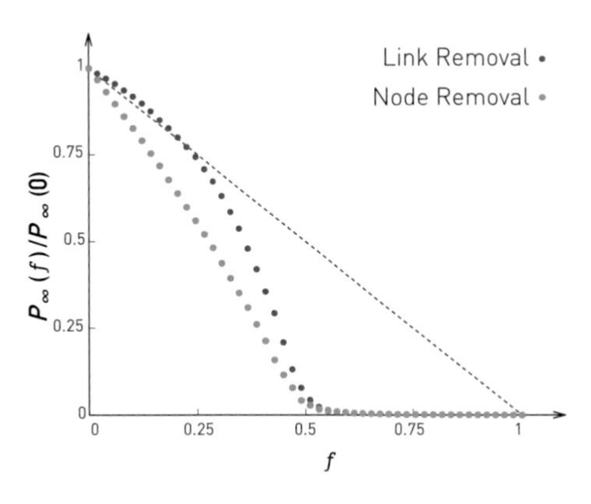

Figure 8.10 **Robustness and Link Removal**
What happens if we randomly remove the links rather than the nodes? The calculations predict that the critical threshold f_c is the same for random link and node removal [16, 239]. To illustrate this, we compare the impact of random link and node removal on a random network with $\langle k \rangle = 2$. The plot indicates that the network falls apart at the same critical threshold $f_c \simeq 0.5$. The difference is in the shape of the two curves. Indeed, the removal of a fraction f of nodes leaves us with a smaller giant component than the removal of a fraction f of links. This is not unexpected: on average each node removes $\langle k \rangle$ links. Hence the removal of a fraction f of nodes is equivalent to the removal of a fraction $f\langle k \rangle$ of links, which clearly makes more damage than the removal of a fraction f of links.

In summary, in this section we encountered a fundamental property of real networks: their robustness to random failures. Equation **(8.7)** predicts that the breakdown threshold of a network depends on $\langle k \rangle$ and $\langle k^2 \rangle$, which in turn are uniquely determined by the network's degree distribution. Therefore random networks have a finite threshold, but for scale-free networks with $\gamma < 3$ the breakdown threshold converges to one. In other words, we need to remove all nodes to break a scale-free network apart, indicating that these networks show an extreme robustness to random failures.

The origin of this extreme robustness is the large $\langle k^2 \rangle$ term. Given that for most real networks $\langle k^2 \rangle$ is larger than the random expectation, enhanced robustness is a generic property of many networks. This robustness is rooted in the fact that random failures affect mainly the numerous small nodes, which play only a limited role in maintaning a network's integrity.

8.4 Attack Tolerance

The important role the hubs play in holding together a scale-free network motivates our next question: What if we do not remove the nodes randomly, but go after the hubs? That is, we first remove the highest-degree node, followed by the node with the next highest degree and so on. The likelihood that nodes would break in this particular order under normal conditions is essentially zero. Instead, this process mimics an *attack* on the network, as it assumes detailed knowledge of the network topology, an ability to target the hubs and a desire to deliberately cripple the network [12].

The removal of a single hub is unlikely to fragment a network, as the remaining hubs can still hold the network together. After the removal of a few hubs, however, large chunks of nodes start falling off (**Online resource 8.2**). If the attack continues, it can rapidly break the network into tiny clusters.

The impact of hub removal is quite evident in the case of a scale-free network (**Figure 8.11**): the critical point, which is absent under random failures, re-emerges under attacks. Not only re-emerges, but also has a remarkably low value. Therefore, the removal of a small fraction of the hubs is sufficient to break a scale-free network into tiny clusters. The goal of this section is to quantify this attack vulnerability.

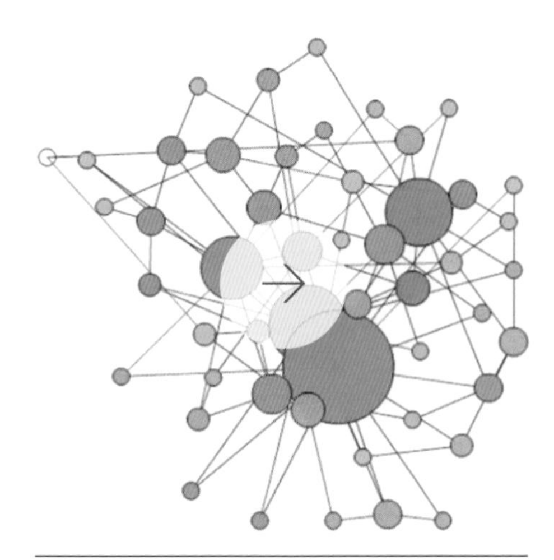

Online Resource 8.2

Scale-Free Networks Under Attack

During an attack we aim to inflict maximum damage on a network. We can do this by removing first the highest-degree node, followed by the next highest degree and so on. As the movie illustrates, it is sufficient to remove only a few hubs to break a scale-free network into disconnected components. Compare this with the network's refusal to break apart under random node failures, as shown in **Online resource 8.1.** Visualization by Dashun Wang.

8.4.1 Critical Threshold Under Attack

An attack on a scale-free network has two consequences (**Figure 8.11**):

- The critical threshold f_c is smaller than $f_c = 1$, indicating that under attacks a scale-free network can be fragmented by the removal of a finite fraction of its hubs.

- The observed f_c is remarkably low, indicating that we need to remove only a tiny fraction of the hubs to cripple the network.

To quantify this process we need to analytically calculate f_c for a network under attack. To do this we rely on the fact that hub removal changes the network in two ways [240]:

- It changes the maximum degree of the network from k_{max} to k'_{max}, as all nodes with degree larger than k'_{max} have been removed.
- The degree distribution of the network changes from p_k to $p'_{k'}$, as nodes connected to the removed hubs will lose links, altering the degrees of the remaining nodes.

By combining these two changes we can map the attack problem into the robustness problem discussed in the previous section. In other words, we can view an attack as random node removal from a network with adjusted k'_{max} and $p'_{k'}$. The calculations predict that the critical threshold f_c for attacks on a scale-free network is the solution of the equation [240] (**Advanced topics 8.F**)

$$f_c^{\frac{2-\gamma}{1-\gamma}} = 2 + \frac{2-\gamma}{3-\gamma} k_{min} \left(f_c^{\frac{3-\gamma}{1-\gamma}} - 1 \right). \quad (8.12)$$

Figure 8.12 shows the numerical solution of **(8.12)** as a function of the degree exponent γ, allowing us to draw several conclusions:

- While f_c for failures decreases monotonically with γ, f_c for attacks can have a non-monotonic behavior: it increases for small γ and decreases for large γ.
- f_c for attacks is always smaller than f_c for random failures.
- For large γ a scale-free network behaves like a random network. As a random network lacks hubs, the impact of an attack is similar to the impact of random node removal. Consequently, the failure and the attack thresholds converge to each other for large γ. Indeed, if $\gamma \to \infty$ then $p_k \to \delta(k - k_{min})$, meaning that all nodes have the same degree k_{min}. Therefore, random failures and targeted attacks become indistinguishable in the $\gamma \to \infty$ limit, obtaining

$$f_c \to 1 - \frac{1}{k_{min} - 1} \quad (8.13)$$

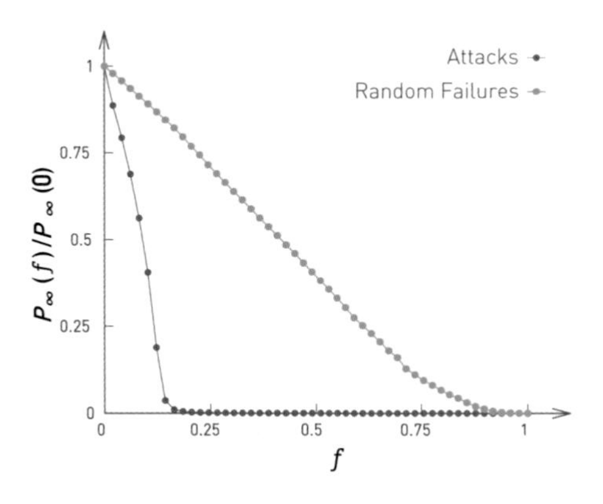

Figure 8.11 **Scale-Free Network Under Attack**
The probability that a node belongs to the largest connected component in a scale-free network under attack (purple) and under random failures (green). For an attack, we remove the nodes in decreasing order of their degree: we start with the biggest hub, followed by the next biggest and so on. In the case of failures, the order in which we choose the nodes is random, independent of the node's degree. The plot illustrates a scale-free network's extreme fragility to attacks: f is small, implying that the removal of only a few hubs can disintegrate the network. The initial network has degree exponent $\gamma = 2.5$, $k_{min} = 2$ and $N = 10,000$.

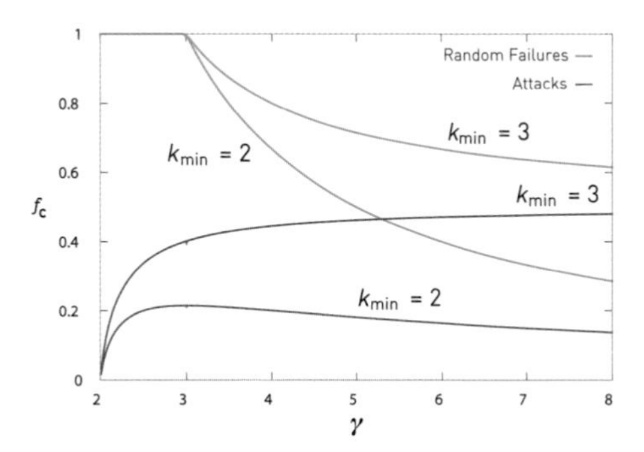

Figure 8.12 **Critical Threshold Under Attack**
The dependence of the breakdown threshold, f, on the degree exponent, γ, for scale-free networks with $k_{min} = 2, 3$. The curves are predicted by **(8.12)** for attacks (purple) and by **(8.7)** for random failures (green).

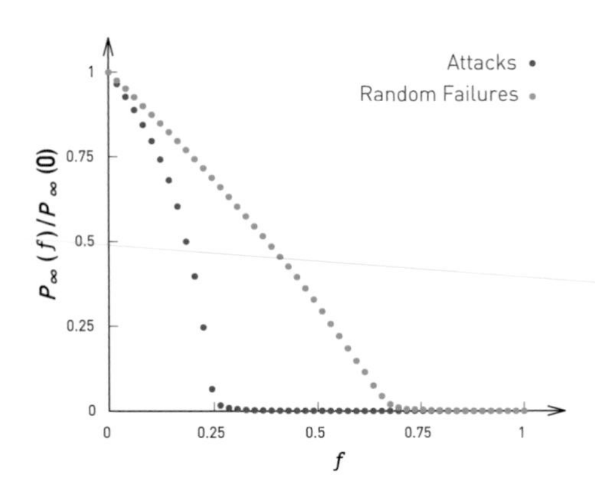

Figure 8.13 **Attacks and Failures in Random Networks**
The fraction of nodes that belong to the giant component in a random network if a fraction f of nodes are randomly removed (green) and in decreasing order of their degree (purple). Both curves indicate the existence of a finite threshold, in contrast with scale-free networks, for which $f_c \to 1$ under random failures. The simulations were performed for random networks with $N = 10,000$ and $\langle k \rangle = 3$.

- As **Figure 8.13** shows, a random network has a finite percolation threshold under both random failures and attacks, as predicted by **Figure 8.12** and **(8.13)** for large γ.

The airport analogy helps us understand the fragility of scale-free networks to attacks: the closing of two large airports, like Chicago's O'Hare Airport or the Atlanta International Airport, for only a few hours would be headline news, altering travel throughout the USA. Should some series of events lead to the simultaneous closure of the Atlanta, Chicago, Denver, and New York airports, the biggest hubs, air travel within the North American continent would come to a halt within hours.

In summary, while random node failures do not fragment a scale-free network, an attack that targets the hubs can easily destroy such a network. This fragility is bad news for the Internet, as it indicates that it is inherently vulnerable to deliberate attacks (**Box 8.2**). It can be good news in medicine, as the vulnerability of bacteria to the removal of their hub proteins offers avenues to design drugs that kill unwanted bacteria.

8.5 Cascading Failures

Throughout this chapter we have assumed that each node failure is a random event, hence the nodes of a network fail independently of each other. In reality, in a network the activity of each node depends on the activity of its neighboring nodes. Consequently, the failure of a node can induce the failure of the nodes connected to it. Let us consider a few examples:

- **Blackouts (Power Grid)**
  After the failure of a node or a link, the electric currents are instantaneously reorganized on the rest of the power grid. For example, on August 10, 1996, a hot day in Oregon, a line carrying 1,300 megawatts sagged close to a tree and snapped. Because electricity cannot be stored, the current it carried was automatically shifted to two lower-voltage lines. As these were not designed to carry the excess current, they too failed. Seconds later the excess current led to the malfunction of 13 generators, eventually causing a blackout in 11 US states and two Canadian provinces [242].

Box 8.2 Paul Baran and the Internet

In 1959 RAND, a Californian think-tank, assigned Paul Baran, a young engineer at that time, to develop a communication system that could survive a Soviet nuclear attack. As a nuclear strike handicaps all equipment within the range of the detonation, Baran had to design a system whose users outside this range would not lose contact with one another. He described the communication network of his time as a "hierarchical structure of a set of stars connected in the form of a larger star," offering an early description of what we call today a scale-free network [241]. He concluded that this topology is too centralized to be viable under attack. He also discarded the hub-and-spoke topology shown in **Figure 8.14a**, noting that the "centralized network is obviously vulnerable as destruction of a single central node destroys communication between the end stations."

Baran decided that the ideal survivable architecture was a decentralized mesh-like network (**Figure 8.14c**). This network is sufficiently redundant so that even if some of its nodes fail, alternative paths can connect the remaining nodes. Baran's ideas were ignored by the military, so when the Internet was born a decade later, it relied on distributed protocols that allowed each node to decide where to link. This decentralized philosophy paved the way to the emergence of a scale-free Internet, rather than the uniform mesh-like topology envisioned by Baran.

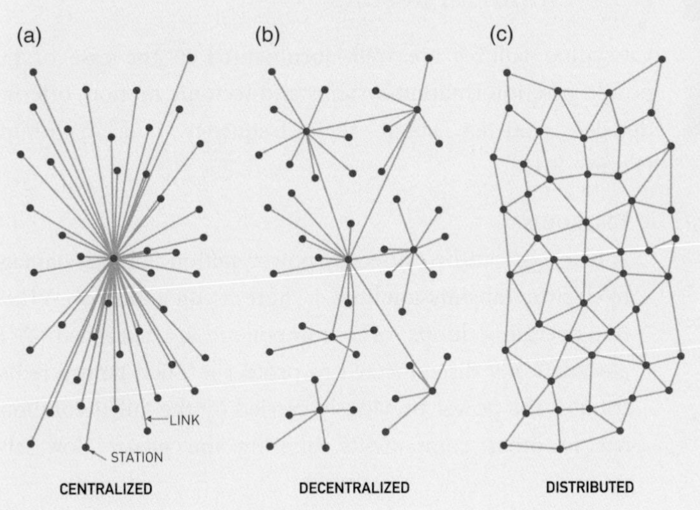

Figure 8.14 **Baran's Network**
Possible configurations of communication networks, as envisioned by Paul Baran in 1959.
After [241].

Table 8.2 **Avalanche Exponents in Real Systems**

The reported avalanche exponents of the power-law distribution (**8.14**) for energy loss in various countries [248], twitter cascades [249] and earthquake sizes [250]. The third column indicates the nature of the measured cascade size *s*, corresponding to power or energy not served, the number of re-tweets generated by a typical tweet and the amplitude of the seismic wave.

Source	Exponent	Cascade
Power grid (North America)	2.0	Power
Power grid (Sweden)	1.6	Energy
Power grid (Norway)	1.7	Power
Power grid (New Zealand)	1.6	Energy
Power grid (China)	1.8	Energy
Twitter cascades	1.75	Re-tweets
Earthquakes	1.67	Seismic wave

Figure 8.15 **Domino Effect**

The *domino effect* is the fall of a series of dominos induced by the fall of the first domino. The term is often used to refer to a sequence of events induced by a local change that propagates throughout the whole system. Hence the domino effect represents perhaps the simplest illustration of cascading failures, the topic of this section.

- **Denial of Service Attacks (Internet)**

  If a router fails to transmit the packets received by it, the Internet protocols will alert the neighboring routers to avoid the troubled equipment by re-routing the packets using alternative routes. Consequently, a failed router increases traffic on other routers, potentially inducing a series of denial of service attacks throughout the Internet [243].

- **Financial Crises**

  Cascading failures are common in economic systems. For example, the drop in house prices in 2008 in the USA spread along the links of the financial network, inducing a cascade of failed banks, companies and even nations [244, 245, 246]. It eventually caused the worst global financial meltdown since the 1930s Great Depression.

While they cover different domains, these examples have several common characteristics. First, the initial failure had only a limited impact on the network structure. Second, the initial failure did not stay localized, but spread along the links of the network, inducing additional failures. Eventually, multiple nodes lost their ability to carry out their normal functions. Consequently, each of these systems experienced *cascading failures*, a dangerous phenomenon in most networks [247]. In this section we discuss the empirical patterns governing such cascading failures. The modeling of these events is the topic of the next section.

8.5.1 Empirical Results

Cascading failures are well documented in the case of the power grid, information systems and tectonic motion, offering detailed statistics about their frequency and magnitude (**Figure 8.15**).

- **Blackouts**

  A blackout can be caused by power station failures, damage to electric transmission lines, a short circuit and so on. When the operating limits of a component are exceeded, it is automatically disconnected to protect it. Such failure redistributes the power previously carried by the failed component to other components, altering the power flow, the

Figure 8.16 **Cascade Size Distributions**

(**a**) The distribution of energy loss for all North American blackouts between 1984 and 1998, as documented by the North American Electrical Reliability Council. The distribution is typically fitted to **(8.14)**. The reported exponents for different countries are listed in **Table 8.2**. After [248].

(**b**) The distribution of cascade sizes on Twitter. While most tweets go unnoticed, a tiny fraction of tweets are shared thousands of times. Overall, the re-tweet numbers are well approximated by **(8.14)** with $\alpha \simeq 1.75$. After [249].

(**c**) The cumulative distribution of earthquake amplitudes recorded between 1977 and 2000. The dashed lines indicate the power-law fit **(8.14)** used by seismologists to characterize the distribution. The earthquake magnitude shown on the horizontal axis is the logarithm of s, which is the amplitude of the observed seismic waves. After [300].

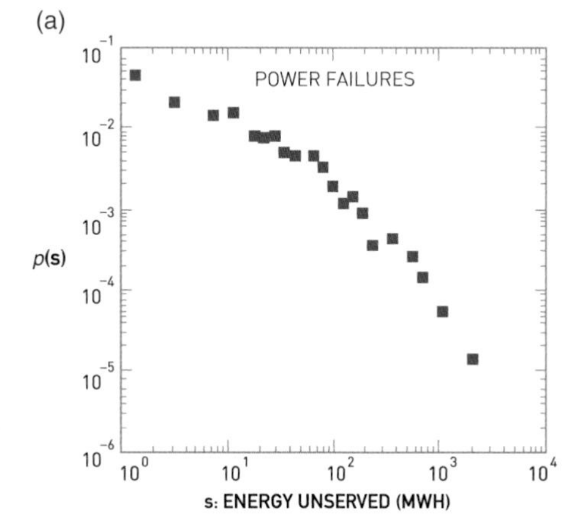

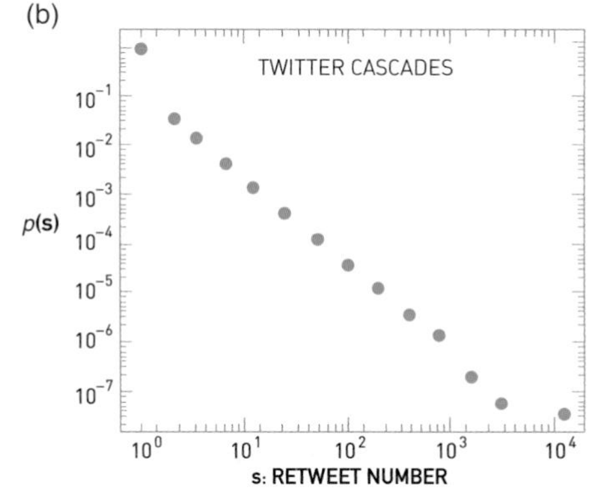

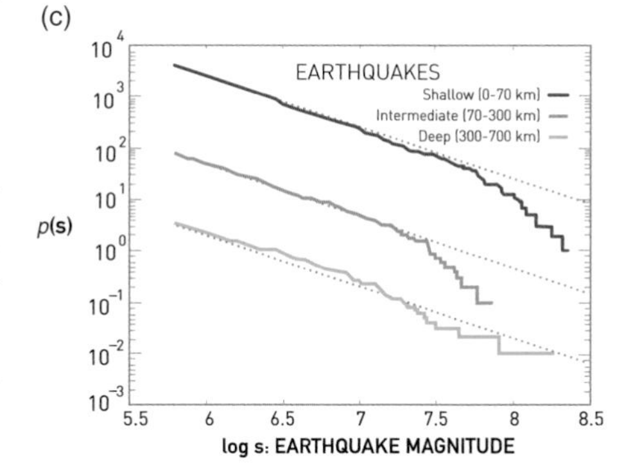

frequency, the voltage and the phase of the current, and the operation of the control, monitoring and alarm systems. These changes can in turn disconnect other components as well, starting an avalanche of failures.

A frequently recorded measure of blackout size is the energy unserved. **Figure 8.16a** shows the probability distribution $p(s)$ of energy unserved in all North American blackouts between 1984 and 1998. Electrical engineers approximate the obtained distribution with the power law [248],

$$p(s) \sim s^{-a}, \qquad (8.14)$$

where the *avalanche exponent* α is listed in **Table 8.2** for several countries. The power-law nature of this distribution indicates that most blackouts are rather small, affecting only a few consumers. These coexist, however, with occasional major blackouts, when millions of consumers lose power (**Figure 8.17**).

- **Information Cascades**

  Modern communication systems, from email to Facebook or Twitter, facilitate the cascade-like spreading of information along the links of the social network. As the events pertaining to the spreading process often leave digital traces, these platforms allow researchers to detect the underlying cascades.

Figure 8.17 **Northeast Blackout of 2003**
One of the largest blackouts in North America took place on August 14, 2003, just before 4:10 p.m. Its cause was a software bug in the alarm system at a control room of the First Energy Corporation in Ohio. Missing the alarm, the operators were unaware of the need to redistribute the power after an overloaded transmission line hit a tree. Consequently, a normally manageable local failure began a cascading failure that shut down more than 508 generating units at 265 power plants, leaving an estimated 10 million people without electricity in Ontario and 45 million in eight US states. The figure highlights the states affected by the blackout. For a satellite image of the blackout, see **Figure 1.1**.

The micro-blogging service Twitter has been particularly studied in this context. On Twitter, the network of who follows whom can be reconstructed by crawling the service's follower graph. As users frequently share web-content using URL shorteners, one can also track each spreading/sharing process. A study tracking 74 million such events over two months followed the diffusion of each URL from a particular seed node through its re-posts until the end of a cascade (**Figure 8.18**). As **Figure 8.16b** indicates, the size distribution of the observed cascades follows the power law **(8.14)** with an avalanche exponent $\alpha \approx 1.75$ [249]. The power law indicates that the vast majority of posted URLs do not spread at all, a conclusion supported by the fact that the average cascade size is only $\langle s \rangle = 1.14$. Yet, a small fraction of URLs are re-posted thousands of times.

- **Earthquakes**
Geological fault surfaces are irregular and sticky, prohibiting their smooth slide against each other. Once a fault has locked, the continued relative motion of the tectonic plates accumulates an increasing amount of strain energy around the fault surface. When the stress becomes sufficient to break through the asperity, a sudden slide releases the stored energy, causing an earthquake. Earthquakes can also be induced by the natural rupture of geological faults, by volcanic activity, landslides, mine blasts and even nuclear tests.

Each year around 500,000 earthquakes are detected with instrumentation. Only about 100,000 of these are sufficiently strong to be felt by humans. Seismologists approximate the distribution of earthquake amplitudes with the power law **(8.14)** with $\alpha \approx 1.67$ (**Figure 8.16c**) [250].

Earthquakes are rarely considered a manifestly network phenomenon, given the difficulty of mapping out the precise network of interdependencies that causes them. Yet, the resulting cascading failures bear many similarities to network-based cascading events, suggesting common mechanisms.

The power-law distribution **(8.14)** followed by blackouts, information cascades and earthquakes indicates that most cascading failures are relatively small. These small cascades capture the loss of electricity in a few houses, tweets of little interest to most users or earthquakes so small that one needs sensitive instruments to detect them. Equation **(8.14)** predicts that these numerous small events coexist with a few exceptionally large events. Examples of such major cascades include the 2003 power outage in North America (**Figure 8.17**), the tweet *Iran Election Crisis: 10 Incredible YouTube Videos* (http://bit.ly/vPDLo) that was shared 1,399 times [251] or the January 2010 earthquake in Haiti, with over 200,000 victims. Interestingly, the avalanche exponents reported by electrical engineers, media researches and seismologists are surprisingly close to each other, varying between 1.6 and 2 (**Table 8.2**).

Cascading failures are documented in many other environments:

- The consequences of bad weather or mechanical failures can cascade through airline schedules, delaying multiple flights and stranding thousands of passengers (**Box 8.3**) [252].
- The disappearance of a species can cascade through the food web of an ecosystem, inducing the extinction of numerous species and altering the habitat of others [253–256].
- The shortage of a particular component can cripple supply chains. For example, the 2011 floods in Thailand resulted in a chronic shortage of car components that disrupted the production chain of more than 1,000 automotive factories worldwide. Therefore, the damage was not limited to the flooded factories, but resulted in worldwide insurance claims reaching $20 billion [257].

In summary, cascading effects are observed in systems of rather different nature. Their size distribution is well approximated by the power law **(8.14)**, implying that most cascades are too small to be noticed; a few, however, are huge, having a global impact. The goal of the next section is to understand the origin of these phenomena and to build models that can reproduce their salient features.

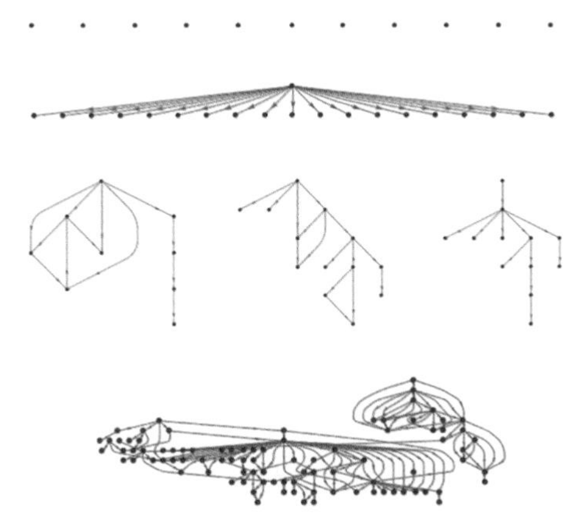

Figure 8.18 **Information Cascades**
Examples of information cascades on Twitter. Nodes denote Twitter accounts, the top node corresponding to the account that first posted a certain shortened URL. The links correspond to those who re-tweeted it. These cascades capture the heterogeneity of information avalanches: most URLs are not re-tweeted at all, appearing as single nodes in the figure. Some, however, start major re-tweet avalanches, like the one seen at the bottom panel. After [249].

Box 8.3 Cascading flight Congestions

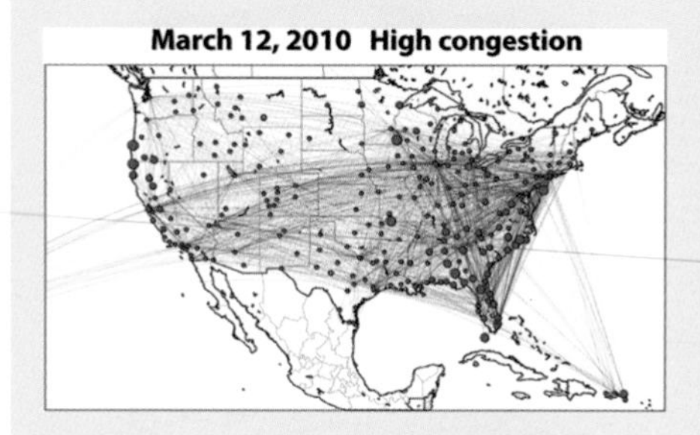

Figure 8.19 **Clusters of Congested Airports**

US aviation map showing congested airports as purple nodes and those with normal traffic as green nodes. The lines correspond to the direct flights between them on March 12, 2010. The clustering of the congested airports indicates that the delays are not independent of each other, but cascade through the airport network. After [252].

Flight delays in the USA have an economic impact of over $40 billion per year [258], caused by the need for enhanced operations, passenger loss of time, decreased productivity and missed business and leisure opportunities. A flight delay is the time difference between the expected and actual departure/arrival times of a flight. Airline schedules include a buffer period between consecutive flights to accommodate short delays. When a delay exceeds this buffer, subsequent flights that use the same aircraft, crew or gate are also delayed. Consequently, a delay can propagate in a cascade-like fashion through the airline network.

While most flights in 2010 were on time, 37.5% arrived or departed late [252]. The delay distribution follows (8.14), implying that while most flights were delayed by just a few minutes, a few were hours behind schedule. These long delays induce correlated delay patterns, a signature of cascading congestions in the air transportation system (**Figure 8.19**).

8.6 Modeling Cascading Failures

The emergence of a cascading event depends on many variables, from the structure of the network on which the cascade propagates to the nature of the propagation process and the breakdown criteria of each individual component. The empirical results indicate that despite the diversity of these variables, the size distribution of the observed avalanches is universal, being independent of the particularities of the system. The purpose of this section is to understand the mechanisms governing cascading phenomena and to explain the power-law nature of the avalanche size distribution.

Numerous models have been proposed to capture the dynamics of cascading events [248, 259–265]. While these models differ in the degree of fidelity they employ to capture specific phenomena, they indicate that systems that develop cascades share three key ingredients:

(i) The system is characterized by some flow over a network, like the flow of electric current in the power grid or the flow of information in communication systems.

(ii) Each component has a local breakdown rule that determines when it contributes to a cascade, either by failing (power grid, earthquakes) or by choosing to pass on a piece of information (Twitter).
(iii) Each system has a mechanism to redistribute the traffic to other nodes upon the failure or the activation of a component.

Next, we discuss two models that predict the characteristics of cascading failures at different levels of abstraction.

8.6.1 Failure Propagation Model

Introduced to model the spread of ideas and opinions [260], the failure propagation model is frequently used to describe cascading failures as well [265]. The model is defined as follows.

Consider a network with an arbitrary degree distribution, where each node contains an agent. An agent i can be in the state 0 (*active* or *healthy*) or 1 (*inactive* or *failed*), and is characterized by a breakdown threshold that has the same value $\varphi_i = \varphi$ for all nodes i.

All agents are initially in the healthy state 0. At time $t = 0$ one agent switches to state 1, corresponding to an initial component failure or to the release of a new piece of information. In each subsequent time step we randomly pick an agent and update its state following a threshold rule:

- If the selected agent i is in state 0, it inspects the state of its k_i neighbors. The agent i adopts state 1 (i.e. it also fails) if at least a fraction φ of its k_i neighbors are in state 1, otherwise it retains its original state 0.
- If the selected agent i is in state 1, it does not change its state.

In other words, a healthy node i changes its state if a fraction φ of its neighbors have failed. Depending on the local network topology, an initial perturbation can die out immediately, failing to induce the failure of any other node. It can also lead to the failure of multiple nodes, as illustrated in **Figure 8.20a,b**. The simulations document three regimes with distinct avalanche characteristics (**Figure 8.20c**):

- **Subcritical Regime**
  If $\langle k \rangle$ is high, changing the state of a node is unlikely to move other nodes over their threshold, as the healthy nodes have many healthy neighbors. In this regime cascades die out quickly and their sizes follow an exponential distribution. Hence the system is unable to support large global cascades (blue symbols, **Figure 8.20c,d**).
- **Supercritical Regime**
  If $\langle k \rangle$ is small, flipping a single node can put several of its neighbors over the threshold, triggering a global cascade. In this regime perturbations induce major breakdowns (purple symbols, **Figure 8.20c,d**).
- **Critical Regime**
  At the boundary of the subcritical and supercritical regimes, the avalanches have widely different sizes. Numerical simulations indicate that in this regime the avalanche sizes s follow **(8.14)** (green and orange symbols, **Figure 8.21d**) with $\alpha = 3/2$ if the underlying network is random.

(a)

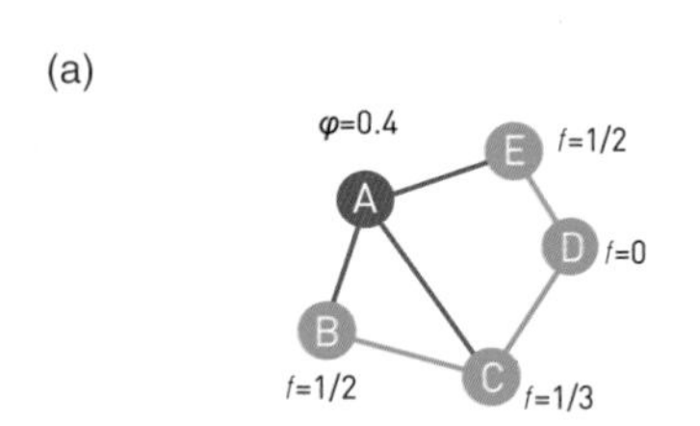

(b)

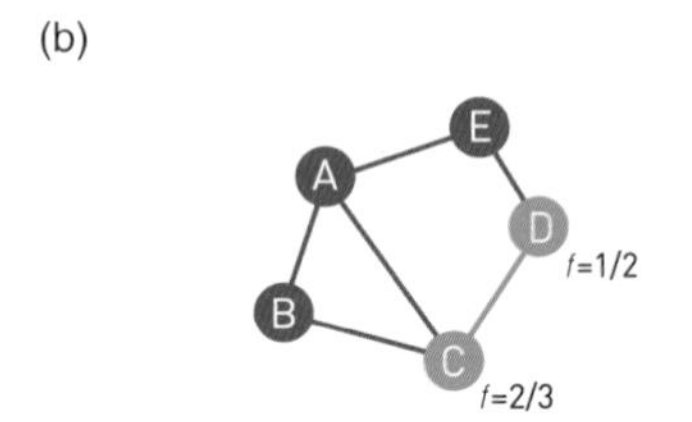

(c)

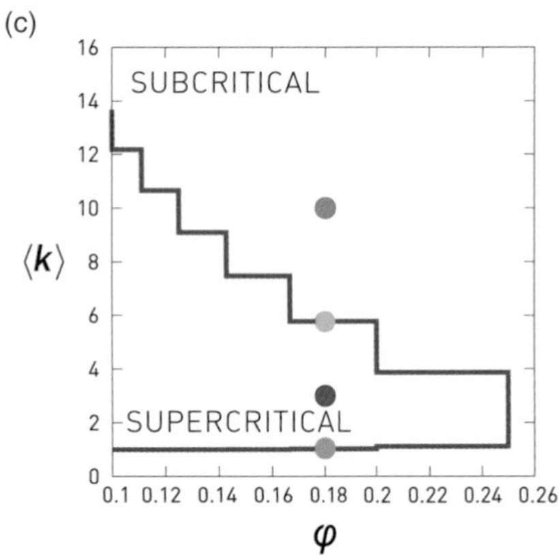

(d)

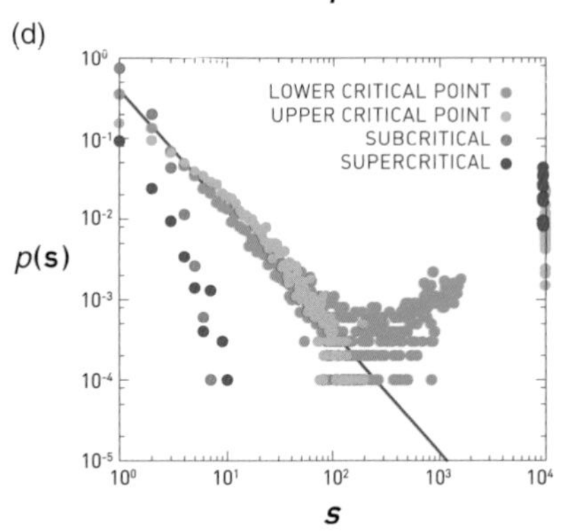

Figure 8.20 **Failure Propagation Model**
(**a,b**) The development of a cascade in a small network in which each node has the same breakdown threshold $\varphi = 0.4$. Initially all nodes are in state 0, shown as green circles. After node A changes its state to 1 (purple), its neighbors B and E will have a fraction $f = 1/2 > 0.4$ of their neighbors in state 1. Consequently they also fail, changing their state to 1, as shown in (b). In the next time step C and D will also fail, as both have $f > 0.4$. Consequently the cascade sweeps the whole network, reaching a size $s = 5$. One can check that if we initially flip node B, it will not induce an avalanche.
(**c**) The phase diagram of the failure propagation model in terms of the threshold function φ and the average degree $\langle k \rangle$ of the network on which the avalanche propagates. The continuous line encloses the region of the $(\langle k \rangle, \varphi)$ plane in which the cascades can propagate in a random graph.
(**d**) Cascade size distributions for $N = 10{,}000$ and $\varphi = 0.18$, $\langle k \rangle = 1.05$ (green), $\langle k \rangle = 3.0$ (purple), $\langle k \rangle = 5.76$ (orange) and $\langle k \rangle = 10.0$ (blue). At the lower critical point we observe a power law $p(s)$ with exponent $\alpha = 3/2$. In the supercritical regime we have only a few small avalanches, as most cascades are global. In the upper critical and subcritical regime we see only small avalanches. After [260].

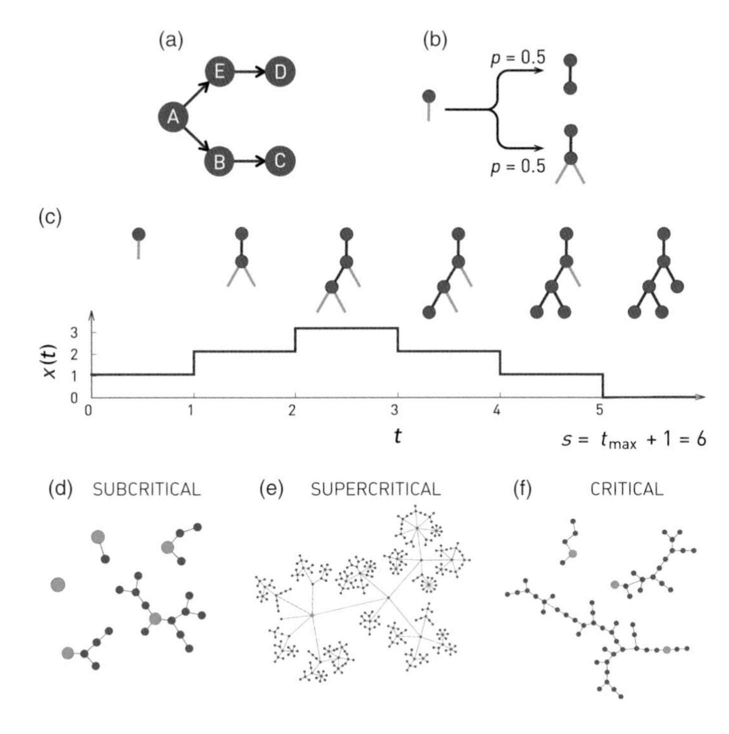

Figure 8.21 **Branching Model**

(a) The branching process mirroring the propagation of the failure shown in **Figure 8.20a,b**. The perturbation starts from node *A*, whose failure flips *B* and *E*, which in turn flip *C* and *D*, respectively.

(b) An elementary branching process. Each active link (green) can become inactive with probability $p_0 = 1/2$ (top) or give birth to two new active links with probability $p_2 = 1/2$ (bottom).

(c) To analytically calculate $p(s)$ we map the branching process into a diffusion problem. For this we show the number of active sites, $x(t)$, as a function of time t. A nonzero $x(t)$ means that the avalanche persists. When $x(t)$ becomes zero, we lose all active sites and the avalanche ends. In the example shown in the image this happens at $t = 5$, hence the size of the avalanche is $t_{max} + 1 = 6$.

An exact mapping between the branching model and a one-dimensional random walk helps us calculate the avalanche exponent. Consider a branching process starting from a stub with one active end. When the active site becomes inactive, it decreases the number of its active sites, i.e. $x \to x - 1$. When the active site branches, it creates two active sites, i.e. $x \to x + 1$. This maps the avalanche size s to the time it takes for the walk that starts at $x = 1$ to reach $x = 0$ for the first time. This is a much-studied process in random walk theory, predicting that the return time distribution follows a power law with exponent 3/2 [262]. For a branching process corresponding to scale-free p_k, the avalanche exponent depends on γ, as shown in **Figure 8.22**.

(d)–(f) Typical avalanches generated by the branching model in the subcritical (d), supercritical (e) and critical regime (f). The green node in each cascade marks the root of the tree, representing the first perturbation. In (d) and (f) we show multiple trees, while in (e) we show only one, as each tree (avalanche) grows indefinitely.

8.6.2 Branching Model

Given the complexity of the failure propogation model, it is hard to analytically predict the scaling behavior of the obtained avalanches. To understand the power-law nature of $p(s)$ and to calculate the avalanche exponent α, we turn to the branching model. This is the simplest model that still captures the basic features of a cascading event.

The model builds on the observation that each cascading failure follows a branching process. Indeed, let us call the node whose initial failure triggers the avalanche the *root of the tree.* The branches of the tree are the nodes whose failure was triggered by this initial failure. For example, in **Figure 8.20a, b**, the breakdown of node *A* starts the avalanche, hence *A* is the root of the tree. The failure of *A* leads to the failure of *B* and *E*, representing the two branches of the tree. Subsequently, *E* induces the failure of *D* and *B* leads to the failure of *C* (**Figure 8.21a**).

The branching model captures the essential features of avalanche propagation (**Figure 8.21**). The model starts with a single active node. In the next time step each active node produces k offspring, where k is selected from a p_k distribution. If a node selects $k = 0$, that branch dies out (**Figure 8.21b**). If it selects $k > 0$, it will have k new active sites. The size of an avalanche corresponds to the size of the tree when all active sites have died out (**Figure 8.21c**).

The branching model predicts the same phases as those observed in the cascading failures model. The phases are now determined only by $\langle k \rangle$, hence by the p_k distribution.

- **Subcritical Regime:**
  For $\langle k \rangle < 1$ on average each branch has less than one offspring. Consequently, each tree will terminate quickly (**Figure 8.21d**). In this regime the avalanche sizes follow an exponential distribution.
- **Supercritical Regime:** $\langle k \rangle > 1$
  For $\langle k \rangle > 1$ on average each branch has more than one offspring. Consequently, the tree will continue to grow indefinitely (**Figure 8.21e**). Hence in this regime all avalanches are global.
- **Critical Regime:** $\langle k \rangle = 1$
  For $\langle k \rangle = 1$ on average each branch has exactly one offspring. Consequently, some trees are large and others die out shortly (**Figure 8.21e**). Numerical simulations indicate that in this regime the avalanche size distribution follows the power law (8.14).

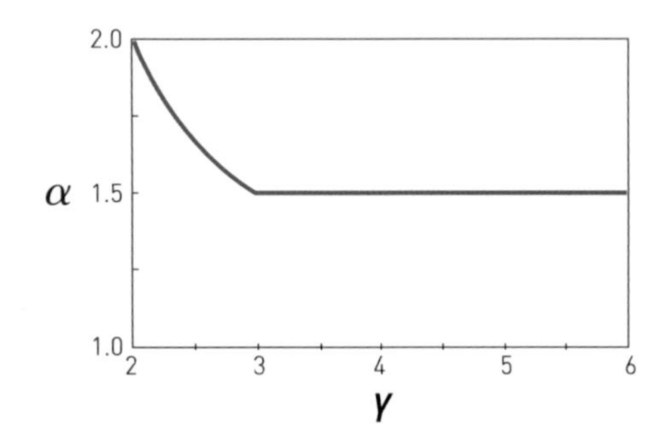

Figure 8.22 **The Avalanche Exponent**
The dependence of the avalanche exponent α on the degree exponent γ of the network on which the avalanche propagates, according to **(8.15)**. The plot indicates that between $2 < \gamma < 3$ the avalanche exponent depends on the degree exponent. Beyond $\gamma = 3$, however, the avalanches behave as they would when spreading on a random network, in which case we have $\alpha = 3/2$.

The branching model can be solved analytically, allowing us to determine the avalanche size distribution for an arbitrary p_k. If p_k is exponentially bounded, that is it has an exponential tail, the calculations predict $\alpha = 3/2$. If, however, p_k is scale-free, then the avalanche exponent depends on the power law exponent γ, following (**Figure 8.22**) [262, 263]

$$\alpha = \begin{cases} 3/2 & \gamma \geq 3 \\ \gamma/(\gamma - 1) & 2 < \gamma < 3 \end{cases}. \tag{8.15}$$

This prediction allows us to revisit **Table 8.2**, finding that the empirically observed avalanche exponents are all between 1.5 and 2, as predicted by **(8.15)**.

In summary, we have discussed two models that capture the dynamics of cascading failures: the failure propagation model and the branching model. In the literature we may also encounter the *overload model*, which is designed to capture power grid failures [248], or the *sandpile model*, that captures the behavior of cascading failures in the critical regime [261, 262]. Other models can also account for the fact that nodes and links have different capacities to carry traffic [264]. These models differ in their realism and the number and nature of their tuning parameters. Yet, they all predict the existence of a critical state, in which the avalanche sizes follow a power law. The avalanche exponent α is uniquely determined by the degree exponent of the network on which the avalanche propagates. The fact that models with rather different propagation dynamics and failure mechanisms predict the same scaling law and avalanche exponent suggests that the underlying phenomenon is universal, that is model independent.

8.7 Building Robustness

Can we enhance a network's robustness? In this section we show that the insights we have gained about the factors that influence robustness allow us to design networks that can simultaneously resist random failures and attacks. We also discuss how to stop a cascading failure, allowing us to enhance a system's dynamical robustness. Finally, we apply the developed tools to the power grid, linking its robustness to its reliability.

8.7.1 Designing Robust Networks

Designing networks that are simultaneously robust to attacks *and* random failures appears to be a conflicting desire [266, 267, 268, 269]. For example, the hub-and-spoke network of **Figure 8.23a** is robust to random failures, as only the failure of its central node can break the network into isolated components. Therefore, the probability that a random failure will fragment the network is $1/N$, which is negligible for large N. At the same time this network is vulnerable to attacks, as the removal of a single node, its central hub, breaks the network into isolated nodes.

We can enhance this network's attack tolerance by connecting its peripheral nodes (**Figure 8.23b**), so that the removal of

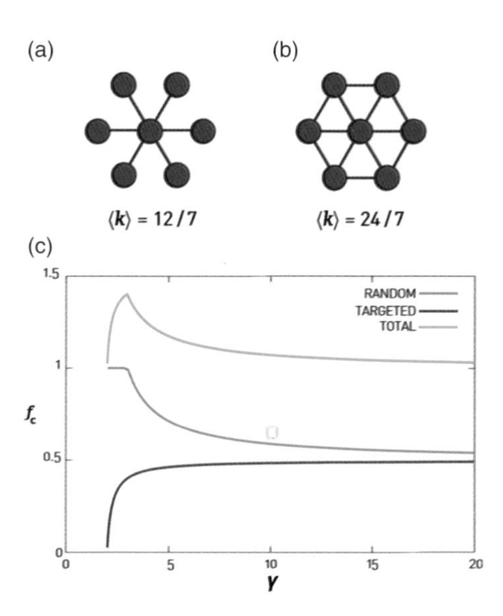

Figure 8.23 **Enhancing Robustness**
(a) A hub-and-spoke network is robust to random failures but has low tolerance to an attack that removes its central hub.
(b) By connecting some of the small-degree nodes, the reinforced network has a higher tolerance to targeted attacks. This increases the cost measured by $\langle k \rangle$, which is higher for the reinforced network.
(c) Random, f_c^{rand}, targeted, f_c^{targ} and total, f_c^{tot} percolation thresholds for scale-free networks as a function of the degree exponent γ for a network with $k_{min} = 3$.

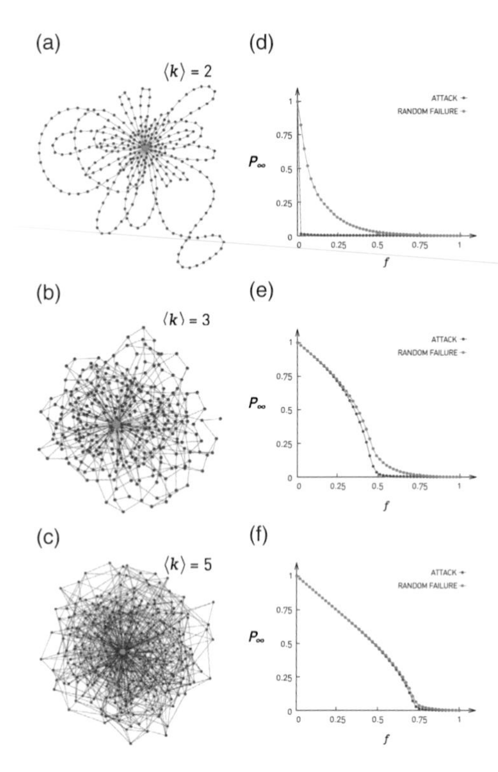

Figure 8.24 **Optimizing Attack and Failure Tolerance**

The figure illustrates the optimal network topologies predicted by (8.16) and (8.17), consisting of a single hub of size (8.18) and the rest of the nodes have the same degree $k_{\min}$ determined by $\langle k\rangle$. The left panels show the network topology for $N = 300$; the right panels show the failure/attack curves for $N = 10,000$.

(a) For small $\langle k\rangle$ the hub holds the network together. Once we remove this central hub the network breaks apart. Hence the attack and error curves are well separated, indicating that the network is robust to random failures but fragile to attacks.

(b) For larger $\langle k\rangle$ a giant component emerges, that exists even without the central hub. Hence, while the hub enhances the system's robustness to random failures, it is no longer essential for the network. In this case both the attack f_c^{targ} and the error f_c^{rand} are large.

(c) For even larger $\langle k\rangle$ the error and the attack curves are indistinguishable, indicating that the network's response to attacks and random failures is indistinguishable. In this case the network is well connected even without its central hub (**Box 8.4**).

the hub does not fragment the network. There is a price, however, for this enhanced robustness: it requires us to double the number of links. If we define the cost to build and maintain a network as proportional to its average degree $\langle k\rangle$, the cost of the network of **Figure 8.23b** is 24/7, double the cost 12/7 of the network of **Figure 8.23a**. The increased cost prompts us to refine our question: Can we maximize the robustness of a network to both random failures and targeted attacks without changing the cost?

A network's robustness against random failures is captured by its percolation threshold f_c, which is the fraction of the nodes we must remove for the network to fall apart. To enhance a network's robustness, we must increase f_c. According to **(8.7)**, f_c depends only on $\langle k\rangle$ and $\langle k^2\rangle$. Consequently, the degree distribution which maximizes f_c needs to maximize $\langle k^2\rangle$ if we wish to keep the cost $\langle k\rangle$ fixed. This is achieved by a bimodal distribution, corresponding to a network with only two kinds of node, with degrees $k_{\min}$ and $k_{\max}$ (**Figure 8.23a,b**).

If we wish to simultaneously optimize the network topology against both random failures and attacks, we search for topologies that maximize the sum (**Figure 8.24c**)

$$f_c^{\mathrm{tot}} = f_c^{\mathrm{rand}} + f_c^{\mathrm{targ}}. \tag{8.16}$$

A combination of analytical arguments and numerical simulations indicates that this too is best achieved by the bimodal degree distribution [266, 267, 268, 269]

$$p_k = (1-r)\,\delta(k-k_{\min}) + r\delta(k-k_{\max}), \tag{8.17}$$

describing a network in which a fraction r of nodes have degree $k_{\max}$ and the remaining fraction $1-r$ have degree $k_{\min}$.

As we show in **Advanced topics 8.G**, the maximum of f_c^{tot} is obtained when $r = 1/N$, that is when there is a single node with degree $k_{\max}$ and the remaining nodes have degree $k_{\min}$. In this case the value of $k_{\max}$ depends on the system size as

$$k_{\max} = AN^{2/3}. \tag{8.18}$$

In other words, a network that is robust to both random failures and attacks has a single hub with degree **(8.18)**, and the rest of the nodes have the same degree $k_{\min}$. This hub-and-spoke

Box 8.4 Halting Cascading Failures

Can we avoid cascading failures? The first instinct is to reinforce the network by adding new links. The problem with reinforcement is that in most real systems the time needed to establish a new link is much larger than the time scale of a cascading failure. For example, thanks to regulatory, financial and legal barriers, building a new transmission line on the power grid can take up to two decades. In contrast, a cascading failure can sweep the power grid in a few seconds.

In a counterintuitive fashion, the impact of cascading failures can be reduced through selective node and link removal [270]. To do so we note that each cascading failure has two parts:

(i) *Initial failure* is the breakdown of the first node or link, representing the source of the subsequent cascade.
(ii) *Propagation* is when the initial failure induces the failure of additional nodes and starts cascading through the network.

Typically the time interval between (i) and (ii) is much shorter than the time scale over which the network could be reinforced. Yet, simulations indicate that the size of a cascade can be reduced if we intentionally remove additional nodes right after the initial failure (i), but before the failure could propagate. Even though the intentional removal of a node or a link causes further damage to the network, the removal of a well-chosen component can suppress the cascade propagation [270]. Simulations indicate that to limit the size of the cascades we must remove nodes with small loads and links with large excess loads in the vicinity of the initial failure. The mechanism is similar to the method used by firefighters, who set a controlled fire in the fire-line to consume the fuel in the path of a wildfire.

A dramatic manifestation of this approach is provided by the *Lazarus effect*, the ability to revive a previously "dead" bacteria (i.e., one that is unable to grow and multiply). This can be achieved through the knockout of a few well-selected genes (**Figure 8.25**) [271]. Therefore, in a counterintuitive fashion, controlled damage can be beneficial to a network.

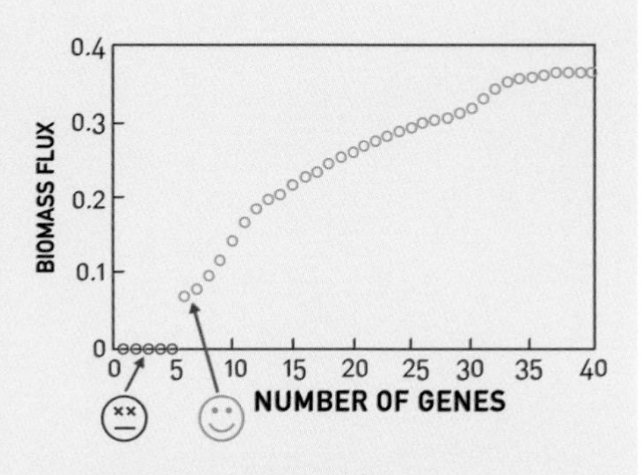

Figure 8.25 **Lazarus Effect**
The growth rate of bacteria is determined by its ability to generate biomass, the molecules it needs to build its cell wall, DNA and other cellular components. If some key genes are missing, the bacteria is unable to generate the necessary biomass. Unable to multiply, it will eventually die. Genes in whose absence the *biomass flux* is zero are called *essential*.

The plot shows the biomass flux for *E. coli*, a type of bacteria frequently studied by biologists. The original mutant is missing an essential gene, hence its biomass flux is zero, as shown on the vertical axis. Consequently, it cannot multiply. Yet, as the figure illustrates, by removing five additional genes we can turn on the biomass flux. Therefore, counterintuitively, we can revive a dead organism through the removal of further genes, a phenomenon called the *Lazarus effect* [271].

topology is obviously robust against random failures as the chance of removing the central hub is $1/N$, which is tiny for large N.

The obtained network may appear to be vulnerable to an attack that removes its hub, but it is not necessarily so. Indeed, the network's giant component is held together by both the central hub as well as the many nodes with degree k_{min} that for $k_{min} > 1$ form a giant component on their own. Hence, while the removal of the k_{max} hub causes a major one-time loss, the remaining low-degree nodes are robust against subsequent targeted removal (**Figure 8.24c**).

8.7.2 Case Study: Estimating Robustness

The European power grid is an ensemble of more than 34 national power grids consisting of over 3,000 generators and substations (nodes) and 200,000 km of transmission lines (**Figure 8.26a–d**). The network's degree distribution can be approximated by (**Figure 8.26e**) [272, 273]

$$p_k = \frac{e^{-k/\langle k \rangle}}{\langle k \rangle}, \tag{8.19}$$

indicating that its topology is characterized by a single parameter, $\langle k \rangle$. Such exponential p_k emerges in growing networks that lack preferential attachment (**Section 5.5**).

By knowing $\langle k \rangle$ for each national power grid, we can predict the respective network's critical threshold f_c^{targ} for attacks. As **Figure 8.26f** shows, for national power grids with $\langle k \rangle > 1.5$ there is reasonable agreement between the observed and the predicted f_c^{targ} (Group 1). However, for power grids with $\langle k \rangle < 1.5$ (Group 2) the predicted f_c^{targ} underestimates the real f_c^{targ}, indicating that these national networks are more robust to attacks than expected based on their degree distribution. As we show next, this enhanced robustness correlates with the reliability of the respective national networks.

To test the relationship between robustness and reliability, we use several quantities, collected and reported for each power failure: (1) energy not supplied; (2) total loss of power; (3) average interruption time, measured in minutes per year. The measurements indicate that Group 1 networks, for which the

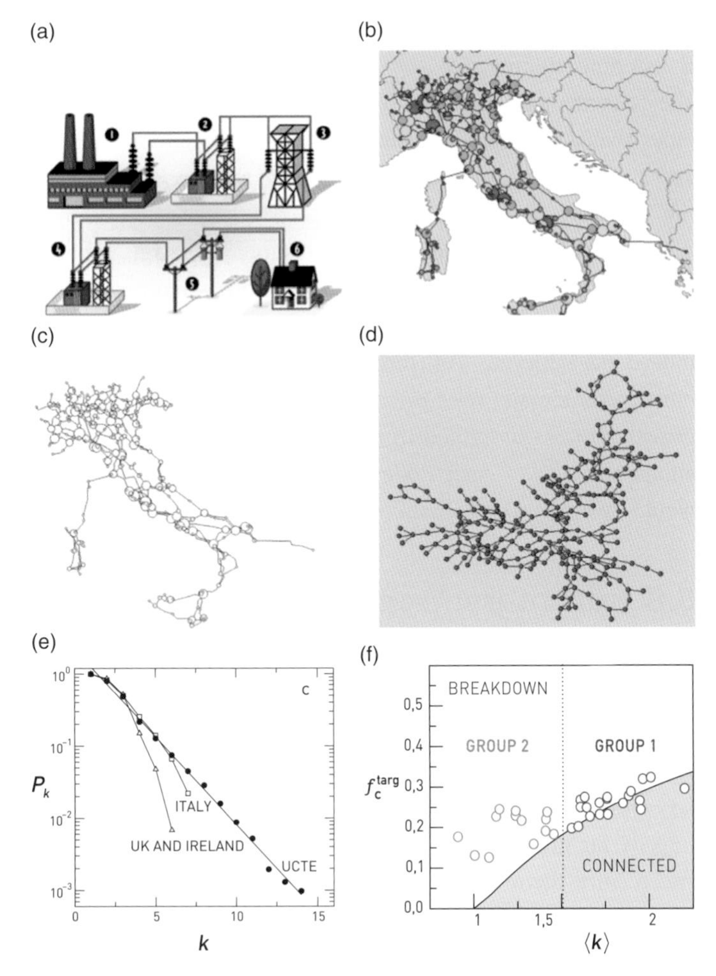

Figure 8.26 **The Power Grid**

(**a**) The power grid is a complex infrastructure consisting of (1) power generators, (2) switching units, (3) the high-voltage transmission grid, (4) transformers, (5) low-voltage lines, (6) consumers, like households or businesses. When we study the network behind the power grid, many of these details are ignored.

(**b**)–(**d**) The Italian power grid with details of production and consumption. Once we strip these details from the network, we obtain the spatial network shown in (c). Once the spatial information is also removed, we arrive at the network (d), which is the typical object of study at the network level.

(**e**) The complementary cumulative degree distribution P_k of the European power grid. The plot shows the data for the full network (UCTE) and separately for Italy, and the joint network of the UK and Ireland, indicating that the national grid's P_k also follows **(8.19)**.

(**f**) The phase space $(f_c^{targ}, \langle k \rangle)$ of exponential uncorrelated networks under attack, where f_c^{targ} is the fraction of hubs we must remove to fragment the network. The continuous curve corresponds to the critical boundary for attacks, below which the network retains its giant component. The plot also shows the estimated $f_c^{targ}(\langle k \rangle)$ for attacks for 33 of the 34 national power grids within the EU, each shown as a separate circle. The plot indicates the presence of two classes of power grids. For countries with $\langle k \rangle > 1.5$ (Group 1), the analytical prediction for $f_c^{targ}(\langle k \rangle)$ agrees with the numerically observed values. For countries with $\langle k \rangle < 1.5$ (Group 2), the analytical prediction underestimates the numerically observed values. Therefore, Group 2 national grids show enhanced robustness to attacks, meaning that they are more robust than expected for a random network with the same degree sequence.

After [272].

real and the theoretical f_c^{targ} agree, represent two-thirds of the full network size and carry almost as much power and energy as the Group 2 networks. Yet, Group 1 accumulates more than five times the average interruption time, more than two times the recorded power losses and almost four times the undelivered energy compared with Group 2 [272]. Hence, the national power grids in Group 1 are significantly more fragile than the power grids in Group 2. This result offers direct evidence that networks which are topologically more robust are also more reliable. At the same time, this finding is rather counterintuitive: one would expect the denser networks to be more robust. We find, however, that the sparser power grids display enhanced robustness.

In summary, a better understanding of the network topology is essential to improve the robustness of complex systems. We can enhance robustness by either designing network topologies that are simultaneously robust to both random failures and attacks, or by interventions that limit the spread of cascading failures.

These results may suggest that we should redesign the topology of the Internet and the power grid to enhance their robustness [274]. Given the opportunity to do so, this could indeed be achieved. Yet, these infrastructural networks were built incrementally over decades, following the self-organized growth process described in the previous chapters. Given the enormous cost of each node and link, it is unlikely that we would ever be given a chance to rebuild them.

8.8 Summary: Achilles' Heel

The masterminds of the September 11, 2001 attack did not choose their targets at random: the World Trade Center in New York, the Pentagon and the White House (an intended target) in Washington DC are the hubs of America's economic, military and political power [50]. Yet, while causing a human tragedy far greater than any other event America has experienced since the Vietnam War, the attacks failed to topple the network. They did offer, however, an excuse to start new wars, like the Iraq and the Afghan wars, triggering a series of cascading events whose impact was far more devastating than the 9/11 terrorist attacks themselves. Yet, all networks, ranging from the economic to the military and the political web, survived. Hence, we can view 9/11 as a tale of robustness and network resilience (**Box 8.5**). The roots of this robustness were uncovered in this chapter: real networks have a whole hierarchy of hubs. Taking out any one of them is not sufficient to topple the underlying network.

The remarkable robustness of real networks represents good news for most complex systems. Indeed, there are uncountable errors in our cells, from misfolding proteins to the late arrival of a transcription factor. Yet, the robustness of the underlying cellular network allows our cells to carry on their normal functions. Network robustness also explains why we rarely notice the effect of router errors on the Internet, or why the

Box 8.5 Robustness, Resilience, Redundancy

Redundancy and resilience are concepts deeply linked to robustness. It is useful to clarify the differences between them.

Robustness
A system is robust if it can maintain its basic functions in the presence of internal and external errors. In a network context, robustness refers to the system's ability to carry out its basic functions even when some of its nodes and links may be missing.

Resilience
A system is resilient if it can adapt to internal and external errors by changing its mode of operation, without losing its ability to function. Hence, resilience is a dynamical property that requires a shift in the system's core activities.

Redundancy
Redundancy implies the presence of parallel components and functions that, if needed, can replace a missing component or function. Networks show considerable redundancy in their ability to navigate information between two nodes, thanks to the multiple independent paths between most node pairs.

disappearance of a species does not result in an immediate environmental catastrophe.

This topological robustness has its price, however: fragility against attacks. As we showed in this chapter, the simultaneous removal of several hubs will break any network. This is bad news for the Internet, as it allows crackers to design strategies that can harm this vital communication system. It is bad news for economic systems, as it indicates that hub removal can cripple the whole economy, as vividly illustrated by the 2009 financial meltdown. Yet, it is good news for drug design, as it suggests that an accurate map of cellular networks can help us develop drugs that can kill unwanted bacteria or cancer cells.

The message of this chapter is simple: network topology, robustness and fragility cannot be separated from one other.

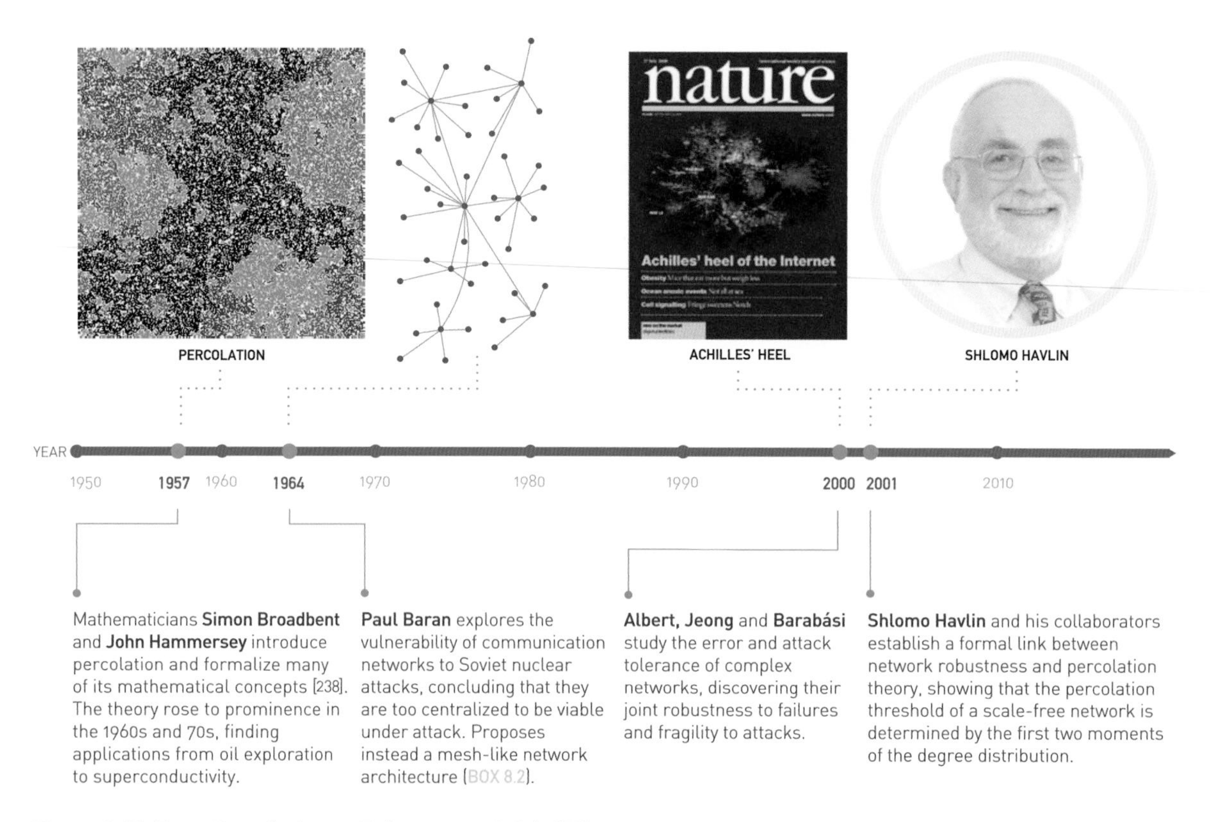

Figure 8.27 **From Percolation to Robustness: A Brief History**
The systematic study of network robustness started with a paper published in *Nature* **(Figure 8.1)** by Réka Albert, Hawoong Jeong and Albert-László Barabási [234], reporting the robustness of scale-free networks to random failures and their fragility to attacks. Yet, the analytical understanding of network robustness relies on percolation theory. In this context, particularly important were the contributions of Shlomo Havlin and collaborators, who established the formal link between robustness and percolation theory and showed that the percolation threshold of a scale-free network is determined by the moments of the degree distribution. A statistical physicist from Israel, Havlin had multiple contributions to the study of networks – from discovering the self-similar nature of real networks [275] to exploring the robustness of layered networks [276].

Rather, each complex system has its own Achilles' heel: the networks behind them are simultaneously robust to random failures but vulnerable to attacks (**Figure 8.27**).

When considering robustness, we cannot ignore the fact that most systems have numerous controls and feedback loops that help them survive in the face of errors and failures. Internet protocols were designed to "route around the trouble," guiding the traffic away from routers that malfunction; cells have numerous mechanisms to dismantle faulty proteins and to shut down malfunctioning genes. This chapter documented a new contribution to robustness: the structure of the underlying network offers a system enhanced failure tolerance.

The robustness of scale-free networks prompts us to ask: Could this enhanced robustness be the reason why many real networks are scale-free? Perhaps real systems have developed a scale-free architecture to satisfy their need for robustness. If this hypothesis is correct, we should be able to set robustness as an optimization criterion and obtain a scale-free network. Yet, as we showed in **Section 8.7**, a network with maximal robustness has a hub-and-spoke topology. Its degree distribution is bimodal, rather than a power law. This suggests that robustness is not the principle that drives the development of real networks. Rather, networks are scale-free thanks to growth and preferential attachment. It so happens that scale-free networks also have enhanced robustness. Yet, they are not the most robust networks we could design.

Box 8.6 At A Glance: Network Robustness

Malloy–Reed Criterion

A giant component exists if

$$\frac{\langle k^2\rangle}{\langle k\rangle} > 2$$

Random Failures

$$f_c = 1 - \frac{1}{\frac{\langle k^2\rangle}{\langle k\rangle} - 1}$$

Random Network

$$f_c^{ER} = 1 - \frac{1}{\langle k\rangle}$$

Enhanced Robustness: Attacks

$$f_c > f_c^{ER}$$

Attacks

$$f_c^{\frac{2-\gamma}{1-\gamma}} = 2 + \frac{2-\gamma}{3-\gamma} k_{min} \left(f_c^{\frac{3-\gamma}{1-\gamma}} - 1 \right)$$

Cascading Failures

$$p(s) \sim s^{-\alpha}$$

$$\alpha = \begin{cases} 3/2 & \gamma > 3 \\ \frac{\gamma}{\gamma - 1} & 2 < \gamma < 3 \end{cases}$$

8.9 Homework

8.9.1 Random Failure: Beyond Scale-Free Networks

Calculate the critical threshold f_c for networks with the following degree distributions:

(a) Power law with exponential cutoff.
(b) Lognormal.
(c) Delta (all nodes have the same degree).

Assume that the networks are uncorrelated and infinite. Refer to **Table 4.2** for the functional form of each distribution and the corresponding first and second moments. Discuss the consequences of the obtained results for network robustness.

8.9.2 Critical Threshold in Correlated Networks

Generate three networks with 10^4 nodes, that are assortative, disassortative and neutral and have a power-law degree distribution with degree exponent $\gamma = 2.2$. Use the Xalvi-Brunet and Sokolov algorithm described in **Section 7.5** to generate the networks. With the help of a computer, study the robustness of the three networks against random failures, and compare their $P_\infty(f)/P_\infty(0)$ ratio. Which network is the most robust? Can you explain why?

8.9.3 Failure of Real Networks

Determine the number of nodes that need to fail to break the networks listed in **Table 4.1**. Assume that each network is uncorrelated.

8.9.4 Conspiracy in Social Networks

In a Big Brother society, the thought police want to follow a "divide-and-conquer" strategy by fragmenting the social network into isolated components. You belong to the resistance and want to foil their plans. There are rumors that the police want to detain individuals who have many friends and individuals whose friends tend to know each other. The resistance puts you in charge to decide which individuals to protect: those whose friendship circle is highly interconnected or those with many friends. To decide, you simulate two different attacks on your network, by removing (i) the nodes that have the highest clustering coefficient and (ii) the nodes that have the largest degree. Study the size of the giant component as a function of the fraction of removed nodes for the two attacks on the following networks:

(a) A network with $N = 10^4$ nodes generated with the configuration model (**Section 4.8**) and power law degree distribution with $\gamma = 2.5$.
(b) A network with $N = 10^4$ nodes generated with the hierarchical model described in **Figure 9.16** and **Advanced topic 9.B**.

Which is the most sensitive topological information, clustering coefficient or degree? Which, if protected, limits the damage best? Would it be better if all individuals' information (clustering coefficient, degree, etc.) could be kept secret? Why?

8.9.5 Avalanches in Networks

Generate a random network with the Erdős–Rényi $G(N,p)$ model and a scale-free network with the configuration model, with $N = 10^3$ nodes and average degree $\langle k \rangle = 2$. Assume that on each node there is a bucket which can hold as many sand grains as the node degree. Simulate then the following process:

(a) At each time step add a grain to a randomly chosen node i.
(b) If the number of grains at node i reaches or exceeds its bucket size, then it becomes unstable and all the grains at the node topple to the buckets of its adjacent nodes.
(c) If this toppling causes any of the adjacent nodes' buckets to be unstable, subsequent topplings follow on those nodes, until there is no unstable bucket left. We call this sequence of toppings an avalanche, its size s being equal to the number of nodes that turned unstable following an initial perturbation (adding one grain).

Repeat (a)–(c) 10^4 times. Assume that at each time step a fraction 10^{-4} of sand grains is lost in the transfer, so that the network buckets do not become saturated with sand. Study the avalanche distribution $p(s)$.

8.10 Advanced Topics 8.A Percolation in Scale-Free Networks

To understand how a scale-free network breaks apart as we approach the threshold **(8.7)**, we need to determine the corresponding critical exponents γ_p, β_p and ν. The calculations indicate that the scale-free property alters the value of these exponents, leading to systematic deviations from the exponents that characterize random networks (**Section 8.2**).

Let us start with the probability P_∞ that a randomly selected node belongs to the giant component. According to **(8.2)** this follows a power law near p_c (or f_c in the case of node removal). The calculations predict that for a scale-free network the exponent β_p [16, 277–280] depends on the degree exponent γ:

$$\beta_p = \begin{cases} \dfrac{1}{3-\gamma} & 2 < \gamma < 3, \\ \dfrac{1}{\gamma - 3} & 3 < \gamma < 4, \\ 1 & \gamma > 4. \end{cases} \tag{8.20}$$

Hence, while for a random network (corresponding to $\gamma > 4$) we have $\beta_p = 1$, for most scale-free networks of practical interest $\beta_p > 1$. Therefore, the giant component collapses faster in the vicinity of the critical point in a scale-free network than in a random network.

The exponent characterizing the average component size near p_c follows [277]

$$\gamma_P = \begin{cases} 1 & \gamma > 3 \\ -1 & 2 < \gamma < 3 \end{cases}. \tag{8.21}$$

The negative γ_p for $\gamma < 3$ may appear surprising. Note, however, that for $\gamma < 3$ we always have a giant component. Hence, the divergence **(8.1)** cannot be observed in this regime.

For a randomly connected network with arbitrary degree distribution, the size distribution of the finite clusters follows [277, 279, 280]

$$n_s \sim s^{-\tau} e^{-s/s^*} \tag{8.22}$$

Here, n_s is the number of clusters of size s and s^* is the crossover cluster size. At criticality,

$$s^* \sim |p - p_c|^{-\sigma} \tag{8.23}$$

The critical exponents are

$$\tau = \begin{cases} \frac{5}{2} & \gamma > 4 \\ \frac{2\gamma - 3}{\gamma - 2} & 2 < \gamma < 4, \end{cases} \tag{8.24}$$

$$\sigma = \begin{cases} \frac{3-\gamma}{\gamma - 2} & 2 < \gamma < 3 \\ \frac{\gamma - 3}{\gamma - 2} & 3 < \gamma < 4 \\ \frac{1}{2} & \gamma > 4. \end{cases} \tag{8.25}$$

Once again, the random network values $\tau = 5/2$ and $\sigma = 1/2$ are recovered for $\gamma > 4$.

In summary, the exponents describing the breakdown of a scale-free network depend on the degree exponent γ. This is true even in the range $3 < \gamma < 4$, where the percolation transition occurs at a finite threshold f_c. The mean-field behavior predicted for percolation in infinite dimensions, capturing the response of a random network to random failures, is recovered only for $\gamma > 4$.

8.11 Advanced Topics 8.B Molloy–Reed Criterion

The purpose of this section is to derive the Molloy–Reed criterion, which allows us to calculate the percolation threshold of an arbitrary network. For a giant component to exist, each node that belongs to it must be connected to at least two other nodes on average (**Figure 8.8**). Therefore, the average degree k_i of a randomly chosen node i that is part of the giant component should be at least 2. Denote by $P(k_i|i \leftrightarrow j)$ the conditional probability that a node in a network with degree k_i is connected to a node j that is part of the giant component. This conditional probability allows us to determine the expected degree of node i as [280]

$$\langle k_i|i \leftrightarrow j\rangle = \sum_{k_i} k_i P(k_i|i \leftrightarrow j) = 2. \tag{8.26}$$

In other words, $\langle k_i|i \leftrightarrow j\rangle$ should be equal to or exceed two, the condition for node i to be part of the giant component. We can write the probability appearing in the sum **(8.26)** as

$$P(k_i \,|\, i \leftrightarrow j) = \frac{P(k_i, i \leftrightarrow j)}{P(i \leftrightarrow j)} = \frac{P(i \leftrightarrow j \,|\, k_i)p(k_i)}{P(i \leftrightarrow j)}, \tag{8.27}$$

where we used Bayes' theorem in the last term. For a network with degree distribution p_k, in the absence of degree correlations, we can write

$$P(i \leftrightarrow j) = \frac{2L}{N(N-1)} = \frac{\langle k\rangle}{N-1}, \quad P(i \leftrightarrow j \,|\, k_i) = \frac{k_i}{N-1}, \tag{8.28}$$

which expresses the fact that we can choose between $N-1$ nodes to link to, each with probability $1/(N-1)$, and that we can try this k_i times. We can now return to **(8.26)**, obtaining

$$\sum_{k_i} k_i P(k_i \,|\, i \leftrightarrow j) = \sum_{k_i} k_i \frac{P(i \leftrightarrow j \,|\, k_i)p(k_i)}{P(i \leftrightarrow j)}$$

$$= \sum_{k_i} k_i \frac{k_i p(k_i)}{\langle k\rangle} = \frac{\sum_{k_i} k_i^2 p(k_i)}{\langle k\rangle}. \tag{8.29}$$

With that we arrive at the Molloy–Reed criterion **(8.4)**, providing the condition to have a giant component:

$$k = \frac{\langle k^2 \rangle}{\langle k \rangle} > 2. \tag{8.30}$$

8.12 Advanced Topics 8.C Critical Threshold Under Random Failures

The purpose of this section is to derive **(8.7)**, which provides the critical threshold for random node removal [16, 280]. The random removal of a fraction f of nodes has two consequences:

- It alters the degree of some nodes, as nodes that were previously connected to the removed nodes will lose some links $[k \rightarrow k' \leq k]$.
- Consequently, it changes the degree distribution, as the neighbors of the missing nodes will have an altered degree $[p_k \rightarrow p'_k]$.

To be specific, after we randomly remove a fraction f of nodes, a node with degree k becomes a node with degree k' with probability

$$\binom{k}{k'} f^{k-k'} (1-f)^{k'} \qquad k' \leq k. \tag{8.31}$$

The first f-dependent term in **(8.31)** accounts for the fact that the selected node lost $(k - k')$ links, each with probability f; the next term accounts for the fact that node removal leaves k' links untouched, each with probability $(1 - f)$.

The probability that we have a degree-k node in the original network is p_k; the probability that we have a new node with degree k' in the new network is

$$p'_{k'} = \sum_{k=k'}^{\infty} p_k \binom{k}{k'} f^{k-k'} (1-f)^{k'}. \tag{8.32}$$

Let us assume that we know $\langle k \rangle$ and $\langle k^2 \rangle$ for the original degree distribution p_k. Our goal is to calculate $\langle k' \rangle$, $\langle k'^2 \rangle$ for the new degree distribution $p'_{k'}$, obtained after we randomly removed a fraction f of the nodes. For this we write

$$\begin{aligned}\langle k' \rangle_f &= \sum_{k'=0}^{\infty} k' p'_{k'} \\ &= \sum_{k'=0}^{\infty} k' \sum_{k'=k}^{\infty} p_k \left(\frac{k!}{k'!(k-k')!} \right) f^{k-k'} (1-f)^{k'} \\ &= \sum_{k'=0}^{\infty} \sum_{k=k'}^{\infty} p_k \frac{k(k-1)!}{(k'-1)!(k-k')!} f^{k-k'} (1-f)^{k'-1} (1-f).\end{aligned} \tag{8.33}$$

The sum above is performed over the triangle shown in **Figure 8.28**. We can check that we are performing the same sum if we change the order of summation together with the limits of the sums as

$$\sum_{k'=0}^{\infty} \sum_{k=k'}^{\infty} = \sum_{k=0}^{\infty} \sum_{k'=0}^{k}. \tag{8.34}$$

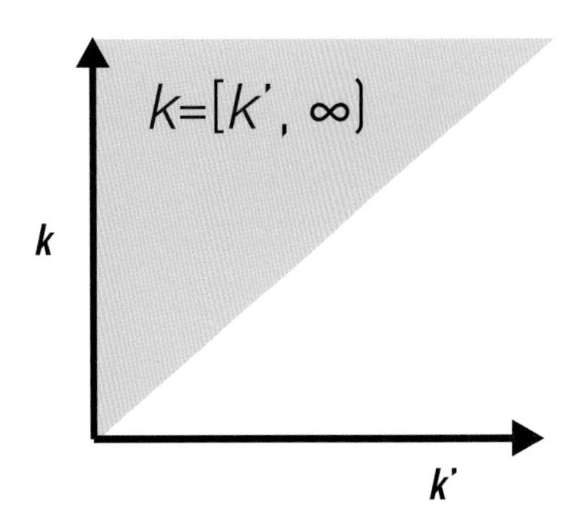

Figure 8.28 **The Integration Domain**
In **(8.34)** we change the integration order, that is the order of the two sums. We can do so because both sums are defined over the triangle shown in purple in the figure.

Hence we obtain

$$\begin{aligned}\langle k' \rangle_f &= \sum_{k=0}^{\infty} k' \sum_{k'=0}^{k} p_k \frac{k(k-1)!}{(k'-1)!(k-k')!} f^{k-k'} (1-f)^{k'-1} (1-f) \\ &= \sum_{k=0}^{\infty} (1-f) k p_k \sum_{k'=0}^{k} \frac{(k-1)!}{(k'-1)!(k-k')!} f^{k-k'} (1-f)^{k'-1} \\ &= \sum_{k=0}^{\infty} (1-f) k p_k \sum_{k'=0}^{k} \binom{k-1}{k'-1} f^{k-k'} (1-f)^{k'-1} \\ &= \sum_{k=0}^{\infty} (1-f) k p_k \\ &= (1-f)\langle k \rangle.\end{aligned} \tag{8.35}$$

This connects $\langle k' \rangle$ to the original $\langle k \rangle$ after the random removal of a fraction f of nodes.

We perform a similar calculation for $\langle k'^2 \rangle$:

$$\begin{aligned}\langle k'^2 \rangle_f &= \langle k'(k'-1) + k' \rangle \\ &= \langle k'(k'-1) \rangle_f + \langle k' \rangle_f \\ &= \sum_{k'=0}^{\infty} k'(k'-1) p'_{k'} + \langle k' \rangle_f.\end{aligned} \tag{8.36}$$

Again, we change the order of the sums (**Figure 8.28**), obtaining

$$
\begin{aligned}
&\langle k'(k'-1)\rangle_f \\
&= \sum_{k'=0}^{\infty} k'(k'-1)p'_{k'} \\
&= \sum_{k'=0}^{\infty} k'(k'-1)\sum_{k'=0}^{\infty} \frac{k'(k'-1)}{k'!(k-k')!} f^{k-k'}(1-f)^{k'} \\
&= \sum_{k=0}^{\infty} k'(k'-1)\sum_{k'=0}^{k} p_k \frac{k'(k'-1)}{k'!(k-k')!} f^{k-k'}(1-f)^{k'} \\
&= \sum_{k=0}^{\infty}\sum_{k'=0}^{k} p_k \frac{k'!}{(k'-2)!(k-k')!} f^{k-k'}(1-f)^{k'-2}(1-f)^2 \\
&= \sum_{k=0}^{\infty}(1-f)^2 k(k-1)p_k \sum_{k'=0}^{k} \frac{(k'-2)!}{(k'-2)!(k-k')!} f^{k-k'}(1-f)^{k'-2} \\
&= \sum_{k=0}^{\infty}(1-f)^2 k(k-1)p_k \sum_{k'=0}^{k} \binom{k-2}{k'-2} f^{k-k'}(1-f)^{k'-2} \\
&= \sum_{k=0}^{\infty}(1-f)^2 k(k-1)p_k \\
&= (1-f)^2\langle k(k-1)\rangle. \qquad (8.37)
\end{aligned}
$$

Hence, we obtain

$$
\begin{aligned}
\langle k'^2\rangle_f &= \langle k'(k'-1)+k'\rangle_f \\
&= \langle k'(k'-1)\rangle_f + \langle k'\rangle_f \\
&= (1-f)^2\langle k(k-1)\rangle + (1-f)\langle k\rangle \\
&= (1-f)^2(\langle k^2\rangle - \langle k\rangle) + (1-f)\langle k\rangle \\
&= (1-f)^2\langle k^2\rangle - (1-f)^2\langle k\rangle + (1-f)\langle k\rangle \\
&= (1-f)^2\langle k^2\rangle - (-f^2+2f-1+1-f)\langle k\rangle \\
&= (1-f)^2\langle k^2\rangle + f(1-f)\langle k\rangle. \qquad (8.38)
\end{aligned}
$$

This connects $\langle k'^2\rangle$ to the original $\langle k^2\rangle$ after the random removal of a fraction f of nodes. Let us put the results **(8.35)** and **(8.38)** together:

$$\langle k'\rangle_f = (1-f)\langle k\rangle, \qquad (8.39)$$

$$\langle k'\rangle_f = (1-f)^2\langle k^2\rangle + f(1-f)\langle k\rangle. \qquad (8.40)$$

According to the Molloy–Reed criterion **(8.4)**, the breakdown threshold is given by

$$\kappa = \frac{\langle k'^2 \rangle_f}{\langle k' \rangle_f} = 2. \tag{8.41}$$

Inserting **(8.38)** and **(8.40)** into **(8.41)**, we obtain our final result **(8.7)**,

$$f_c = 1 - \frac{1}{\frac{\langle k^2 \rangle}{\langle k \rangle} - 1} \tag{8.42}$$

providing the breakdown threshold of networks with arbitrary p_k under random node removal.

8.13 Advanced Topics 8.D Breakdown of a Finite Scale-Free Network

In this section we derive the dependence **(8.10)** of the breakdown threshold of a scale-free network on the network size N. We start by calculating the mth moment of a power-law distribution

$$\begin{aligned}\langle k^m \rangle &= (\gamma - 1) k_{\text{min}}^{\gamma-1} \int_{k_{\text{min}}}^{k_{\text{max}}} k^{m-\gamma} dk \\ &= \frac{(\gamma - 1)}{(m - \gamma + 1)} k_{\text{min}}^{\gamma-1} [k^{m-\gamma+1}]_{k_{\text{min}}}^{k_{\text{max}}}.\end{aligned} \tag{8.43}$$

Using **(4.18)**,

$$k_{\text{max}} = k_{\text{min}} N^{\frac{1}{\gamma-1}} \tag{8.44}$$

we obtain

$$\langle k^m \rangle = \frac{(\gamma - 1)}{(m - \gamma + 1)} k_{\text{min}}^{\gamma-1} [k_{\text{max}}^{m-\gamma+1} - k_{\text{min}}^{m-\gamma+1}]. \tag{8.45}$$

To calculate f_c we need to determine the ratio

$$\kappa = \frac{\langle k^2 \rangle}{\langle k \rangle} = \frac{(2 - \gamma) k_{\text{max}}^{3-\gamma} - k_{\text{min}}^{3-\gamma}}{(3 - \gamma) k_{\text{max}}^{2-\gamma} - k_{\text{min}}^{2-\gamma}}, \tag{8.46}$$

which for large N (and hence for large k_{max}) depends on γ as

$$\kappa = \frac{\langle k^2 \rangle}{\langle k \rangle} = \left| \frac{2 - \gamma}{3 - \gamma} \right| \begin{cases} k_{\text{min}} & \gamma > 3 \\ k_{\text{max}}^{3-\gamma} k_{\text{min}}^{\gamma-2} & 3 > \gamma > 2. \\ k_{\text{max}} & 2 > \gamma > 1 \end{cases} \tag{8.47}$$

The breakdown threshold is given by **(8.7)**

$$f_c = 1 - \frac{1}{\kappa - 1}, \tag{8.48}$$

where κ is given by **(8.46)**. Inserting **(8.43)** into **(8.42)** and **(8.47)**, we obtain

$$f_c \approx 1 - \frac{C}{N^{\frac{3-\gamma}{\gamma-1}}}, \tag{8.49}$$

which is **(8.10)**.

8.14 Advanced Topics 8.E Attack and Error Tolerance of Real Networks

In this section we explore the attack and error curves for the ten reference networks discussed in **Table 4.1** and **(8.2)**. The corresponding curves are shown in **Figure 8.29**. Their inspection reveals several patterns, confirming the results discussed in this chapter:

- For all networks the error and attack curves separate, confirming the Achilles' heel property (**Section 8.8**): real networks are robust to random failures but are fragile to attacks.
- The separation between the error and attack curves depends on the average degree and the degree heterogeneity of each network. For example, for the citation and the actor networks, f_c for the attacks is in the vicinity of 0.5 and 0.75, respectively – rather large values. This is because these networks are rather dense, with $\langle k \rangle = 20.8$ for the citation network and $\langle k \rangle = 83.7$ for the actor network. Hence, these networks can survive the removal of a very high fraction of their hubs.

8.15 Advanced Topics 8.F Attack Threshold

The goal of this section is to derive **(8.12)**, providing the attack threshold of a scale-free network. We aim to calculate f_c for

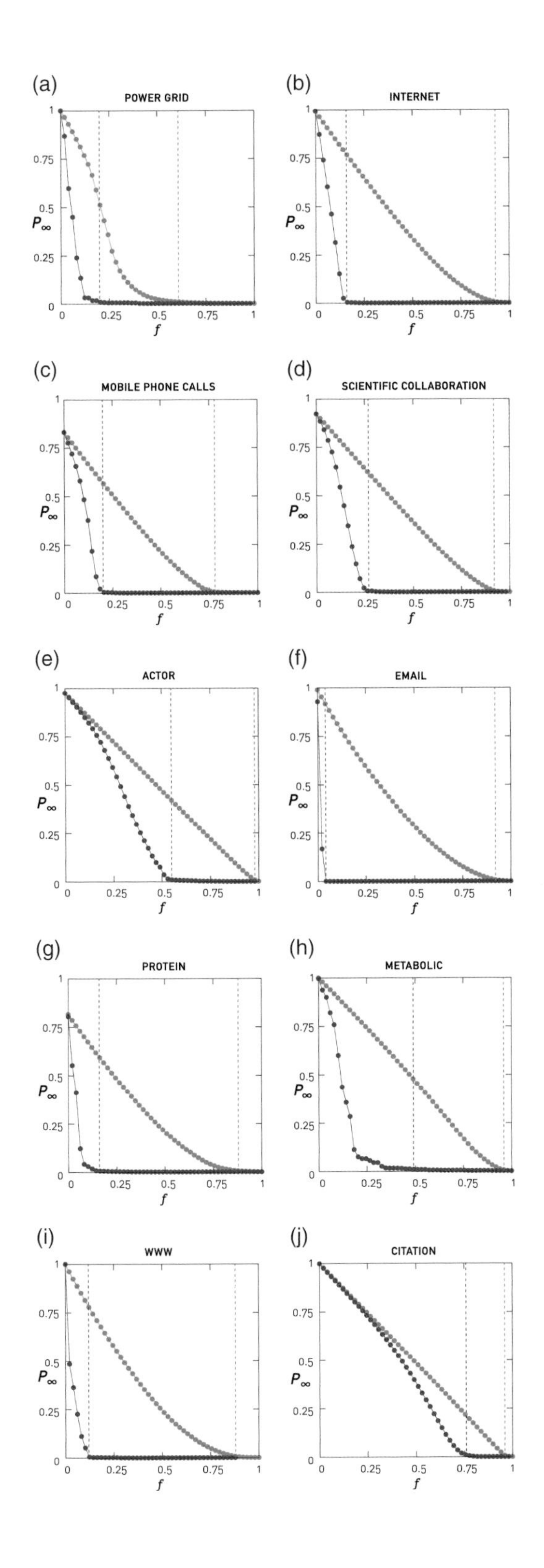

Figure 8.29 **Error and Attack Curves**

The error (green) and attack (purple) curves for the ten reference networks listed in **Table 4.1**. The green vertical line corresponds to the estimated f_c^{rand} for errors, while the purple vertical line corresponds to f_c^{targ} for attacks. The estimated f_c corresponds to the point where the giant component first drops below 1% of its original size. In most systems this procedure offers a good approximation for f_c. The only exception is the metabolic network, for which $f_c^{\text{targ}} < 0.25$, but a small cluster persists, pushing the reported f_c^{targ} to $f_c^{\text{targ}} \simeq 0.5$.

an uncorrelated scale-free network, generated by the configuration model with $p_k = c \cdot k^{-\gamma}$, where $k = k_{\min}, \ldots, k_{\max}$ and $c \approx (\gamma - 1)/(k_{\min}^{-\gamma+1} - k_{\max}^{-\gamma+1})$.

The removal of a fraction f of nodes in a decreasing order of their degree (hub removal) has two effects [240, 280]:

(i) The maximum degree of the network changes from $k_{\max}$ to $k'_{\max}$.

(ii) The links connected to the removed hubs are also removed, changing the degree distribution of the remaining network.

The resulting network is still uncorrelated, therefore we can use the Molloy–Reed criterion to determine the existence of a giant component.

We start by considering the impact of (i). The new upper cutoff, $k'_{\max}$, is given by

$$f = \int_{k'_{\max}}^{k_{\max}} p_k dk = \frac{\gamma - 1}{\gamma - 1} \frac{k'^{-\gamma+1}_{\max} - k^{-\gamma+1}_{\max}}{k^{-\gamma+1}_{\min} - k^{-\gamma+1}_{\max}}. \tag{8.50}$$

If we assume that $k_{\max} \gg k'_{\max}$ and $k_{\max} \gg k_{\min}$ (true for large scale-free networks with natural cutoff), we can ignore the $k_{\max}$ terms, obtaining

$$f = \left(\frac{k'_{\max}}{k_{\min}} \right)^{-\gamma+1}, \tag{8.51}$$

which leads to

$$k'_{\max} = k_{\min} f^{\frac{1}{1-\gamma}}. \tag{8.52}$$

Equation **(8.52)** provides the new maximum degree of the network after we remove a fraction f of the hubs.

Next we turn to (ii), accounting for the fact that hub removal changes the degree distribution $p_k \to p'_k$. In the absence of degree correlations, we assume that the links of the removed hubs connect to randomly selected stubs. Consequently, we calculate the fraction of links removed "randomly," $\tilde{f}$, as a consequence of removing a fraction f of the hubs:

$$\tilde{f} = \frac{\int_{k'_{max}}^{k_{max}} kp_k dk}{\langle k \rangle} = \frac{1}{\langle k \rangle} c \int_{k'_{max}}^{k_{max}} k^{-\gamma+1} dk$$

$$= \frac{1}{\langle k \rangle} \frac{1-\gamma}{2-\gamma} \frac{k'^{-\gamma+2}_{max} - k^{-\gamma+2}_{max}}{k^{-\gamma+1}_{min} - k^{-\gamma+2}_{max}}. \quad (8.53)$$

Ignoring the k_{max} term again and using $\langle k \rangle \approx \frac{\gamma-1}{\gamma-2} k_{min}$, we obtain

$$\tilde{f} = \left(\frac{k'_{max}}{k_{min}} \right)^{-\gamma+2}. \quad (8.54)$$

Using (8.51), we obtain

$$\tilde{f} = f^{\frac{2-\gamma}{1-\gamma}}. \quad (8.55)$$

For $\gamma \to 2$ we have $\tilde{f} \to 1$, which means that the removal of a tiny fraction of the hubs removes all links, potentially destroying the network. This is consistent with the finding of **Chapter 4** that for $\gamma = 2$ the hubs dominate the network.

In general, the degree distribution of the remaining network is

$$p'_{k'} = \sum_{k=k_{min}}^{k'_{max}} \binom{k}{k'} \tilde{f}^{k-k'} (1-\tilde{f})^{k'} p_k. \quad (8.56)$$

Note that we obtained the degree distribution **(8.32)** in **Advanced topics 8.C**. This means that we can now proceed with the calculation method developed there for random node removal. To be specific, we calculate κ for a scale-free network with k_{min} and k'_{max} using **(8.45)**:

$$\kappa = \frac{2-\gamma}{3-\gamma} \frac{k'^{3-\gamma}_{max} - k^{3-\gamma}_{min}}{k'^{2-\gamma}_{max} - k^{2-\gamma}_{min}}. \quad (8.57)$$

Substituting this into (8.52), we have

$$\kappa = \frac{2-\gamma}{3-\gamma} \frac{k^{3-\gamma}_{min} f^{(3-\gamma)/(1-\gamma)} - k^{3-\gamma}_{min}}{k^{2-\gamma}_{min} f^{(2-\gamma)/(1-\gamma)} - k^{2-\gamma}_{min}} = \frac{2-\gamma}{3-\gamma} k_{min} \frac{f^{(3-\gamma)/(1-\gamma)} - 1}{f^{(2-\gamma)/(1-\gamma)} - 1}. \quad (8.58)$$

After simple transformations we obtain

$$f_c^{\frac{2-\gamma}{1-\gamma}} = 2 + \frac{2-\gamma}{3-\gamma} k_{\min} \left(f_c^{\frac{3-\gamma}{1-\gamma}} - 1 \right). \tag{8.59}$$

8.16 Advanced Topics 8.G The Optimal Degree Distribution

In this section we derive the bimodal degree distribution that simultaneously optimizes a network's topology against attacks and failures, as discussed in **Section 8.7** [267]. Let us assume, as we did in **(8.17)**, that the degree distribution is bimodal, consisting of two delta functions:

$$p_k = (1-r)\delta(k-k_{\min}) + r\delta(k-k_{\max}). \tag{8.60}$$

We start by calculating the total threshold, f^{tot}, as a function of r and $k_{\max}$ for a fixed $\langle k \rangle$. To obtain analytical expressions for f_c^{rand} and f_c^{targ} we calculate the moments of the bimodal distribution **(8.60)**,

$$\begin{aligned} \langle k \rangle &= (1-r)k_{\min} + rk_{\max}, \\ \langle k^2 \rangle &= (1-r)k_{\min}^2 + rk_{\max}^2 = \frac{(\langle k \rangle - rk_{\max})^2}{1-r} + rk_{\max}^2. \end{aligned} \tag{8.61}$$

Inserting these into **(8.7)** we obtain

$$f_c^{\text{rand}} = \frac{\langle k \rangle^2 - 2r\langle k \rangle k_{\max} - 2(1-r)\langle k \rangle + rk_{\max}^2}{\langle k \rangle^2 - 2r\langle k \rangle k_{\max} - (1-r)\langle k \rangle + rk_{\max}^2}. \tag{8.62}$$

To determine the threshold for targeted attack, we must consider the fact that we have only two types of node: a fraction r of nodes with degree $k_{\max}$ and a fraction $1-r$ with degree $k_{\min}$. Hence, hub removal can either remove all hubs (case (i)) or only some fraction of them (case (ii)).

(i) $f_c^{\text{targ}} > r$: In this case all hubs have been removed, hence the nodes left after the targeted attack have degree $k_{\min}$. We therefore obtain

$$f_c^{\text{targ}} = r + \frac{1-r}{\langle k \rangle - rk_{\max}} \left\{ \langle k \rangle \frac{\langle k \rangle - rk_{\max} - 2(1-r)}{\langle k \rangle - rk_{\max} - (1-r)} - rk_{\max} \right\}. \tag{8.63}$$

(ii) $f_c^{targ} < r$: In this case the removed nodes are all from the high-degree group, leaving behind some k_{max} nodes. Hence we obtain

$$f_c^{targ} = \frac{\langle k \rangle^2 - 2r\langle k \rangle k_{max} - 2(1-r)\langle k \rangle}{k_{max}(k_{max}-1)(1-r)}. \tag{8.64}$$

With the thresholds **(8.62)–(8.64)** we can now evaluate the total threshold f_c^{tot} **(8.16)**. To obtain an expression for the optimal value of k_{max} as a function of r, we determine the value of k for which f_c^{tot} is maximal. Using **(8.62)** and **(8.64)**, we find that for small r the optimal value of k_{max} can be approximated by

$$k_{max} \sim \left\{ \frac{2\langle k \rangle^2 (\langle k \rangle - 1)^2}{2\langle k \rangle - 1} \right\}^{1/3} r^{-2/3} = Ar^{-2/3}. \tag{8.65}$$

Using this result and (8.16), for small r we have

$$f_c^{tot} = 2 - \frac{1}{\langle k \rangle - 1} - \frac{3\langle k \rangle}{A^2} r^{1/3} + O(r^{2/3}). \tag{8.66}$$

Thus, f_c^{tot} approaches the theoretical maximum when r approaches zero. For a network of N nodes the maximum value of f_c^{tot} is obtained when $r = 1/N$, this being the smallest value consistent with having at least one node of degree k_{max}. Given this r, the equation determining the optimal k_{max}, representing the size of the central hubs, is [267]

$$k_{max} = AN^{2/3} \tag{8.67}$$

where A is defined in **(8.65)**.

Figure 9.0 **Art & Networks: K-Mode**

K-Mode is an art collective that publishes trend reports with an unusual take on various concepts. The image shows a page from *Youth Mode: A Report on Freedom,* discussing the subtle shift in the origins and meaning of communities, the topic of this chapter [281].

CHAPTER 9

Communities

9.1 Introduction

Belgium appears to be the model bi-cultural society: 59% of its citizens are Flemish, speaking Dutch and 40% are Walloons, speaking French. As multi-ethnic countries break up all over the world, we must ask: How has this country fostered the peaceful coexistence of these two ethnic groups since 1830? Is Belgium a densely knitted society, where it does not matter if one is Flemish or Walloon? Or do we have two nations within the same borders that have learned to minimize contact with each other?

The answer was provided by Vincent Blondel and his students in 2007, who developed an algorithm to identify the country's community structure. They started from the mobile call network, placing individuals next to whom they regularly called on their mobile phone [282]. The algorithm revealed that Belgium's social network is broken into two large clusters of

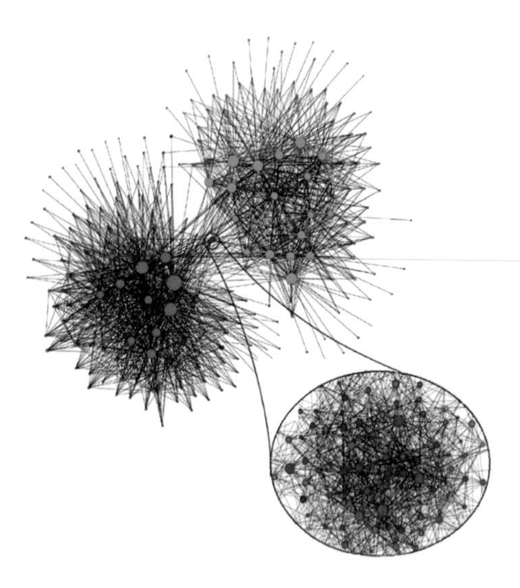

Figure 9.1 Communities in Belgium
Communities extracted from the call pattern of the consumers of the largest Belgian mobile phone company. The network has about two million mobile phone users. The nodes correspond to communities, the size of each node being proportional to the number of individuals in the corresponding community. The color of each community on a red–green scale represents the language spoken in the particular community, red for French and green for Dutch. Only communities of more than 100 individuals are shown. The community that connects the two main clusters consists of several smaller communities with less obvious language separation, capturing the culturally mixed Brussels – the country's capital.
After [282].

communities and that individuals in one of these clusters rarely talk with individuals from the other cluster (**Figure 9.1**). The origin of this separation became obvious once they assigned to each node the language spoken by each individual, learning that one cluster consisted almost exclusively of French speakers and the other of Dutch speakers.

In network science we call a *community* a group of nodes that have a higher likelihood of connecting to each other than to nodes from other communities. To gain intuition about community organization, next we discuss two areas where communities play a particularly important role:

- **Social Networks**
  Social networks are full of easy-to-spot communities, something that scholars noticed decades ago [283–286]. Indeed, the employees of a company are more likely to interact with their coworkers than with employees of other companies [283]. Consequently, work places appear as densely interconnected communities within the social network. Communities could also represent circles of friends, or a group of individuals who pursue the same hobby together, or individuals living in the same neighborhood.
  A social network that has received particular attention in the context of community detection is known as *Zachary's Karate Club* (**Figure 9.2**) [286], capturing the links between 34 members of a karate club. Given the club's small size, each club member knew everyone else. To uncover the true relationships between club members, sociologist Wayne Zachary documented 78 pairwise links between members who regularly interacted outside the club (**Figure 9.2a**).
  Interest in the dataset is driven by a singular event: a conflict between the club's president and the instructor split the club in two. About half of the members followed the instructor and the other half the president, a breakup that unveiled the ground truth, representing the club's underlying community structure (**Figure 9.2a**). Today, community-finding algorithms are often tested based on their ability to infer these two communities from the structure of the network before the split.
- **Biological Networks**
  Communities play a particularly important role in our understanding of how specific biological functions are

encoded in cellular networks. Two years before receiving the Nobel Prize in Medicine, Lee Hartwell argued that biology must move beyond its focus on single genes. It must explore instead how groups of molecules form functional modules to carry out specific cellular functions [288]. Ravasz and collaborators [289] made the first attempt to systematically identify such modules in metabolic networks. They did so by building an algorithm to identify groups of molecules that form locally dense communities (**Figure 9.3**).

Communities play a particularly important role in understanding human diseases. Indeed, proteins that are involved in the same disease tend to interact with each other [290]. This finding inspired the *disease module* hypothesis [29], stating that each disease can be linked to a well-defined neighborhood of the cellular network.

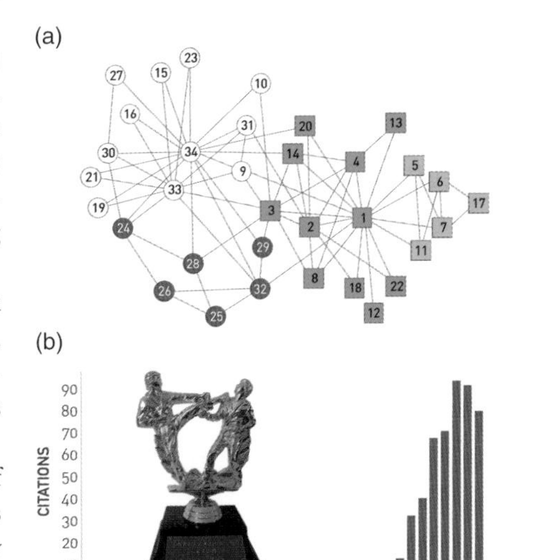

Figure 9.2 **Zachary's Karate Club**

(a) The connections between the 34 members of Zachary's Karate Club. Links capture interactions between the club members *outside the club*. The circles and the squares denote the two fractions that emerged after the club split in two. The colors capture the best community partition predicted by an algorithm that optimizes the modularity coefficient M (**Section 9.4**). The community boundaries closely follow the split: the white and purple communities capture one fraction and the green and orange communities the other. After [287].

(b) The citation history of the Zachary Karate Club paper [286] mirrors the history of community detection in network science. Indeed, there was virtually no interest in Zachary's paper until Girvan and Newman used it as a benchmark for community detection in 2002 [47]. After that the number of citations of the paper exploded, reminiscent of the citation explosion of Erdős and Rényi's work following the discovery of scale-free networks (**Figure 3.15**).

The frequent use of Zachary's karate club network as a benchmark in community detection inspired the Zachary Karate Club Club, whose tongue-in-cheek statute states: "The first scientist at any conference on networks who uses Zachary's karate club as an example is inducted into the Zachary Karate Club Club, and awarded a prize."

Hence the prize is not based on merit, but on the simple act of participation. Yet, its recipients are prominent network scientists (http://networkkarate.tumblr.com/). The figure shows the Zachary Karate Club trophy, which is always held by the latest inductee. Photo courtesy of Marián Boguñá.

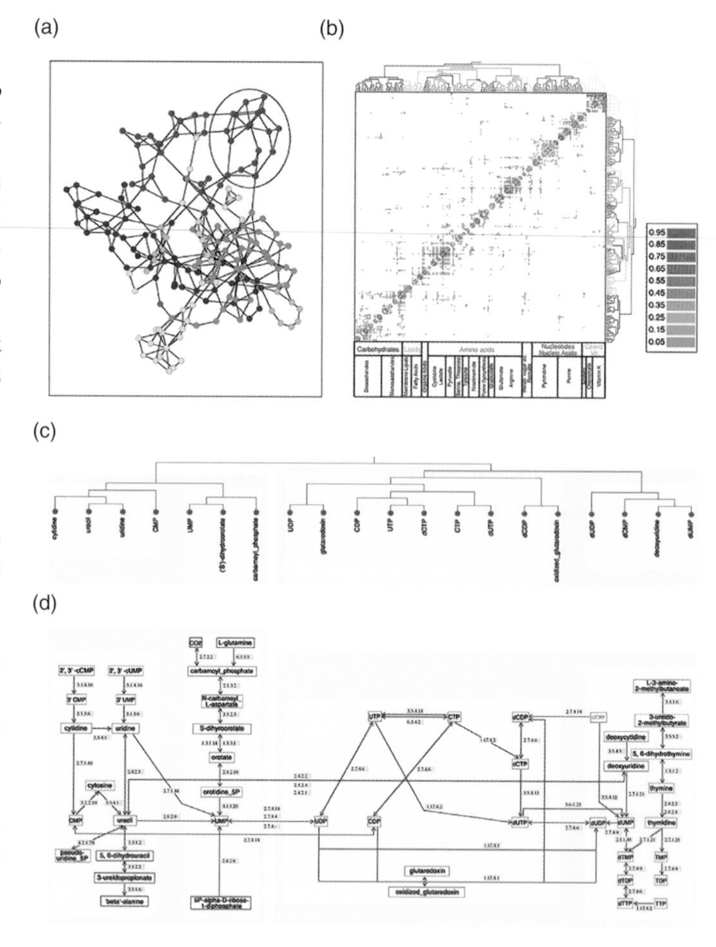

Figure 9.3 **Communities in Metabolic Networks**

The *E. coli* metabolism offers fertile ground to investigate the community structure of biological systems [289].

(a) The biological modules (communities) identified by the Ravasz algorithm [289] (**Section 9.3**). The color of each node, capturing the predominant biochemical class to which it belongs, indicates that different functional classes are segregated in distinct network neighborhoods. The highlighted region selects the nodes that belong to the pyrimidine metabolism, one of the predicted communities.

(b) The topological overlap matrix of the *E. coli* metabolism and the corresponding dendrogram allowing us to identify the modules shown in (a). The color of the branches reflects the predominant biochemical role of the participating molecules, like carbohydrates (blue), nucleotide and nucleic acid metabolism (red) and lipid metabolism (cyan).

(c) The red right branch of the dendrogram tree shown in (b), highlighting the region corresponding to the pyrimidine module.

(d) The detailed metabolic reactions within the pyrimidine module. The boxes around the reactions highlight the communities predicted by the Ravasz algorithm.

After [289].

The examples discussed above illustrate the diverse motivations that drive community identification. The existence of communities is rooted in who connects to whom, hence they cannot be explained based on the degree distribution alone. To extract communities we must therefore inspect a network's detailed wiring diagram. These examples inspire the starting hypothesis of this chapter:

H1: Fundamental Hypothesis

A network's community structure is uniquely encoded in its wiring diagram.

According to the fundamental hypothesis, there is a ground truth about a network's community organization that can be uncovered by inspecting A_{ij}.

The purpose of this chapter is to introduce the concepts necessary to understand and identify the community structure of a complex network. We will ask how to define communities, explore the various community characteristics and introduce a series of algorithms, relying on different principles, for community identification.

9.2 Basics of Communities

What do we really mean by a community? How many communities are in a network? How many different ways can we partition a network into communities? In this section we address these frequently emerging questions in community identification.

9.2.1 Defining Communities

Our sense of communities rests on a second hypothesis (**Figure 9.4**):

H2: Connectedness and Density Hypothesis
A community is a locally dense connected subgraph in a network.

In other words, all members of a community must be reached through other members of the same community (*connectedness*). At the same time we expect nodes that belong to a community to have a higher probability of linking to other members of that community than to nodes that do not belong to the same community (*density*). While this hypothesis considerably narrows what would be considered a community, it does not uniquely define it. Indeed, as we discuss below, several community definitions are consistent with H2.

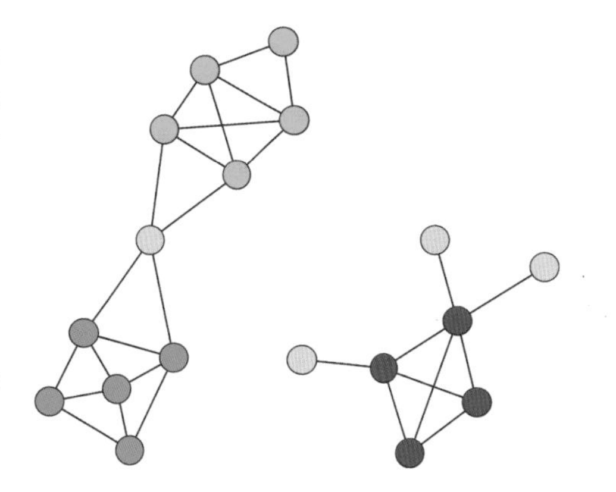

Figure 9.4 Connectedness and Density Hypothesis
Communities are locally dense connected subgraphs in a network. This expectation relies on two distinct hypotheses:
Connectedness Hypothesis
Each community corresponds to a connected subgraph, like the subgraphs formed by the orange, the green or the purple nodes. Consequently, if a network consists of two isolated components, each community is limited to only one component. The hypothesis also implies that on the same component a community cannot consist of two subgraphs that do not have a link to each other. Consequently, the orange and the green nodes form separate communities.
Density Hypothesis
Nodes in a community are more likely to connect to other members of the same community than to nodes in other communities. The orange, the green and the purple nodes satisfy this expectation.

Maximum Cliques

One of the first papers on community structure, published in 1949, defined a community as a group of individuals whose members all know each other. In graph-theoretic terms this means that a community is a *complete subgraph* or a *clique*. A clique automatically satisfies H2: it is a connected subgraph with maximal link density. Yet, viewing communities as cliques has several drawbacks:

- While triangles are frequent in networks, larger cliques are rare.
- Requiring a community to be a complete subgraph may be too restrictive, missing many other legitimate communities. For example, none of the communities of **Figures 9.2 and 9.3** correspond to complete subgraphs.

Strong and Weak Communities

To relax the rigidity of cliques, consider a connected subgraph C of N_C nodes in a network. The *internal degree* k_i^{int} of node i is the number of links that connect i to other nodes in C. The *external degree* k_i^{ext} is the number of links that connect i to the rest of the network. If $k_i^{\text{ext}} = 0$, each neighbor of i is within C, hence C is a good community for node i. If $k_i^{\text{int}} = 0$, then node i should be assigned to a different community. These definitions allow us to distinguish two kinds of community (**Figure 9.5**):

- **Strong Community**

  C is a *strong community* if each node within C has more links within the community than with the rest of the graph [291, 292]. Specifically, a subgraph C forms a strong community if for each node $i \in C$,

$$k_i^{\text{int}}(C) > k_i^{\text{ext}}(C). \tag{9.1}$$

- **Weak Community**

  C is a *weak community* if the total internal degree of a subgraph exceeds its total external degree [292]. Specifically, a subgraph C forms a weak community if

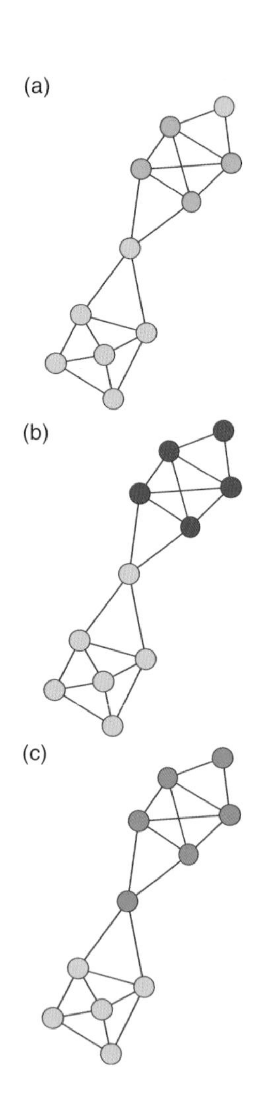

Figure 9.5 **Defining Communities**

(a) **Cliques**

A *clique* corresponds to a complete subgraph. The highest-order clique of this network is a square, shown in orange. There are several three-node cliques in this network. Can you find them?

(b) **Strong Communities**

A *strong community*, defined in (**9.1**), is a connected subgraph whose nodes have more links to other nodes in the same community than to nodes that belong to other communities. Such a strong community is shown in purple. There are additional strong communities in the graph – can you find at least two more?

(c) **Weak Communities**

A *weak community*, defined in (**9.2**), is a subgraph whose nodes' total internal degree exceeds their total external degree. The green nodes represent one of the several possible weak communities of this network.

$$\sum_{i \in C} k_i^{\text{int}}(C) > \sum_{i \in C} k_i^{\text{ext}}(C). \tag{9.2}$$

A weak community relaxes the strong community requirement by allowing some nodes to violate **(9.1)**. In other words, the inequality **(9.2)** applies to the community as a whole rather than to each node individually.

Note that each clique is a strong community, and each strong community is a weak community. The converse is generally not true (**Figure 9.5**).

The community definitions discussed above (cliques, strong and weak communities) refine our notions of communities. At the same time they indicate that we do have some freedom in defining communities.

9.2.2 Number of Communities

How many ways can we group the nodes of a network into communities? To answer this question consider the simplest community-finding problem, called *graph bisection*. We aim to divide a network into two non-overlapping subgraphs, such that the number of links between the nodes in the two groups, called the *cut size*, is minimized (**Box 9.1**).

Graph Partitioning

We can solve the graph-bisection problem by inspecting all possible divisions into two groups and choosing the one with the smallest cut size. To determine the computational cost of this brute-force approach we note that the number of distinct ways we can partition a network of N nodes into groups of N_1 and N_2 nodes is

$$\frac{N!}{N_1!N_2!}. \tag{9.3}$$

Using Stirling's formula $n! \simeq \sqrt{2\pi n}(n/e)^n$ we can write **(9.3)** as

$$\begin{aligned}\frac{N!}{N_1!N_2!} &\simeq \frac{\sqrt{2\pi N}(N/e)^N}{\sqrt{2\pi N_1}(N_1/e)^{N_1}\sqrt{2\pi N_2}(N_2/e)^{N_2}} \\ &\sim \frac{N^{N+1/2}}{N_1^{N_1+1/2}N_2^{N_2+1/2}}.\end{aligned} \tag{9.4}$$

Box 9.1 Graph Partitioning

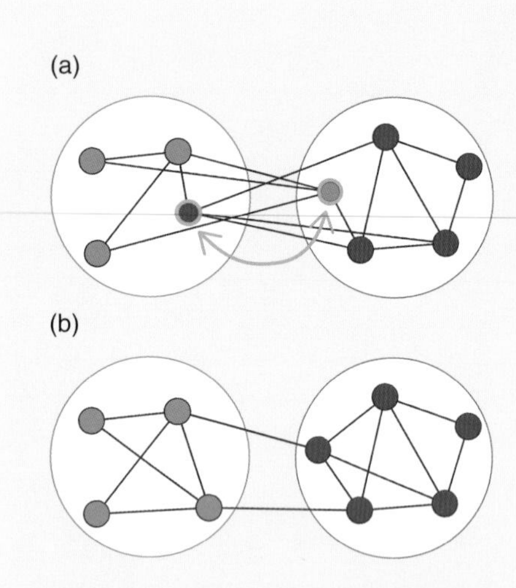

Figure 9.6 **Kerninghan–Lin Algorithm**
The best-known algorithm for graph partitioning was proposed in 1970 [294]. We illustrate this with graph bisection, which starts by randomly partitioning the network into two groups of pre-defined sizes. Next we select a node pair (i,j), where i and j belong to different groups, and swap them, recording the resulting change in the cut size. By testing all (i,j) pairs we identify the pair that results in the largest reduction of the cut size, like the pair highlighted in (a). By swapping them we arrive at the partition shown in (b). In some implementations of the algorithm, if no pair reduces the cut size, we swap the pair that increases the cut size the least.

Chip designers face a problem of exceptional complexity: they need to place on a chip 2.5 billion transistors such that their wires do not intersect. To simplify the problem they first partition the wiring diagram of an integrated circuit (IC) into smaller subgraphs, chosen such that the number of links between them is minimal. Then they lay out different blocks of an IC individually, and reconnect these blocks. A similar problem is encountered in parallel computing, where a large computational problem is partitioned into subtasks and assigned to individual chips. The assignment must minimize the typically slow communication between processors.

The problem faced by chip designers or software engineers is called *graph partitioning* in computer science [293]. The algorithms developed for this purpose, like the widely used Kerninghan–Lin algorithm (**Figure 9.6**), are the predecessors of the community-finding algorithms discussed in this chapter.

There is an important difference between graph partitioning and community detection: graph partitioning divides a network into a pre-defined number of smaller subgraphs. In contrast, community detection aims to uncover the inherent community structure of a network. Consequently, in most community-detection algorithms the number and size of the communities are not predefined, but need to be discovered by inspecting the network's wiring diagram.

To simplify the problem, let us set the goal of dividing the network into two equal sizes $N_1 = N_2 = N/2$. In this case **(9.4)** becomes

$$\frac{2^{N+1}}{\sqrt{N}} = e^{(N+1)\ln 2 - \frac{1}{2}\ln N}, \qquad (9.5)$$

indicating that the number of bisections increases exponentially with the size of the network.

To illustrate the implications of **(9.5)**, consider a network with ten nodes which we bisect into two subgraphs of size $N_1 = N_2 = 5$. According to **(9.3)**, we need to check 252 bisections to find the one with the smallest cut size. Let us assume that our computer can inspect these 252 bisections in one millisecond (10^{-3} sec). If we next wish to bisect a network with 100 nodes into groups of size $N_1 = N_2 = 50$, according to **(9.3)** we need to check approximately 10^{29} divisions, requiring about 10^{16} years on the same computer. Therefore, our brute-force strategy is bound to fail, being impossible to inspect all bisections for even a modest-sized network.

Community Detection

While in graph partitioning the number and size of communities are pre-defined, in community detection both parameters are unknown. We call a partition a division of a network into an arbitrary number of groups, such that each node belongs to one and only one group. The number of possible partitions follows [295–298]

$$B_N = \frac{1}{e}\sum_{j=0}^{\infty}\frac{j^N}{j!}. \tag{9.6}$$

As **Figure 9.7** indicates, B_N grows faster than exponentially with the network size for large N.

Equations **(9.5)** and **(9.6)** signal the fundamental challenge of community identification: the number of possible ways we can partition a network into communities grows exponentially or faster with the network size N. Therefore, it is impossible to inspect all partitions of a large network (**Box 9.2**).

In summary, our notion of communities rests on the expectation that each community corresponds to a locally dense connected subgraph. This hypothesis leaves room for numerous community definitions, from cliques to weak and strong communities. Once we adopt a definition, we could identify communities by inspecting all possible partitions of a network, selecting the one that best satisfies our definition. Yet, the number of partitions grows faster than exponentially with the network size, making such brute-force approaches computationally infeasible. We therefore need algorithms that can identify communities without inspecting all partitions. This is the subject of the next sections.

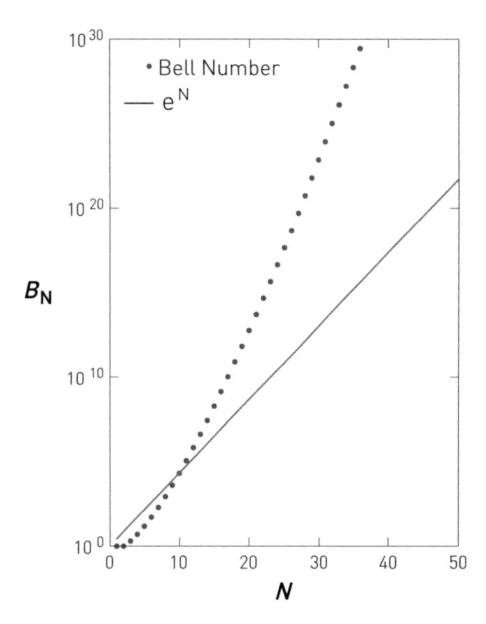

Figure 9.7 **Number of Partitions**

The number of partitions of a network of size N is provided by the Bell number **(9.6)**. The figure compares the Bell number to an exponential function, illustrating that the number of possible partitions grows faster than exponentially. Given that there are over 10^{40} partitions for a network of size $N = 50$, brute-force approaches that aim to identify communities by inspecting all possible partitions are computationally infeasible.

Box 9.2 NP Completeness

Figure 9.8 **Night at the Movies**

Traveling Salesman is a 2012 intellectual thriller about four mathematicians who have solved the *P* vs. *NP* problem and are now struggling with the implications of their discovery. The *P vs. NP problem* asks whether every problem whose solution can be verified in polynomial time can also be solved in polynomial time. This is one of the seven Millennium Prize Problems, hence a $1,000,000 prize waits for the first correct solution. The *Traveling Salesman* refers to a salesman who tries to find the shortest route to visit several cities exactly once, at the end returning to his starting city. While the problem appears simple, it is in fact NP-complete – we need to try all combinations to find the shortest path.

How long does it take to execute an algorithm? The answer is not given in minutes and hours, as the execution time depends on the speed of the computer on which we run the algorithm. We count instead the number of computations the algorithm performs. For example, an algorithm that aims to find the largest number in a list of N numbers has to compare each number in the list with the maximum found so far. Consequently, its execution time is proportional to N. In general, we call an algorithm *polynomial* if its execution time follows N^x.

An algorithm whose execution time is proportional to N^3 is slower on any computer than an algorithm whose execution time is N. But this difference dwindles in significance compared with an exponential algorithm, whose execution time increases as 2^N. For example, if an algorithm whose execution time is proportional to N takes a second for $N = 100$ elements, then an N^3 algorithm takes almost three hours on the same computer. Yet an exponential algorithm (2^N) will take 10^{20} years to complete.

The problem that an algorithm can solve in polynomial time is called a *class P* problem. Several computational problems encountered in network science have no known polynomial-time algorithms, with the available algorithms requiring exponential running time. Yet, the correctness of the solution can be checked quickly – that is, in polynomial time. Such problems, called *NP-complete*, include the traveling salesman problem (**Figure 9.8**), the graph coloring problem, maximum clique identification, partitioning a graph into subgraphs of specific type and the vertex cover problem (**Box 7.4**).

The ramifications of NP-completeness have captured the fascination of the popular media as well. Charlie Epps, the main character of the CBS TV series *Numbers*, spends the last three months of his mother's life trying to solve an NP-complete problem, convinced that the solution will cure her disease. Similarly, the motive for a double homicide in the CBS TV series *Elementary* is the search for a solution of an NP-complete problem, driven by its enormous value for cryptography.

9.3 Hierarchical Clustering

To uncover the community structure of large real networks we need algorithms whose running time grows polynomially with N. *Hierarchical clustering*, the topic of this section, helps us achieve this goal.

The starting point of hierarchical clustering is a *similarity matrix*, whose elements x_{ij} indicate the distance of node i from node j. In community identification the similarity is extracted from the relative position of nodes i and j within the network.

Once we have x_{ij}, hierarchical clustering iteratively identifies groups of nodes with high similarity. We can use two different procedures to achieve this: *agglomerative algorithms* merge nodes with high similarity into the same community, while *divisive algorithms* isolate communities by removing low-similarity links that tend to connect communities. Both procedures generate a hierarchical tree, called a dendrogram, that predicts the possible community partitions. Next we explore the use of agglomerative and divisive algorithms to identify communities in networks.

9.3.1 Agglomerative Procedures: The Ravasz Algorithm

We illustrate the use of *agglomerative hierarchical clustering* for community detection by discussing the *Ravasz algorithm*, proposed to identify functional modules in metabolic networks [289]. The algorithm consists of the following steps:

Step 1: Define the Similarity Matrix

In an agglomerative algorithm, similarity should be high for node pairs that belong to the same community and low for node pairs that belong to different communities. In a network context nodes that connect to each other and share neighbors likely belong to the same community, hence their x_{ij} should be large. The topological overlap matrix (**Figure 9.9**)

$$x_{ij}^0 = \frac{J(i,j)}{\min(k_i, k_j) + 1 - \Theta(A_{ij})} \tag{9.7}$$

captures this expectation. Here, $\Theta(x)$ is the Heaviside step function, which is zero for $x \leq 0$ and one for $x > 0$; $J(i, j)$ is the number of common neighbors of node i and j, to which we add one (+1) if there is a direct link between i and j; $\min(k_i, k_j)$ is the smaller of the degrees k_i and k_j. Consequently: $x_{ij}^0 = 1$ if

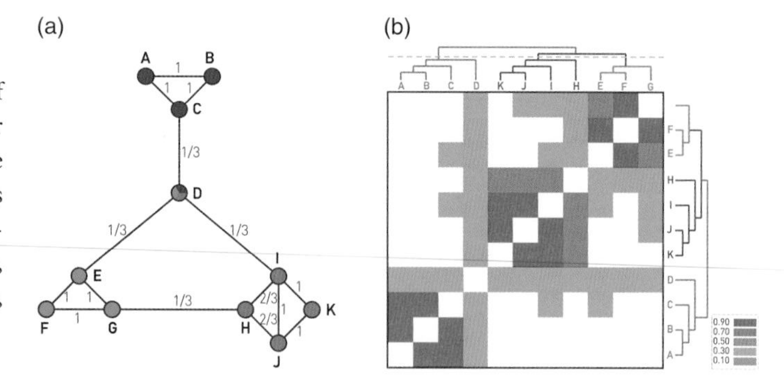

Figure 9.9
(a) The Ravasz Algorithm
A small network illustrating the calculation of the topological overlap x^0_{ij}. For each node pair (i,j) we calculate the overlap **(9.7)**. The obtained x^0_{ij} for each connected node pair is shown on each link. Note that x^0_{ij} can be non-zero for nodes that do not link to each other, but have a common neighbor. For example, $x^0_{ij} = 1/3$ for C and E.
(b) Topological Overlap Matrix
The topological overlap matrix x_{ij} for the network shown in (a). The rows and columns of the matrix were reordered after applying average linkage clustering, placing next to each other nodes with the highest topological overlap. The colors denote the degree of topological overlap between each node pair, as calculated in (a). By cutting the dendrogram with the orange line, it recovers the three modules built into the network. The dendogram indicates that the EFG and the HIJK modules are closer to each other than they are to the ABC module. After [289].

nodes i and j have a link to each other and have the same neighbors, like A and B in **Figure 9.9a**.
$x^0_{ij} = 0$ if i and j do not have common neighbors, nor do they link to each other, like A and E.
Members of the same dense local network neighborhood have high topological overlap, like nodes H, I, J, K or E, F, G.

Step 2: Decide Group Similarity

As nodes are merged into small communities, we must measure how similar two communities are. Three approaches, called *single, complete* and *average cluster similarity*, are frequently used to calculate the community similarity from the node-similarity matrix x_{ij} (**Figure 9.10**). The Ravasz algorithm uses the *average cluster similarity* method, defining the similarity of two communities as the average of x_{ij} over all node pairs i and j that belong to distinct communities (**Figure 9.10d**).

Step 3: Apply Hierarchical Clustering

The Ravasz algorithm uses the following procedure to identify the communities:

1. Assign each node to a community of its own and evaluate x_{ij} for all node pairs.
2. Find the community pair or the node pair with the highest similarity and merge them into a single community.
3. Calculate the similarity between the new community and all other communities.
4. Repeat steps 2 and 3 until all nodes form a single community.

Step 4: Dendrogram

The pairwise mergers of step 3 will eventually pull all nodes into a single community. We can use a dendrogram to extract the underlying community organization.

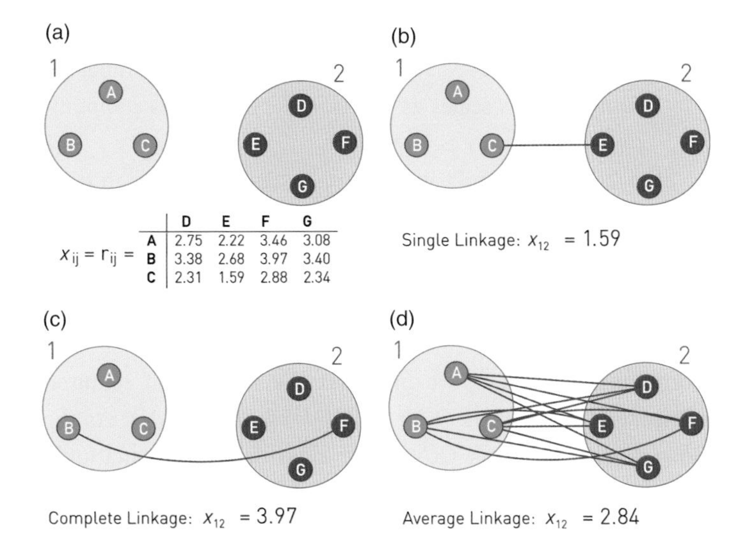

Figure 9.10 **Cluster Similarity**

In agglomerative clustering we need to determine the similarity of two communities from the node-similarity matrix x_{ij}. We illustrate this procedure for a set of points whose similarity x_{ij} is the physical distance r_{ij} between them. In networks, x_{ij} corresponds to some network-based distance measure, like x_{ij}^0 defined in **(9.7)**.

(a) **Similarity Matrix**

Seven nodes forming two distinct communities. The table shows the distance r_{ij} between each node pair, acting as the similarity x_{ij}.

(b) **Single-Linkage Clustering**

The similarity between communities 1 and 2 is the smallest of all x_{ij}, where i and j are in different communities. Hence, the similarity is $x_{12} = 1.59$, corresponding to the distance between nodes C and E.

(c) **Complete-Linkage Clustering**

The similarity between two communities is the maximum of x_{ij}, where i and j are in distinct communities. Hence, $x_{12} = 3.97$.

(d) **Average-Linkage Clustering**

The similarity between two communities is the average of x_{ij} over all node pairs i and j that belong to different communities. This is the procedure implemented in the Ravasz algorithm, providing $x_{12} = 2.84$.

The dendrogram visualizes the order in which the nodes are assigned to specific communities. For example, the dendrogram of **Figure 9.9b** tells us that the algorithm first merged nodes A with B, K with J and E with F, as each of these pairs have $x_{ij}^0 = 1$. Next, node C was added to the (A, B) community, I to (K, J) and G to (E, F).

To identify the communities we must cut the dendrogram. Hierarchical clustering does not tell us where that cut should be. Using, for example, the cut indicated as a dashed line in **Figure 9.9b**, we recover the three obvious communities (ABC, EFG and HIJK).

Applied to the *E. coli* metabolic network (**Figure 9.3a**), the Ravasz algorithm identifies the nested community structure of the bacterial metabolism. To check the biological relevance of these communities, we color-coded the branches of the dendrogram according to the known biochemical classification of each metabolite. As shown in **Figure 9.3b**, substrates with similar biochemical role tend to be located on the same branch of the tree. In other words, the known biochemical classification of these metabolites confirms the biological relevance of the communities extracted from the network topology.

Computational Complexity

How many computations do we need to run the Ravasz algorithm? The algorithm has four steps, each with its own computational complexity:

(a)

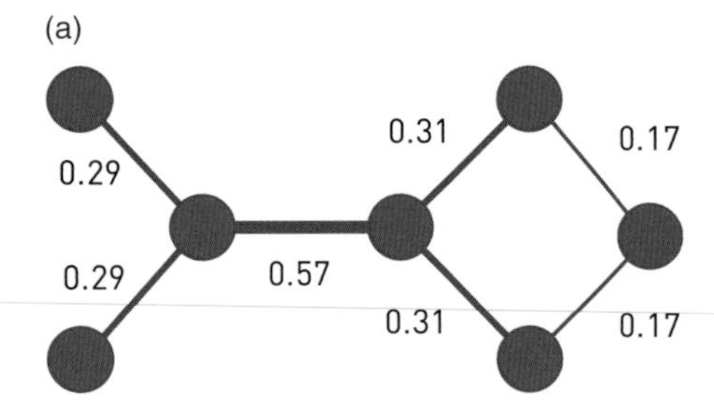

(b)

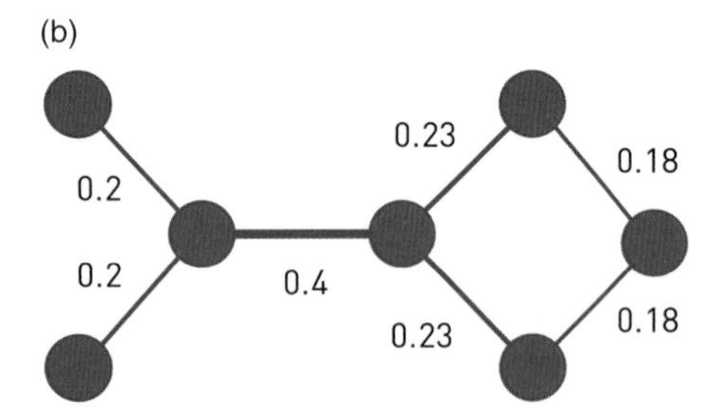

Figure 9.11 **Centrality Measures**

Divisive algorithms require a centrality measure that is high for nodes that belong to different communities and low for node pairs in the same community. Two frequently used measures can achieve this:

(a) **Link Betweenness**

Link betweenness captures the role of each link in information transfer. Hence, x_{ij} is proportional to the number of shortest paths between all node pairs that run along the link (i,j). Consequently, inter-community links, like the central link in the figure with $x_{ij} = 0.57$, have large betweenness. The calculation of link betweenness scales as $O(LN)$, or $O(N^2)$ for a sparse network [299].

(b) **Random-Walk Betweenness**

A pair of nodes m and n are chosen at random. A walker starts at m, following each adjacent link with equal probability until he reaches n. *Random-walk betweenness* x_{ij} is the probability that the link $i \rightarrow j$ was crossed by the walker after averaging over all possible choices for the starting nodes m and n. The calculation requires the inversion of an $N \times N$ matrix, with $O(N^3)$ computational complexity and averaging the flows over all node pairs, with $O(LN^2)$. Hence the total computational complexity of random-walk betweenness is $O[(L+N)N^2]$, or $O(N^3)$ for a sparse network.

Step 1: The calculation of the similarity matrix x_{ij}^0 requires us to compare N^2 node pairs, hence the number of computations scales as N^2. In other words, its *computational complexity* is $O(N^2)$.

Step 2: Group similarity requires us to determine in each step the distance of the new cluster from all other clusters. Doing this N times requires $O(N^2)$ calculations.

Steps 3 & 4: The construction of the dendrogram can be performed in $O(N \log N)$ steps.

Combining steps 1–4, we find that the number of required computations scales as $O(N^2) + O(N^2) + O(N \log N)$. As the slowest step scales as $O(N^2)$, the algorithm's computational complexity is $O(N^2)$. Hence, hierarchal clustering is much faster than the brute-force approach, which generally scales as $O(e^N)$.

9.3.2 Divisive Procedures: The Girvan–Newman Algorithm

Divisive procedures systematically remove the links connecting nodes that belong to different communities, eventually breaking a network into isolated communities. We illustrate their use by introducing an algorithm proposed by Michelle Girvan and Mark Newman [47, 299], consisting of the following steps:

Step 1: Define Centrality

While in agglomerative algorithms x_{ij} selects node pairs that belong to the same community, in divisive algorithms x_{ij}, called *centrality*, selects node pairs that are in different communities. Hence we want x_{ij} to be high if nodes i and j belong to different communities and low if they are in the same community. Two centrality measures that satisfy this expectation are discussed in **Figure 9.11**. The fastest of the two is *link betweenness*, defining x_{ij} as the number of shortest paths that go through the link (i, j). Links connecting different communities are expected to have large x_{ij}, while links within a community have small x_{ij}.

Step 2: Hierarchical Clustering

The final steps of a divisive algorithm mirror those we used in agglomerative clustering (**Figure 9.12**):

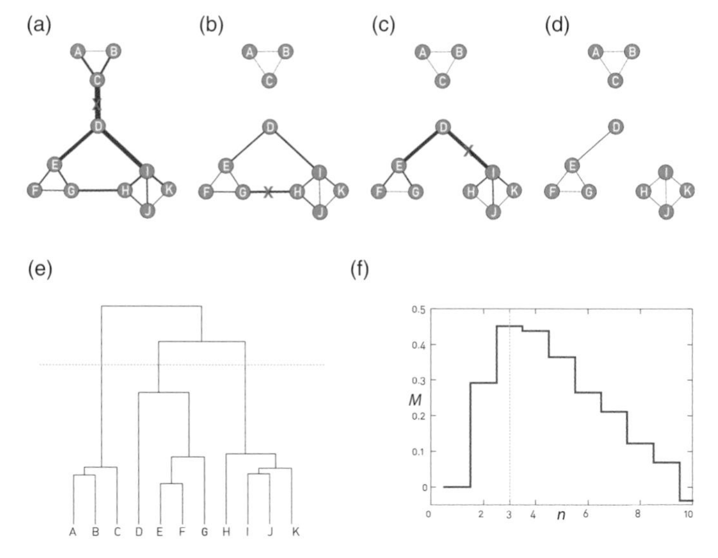

Figure 9.12 **The Girvan-Newman Algorithm**

(a) The divisive hierarchical algorithm of Girvan and Newman uses link betweenness (**Figure 9.11a**) as centrality. In the figure the link weights, assigned proportionally to x_{ij}, indicate that links connecting different communities have the highest x_{ij}. Indeed, each shortest path between these communities must run through them.

(b)–(d) The sequence of images illustrates how the algorithm removes one by one the three highest x_{ij} links, leaving three isolated communities behind. Note that betweenness needs to be recalculated after each link removal.

(e) The dendrogram generated by the Girvan–Newman algorithm. The cut at level 3, shown as an orange dotted line, reproduces the three communities present in the network.

(f) The modularity function, M, introduced in **Section 9.4**, helps us select the optimal cut. Its maximum agrees with our expectation that the best cut is at level 3, as shown in (e).

1. Compute the centrality x_{ij} of each link.
2. Remove the link with the largest centrality. In case of a tie, choose one link randomly.
3. Recalculate the centrality of each link for the altered network.
4. Repeat steps 2 and 3 until all links are removed.

Girvan and Newman applied their algorithm to Zachary's Karate Club (**Figure 9.2a**), finding that the predicted communities matched almost perfectly the two groups after the break-up. Only node 3 was classified incorrectly.

Computational Complexity

The rate-limiting step of divisive algorithms is the calculation of centrality. Consequently, the algorithm's computational complexity depends on which centrality measure we use. The most efficient is link betweenness, with $O(LN)$ [300, 301, 302] (**Figure 9.11a**). Step 3 of the algorithm introduces an additional factor L in the running time, hence the algorithm scales as $O(L^2N)$, or $O(N^3)$ for a sparse network.

9.3.3 Hierarchy in Real Networks

Hierarchical clustering raises two fundamental questions:

Nested Communities

First, it assumes that small modules are nested into larger ones. These *nested communities* are well captured by the dendrogram (**Figures 9.9b** and **9.12e**). How do we know, however, if such a

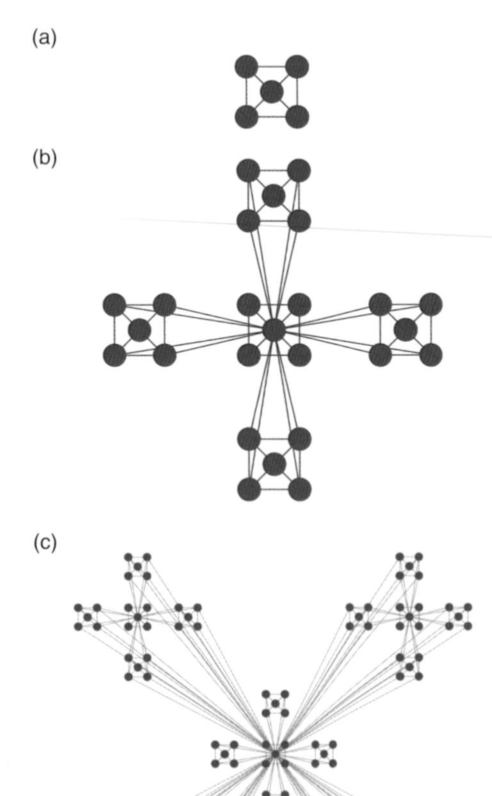

Figure 9.13 **Hierarchical Network**
The iterative construction of a deterministic hierarchical network.
(a) Start from a fully connected module of five nodes. Note that the diagonal nodes are also connected, but the links are not visible.
(b) Create four identical replicas of the starting module and connect the peripheral nodes of each module to the central node of the original module. This way we obtain a network with $N = 25$ nodes.
(c) Create four replicas of the 25-node module and connect the peripheral nodes again to the central node of the original module, obtaining an $N = 125$-node network. This process is continued indefinitely.
After [303].

hierarchy is indeed present in a network? Could this hierarchy be imposed by our algorithms, whether or not the underlying network has a nested community structure?

Communities and the Scale-Free Property
Second, the density hypothesis states that a network can be partitioned into a collection of subgraphs that are only weakly linked to other subgraphs. How can we have isolated communities in a scale-free network if the hubs inevitably link multiple communities?

The *hierarchical network model*, whose construction is shown in **Figure 9.13**, resolves the conflict between communities and the scale-free property and offers intuition about the structure of nested hierarchical communities. The network obtained has several key characteristics:

Scale-Free Property
The hierarchical model generates a scale-free network with degree exponent (**Figure 9.14a, Advanced topics 9.A**)

$$\gamma = 1 + \frac{\ln 5}{\ln 4} = 2.161.$$

Size-Independent Clustering Coefficient
While for the Erdős–Rényi and the Barabási–Albert models the clustering coefficient decreases with N (**Section 5.9**), for the hierarchical network we have $C = 0.743$ independent of the network size (**Figure 9.14c**). Such an N-independent clustering coefficient has been observed in metabolic networks [289].

Hierarchical Modularity
The model consists of numerous small communities that form larger communities, which again combine into ever larger communities. The quantitative signature of this nested hierarchical modularity is the dependence of a node's clustering coefficient on the node's degree [289, 303, 304]

$$C(k) \sim k^{-1}. \tag{9.8}$$

In other words, the higher a node's degree, the smaller is its clustering coefficient.

Equation **(9.8)** captures the way the communities are organized in a network. Indeed, small-degree nodes have high C because they reside in dense communities. High-degree nodes have small C because they connect to different communities.

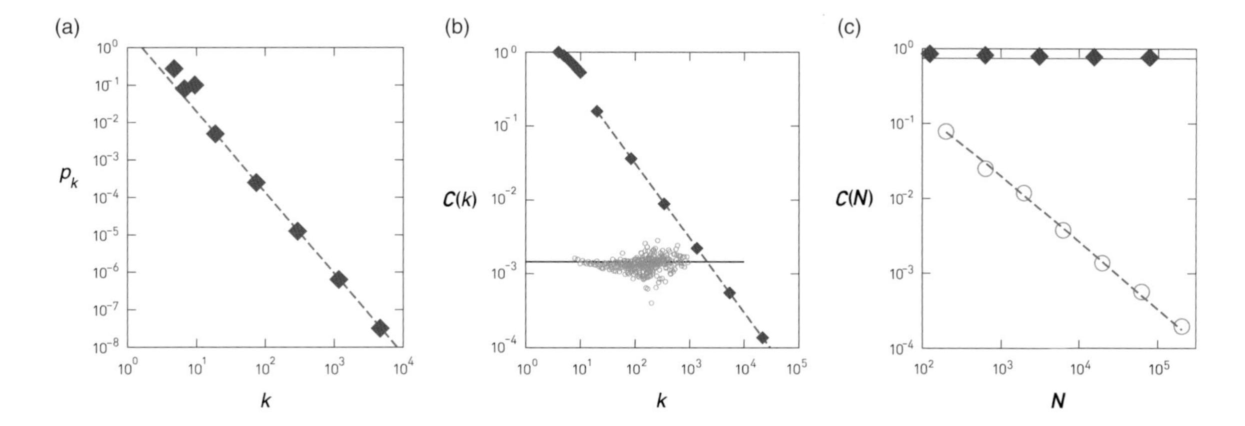

Figure 9.14 **Scaling in Hierarchical Networks**

Three quantities characterize the hierarchical network shown in **Figure 9.13**:

(a) **Degree Distribution**
The scale-free nature of the generated network is illustrated by the scaling of p_k with slope $\gamma = \ln 5/\ln 4$, shown as a dashed line. See **Advanced topics 9.A** for the derivation of the degree exponent.

(b) **Hierarchical Clustering**
$C(k)$ follows **(9.8)**, shown as a dashed line. The circles show $C(k)$ for a randomly wired scale-free network, obtained from the original model by degree-preserving randomization. The lack of scaling indicates that the hierarchical architecture is lost under rewiring. Hence $C(k)$ captures a property that goes beyond the degree distribution.

(c) **Size-Independent Clustering Coefficient**
The dependence of the clustering coefficient C on the network size N. For the hierarchical model, C is independent of N (filled symbols) while for the Barabási–Albert model, $C(N)$ decreases (empty symbols).
After [303].

For example, in **Figure 9.13c** the nodes at the center of the five-node modules have $k = 4$ and clustering coefficient $C = 4$. Those at the center of the 25-node modules have $k = 20$ and $C = 3/19$. Those at the center of the 125-node modules have $k = 84$ and $C = 3/83$. Hence, the higher the degree of a node, the smaller is its C.

The hierarchical network model suggests that inspecting $C(k)$ allows us to decide if a network is hierarchical. For the Erdős–Rényi and the Barabási–Albert models, $C(k)$ is independent of k, indicating that they do not display hierarchical modularity. To see if hierarchical modularity is present in real systems, we calculated $C(k)$ for ten reference networks, finding that (**Figure 9.36**):

- Only the power grid lacks hierarchical modularity, its $C(k)$ being independent of k (**Figure 9.36a**).
- For the remaining nine networks, $C(k)$ decreases with k. Hence, in these networks small nodes are part of small dense communities, while hubs link disparate communities to each other.
- For the scientific collaboration, metabolic and citation networks, $C(k)$ follows **(9.8)** in the high-k region. The form of $C(k)$ for the Internet, mobile, email, protein interaction and WWW networks needs to be derived individually, as for those $C(k)$ does not follow **(9.8)**. More detailed network models predict $C(k) \sim k^{-\beta}$ where β is between 0 and 2 [303, 304].

In summary, in principle hierarchical clustering does not require preliminary knowledge about the number and size of communities. In practice it generates a dendrogram that offers a family of community partitions characterizing the studied

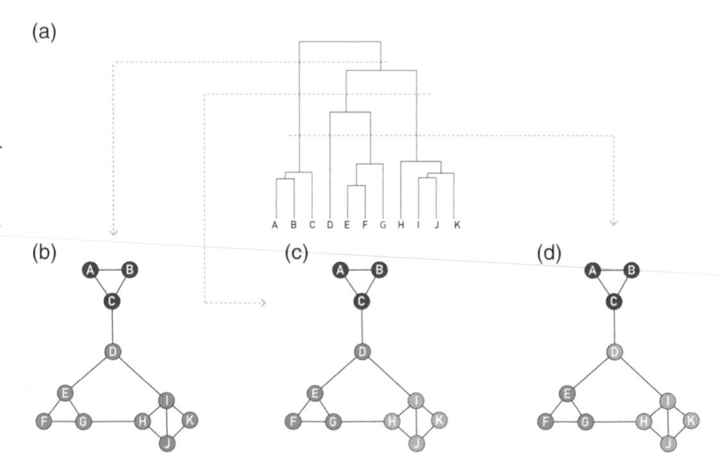

Figure 9.15 **Ambiguity in Hierarchical Clustering**
Hierarchical clustering does not tell us where to cut a dendrogram. Indeed, depending on where we make the cut in the dendrogram of **Figure 9.9a**, we obtain (**b**) two, (**c**) three or (**d**) four communities. While for a small network we can visually decide which cut captures best the underlying community structure, it is impossible to do so in larger networks. In the next section we discuss modularity, which helps us select the optimal cut.

network. This dendrogram does not tell us which partition captures best the underlying community structure. Indeed, any cut of the hierarchical tree offers a potentially valid partition (**Figure 9.15**). This is at odds with our expectation that in each network there is a ground truth, corresponding to a unique community structure.

While there are multiple notions of hierarchy in networks [305, 306], inspecting $C(k)$ helps us decide if the underlying network has hierarchical modularity. We find that $C(k)$ decreases in most real networks, indicating that most real systems display hierarchical modularity. At the same time, $C(k)$ is independent of k for the Erdős–Rényi or Barabási–Albert models, indicating that these canonical models lack a hierarchical organization.

9.4 Modularity

In a randomly wired network the connection pattern between the nodes is expected to be uniform, independent of the network's degree distribution. Consequently, these networks are not expected to display systematic local density fluctuations that we could interpret as communities. This expectation inspired the third hypothesis of community organization:

H3: Random Hypothesis
Randomly wired networks lack an inherent community structure.

This hypothesis has some actionable consequences: by comparing the link density of a community with the link density obtained for the same group of nodes for a randomly rewired network, we could decide if the original community corresponds to a dense subgraph, or if its connectivity pattern emerged by chance.

In this section we show that systematic deviations from a random configuration allow us to define a quantity called *modularity*, which measures the quality of each partition. Hence, modularity allows us to decide if a particular community partition is better than some other one. Finally, modularity optimization offers a novel approach to community detection.

9.4.1 Modularity

Consider a network with N nodes and L links and a partition into n_c communities, each community having N_c nodes connected to each other by L_c links, where $c = 1, \ldots, n_c$. If L_c is larger than the expected number of links between the N_c nodes given the network's degree sequence, then the nodes of the subgraph C_c could indeed be part of a true community, as expected based on hypothesis H2 (**Figure 9.2**). We therefore measure the difference between the network's real wiring diagram (A_{ij}) and the expected number of links between i and j if the network is randomly wired (p_{ij}),

$$M_c = \frac{1}{2L} \sum_{(i,j) \in Cc} (A_{ij} - p_{ij}). \tag{9.9}$$

Here, p_{ij} can be determined by randomizing the original network, while keeping the expected degree of each node unchanged. Using the degree-preserving null model **(7.1)**, we have

$$p_{ij} = \frac{k_i k_j}{2L}. \tag{9.10}$$

If M_c is positive, then the subgraph C_c has more links than expected by chance, hence it represents a potential community. If M_c is zero, then the connectivity between the N_c nodes is random, fully explained by the degree distribution. Finally, if M_c is negative, then the nodes of C_c do not form a community.

Using **(9.10)** we can derive a simpler form for the modularity **(9.9)** (**Advanced topics 9.B**)

$$M_c = \frac{L_c}{L} - \left(\frac{k_c}{2L}\right)^2, \tag{9.11}$$

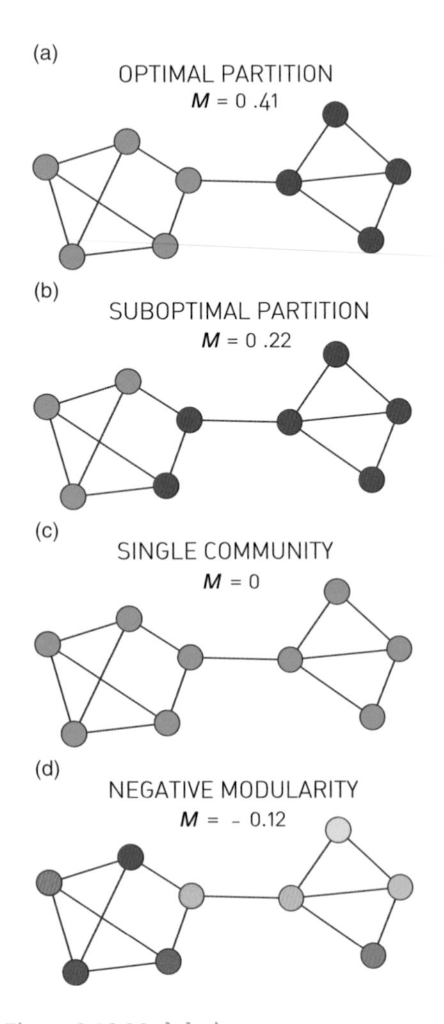

Figure 9.16 **Modularity**
To better understand the meaning of modularity, we show M defined in **(9.12)** for several partitions of a network with two obvious communities.
(a) **Optimal Partition**
The partition with maximal modularity $M = 0.41$ closely matches the two distinct communities.
(b) **Sub-optimal Partition**
A partition with a sub-optimal but positive modularity, $M = 0.22$, fails to correctly identify the communities present in the network.
(c) **Single Community**
If we assign all nodes to the same community we obtain $M = 0$, independent of the network structure.
(d) **Negative Modularity**
If we assign each node to a different community, modularity is negative, obtaining $M = -0.12$.

where L_c is the total number of links within the community C_c and k_c is the total degree of the nodes in this community.

To generalize these ideas to a full network, consider the complete partition that breaks the network into n_c communities. To see if the local link density of the subgraphs defined by this partition differs from the expected density in a randomly wired network, we define the partition's *modularity* by summing **(9.11)** over all n_c communities [299]

$$M_c = \sum_{c=1}^{n_c} \left[\frac{L_c}{L} - \left(\frac{k_c}{2L} \right)^2 \right]. \tag{9.12}$$

Modularity has several key properties:

- **Higher Modularity Implies Better Partition.**
  The higher is M for a partition, the better is the corresponding community structure. Indeed, in **Figure 9.16a** the partition with the maximum modularity ($M = 0.41$) accurately captures the two obvious communities. A partition with a lower modularity clearly deviates from these communities (**Figure 9.16b**). Note that the modularity of a partition cannot exceed one [307, 308].
- **Zero and Negative Modularity.**
  By taking the whole network as a single community we obtain $M = 0$, as in this case the two terms in parentheses in **(9.12)** are equal (**Figure 9.16c**). If each node belongs to a separate community, we have $L_c = 0$ and the sum **(9.12)** has n_c negative terms, hence M is negative (**Figure 9.16d**).

We can use modularity to decide which of the many partitions predicted by a hierarchical method offers the best community structure, selecting the one for which M is maximal. This is illustrated in **Figure 9.12f**, which shows M for each cut of the dendrogram, finding a clear maximum when the network breaks into three communities.

9.4.2 The Greedy Algorithm

The expectation that partitions with higher modularity correspond to partitions that more accurately capture the underlying community structure prompts us to formulate our final hypothesis:

H4: Maximal Modularity Hypothesis
For a given network, the partition with maximum modularity corresponds to the optimal community structure.

The hypothesis is supported by the inspection of small networks, for which the maximum M agrees with the expected communities (**Figures 9.12 and 9.16**).

The maximum modularity hypothesis is the starting point of several community-detection algorithms, each seeking the partition with the largest modularity. In principle, we could identify the best partition by checking M for all possible partitions, selecting the one for which M is largest. Given, however, the exceptionally large number of partitions, this brute-force approach is computationally not feasible. Next we discuss an algorithm that finds partitions with close to maximal M, while bypassing the need to inspect all partitions.

Greedy Algorithm

The first modularity-maximization algorithm, proposed by Newman [309], iteratively joins pairs of communities if the move increases the partition's modularity. The algorithm follows these steps:

1. Assign each node to a community of its own, starting with N communities of single nodes.
2. Inspect each community pair connected by at least one link and compute the modularity difference ΔM obtained if we merge them. Identify the community pair for which ΔM is the largest and merge them. Note that modularity is always calculated for the full network.
3. Repeat step 2 until all nodes merge into a single community, recording M for each step.
4. Select the partition for which M is maximal.

To illustrate the predictive power of the greedy algorithm, consider the collaboration network between physicists, consisting of $N = 56,276$ scientists in all branches of physics who posted papers on arxiv.org (**Figure 9.17**). The greedy algorithm predicts about 600 communities with peak modularity $M = 0.713$. Four of these communities are very large, together containing 77% of all nodes (**Figure 9.17a**). In the largest community, 93% of the authors publish in condensed matter

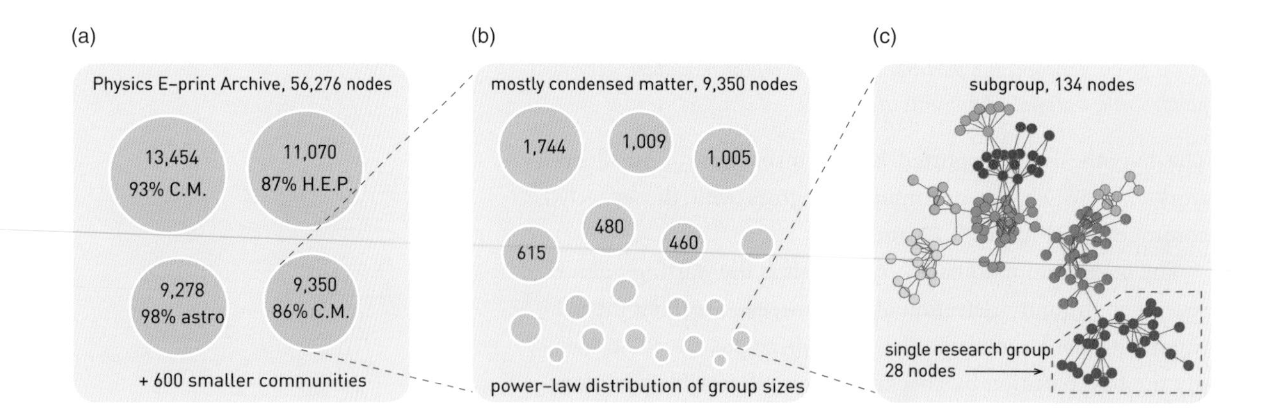

Figure 9.17 **The Greedy Algorithm**

(a) **Clustering Physicists**
The community structure of the collaboration network of physicists. The greedy algorithm predicts four large communities, each composed primarily of physicists of similar interests. To see this on each cluster we show the percentage of members who belong to the same subfield of physics. Specialties are determined by the subsection(s) of the e-print archive in which individuals post papers. C.M. indicates condensed matter, H.E.P. high-energy physics and astro astrophysics. These four large communities coexist with 600 smaller communities, resulting in an overall modularity $M = 0.713$.
(b) **Identifying Subcommunities**
We can identify subcommunities by applying the greedy algorithm to each community, treating them as separate networks. This procedure splits the condensed matter community into many smaller subcommunities, increasing the modularity of the partition to $M = 0.807$.
(c) **Research Groups**
One of these smaller communities is further partitioned, revealing individual researchers and the research groups they belong to.
After [309].

physics while 87% of the authors in the second largest community publish in high-energy physics, indicating that each community contains physicists of similar professional interests. The accuracy of the greedy algorithm is also illustrated in **Figure 9.2a**, showing that the community structure with the highest M for the Zachary Karate Club accurately captures the club's subsequent split.

Computational Complexity

Since the calculation of each ΔM can be done in constant time, step 2 of the greedy algorithm requires $O(L)$ computations. After deciding which communities to merge, the update of the matrix can be done in a worst-case time $O(N)$. Since the algorithm requires $N - 1$ community mergers, its complexity is $O[(L + N)N]$, or $O(N^2)$ on a sparse graph. Optimized implementations reduce the algorithm's complexity to $O(N\log^2 N)$ (**Online resource 9.1**).

9.4.3 Limits of Modularity

Given the important role modularity plays in community identification, we must be aware of some of its limitations.

Resolution Limit

Modularity maximization forces small communities into larger ones. Indeed, if we merge communities A and B into a single community, the network's modularity changes with (**Advanced topics 9.B**)

$$\Delta M_{AB} = \frac{l_{AB}}{L} - \frac{k_A k_B}{2L^2}, \tag{9.13}$$

where l_{AB} is the number of links that connect the nodes in community A with total degree k_A to the nodes in community B with total degree k_B. If A and B are distinct communities, they should remain distinct when M is maximized. As we show next, this is not always the case.

Consider the case $k_A k_B/2L < 1$, when **(9.13)** predicts $\Delta M_{AB} > 0$ if there is at least one link between the two communities ($l_{AB} \geq 1$). Hence we must merge A and B to maximize modularity. Assuming for simplicity that $k_A \sim k_B = k$, if the total degree of the communities satisfies

$$k \leq \sqrt{2L} \tag{9.14}$$

then modularity increases by merging A and B into a single community, even if A and B are otherwise distinct communities. This is an artifact of modularity maximization: if k_A and k_B are under the threshold **(9.14)**, the *expected* number of links between them is smaller than one. Hence, even a single link between them will force the two communities together when we maximize M. This resolution limit has several consequences:

- Modularity maximization cannot detect communities that are smaller than the resolution limit **(9.14)**. For example, for the WWW sample with $L = 1{,}497{,}134$ (**Table 2.1**), modularity maximization will have difficulty resolving communities with total degree $k_C \lesssim 1{,}730$.
- Real networks contain numerous small communities [311, 312]. Given the resolution limit **(9.14)**, these small communities are systematically forced into larger communities, offering a misleading characterization of the underlying community structure.

To avoid the resolution limit we can further subdivide the large communities obtained by modularity optimization [309, 313]. For example, treating the smaller of the two condensed-matter groups of **Figure 9.17a** as a separate network and feeding it again into the greedy algorithm, we obtain about 100 smaller communities with an increased modularity $M = 0.807$ (**Figure 9.17b**) [309].

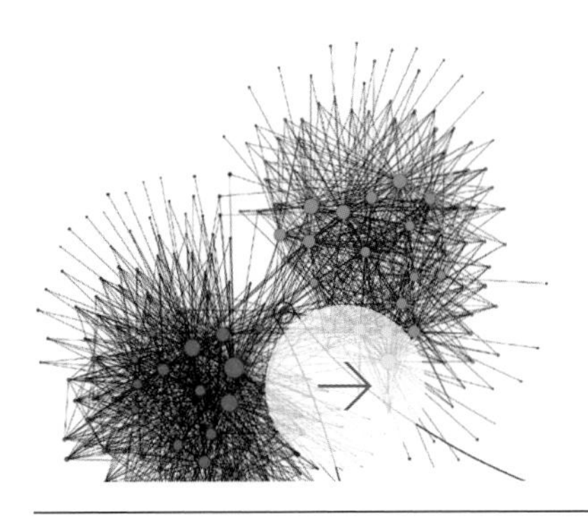

Online Resource 9.1
Modularity-Based Algorithms

There are several widely used community-finding algorithms that maximize modularity.

Optimized Greedy Algorithm

The use of data structures for sparse matrices can decrease the greedy algorithm's computational complexity to $O(N\log^2 N)$ [310]. See http://cs.unm.edu/~aaron/research/fastmodularity.htm for the code.

Louvain Algorithm

The modularity optimization algorithm achieves a computational complexity of $O(L)$ [282]. Hence it allows us to identify communities in networks with millions of nodes, as illustrated in **Figure 9.1**. The algorithm is described in **Advanced topics 9.C**. See https://sites.google.com/site/findcommunities/ for the code.

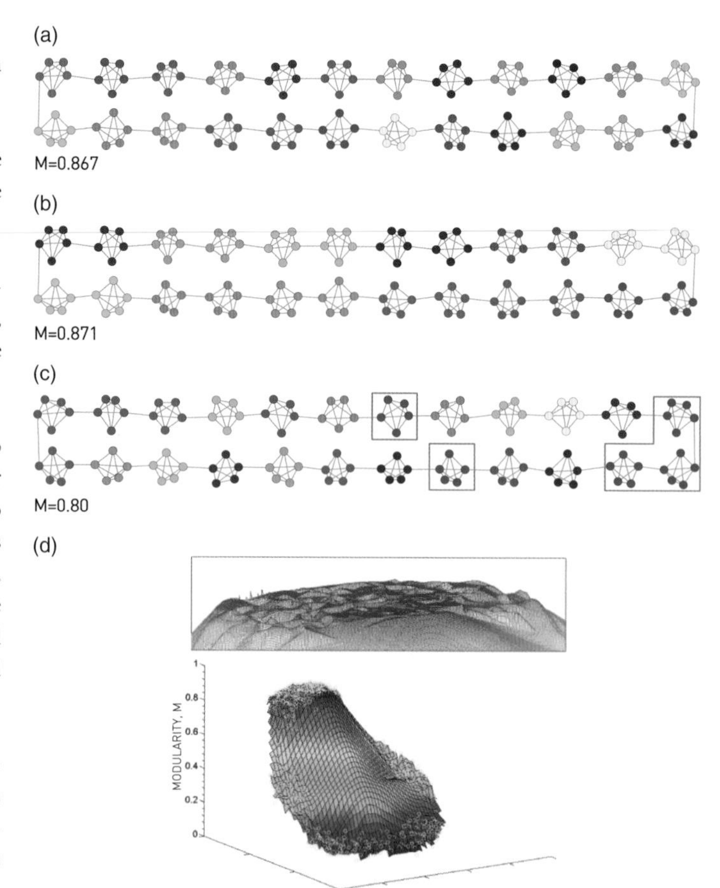

Figure 9.18 **Modularity Maxima**

A ring network consisting of 24 cliques, each made up of five nodes.

(a) **The Intuitive Partition**

The best partition should correspond to the configuration where each cluster is a separate community. This partition has $M = 0.867$.

(b) **The Optimal Partition**

If we combine the clusters into pairs, as illustrated by the node colors, we obtain $M = 0.871$, higher than M obtained for the intuitive partition (a).

(c) **Random Partition**

Partitions with comparable modularity tend to have rather distinct community structure. For example, if we assign each cluster randomly to communities, even clusters that have no links to each other, like the five highlighted clusters, may end up in the same community. The modularity of this random partition is still high, $M = 0.80$, not too far from the optimal $M = 0.87$.

(d) **Modularity Plateau**

The modularity function of the network reconstructed from 997 partitions. The vertical axis gives the modularity M, revealing a high-modularity plateau that consists of numerous low-modularity partitions. We lack, therefore, a clear modularity maxima – instead, the modularity function is highly degenerate.

After [314].

Modularity Maxima

All algorithms based on maximal modularity rely on the assumption that a network with a clear community structure has an optimal partition with a maximal M [314]. In practice, we hope that M_{max} is an easy-to-find maximum and that the communities predicted by all other partitions are distinguishable from those corresponding to M_{max}. Yet, as we show next, this optimal partition is difficult to identify among a large number of close-to-optimal partitions.

Consider a network composed of n_c subgraphs with comparable link densities $k_C \approx 2L/n_c$. The best partition should correspond to the one where each cluster is a separate community (**Figure 9.18a**), in which case $M = 0.867$. Yet, if we merge the neighboring cluster pairs into a single community, we obtain a higher modularity $M = 0.87$ (**Figure 9.18b**). In general, **(9.13)** and **(9.14)** predict that if we merge a pair of clusters, we change modularity with

$$\Delta M = \frac{l_{AB}}{L} - \frac{2}{n_c^2}. \tag{9.15}$$

In other words, the drop in modularity is less than $\Delta M = -2/n_c$. For a network with $n_c = 20$ communities, this change is at most $\Delta M = -0.005$, tiny compared with the maximal modularity $M \simeq 0.87$ (**Figure 9.18b**). As the number of groups increases, ΔM_{ij} goes to zero, hence it becomes increasingly difficult to distinguish the optimal partition from the numerous sub-optimal alternatives whose modularity is practically indistinguishable from $M_{\max}$. In other words, the modularity function is not peaked around a single optimal partition, but has a high modularity plateau (**Figure 9.18d**).

In summary, modularity offers a first-principle understanding of a network's community structure. Indeed, **(9.16)** incorporates in a compact form a number of essential questions, like what we mean by a community, how we choose the appropriate null model and how we measure the goodness of a particular partition. Consequently, modularity optimization plays a central role in the community-finding literature.

At the same time, modularity has several well-known limitations. First, it forces together small, weakly connected communities. Second, networks lack clear modularity maxima, developing instead a modularity plateau containing many partitions with hard-to-distinguish modularity. This plateau explains why numerous modularity-maximization algorithms can rapidly identify a high-M partition: they identify one of the numerous partitions with close-to-optimal M. Finally, analytical calculations and numerical simulations indicate that even random networks contain high-modularity partitions, at odds with hypothesis H3, which motivated the concept of modularity [315, 316, 317].

Modularity optimization is a special case of a larger problem: finding communities by optimizing some quality function Q. The greedy algorithm and the Louvain algorithm described in **Advanced topics 9.C** assume that $Q = M$, seeking partitions with maximal modularity. In **Advanced topics 9.C** we also describe the Infomap algorithm, which finds communities by minimizing the map equation L, an entropy-based measure of the partition quality [318, 319, 320].

9.5 Overlapping Communities

A node is rarely confined to a single community. Consider a scientist, who belongs to the community of scientists that share his professional interests. Yet, he also belongs to a community consisting of family members and relatives, and perhaps another community of individuals sharing his hobby (**Figure 9.19**). Each of these communities consists of individuals who are members of several other communities, resulting in a complicated web of nested and overlapping communities [311]. Overlapping communities are not limited to social systems: the same genes are often implicated in multiple diseases, an indication that disease modules of different disorders overlap [29].

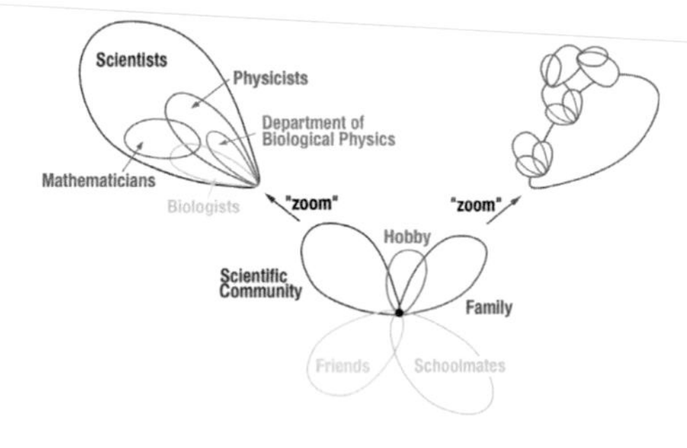

Figure 9.19 **Overlapping Communities**
Schematic representation of the communities surrounding Tamás Vicsek, who introduced the concept of overlapping communities. A zoom into the scientific community illustrates the nested and overlapping structure of the community characterizing his scientific interests.
After [311].

While the existence of a nested community structure has long been appreciated by sociologists [321] and by the engineering community interested in graph partitioning, the algorithms discussed so far force each node into a single community. A turning point was the work of Tamás Vicsek and collaborators [311, 322], who proposed an algorithm to identify overlapping communities, bringing the problem to the attention of the network science community. In this section we discuss two algorithms to detect overlapping communities, clique percolation and link clustering.

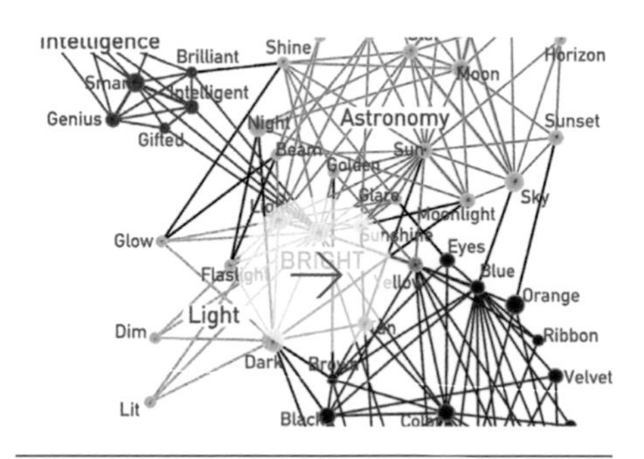

Online Resource 9.2

CFinder

The CFinder software, allowing us to identify overlapping communities, can be downloaded from www.cfinder.org.

9.5.1 Clique Percolation

The *clique percolation algorithm*, often called *CFinder* (**Online resource 9.2**), views a community as the union of overlapping cliques [311]:

- Two k-cliques are considered adjacent if they share $k-1$ nodes (**Figure 9.20b**).

  A k-clique community is the largest connected subgraph obtained by the union of all adjacent k-cliques (**Figure 9.20c**).
- k-cliques that cannot be reached from a particular k-clique belong to other k-clique communities (**Figure 9.20c,d**).

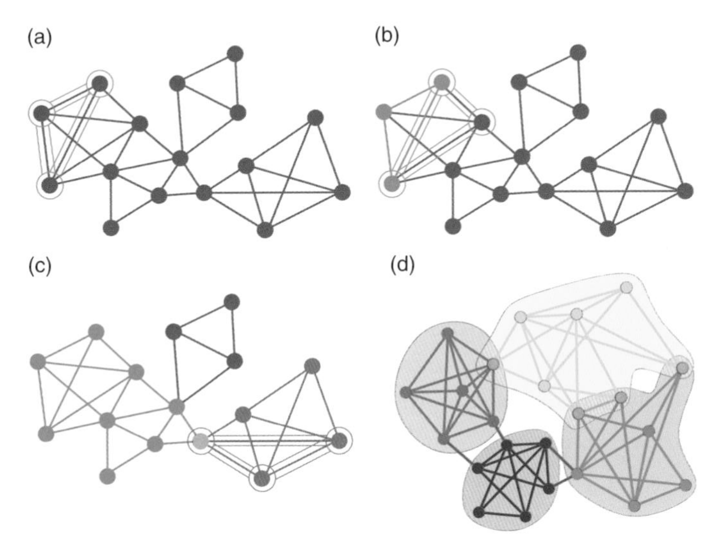

Figure 9.20 **The Clique Percolation Algorithm (CFinder)**

To identify $k = 3$-clique communities we roll a triangle across the network, such that each subsequent triangle shares one link (two nodes) with the previous triangle.

(a,b) **Rolling Cliques**

Starting from the triangle shown in green in (a), (b) illustrates the second step of the algorithm.

(c) **Clique Communities for $k = 3$**

The algorithm pauses when the final triangle of the green community is added. As no more triangles share a link with the green triangles, the green community has been completed. Note that there can be multiple k-clique communities in the same network. We illustrate this by showing a second community in blue. The figure highlights the moment when we add the last triangle of the blue community. The blue and green communities overlap, sharing the orange node.

(d) **Clique Communities for $k = 4$**

$k = 4$ community structure of a small network, consisting of complete four-node subgraphs that share at least three nodes. Orange nodes belong to multiple communities.

Images courtesy of Gergely Palla.

The CFinder algorithm identifies all cliques and then builds an $N_{clique} \times N_{clique}$ clique–clique overlap matrix O, where N_{clique} is the number of cliques and O_{ij} is the number of nodes shared by cliques i and j (**Figure 9.39**). A typical output of the CFinder algorithm is shown in **Figure 9.21**, displaying the community structure of the word *bright*. In the network, two words are linked to each other if they have a related meaning. We can easily check that the overlapping communities identified by the algorithm are meaningful: the word *bright* simultaneously belongs to a community containing light-related words, like *glow* or *dark*; to a community capturing colors (*yellow*, *brown*); to a community consisting of astronomical terms (*sun*, *ray*); and to a community linked to intelligence (*gifted*, *brilliant*). The example also illustrates the difficulty the earlier algorithms would have in identifying communities of this network: they would force *bright* into one of the four communities and remove it from the other three. Hence, communities would

Figure 9.21 **Overlapping Communities**

Communities containing the word *bright* in the South Florida Free Association network, whose nodes are words, connected by a link if their meaning is related. The community structure identified by the CFinder algorithm accurately describes the multiple meanings of *bright*, a word that can be used to refer to light, color, astronomical terms, or intelligence.

After [311].

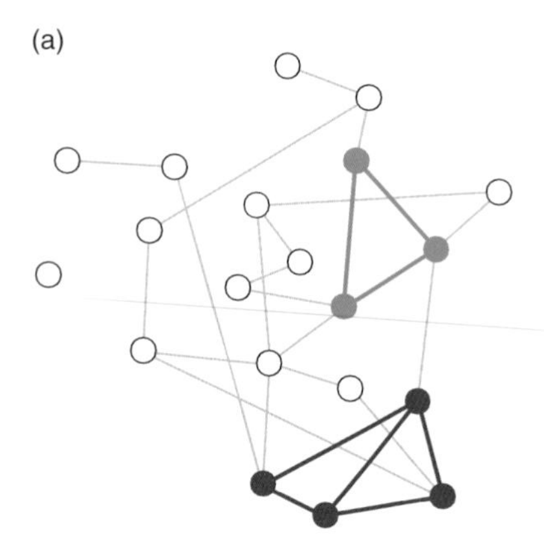

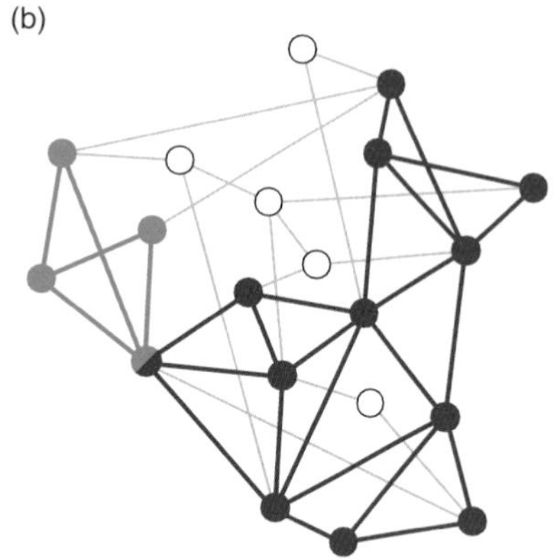

Figure 9.22 The Clique Percolation Algorithm (CFinder)
Random networks built with probabilities $p = 0.13$ (a) and $p = 0.22$ (b). As both p's are larger than the link percolation threshold ($p_c = 1/N = 0.05$ for $N = 20$), in both cases most nodes belong to a giant component.
(a) **Subcritical Communities**
The 3-clique (triangle) percolation threshold is $p_c(3) = 0.16$ according to **(9.16)**, hence at $p = 0.13$ we are below it. Therefore, only two small 3-clique percolation clusters are observed, which do not connect to each other.
(b) **Supercritical Communities**
For $p = 0.22$ we are above $p_c(3)$, hence we observe multiple 3-cliques that form a giant 3-clique percolation cluster (purple). This network also has a second overlapping 3-clique community, shown in green.
After [322].

be stripped of a key member, leading to outcomes that are difficult to interpret.

Could the communities identified by CFinder emerge by chance? To distinguish the real k-clique communities from communities that are a pure consequence of high link density, we explore the percolation properties of k-cliques in a random network [322]. As we discussed in **Chapter 3**, if a random network is sufficiently dense, it has numerous cliques of varying order. A large k-clique community emerges in a random network only if the connection probability p exceeds the threshold (**Advanced topics 9.D**)

$$p_c(k) = \frac{1}{[(k-1)N]^{1/(k-1)}}. \tag{9.16}$$

Under $p_c(k)$ we expect only a few isolated k-cliques (**Figure 9.22a**). Once p exceeds $p_c(k)$, we observe numerous cliques that form k-clique communities (**Figure 9.22b**). In other words, each k-clique community has its own threshold:

- For $k = 2$ the k-cliques are links and (**9.16**) reduces to $p_c(k) \sim 1/N$, which is the condition for the emergence of a giant connected component in Erdős–Rényi networks.
- For $k = 3$ the cliques are triangles (**Figure 9.22a,b**) and (**9.16**) predicts $p_c(k) \sim 1/\sqrt{2N}$.

In other words, k-clique communities naturally emerge in sufficiently dense networks. Consequently, to interpret the overlapping community structure of a network, we must compare it with the community structure obtained for the degree-randomized version of the original network.

Computational Complexity

Finding cliques in a network requires algorithms whose running time grows exponentially with N. Yet, the CFinder community definition is based on cliques instead of maximal cliques, which can be identified in polynomial time [323]. If, however, there are large cliques in the network, it is more efficient to identify all cliques using an algorithm with $O(e^N)$ complexity [311]. Despite this high computational complexity, the algorithm is relatively fast, processing the mobile call network of 4 million mobile phone users in less then one day [324] (see also **Figure 9.28**).

9.5.2 Link Clustering

While nodes often belong to multiple communities, links tend to be community specific, capturing the precise relationship that defines a node's membership in a community. For example, a link between two individuals may indicate that they are in the same family, or that they work together, or that they share a hobby, designations that only rarely overlap. Similarly, in biology each binding interaction of a protein is responsible for a different function, uniquely defining the role of the protein in the cell. This specificity of links has inspired the development of community-finding algorithms that cluster links rather than nodes [325, 326].

The *link-clustering algorithm* proposed by Ahn, Bagrow and Lehmann [325] consists of the following steps:

Step 1: Define Link Similarity
The similarity of a link pair is determined by the neighborhood of the nodes connected by them. Consider for example the links (i,k) and (j,k), connected to the same node k. Their similarity is defined as (**Figure 9.23a-c**)

$$S((i,k),(j,k)) = \frac{|n_+(i) \cap n_+(j)|}{|n_+(i) \cup n_+(j)|}, \qquad (9.17)$$

where $n_+(i)$ is the list of the neighbors of node i, including itself. Hence S measures the relative number of common neighbors i and j have. Consequently, $S = 1$ if i and j have the same neighbors (**Figure 9.23c**). The less is the overlap between the neighborhood of the two links, the smaller is S (**Figure 9.23b**).

Step 2: Apply Hierarchical Clustering
The similarity matrix S allows us to use hierarchical clustering to identify link communities (**Section 9.3**). We use a single-linkage procedure, iteratively merging communities with the largest similarity link pairs (**Figure 9.10**).

Taken together, for the network of **Figure 9.23e, (9.17)** provides the similarity matrix shown in (d). The single-linkage hierarchical clustering leads to the dendrogram shown in (d),

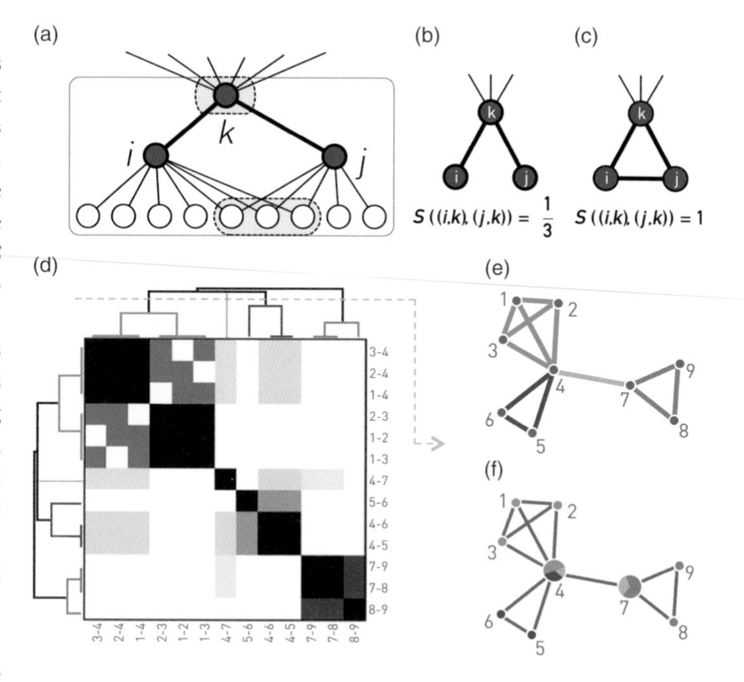

Figure 9.23 **Identifying Link Communities**
The link clustering algorithm identifies links with a similar topological role in a network. It does so by exploring the connectivity patterns of the nodes at the two ends of each link. Inspired by the similarity function of the Ravasz algorithm [289] (**Figure 9.19**), the algorithm aims to assign to high similarity S the links that connect to the same group of nodes.
(a) The similarity S of the (i,k) and (j,k) links connected to node k detects if the two links belong to the same group of nodes. Denoting with $n_+(i)$ the list of neighbors of node i, including itself, we obtain $|n_+(i) \cup n_+(j)| = 12$ and $|n_+(i) \cap n_+(j)| = 4$, resulting in $S = 1/3$ according to **(9.17)**.
(b) For an isolated ($k_i = k_j = 1$) connected triple we obtain $S = 1/3$.
(c) For a triangle we have $S = 1$.
(d) The link similarity matrix for the network shown in (e) and (f). Darker entries correspond to link pairs with higher similarity S. The figure also shows the resulting link dendrogram.
(e) The *link community structure* predicted by the cut of the dendrogram shown as an orange dashed line in (d).
(f) The *overlapping node communities* derived from the link communities shown in (e).
After [325].

whose cuts result in the link communities shown in (e) and the overlapping node communities shown in (f).

Figure 9.24 illustrates the community structure of the characters of Victor Hugo's novel *Les Miserables* identified using the link-clustering algorithm. Anyone familiar with the novel can convince themselves that the communities accurately represent the role of each character. Several characters are placed in multiple communities, reflecting their overlapping roles in the novel. Links, however, are unique to each community.

Computational Complexity

The link-clustering algorithm involves two time-limiting steps: similarity calculation and hierarchical clustering. Calculating the similarity **(9.17)** for a link pair with degrees k_i and k_j requires $\max(k_i,k_j)$ steps. For a scale-free network with degree exponent γ the calculation of similarity has complexity $O(N^{2/(\gamma^{-1})})$, determined by the size of the largest node, k_{max}. Hierarchical clustering requires $O(L)$ time steps. Hence, the algorithm's total computational complexity is $O(N^{2/(\gamma^{-1})})+O(L^2)$. For sparse graphs the latter term dominates, leading to $O(N^2)$.

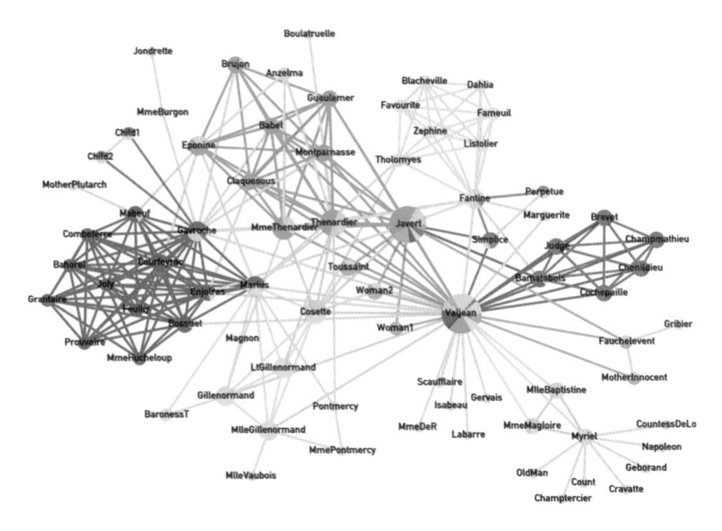

Figure 9.24 **Link Communities**

The network of characters in Victor Hugo's 1862 novel *Les Miserables*. Two characters are connected if they interact directly with each other in the story. The link colors indicate the clusters, light gray nodes corresponding to single-link clusters. Nodes that belong to multiple communities are shown as pie charts, illustrating their membership in each community. Not surprisingly, the main character, Jean Valjean, has the most diverse community membership.

After [325].

The need to detect overlapping communities has inspired numerous algorithms [327]. For example, the CFinder algorithm has been extended to the analysis of weighted [328], directed and bipartite graphs [329, 330]. Similarly, one can derive quality functions for link clustering [326], like the modularity function discussed in **Section 9.4**.

In summary, the algorithms discussed in this section acknowledge the fact that nodes naturally belong to multiple communities. Therefore, by forcing each node into a single community, as we did in the previous sections, we obtain a misleading characterization of the underlying community structure. Link communities recognize the fact that each link accurately captures the nature of the relationship between two nodes. As a bonus, link clustering also predicts the overlapping community structure of a network.

9.6 Testing Communities

Community-identification algorithms offer a powerful diagnosis tool, allowing us to characterize the local structure of real networks. Yet, to interpret and use the predicted communities, we must understand the accuracy of our algorithms. Similarly, the need to diagnose large networks prompts us to address the computational efficiency of our algorithms. In this section we focus on the concepts needed to assess the accuracy and the speed of community finding.

9.6.1 Accuracy

If the community structure is uniquely encoded in the network's wiring diagram, each algorithm should predict precisely the same communities. Yet, given the different hypotheses the various algorithms embody, the partitions uncovered by them can differ, prompting the question: Which community-finding algorithm should we use?

To assess the performance of community-finding algorithms we need to measure an algorithm's *accuracy*, that is its ability to uncover communities in networks whose community structure is known. We start by discussing two *benchmarks*, which are networks with pre-defined community structure, that we can use to test the accuracy of a community-finding algorithm.

Girvan–Newman (GN) Benchmark

The Girvan–Newman benchmark consists of $N = 128$ nodes partitioned into $n_c = 4$ communities of size $N_c = 32$ [47, 331]. Each node is connected with probability p^{int} to the N –1 nodes in its community and with probability p^{ext} to the $3N_c$ nodes in the other three communities. The control parameter

$$\mu = \frac{k^{ext}}{k^{ext} + k^{int}}. \tag{9.18}$$

We expect community-finding algorithms to perform well for small μ (**Figure 9.25a**), when the probability of connecting to nodes within the same community exceeds the probability of connecting to nodes in different communities. The performance of all algorithms should drop for large μ (**Figure 9.25b**), when the link density within the communities becomes comparable to the link density in the rest of the network.

Lancichinetti–Fortunato–Radicchi (LFR) Benchmark

The GN benchmark generates a random graph in which all nodes have comparable degree and all communities have identical size. Yet, the degree distribution of most real networks is fat tailed, and so is the community size distribution (**Figure 9.29**). Hence an algorithm that performs well on the GN benchmark may not do well on real networks. To avoid this limitation, the LFR benchmark (**Figure 9.26**) builds

Figure 9.25 **Testing Accuracy with the GN Benchmark**
The position of each node in (a) and (c) shows the planted communities of the GN benchmark, illustrating the presence of four distinct communities, each with $N_c = 32$ nodes.
(a) The node colors represent the partitions predicted by the Ravasz algorithm for mixing parameter $\mu = 0.40$ given by **(9.18)**. As in this case the communities are well separated, we have an excellent agreement between the planted and the detected communities.
(b) The normalized mutual information as a function of the mixing parameter μ for the Ravasz algorithm. For small μ we have $I_n = 1$ and $n_c = 4$, indicating that the algorithm can easily detect well-separated communities, as illustrated in (a). As we increase μ, the link density difference within and between communities becomes less pronounced. Consequently, the communities are increasingly difficult to identify and I_n decreases.
(c) For $\mu = 0.50$ the Ravasz algorithm misplaces a notable fraction of the nodes, as in this case the communities are not well separated, making it harder to identify the correct community structure.
Note that the Ravasz algorithm generates multiple partitions, hence for each μ we show the partition with the largest modularity, M. Next to (a) and (c), we show the normalized mutual information associated with the corresponding partition and the number of detected communities n_c. The normalized mutual information **(9.23)**, developed for non-overlapping communities, can be extended to overlapping communities as well [333].

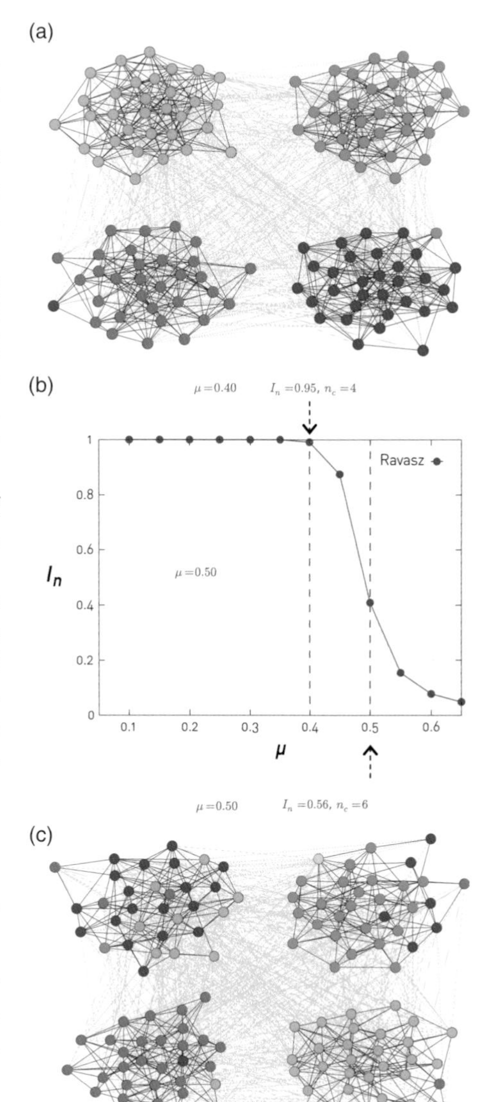

networks for which both the node degrees and the planted community sizes follow power laws [332].

Having built networks with known community structure, next we need tools to measure the accuracy of the partition predicted by a particular community-finding algorithm. As we do so, we must keep in mind that the two benchmarks discussed above correspond to a particular definition of communities. Consequently, algorithms based on clique percolation or link clustering, which embody a different notion of communities, may not fare so well on these.

Measuring Accuracy

To compare the predicted communities with those planted in the benchmark, consider an arbitrary partition into non-overlapping communities. In each step we randomly choose a node and record the label of the community it belongs to. The result is a random string of community labels that follow a $p(C)$

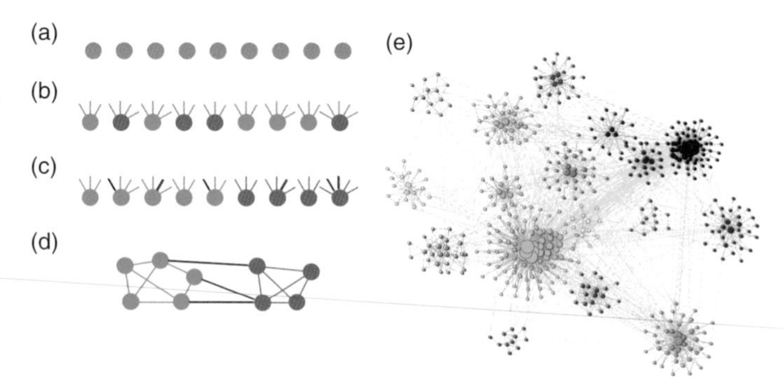

Figure 9.26 **LFR Benchmark**
The construction of the LFR benchmark, which generates networks in which both the node degrees and community sizes follow a power law. The benchmark is built as follows [332]:
(a) Start with N isolated nodes.
(b) Assign each node to a community of size N_c, where N_c follows the power-law distribution $P_{N_c} \sim N_c^{-\zeta}$ with community exponent ζ. Also, assign each node i a degree k_i selected from the power-law distribution $p_k \sim k^{-\gamma}$ with degree exponent γ.
(c) Each node i of a community receives an internal degree $(1 \sim \mu)k_i$, shown as links whose color agrees with the node color. The remaining μk_i degrees, shown as black links, connect to nodes in other communities.
(d) All stubs of nodes of the same community are randomly attached to each other, until no more stubs are "free.' In this way we maintain the sequence of internal degrees of each node in its community. The remaining μk_i stubs are randomly attached to nodes from other communities.
(e) A typical network and its community structure generated by the LFR benchmark with $N = 500$, $\gamma = 2.5$ and $\zeta = 2$.

distribution, representing the probability that a randomly selected node belongs to the community C.

Consider two partitions of the same network, one being the benchmark (ground truth) and the other the partition predicted by a community-finding algorithm. Each partition has its own $p(C_1)$ and $p(C_2)$ distribution.

The joint distribution, $p(C_1, C_2)$, is the probability that a randomly chosen node belongs to community C_1 in the first partition and C_2 in the second. The similarity of the two partitions is captured by the normalized mutual information

$$I_n = \frac{\sum_{C_1, C_2} p(C_1, C_2) \log_2 \frac{p(C_1, C_2)}{p(C_1)p(C_2)}}{\frac{1}{2}H(\{p(C_1)\}) + \frac{1}{2}H(\{p(C_2)\})}. \tag{9.19}$$

The numerator of **(9.19)** is the *mutual information I*, measuring the information shared by the two community assignments: $I = 0$ if C_1 and C_2 are independent of each other; I equals the maximal value $H(\{p(C_1)\}) = H(\{p(C_2)\})$ when the two partitions are identical and

$$H(\{p(C)\}) = -\sum_C p(C) \log_2 p(C) \tag{9.20}$$

is the Shannon entropy.

If all nodes belong to the same community, then we are certain about the next label and $H = 0$, as we do not gain new information by inspecting the community to which the next node belongs. H is maximal if $p(C)$ is the uniform distribution, as in this case we have no idea which community comes next and each new node provides H bits of new information.

In summary, $I_n = 1$ if the benchmark and the detected partitions are identical, and $I_n = 0$ if they are independent of

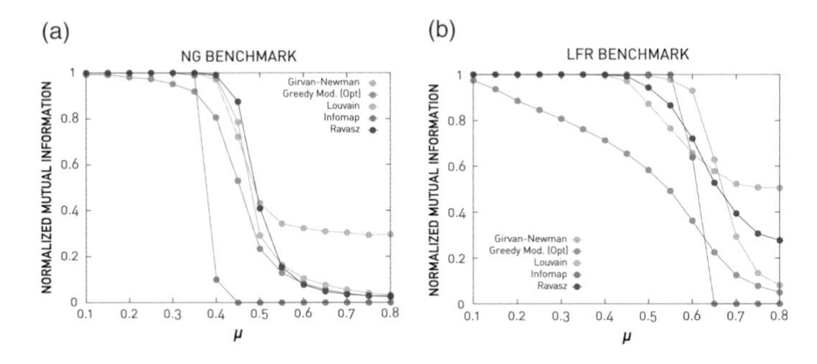

Figure 9.27 **Testing Against Benchmarks**
We tested each community-finding algorithm that predicts non-overlapping communities against the GN and the LFR benchmarks. The plots show the normalized mutual information I_n against μ for five algorithms. For the naming of each algorithm, see **Table 9.1**.

(a) **GN Benchmark**
The horizontal axis shows the mixing parameter **(9.18)**, representing the fraction of links connecting different communities. The vertical axis is the normalized mutual information **(9.19)**. Each curve is averaged over 100 independent realizations.

(b) **LFR Benchmark**
Same as in (a) but for the LFR benchmark. The benchmark parameters are $N = 1{,}000$, $\langle k \rangle = 20$, $\gamma = 2$, $k_{\max} = 50$, $\zeta = 1$, maximum community size: 100, minimum community size: 20. Each curve is averaged over 25 independent realizations.

each other. The utility of I_n is illustrated in **Figure 9.25b**, which shows the accuracy of the Ravasz algorithm for the GN benchmark. In **Figure 9.27** we use I_n to test the performance of each algorithm against the GN and LFR benchmarks. The results allow us to draw several conclusions:

- We have $I_n = 1$ for $\mu < 0.5$. Consequently, when the link density within communities is high compared with their surroundings, most algorithms accurately identify the planted communities. Beyond $\mu = 0.5$ the accuracy of each algorithm drops.
- The accuracy is benchmarks dependent. For the more realistic LFR benchmark, the Louvain and Ravasz methods offer the best performance and greedy modularity performs poorly.

9.6.2 Speed

As discussed in **Section 9.2**, the number of possible partitions increases faster than exponentially with N, becoming astronomically high for most real networks. While community-identification algorithms do not check all partitions, their computational cost still varies widely, determining their speed and consequently the size of the network they can handle. **Table 9.1** summarizes the computational complexity of the algorithms discussed in this chapter. Accordingly, the most efficient are the Louvain and the Infomap algorithms, both of which scale as $O(N\log N)$. The least efficient is CFinder, with $O(e^N)$.

These scaling laws do not capture the actual running time, however. They only show how the running time scales with N. This scaling matters if we need to find communities in very large networks. To get a sense of the true speed of these algorithms we measured their running time for the protein interaction

Table 9.1 **Algorithmic Complexity**

The computational complexity of the community identification algorithms discussed in this chapter. While computational complexity depends on both N and L, for sparse networks with good approximation we have $L \sim N$. We therefore list computational complexity in terms of N only.

Name	Nature	Complexity	Reference
Ravasz	Hierarchical agglomerative	$O(N^2)$	[289]
Girvan–Newman	Hierarchical divisive	$O(N^3)$	[457]
Greedy modularity	Modularity optimization	$O(N^2)$	[309]
Greedy modularity (optimized)	Modularity optimization	$O(N\log^2 N)$	[310]
Louvain	Modularity optimization	$O(L)$	[282]
Infomap	Flow optimization	$O(N\log N)$	[318]
Clique percolation (CFinder)	Overlapping communities	$\exp(N)$	[322]
Link clustering	Hierarchical agglomerative; overlapping communities	$O(N^2)$	[325]

network (N = 2,018), the power grid (N = 4,941) and the scientific collaboration network (N = 23,133), using the same computer. The results, shown in **Figure 9.28**, indicate that:

- The Louvain method requires the shortest running time for all networks. CFinder is just as fast for the mid-size networks, and its running time is comparable to the other algorithms for the larger collaboration network.
- The GN algorithm is the slowest on each network, in line with its predicted high computational complexity (**Table 9.1**). For example, the algorithm failed to find communities in the scientific collaboration network in seven days.

In summary, benchmarks allow us to compare the accuracy and speed of the available algorithms. Given that the development of the fastest and the most accurate community-detection tool remains an active arms race, those interested in the subject should consult the literature that compares algorithms across multiple dimensions [307, 332, 334, 335].

9.7 Characterizing Communities

Research in network science is driven by the desire to quantify the fundamental principles that govern how networks emerge and how they are organized. These organizing principles

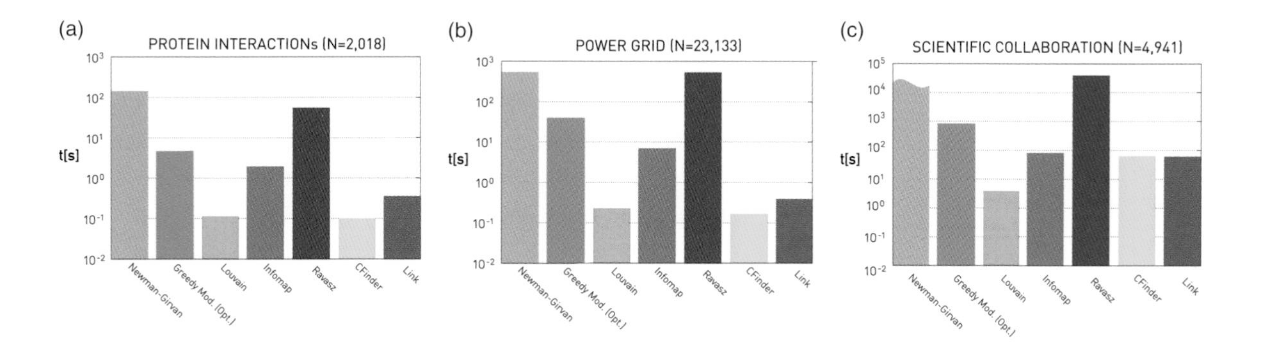

Figure 9.28 **The Running Time**
To compare the speed of community-detection algorithms we used their python implementation, either relying on the versions published by their developers or the available implementation in the *igraph* software package. The Ravasz algorithm was implemented by us, hence it is not optimized, having a larger running time than ideally possible. We ran each algorithm on the same computer. The plots provide their running time in seconds for three real networks. For the science collaboration network the GN algorithm did not finish after seven days, hence we only provide the lower limit of its running time. The higher running time observed for the scientific collaboration network is rooted in the larger size of this network.

impact the structure of communities, as well as our ability to identify them. In this section we discuss community evolution, the characteristics of community size distribution and the role of the link weights in community identification, allowing us to uncover the generic principles of community organization.

9.7.1 Community Size Distribution

According to the fundamental hypothesis (H1), the number and size of communities in a network are uniquely determined by the network's wiring diagram. We must therefore ask: What is the size distribution of these communities?

Many studies report fat-tailed community size distributions, implying that numerous small communities coexist with a few very large ones [292, 309, 310, 311, 334]. To explore how widespread this pattern is, in **Figure 9.29** we show p_{N_c} for three networks, as predicted by various community-finding algorithms. The plots indicate several patterns:

- For the protein interaction and the science collaboration network all algorithms predict an approximately fat-tailed p_{N_c}. Hence in these networks numerous tiny communities coexist with a few large communities.
- For the power grid, different algorithms lead to distinct outcomes. Modularity-based algorithms predict communities with comparable size $N_c \simeq 10^2$. In contrast, the Ravasz algorithm and Infomap predict numerous communities with size $N_c \simeq 10$ and a few larger communities. Finally, clique percolation and link clustering predict an approximately fat-tailed community size distribution.

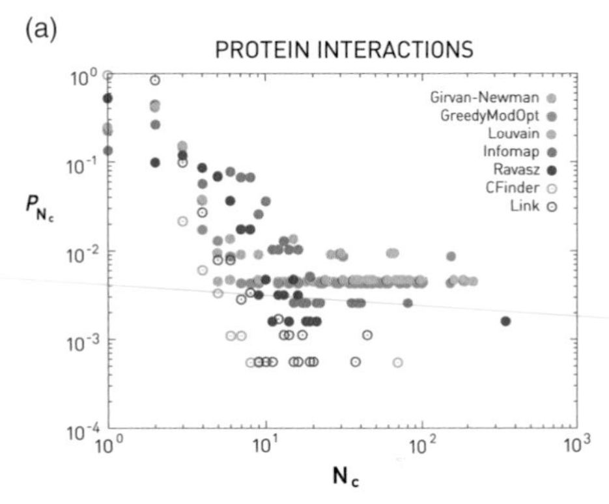

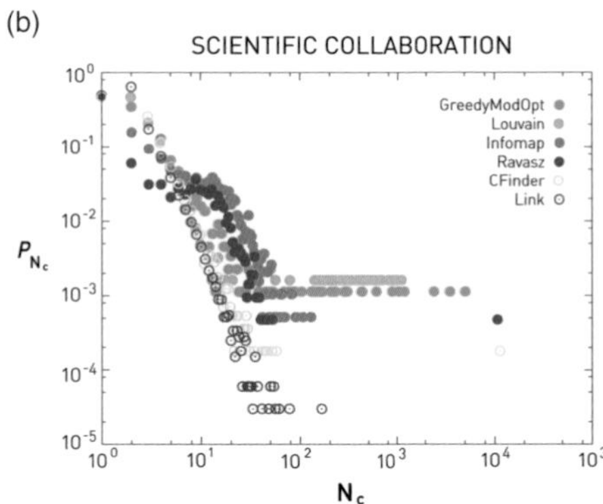

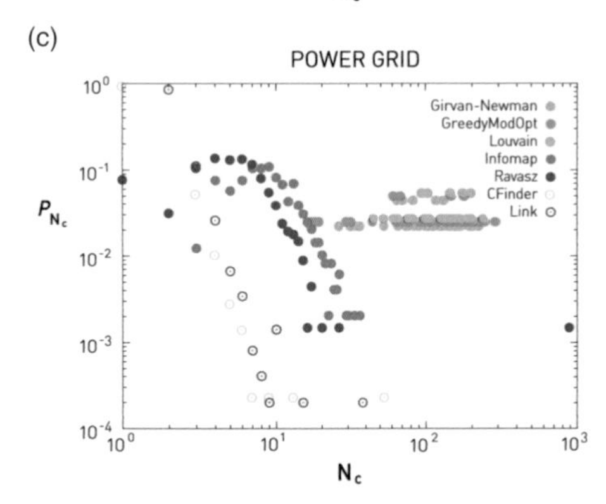

Figure 9.29 **Community Size Distribution**
The community size distribution p_{Nc} predicted by the community-finding algorithms explored in this chapter. The name convention for the algorithms is shown in **Table 9.1**. For the protein interaction (a) and the scientific collaboration network (b) all algorithms predict an approximately fat-tailed community size distribution, hence the predictions are more-or-less consistent with each other. The algorithms offer conflicting results for the power grid, shown in (c).

These differences suggest that the fat-tailed community size distribution is not a byproduct of a particular algorithm. Rather, it is an inherent property of some networks, like the protein and the scientific collaboration networks. The different outcomes for the power grid suggest that this network lacks a unique and detectable community structure.

9.7.2 Communities and Link Weights

Link weights are deeply correlated with the community structure. Yet, as we discuss next, the nature of these correlations is system dependent.

Social Networks

The more time two individuals spend together, the more likely they are to share friends, which increases the chance that they belong to the same community. Consequently, communities in social networks tend to be nucleated around strong ties. Links connecting different communities are weaker in comparison. This pattern, known as the *weak tie hypothesis* [24], is illustrated in **Figure 9.30a** for the mobile call network [115]. We observe that strong ties are indeed predominantly within the numerous small communities, and links connecting communities are visibly weaker.

Transport Systems

The purpose of many technological and biological networks is to transport materials or information. In this case the link weights are expected to correlate with betweenness centrality [336, 337], a proxy of the local traffic carried by a network. As links connecting different communities must transport a considerable amount of traffic, in transport networks strong ties are between communities. In contrast, links within communities are weaker in comparison (**Figure 9.30b**).

The coupling between link weights and community structure suggests that incorporating the link weights could enhance the accuracy of community-finding algorithms. Yet, the different nature of the coupling in social and technological systems serves as a cautionary note: algorithms that aim to place in the same community nodes connected by strong ties may only be effective in social systems. They may offer potentially misleading results in technological and biological systems, where strong ties connect different communities.

9.7.3 Community Evolution

Changes in a network's wiring diagram can have multiple consequences for communities: they can lead to the birth of new communities, the growth or the contraction of existing communities, communities can merge with each other or split into several smaller communities, and finally communities can die (**Figure 9.31**) [324]. Studies focusing on social and communication networks offer several insights into the changes communities experience [324, 338–344]:

Growth
The probability that a node joins a community grows with the number of links the node has to members of that community [344].

Contraction
Nodes with only a few links to members of their community are more likely to leave the community than nodes with multiple links to community members [344]. In weighted networks the probability that a node leaves a community increases with the sum of its link weights to nodes outside the community.

Splitting or Death
The probability that a community disintegrates increases with the aggregate link weights to nodes outside the community.

Age
There is positive correlation between the age of a community and its size, indicating that older communities tend to be larger [324].

Community Stability
The membership of large communities changes faster with time than the membership of smaller communities. Indeed, in social networks large communities often correspond to institutions, companies or schools, that renew themselves by

(a)

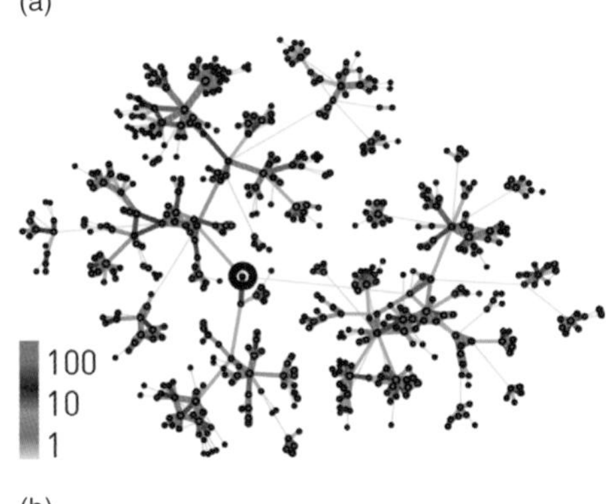

(b)

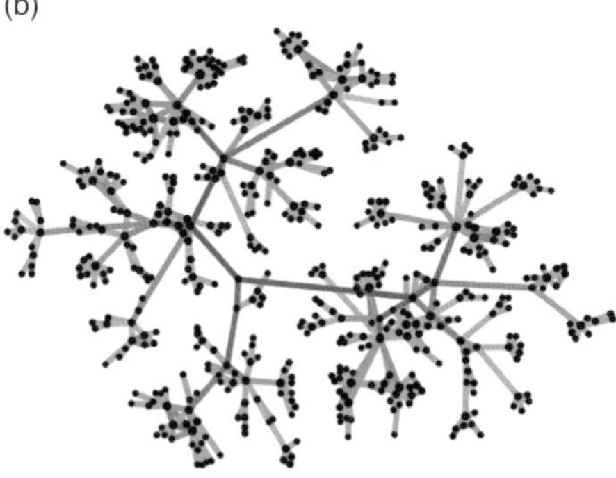

Figure 9.30 Communities and Link Weights
The mobile call network helps us illustrate the relationship between link weights and communities. Links represent mutual calls between users. We show only nodes that are at distance six or less from the individual highlighted as a black circle in (a).
(a) **Real Weights**
The link colors capture the aggregate call duration in minutes (see color bar). In line with the weak-tie hypothesis we find strong ties mainly within communities and weak ties between communities [24].
(b) **Betweenness Centrality**
If the link weights are driven by the need to transport information or materials, as is often the case in technological and biological systems, the weights are well approximated by betweenness centrality (**Figure 9.11**). We colored the links based on each link's betweenness centrality. As the figure indicates, links connecting communities have high betweenness (red), whereas links within communities have low betweenness (green).
After [115].

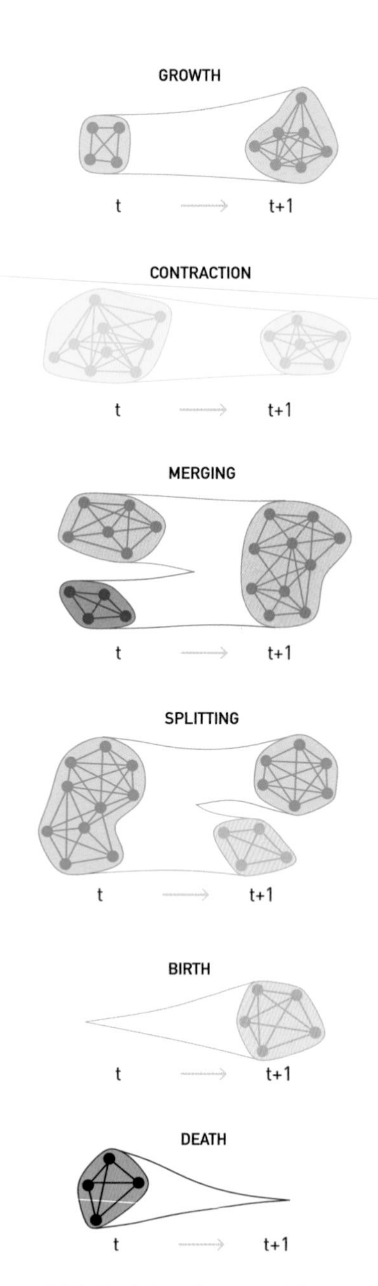

Figure 9.31 **Evolving Communities**
When networks evolve in time, so does the underlying community structure. All changes in community structure are the result of six elementary events in the life of a community, illustrated in the figure: a community can grow or contract; communities can merge or split; new communities are born while others may disappear.
After [324].

accepting new members, hiring new employees or enrolling new students. For small communities, stability requires stable membership [324].

These results were obtained in the context of social systems. Our understanding of the patterns that cover community evolution in technological or biological systems remains limited.

In summary, several recurring patterns characterize the organization and the evolution of communities. The community size distribution is typically fat tailed, indicating the coexistence of many small communities with a few large ones. We also find system-dependent correlations between the community structure and link weights, so that in social systems the strong ties are mainly within communities, while in transport systems they are between communities. Finally, we have gained an increasing understanding of the dynamical patterns that govern community evolution.

9.8 Summary

The ubiquity of communities across different networks has turned community identification into a dynamically developing chapter of network science. Many of the developed algorithms are now available as software packages, allowing their immediate use for network diagnosis. Yet, the efficient use of these algorithms and the interpretation of their predictions requires us to be aware of the assumptions built into them. In this chapter we provided the intellectual and quantitative foundations of community detection, helping us understand the origin and the assumptions behind the most frequently used algorithms.

Despite the successes of community identification, the field is faced with numerous open questions:

Do We Really Have Communities?
Throughout this chapter we avoided a fundamental question: How do we know that there are indeed communities in a particular network? In other words, could we decide that a network has communities without first identifying the communities themselves? The lack of an answer to this question represents

perhaps the most glaring gap in the community-finding literature. Community-finding algorithms are designed to identify communities, whether they are there or not.

Hypotheses or Theorems?
Community identification relies on four hypotheses, summarized in **Box 9.3**. We call them hypotheses because we cannot prove their correctness. Further advances might be able to turn the fundamental, the random and the maximal modularity hypotheses into theorems. Or we may learn about their limitations, as we did in the case of the maximal modularity hypothesis (**Section 9.6**).

Must all Nodes Belong to Communities?
Community-detection algorithms force all nodes into communities. This is likely to be an overkill for most real networks: some nodes belong to a single community, others to multiple communities, and likely many nodes do not belong to any community. Most algorithms used in community identification do not make this distinction, forcing instead all nodes into some community.

Dense vs. Sparse Communities
Most networks explored in this book are sparse. Yet, with improvements in data collection, many real network maps will likely gain numerous links. In dense networks we often see numerous highly overlapping communities, forcing us to re-evaluate the validity of the various hypotheses, and the appropriateness of the community-detection algorithms discussed in this chapter. For example, in highly overlapping communities nodes may have higher external than internal degrees, limiting the validity of the density hypothesis.

Do Communities Matter?
We resort to an example to answer this question. **Figure 9.32a** shows a local neighborhood of the mobile call network, highlighting four communities identified by the link-clustering algorithm (**Section 9.5**). The figure also shows the call frequency at noon (b) and at midnight (c), documenting different calling habits at different parts of the day. We find that the members of the top right community, shown as

Box 9.3 At a Glance: Communities

Community identification rests on several hypotheses, pertaining to the nature of communities.

Fundamental Hypothesis
Communities are uniquely encoded in a network's wiring diagram. They represent a grand truth that remains to be discovered using appropriate algorithms.

Connectedness and Density Hypothesis
A community corresponds to a locally dense connected subgraph.

Random Hypothesis
Randomly wired networks do not have communities.

Maximal Modularity Hypothesis
The partition with the maximum modularity offers the best community structure, where modularity is given by

$$M = \sum_{c=1}^{n_c} \left[\frac{l_c}{L} - \left(\frac{k_c}{2L} \right)^2 \right].$$

(a)

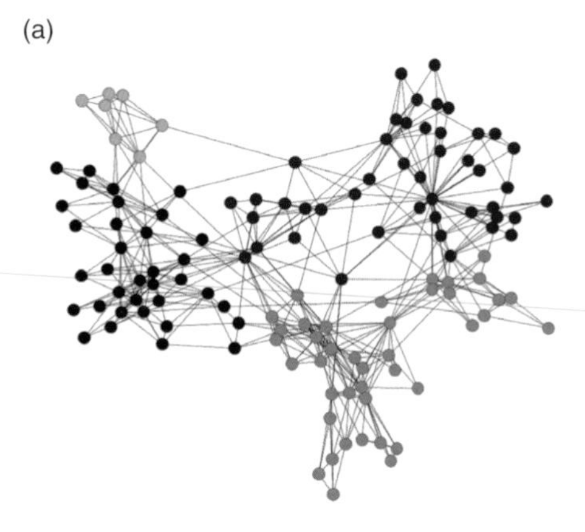

(b)

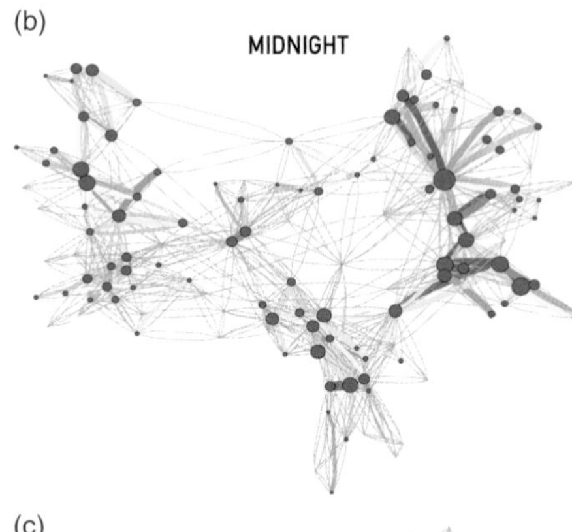

(c)

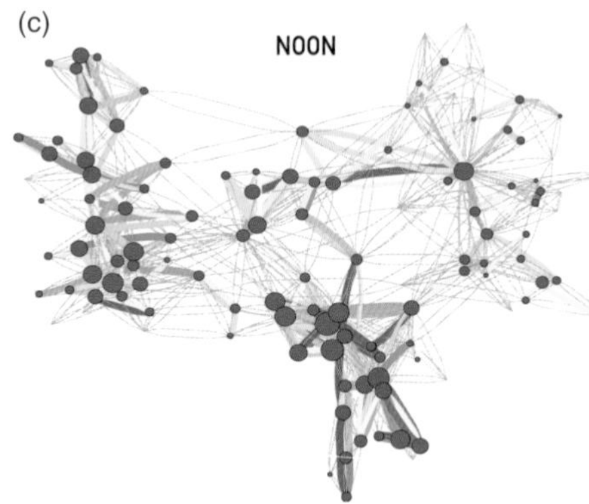

Figure 9.32 **Communities and Call Patterns**
The direct impact of communities on the activity of their members is illustrated by the mobile call network, offering us simultaneous information on community structure and user activity.
(a) **Community Structure**
Four communities of the mobile phone network, each community being colored differently. These communities represent local neighborhoods in the call patterns of over one million consumers, as predicted by the link-clustering algorithm (**Section 9.5**). The rest of the mobile phone network is not shown.
(b) **Midnight Activity**
The calling patterns of the users in the four communities shown in (a). The link colors reflect the frequency of calls in the hour-long interval around midnight. Red links signal numerous calls around midnight; white or missing links imply that the users talked little or did not call each other in this time frame.
(c) **Noon Activity**
The same as in (b) but at noon.
Image courtesy of Sune Lehmann.

brown nodes in (a), are active at midnight (b) but stop calling each other at noon (c). In contrast, the light and dark blue communities are active at noon but sleepy at midnight. This indicates that communities, identified only from the network's wiring diagram, have coherent community-specific activity patterns.

Figure 9.32 suggests that once present, communities have a profound impact on network behavior. Numerous measurements support this conclusion: information travels fast within a community but has difficulty reaching other communities; communities influence the link weights; the presence of communities can lead to degree correlations.

Communities are equally remarkable for their potential applications. For example, strengthening the links between clients that belong to the same community on the WWW can improve the performance of Web-based services [345]. In marketing, community finding can be used to identify customers with similar interests or purchasing habits, helping design efficient product-recommendation systems [346]. Communities are often used to create data structures that can handle queries in a timely fashion [347, 348]. Finally, community-finding algorithms run behind many social network sites, like Facebook,

Twitter or LinkedIn, helping these services discover potential friends, posts of interest and target advertising.

While community finding has deep roots in social and computer science, it is a relatively young chapter of network science (**Box 9.4**). As such, our understanding of community organization continues to develop rapidly, offering increasingly accurate tools to diagnose the local structure of large networks.

9.9 Homework

9.9.1 Hierarchical Networks

Calculate the degree exponent of the hierarchical network shown in **Figure 9.33**.

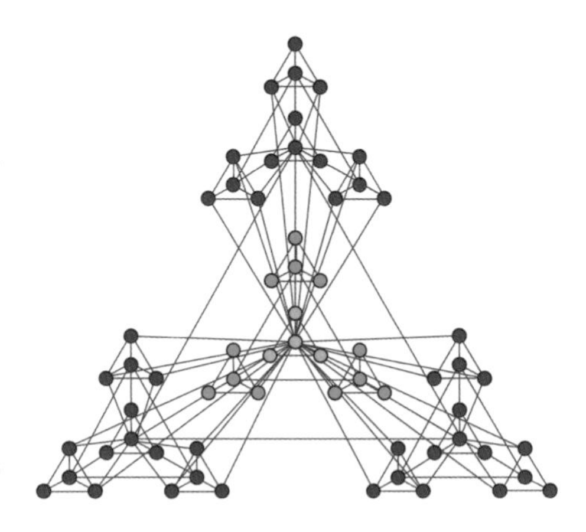

Figure 9.33 **Hierarchical Networks**
The colors represent the subsequent stages of the network's construction.

9.9.2 Communities on a Circle

Consider a one-dimensional lattice with N nodes that form a circle, where each node connects to its two neighbors. Partition the line into n_c consecutive clusters of size $N_c = N/n_c$.

(a) Calculate the modularity of the obtained partition.
(b) According to the maximum modularity hypothesis (**Section 9.4**), the maximum of M_c corresponds to the best partition. Obtain the community size n_c corresponding to the best partition.

9.9.3 Modularity Resolution Limit

Consider a network consisting of a ring of n_c cliques, each clique having N_c nodes and $m(m-1)/2$ links. The neighboring cliques are connected by a single link (**Figure 9.34**). The network has an obvious community structure, each community corresponding to a clique.

(a) Determine the modularity M_{single} of this natural partition, and the modularity M_{pairs} of the partition in which pairs of neighboring cliques are merged into a single community, as indicated by the dotted lines in **Figure 9.34**.

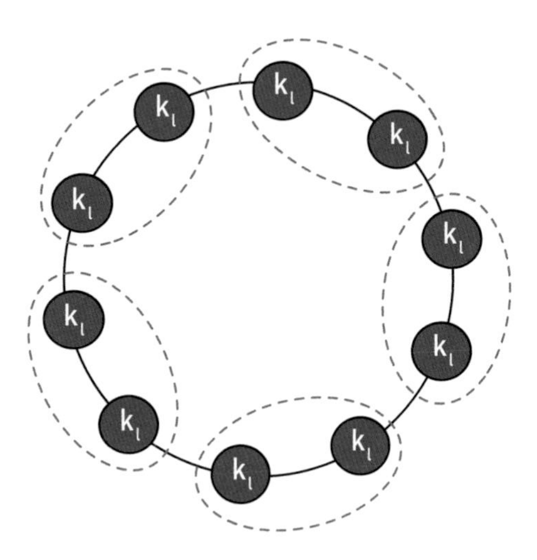

Figure 9.34 **Modularity**

Box 9.4 Community Finding: A Brief History

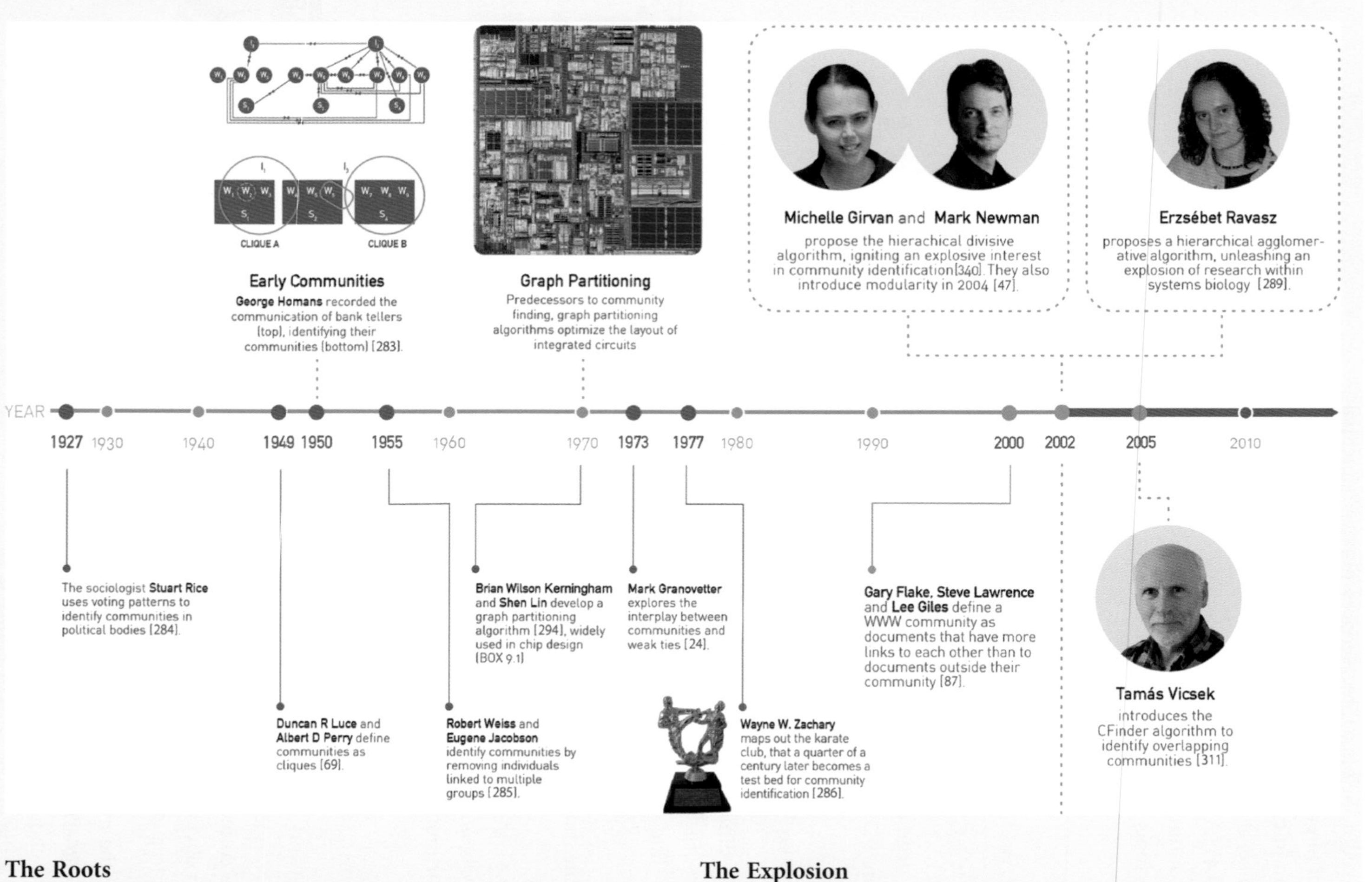

The Roots

Many of the concepts used in community finding have their roots in social and computer science, preceding network science.

The Explosion

The current interest in community finding was ignited by two papers, proposing algorithms to identify communities in social and biological systems.

(b) Show that only for $n_c < \sqrt{2L}$ will the modularity maximum predict the intuitively correct community partition, where

$$L = n_c m(m-1)/2 + n_c$$

(c) Discuss the consequences of violating the above inequality.

9.9.4 Modularity Maximum.

Show that the maximum value of modularity M defined in **(9.12)** cannot exceed one.

9.10 Advanced Topics 9.A Hierarchical Modularity

In this section we discuss the scaling properties of the hierarchical model introduced in **Figure 9.13**. We calculate the degree distribution and the degree-dependent clustering coefficient, deriving **(9.8)**. Finally, we explore the presence of hierarchy in the ten real networks.

9.10.1 Degree Distribution

To compute the model's degree distribution we count the nodes with different degrees. Starting with the five nodes of the first module in **Figure 9.13a**, we label the middle one a *hub* and call the remaining four nodes *peripheral*. All copies of this hub are again called *hubs* and we continue calling copies of peripheral nodes *peripheral* (**Figure 9.35**).

The largest hub at the center of the network acquires 4^n links during the nth iteration. Let us call this central hub H_n and the four copies of this hub H_{n-1} (**Figure 9.35**). We call H_{n-2} the $4 \cdot 5$ leftover module centers whose size equals the size of the network at the $(n-2)$th iteration.

At the nth iteration the degree of the hub H_i follows

$$k_n(H_i) = \sum_{l=1}^{i} 4^l = \frac{4}{3}(4^i - 1), \tag{9.21}$$

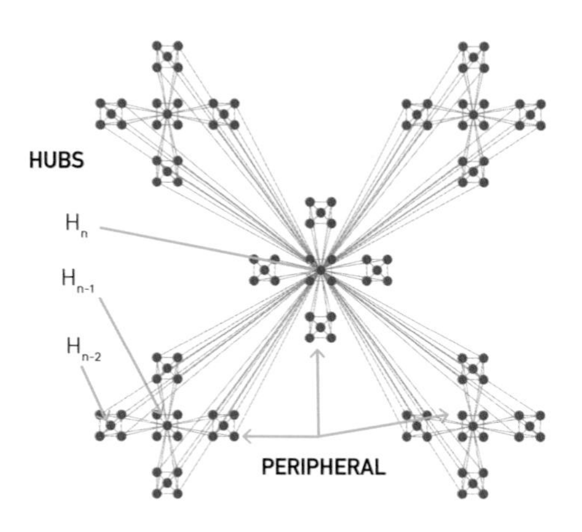

Figure 9.35 **Calculating the Degree Exponent**
The structure of a hierarchical network and the naming convention we use to refer to the hubs. After [289].

where we used

$$\sum_{l=0}^{i} x^l = \frac{x^{i+1}-1}{x-1}, \tag{9.22}$$

or

$$\sum_{l=1}^{i} x^l = \frac{x^{i+1}-1}{x-1} - 1. \tag{9.23}$$

For $i < n$ the number of H_i modules is

$$N_n(H_i) = 4{\cdot}5^{n-i-1}, \tag{9.24}$$

that is there are four modules for $i = n - 1$; $4 \cdot 5$ modules for $i = n - 2$; ... $4 \cdot 5^{n-2}$ for $i = n$. Since we have $4 \cdot 5^{n-i-1}$ H_i-type hubs of degree k_n (H_i), **(9.21)** and **(9.24)** allow us to write

$$\ln N_n(H_i) = C_n - i \cdot \ln 5 \tag{9.25}$$

$$\ln k_n(H_i) \simeq i \cdot \ln 4 + \ln(4/3) \tag{9.26}$$

where

$$C_n = \ln 4 + (n-1)\ln 5. \tag{9.27}$$

Note that in **(9.26)** we used the approximation $4^i - 1 \simeq 4^i$.

For all $k > n + 2$ we can combine **(9.26)** and **(9.27)** to obtain

$$\ln N_n(H_i) = C'_n - \ln k_i \frac{\ln 5}{\ln 4} \tag{9.28}$$

or

$$N_n(H_i) \sim \ln k_i^{-\frac{\ln 5}{\ln 4}}. \tag{9.29}$$

To calculate the degree distribution we need to normalize $N_n(H_i)$ by calculating the ratio

$$p_{k_i} \sim \frac{N_n(H_i)}{k_{i+1} - k_i} \sim k_i^{-\gamma}. \tag{9.30}$$

Using

$$k_{i+1} - k_i = \sum_{l=1}^{i+1} 4^l - \sum_{l=1}^{i} 4^l = 4^{i+1} = 3k_i + 4 \tag{9.31}$$

we obtain

$$p_{k_i} = \frac{k_i^{-\frac{\ln 5}{\ln 4}}}{3k_i + 4} \sim k_i^{-1-\frac{\ln 5}{\ln 4}}. \quad (9.32)$$

In other words, the obtained hierarchical network's degree exponent is

$$\gamma = 1 + \frac{\ln 5}{\ln 4} = 2.16. \quad (9.33)$$

9.10.2 Clustering Coefficient

It is somewhat straightforward to calculate the clustering coefficient of the H_i hubs. Their $\sum_{l=1}^{i} 4^l$ links come from nodes linked in a square, thus the connections between them equal their number. Consequently, the number of links between H_i's neighbors is

$$\sum_{l=1}^{i} 4^l = k_n(H_i), \quad (9.34)$$

providing

$$C(H_i) = \frac{2k_i}{k_i(k_i - 1)} = \frac{2}{k_i - 1}. \quad (9.35)$$

In other words, we obtain

$$C(k) \simeq \frac{2}{k}, \quad (9.36)$$

indicating that $C(k)$ for the hubs scales as k^{-1}, in line with **(9.8)**.

9.10.3 Empirical Results

Figure 9.36 shows the $C(k)$ function for the ten reference networks. We also show $C(k)$ for each network after we applied degree-preserving randomization (green symbols), allowing us to make several observations:

- For small k, all networks have an order of magnitude higher $C(k)$ than their randomized counterpart. Therefore, the small-degree nodes are located in much denser neighborhoods than expected by chance.
- For the scientific collaboration, metabolic and citation networks with a good approximation we have $C(k) \sim k^{-1}$,

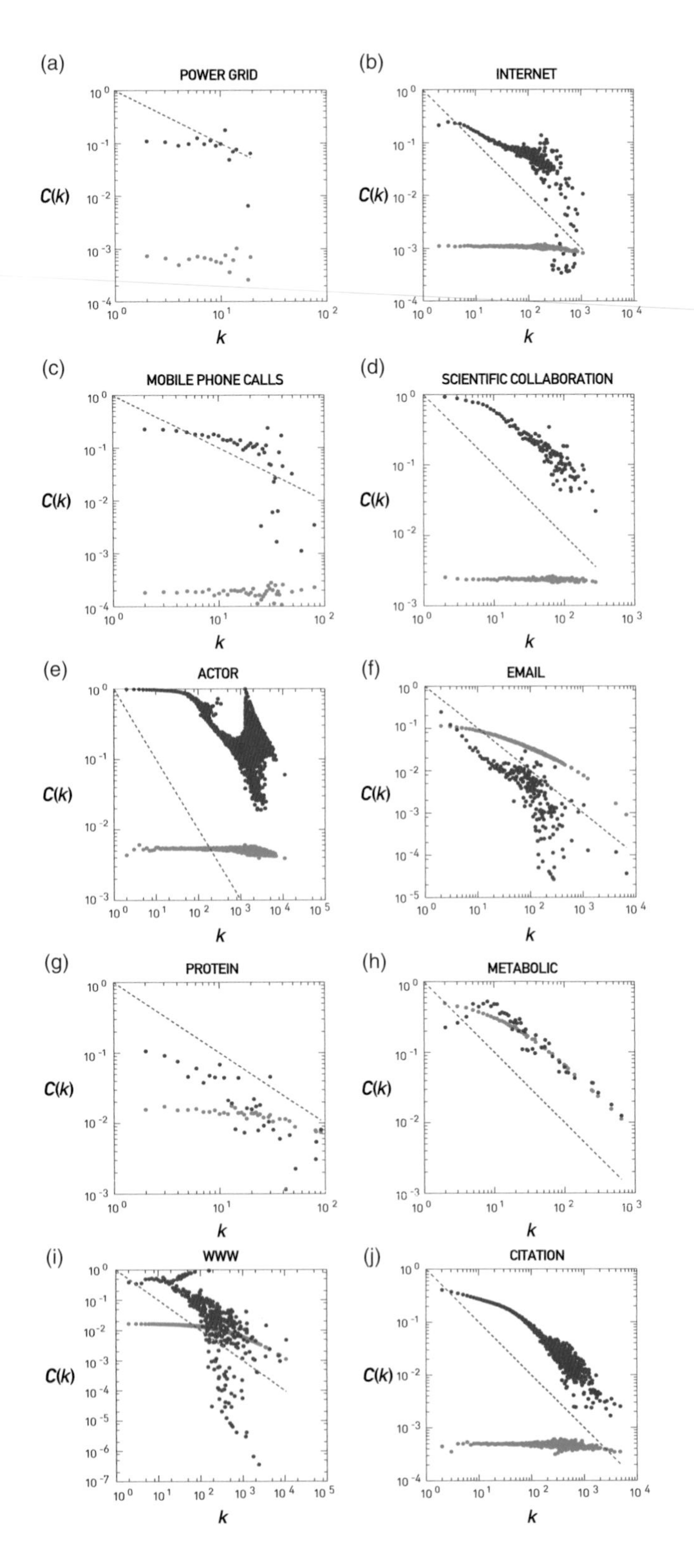

Figure 9.36 **Hierarchy in Real Networks**
The scaling of $C(k)$ with k for the ten reference networks (purple symbols). The green symbols show $C(k)$ obtained after applying degree-preserving randomization to each network, which washes out the local density fluctuations. Consequently, communities and the underlying hierarchy are gone. Directed networks were made undirected to measure $C(k)$. The dashed line in each figure has slope -1, following **(9.8)**, serving as a guide to the eye.

while the randomized $C(k)$ is flat. Hence, these networks display the hierarchical modularity of the model of **Figure 9.13**.

- For the Internet, mobile phone calls, actors, email, protein interactions and the WWW, $C(k)$ decreases with k, while their randomized $C(k)$ is k-independent. Hence, while these networks display a hierarchical modularity, the observed $C(k)$ is not captured by our simple hierarchical model. To fit the $C(k)$ of these systems we need to build models that accurately capture their evolution. Such models predict that, where β can be different from one [303].
- Only for the power grid do we observe a flat, k-independent $C(k)$, indicating the lack of a hierarchical modularity.

Taken together, **Figure 9.36** indicates that most real networks display some nontrivial hierarchical modularity.

9.11 Advanced Topics 9.B Modularity

In this section we derive the expressions **(9.12)** and **(9.13)**, characterizing the modularity fuction and its changes.

9.11.1 Modularity as a Sum Over Communities

Using **(9.9)** and **(9.10)**, we can write the modularity of a full network as

$$M = \frac{1}{2L} \sum_{i,j=1}^{N} \left(A_{ij} - \frac{k_i k_j}{2L} \right) \delta_{C_i, C_j}, \tag{9.37}$$

where C_i is the label of the community to which node i belongs. As only node pairs that belong to the same community contribute to the sum in **(9.37)**, we can rewrite the first term as a sum over communities:

$$\frac{1}{2L} \sum_{i,j=1}^{N} A_{ij} \delta_{C_i, C_j} = \sum_{c=1}^{n_c} \frac{1}{2L} \sum_{i,j \in C_c} A_{ij} = \sum_{c=1}^{n_c} \frac{L_c}{L}, \tag{9.38}$$

where L_c is the number of links within community C_c. The factor 2 disappears because each link is counted twice in A_{ij}.

In a similar fashion, the second term of **(9.37)** becomes

$$\frac{1}{2L}\sum_{i,j=1}^{N}\frac{k_i k_j}{2L}\delta_{C_i,C_j}=\sum_{c=1}^{n_c}\frac{1}{(2L)^2}\sum_{i,j\in C_c}k_i k_j=\sum_{c=1}^{n_c}\frac{k_c^2}{4L^2},\qquad(9.39)$$

where k_c is the total degree of the nodes in community C_c. Indeed, in the configuration model the probability that a stub connects to a randomly chosen stub is $\frac{1}{2L}$, as in total we have $2L$ stubs in the network. Hence, the $2L$ likelihood that our stub connects to a stub inside the module is $\frac{k_c}{2L}$. By repeating this procedure for all k_c stubs within the community C_c and adding 1/2 to avoid double counting, we obtain the last term of **(9.39)**.

Combining **(9.38)** and **(9.39)** leads to **(9.12)**.

9.11.2 Merging Two Communities

Consider communities A and B and denote by k_A and k_B the total degree in these communities (equivalent to k_c above). We wish to calculate the change in modularity after we merge these two communities. Using **(9.12)**, this change can be written as

$$\Delta M_{AB}=\left[\frac{L_{AB}}{L}-\left(\frac{k_{AB}}{2L}\right)^2\right]-\left[\frac{L_A}{L}-\left(\frac{k_A}{2L}\right)^2+\frac{L_B}{L}-\left(\frac{k_B}{2L}\right)^2\right],\qquad(9.40)$$

where

$$L_{AB}=L_A+L_B+l_{AB},\qquad(9.41)$$

l_{AB} is the number of direct links between the nodes of communities A and B, and

$$k_{AB}=k_A+k_B.\qquad(9.42)$$

After inserting **(9.41)** and **(9.42)** into **(9.40)**, we obtain

$$\Delta M_{AB}=\frac{l_{AB}}{L}-\frac{k_A k_B}{2L^2},\qquad(9.43)$$

which is **(9.13)**.

9.12 Advanced Topics 9.C Fast Algorithms for Community Detection

The algorithms discussed in this chapter were chosen to illustrate the fundamental ideas and concepts pertaining to community detection. Consequently, they are not guaranteed to be either the fastest or the most accurate algorithms. Recently two algorithms, called the *Louvain algorithm* and *Infomap*, have gained popularity, as their accuracy is comparable with the accuracy of the algorithms covered in this chapter but offer better scalability. Consequently, we can use them to identify communities in very large networks.

There are many similarities between the two algorithms:

- They both aim to optimize a quality function Q. For the Louvain algorithm Q is modularity, M, and for Infomap Q is an entropy-based measure called the map equation or L.
- Both algorithms use the same optimization procedure.

Given these similarities, we discuss the algorithms together.

9.12.1 The Louvain Algorithm

The $O(N^2)$ computational complexity of the greedy algorithm can be prohibitive for very large networks. A modularity optimization algorithm with better scalability was proposed by Blondel and collaborators [282]. The *Louvain algorithm* consists of two steps that are repeated iteratively (**Figure 9.37**):

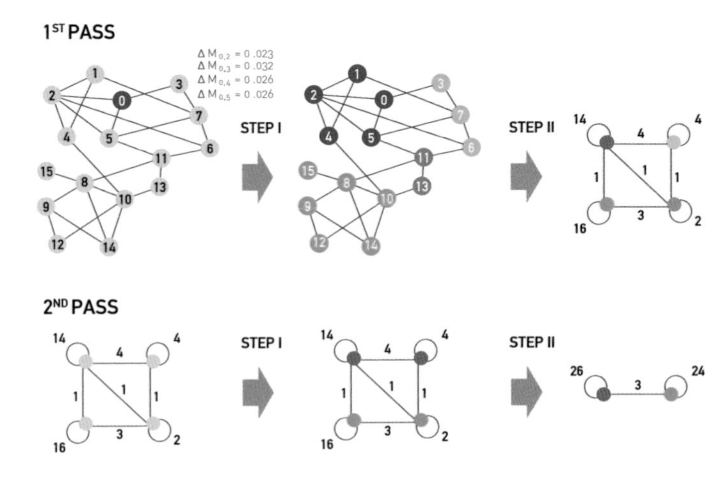

Figure 9.37 **The Louvain Algorithm**
The main steps of the Louvain algorithm. Each *pass* consists of two distinct *steps.*
Step I
Modularity is optimized by local changes. We choose a node and calculate the change in modularity, **(9.44)**, if the node joins the community of its immediate neighbors. The figure shows the expected modularity change $\Delta M_{0,i}$ for node 0. Accordingly, node 0 will join node 3 as the modularity change for this move is the largest, being $\Delta M_{0,3} = 0.032$. This process is repeated for each node, the node colors corresponding to the resulting communities.
Step II
The communities obtained in step I are aggregated, building a new network of communities. Nodes belonging to the same community are merged into a single node, as shown on the top right. This process will generate self-loops, corresponding to links between nodes in the same community that are now merged into a single node.
The sum of steps I and II is called a *pass*. The network obtained after each pass is processed again (pass 2), until no further increase of modularity is possible.
After [282].

Step I

Start with a weighted network of N nodes, initially assigning each node to a different community. For each node i we evaluate the gain in modularity if we place node i in the community of one of its neighbors j. We then move node i into the community for which the modularity gain is the largest, but only if this gain is positive. If no positive gain is found, i stays in its original community. This process is applied to all nodes until no further improvement can be achieved, completing step I.

The modularity change ΔM obtained by moving an isolated node i into a community C can be calculated using

$$\Delta M = \left[\frac{\sum_{in} + 2k_{i,in}}{2W} - \left(\frac{\sum_{tot} + k_i}{2W}\right)^2\right] - \left[\frac{\sum_{in}}{2W} - \left(\frac{\sum_{tot}}{2W}\right)^2 - \left(\frac{k_i}{2W}\right)^2\right] \tag{9.44}$$

where $\sum_{in}$ is the sum of the weights of the links inside C (which is L_C for an unweighted network); $\sum_{tot}$ is the sum of the link weights of all nodes in C; k_i is the sum of the weights of the links incident to node i; $k_{i,in}$ is the sum of the weights of the links from i to nodes in C; W is the sum of the weights of all links in the network.

Note that ΔM is a special case of **(9.13)**, which provides the change in modularity after merging communities A and B. In the current case B is an isolated node. We can use ΔM to determine the modularity change when i is removed from the community it belonged to earlier. For this we calculate ΔM for merging i with the community C after we excluded i from it. The change after removing i is $-\Delta M$.

Step II

We construct a new network whose nodes are the communities identified during step I. The weight of the link between two nodes is the sum of the weight of the links between the nodes in the corresponding communities. Links between nodes of the same community lead to weighted self-loops.

Once step II is completed, we repeat steps I and II, calling their combination a *pass* (**Figure 9.37**). The number of communities decreases with each pass. The passes are repeated until there are no more changes and maximum modularity is attained.

Computational Complexity

The Louvain algorithm is more limited by storage demands than by computational time. The number of computations scales linearly with L for the most time-consuming first pass. With subsequent passes over a decreasing number of nodes and links, the complexity of the algorithm is at most $O(L)$. It therefore allows us to identify communities in networks with millions of nodes.

9.12.2 Infomap

Introduced by Martin Rosvall and Carl T. Bergstrom, Infomap exploits data compression for community identification (**Figure 9.38**) [318, 319, 320]. It does this by optimizing a quality function for community detection in directed and weighted networks, called the *map equation*.

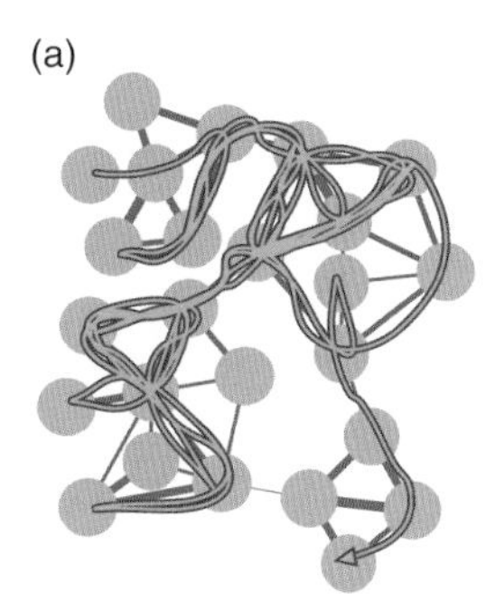

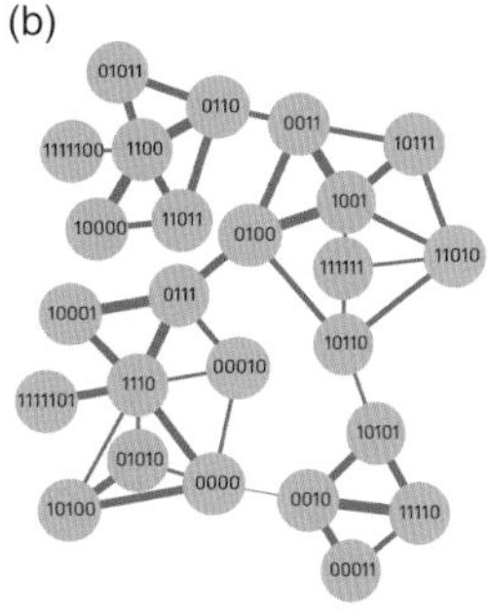

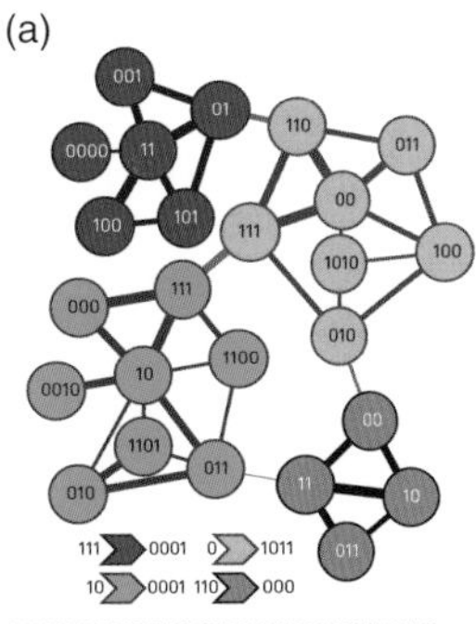

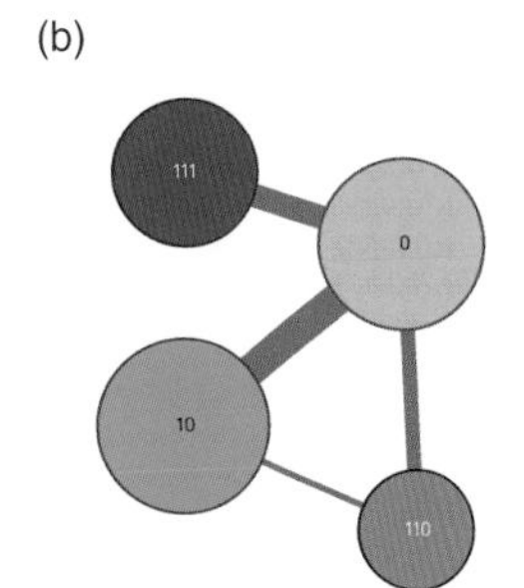

Figure 9.38 **From Data Compression to Communities**

Infomap detect communities by compressing the movement of a random walker on a network.

(a) The orange line shows the trajectory of a random walker on a small network. We want to describe this trajectory with a minimal number of symbols, which we can achieve by assigning repeatedly used structures (communities) short and unique names.

(b) We start by giving a unique name to each node. This is derived using Huffman coding, a data-compression algorithm that assigns each node a code using the estimated probability that the random walk visits that node. The 314 bits under the network describe the sample trajectory of the random walker shown in (a), starting with 1111100 for the first node of the walk in the upper left corner, 1100 for the second node, etc., and ending with 00011 for the last node on the walk in the lower right corner.

(c) The figure shows a two-level encoding of the random walk, in which each community receives a unique name, but the names of nodes within communities are reused. This code yields on average a 32% shorter coding. The codes naming the communities and the codes used to indicate an exit from each community are shown to the left and the right of the arrows under the network, respectively. Using this code, we can describe the walk in (a) by the 243 bits shown under the network in (c). The first three bits 111 indicate that the walk begins in the red community, the code 0000 specifies the first node of the walk, etc.

(d) By reporting only the community names, and not the locations of each node within the communities, we obtain an efficient coarse graining of the network, which corresponds to its community structure.

Consider a network partitioned into n_c communities. We wish to encode in the most efficient fashion the trajectory of a random walker on this network. In other words, we want to describe the trajectory with the smallest number of symbols. The ideal code should take advantage of the fact that the random walker tends to get trapped into communities, staying there for a long time (**Figure 9.38c**).

To achieve this coding we assign:

- One code to each community (index codebook). For example, the purple community in **Figure 9.38c** is assigned the code 111.
- Codewords for each node within each community. For example, the top left node in (c) is assigned 001. Note that the same node code can be reused in different communities.
- Exit codes that mark when the walker leavers a community, like 0001 for the purple community in (c).

The goal, therefore, is to build a code that offers the shortest description of the random walk. Once we have this code, we can identify the network's community structure by reading the index codebook, which is uniquely assigned to each community (**Figure 9.38c**).

The optimal code is obtained by finding the minimum of the *map equation*

$$\mathcal{L} = qH(Q) + \sum_{c=1}^{n_c} p_{\circlearrowright}^{c} H(P_c). \tag{9.45}$$

In a nutshell, the first term of **9.45** gives the average number of bits necessary to describe the movement *between communities*, where q is the probability that the random walker switches communities during a given step.

The second term gives the average number of bits necessary to describe movement *within communities*. Here, $H(P_c)$ is the entropy of within-community movements – including an "exit code" to capture the departure from a community i.

The specific terms of the map equation, and their calculation in terms of the probabilities capturing the movement of a random walker on a network, are somewhat involved. They are described in detail in [318, 319, 320]. **Online resource 9.3** offers an interactive tool to illustrate the mechanism behind **(9.45)** and its use.

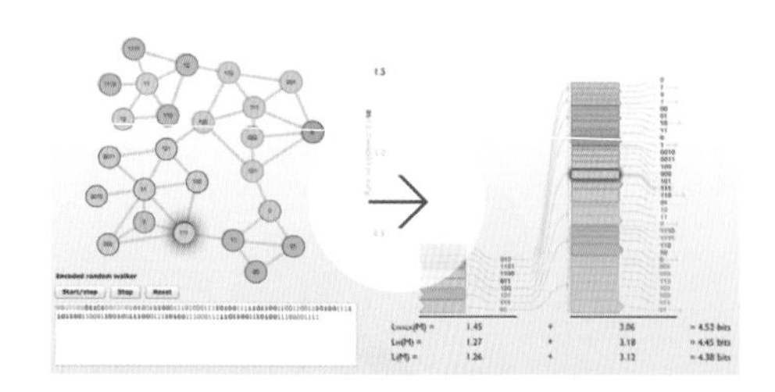

Online Resource 9.3

Map Equation for Infomap

For a dynamic visualization of the mechanism behind the map equation, see http://www.tp.umu.se/~rosvall/livemod/mapequation/.

In the end $\mathcal{L}$ serves as a quality function, taking a specific value for a particular partition of the network into communities. To find the best partition, we must minimize $\mathcal{L}$ over all possible partitions. The popular implementation of this optimization procedure follows steps I and II of the Louvain algorithm: we assign each node to a separate community, and we systematically join neighboring nodes into modules if the move decreases $\mathcal{L}$. After each move, $\mathcal{L}$ is updated using **(9.45)**. The obtained communities are joined into supercommunities, finishing a pass, after which the algorithm is restarted on the new, reduced network.

Computational Complexity

The computational complexity of Infomap is determined by the procedure used to minimize the map equation $\mathcal{L}$. If we use the Louvain procedure, the computational complexity is the same as that of the Louvain algorithm, that is at most $O(L\log L)$ or $O(N\log N)$ for a sparse graph.

In summary, the Louvain algorithm and Infomap offer tools for fast community identification. Their accuracy across benchmarks is comparable with the accuracy of the algorithms discussed throughout this chapter (**Figure 9.28**).

9.13 Advanced Topics 9.D Threshold for Clique Percolation

In this section we derive the percolation threshold **(9.16)** for clique percolation on a random network and discuss the main steps of the CFinder algorithm (**Figure 9.39**).

When we roll a k-clique to an adjacent k-clique by relocating one of its nodes, the expectation value of the number of adjacent k-cliques for the template to roll further should equal exactly one at the percolation threshold (**Figure 9.20**). Indeed, a smaller than one expectation value will result in a premature end of the k-clique percolation clusters, because starting from any k-clique, the rolling would quickly come to a halt. Consequently, the size of the clusters would decay exponentially. A larger than one expectation value, on the contrary, allows the clique community to grow indefinitely, guaranteeing that we have a giant cluster in the system.

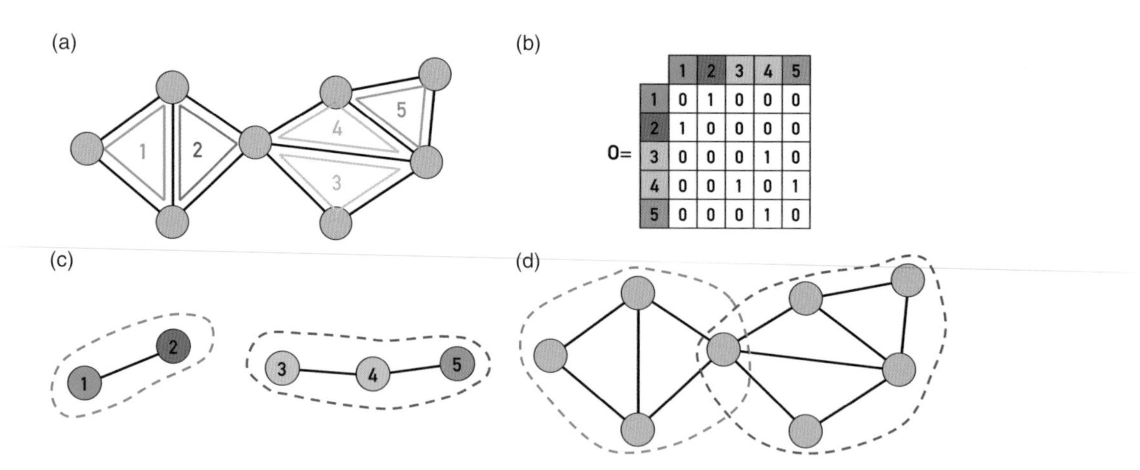

	1	2	3	4	5
1	0	1	0	0	0
2	1	0	0	0	0
3	0	0	0	1	0
4	0	0	1	0	1
5	0	0	0	1	0

Figure 9.39 **CFinder algorithm**
The main steps of the CFinder algorithm.
(**a**) Starting from the network shown in the figure, our goal is to identify all cliques. All five $k = 3$ cliques present in the network are highlighted.
(**b**) The overlap matrix O of the $k = 3$ cliques. This matrix is viewed as an adjacency matrix of a network whose nodes are the cliques of the original network. The matrix indicates that we have two connected components, one consisting of cliques (1, 2) and the other of cliques (3, 4, 5). The connected components of this network map into the communities of the original network.
(**c**) The two clique communities predicted by the adjacency matrix.
(**d**) The two clique communities shown in (c), mapped on the original network.

The above expectation value is provided by

$$(k-1)(N-k-1)^{k-1}, \tag{9.46}$$

where the term $(k-1)$ counts the number of nodes of the template that can be selected for the next relocation; the term $(N-k-1)^{k-1}$ counts the number of potential destinations for this relocation, out of which only the fraction p^{k-1} is acceptable, because each of the new $(k-1)$ edges (associated with the relocation) must exist in order to obtain a new k-clique. For large N, **(9.46)** simplifies to

$$(k-1)Np_c^{k-1} = 1,$$

which leads to **(9.16)**.

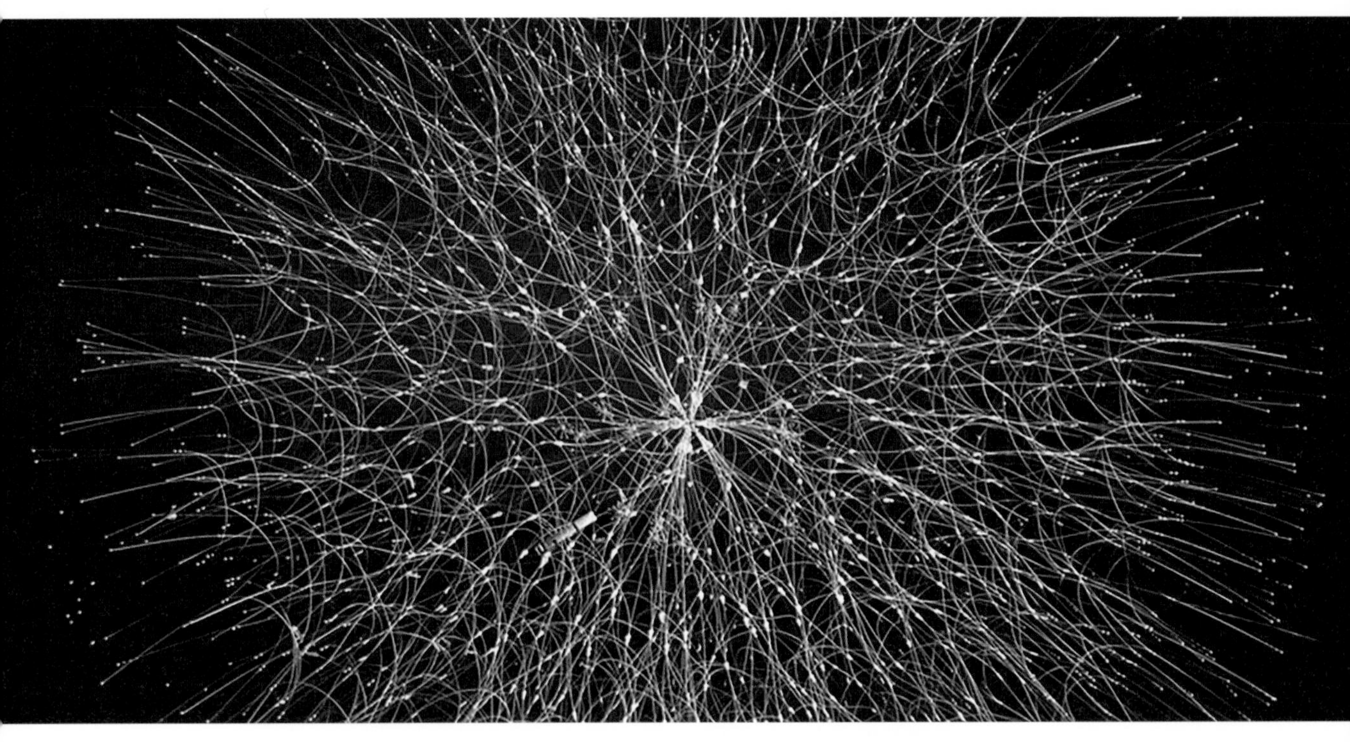

Figure 10.0 **Art & Networks: Bill Smith**
An epidemiological model of the perfect infectious disease (evolved growth system) is an artwork by Bill Smith, an Illinois-based artist (http://www.widicus.org).

CHAPTER 10

Spreading Phenomena

10.1 Introduction

On the night of February 21, 2003 a physician from Guangdong Province in southern China checked into the Metropole Hotel in Hong Kong. He had previously treated patients suffering from a disease that, lacking a clear diagnosis, was called *atypical pneumonia*. Next day, after leaving the hotel, he went to the local hospital, this time as a patient. He died there several days later of atypical pneumonia [349].

The physician did not leave the hotel without a trace: that night 16 other guests of the Metropole Hotel and one visitor also contracted the disease that was eventually renamed Severe Acute Respiratory Syndrome (SARS). These guests carried the SARS virus with them to Hanoi, Singapore and Toronto, sparking outbreaks in each of those cities. Epidemiologists later traced close to half of the 8,100 documented cases of SARS back to the Metropole Hotel. With that, the physician who

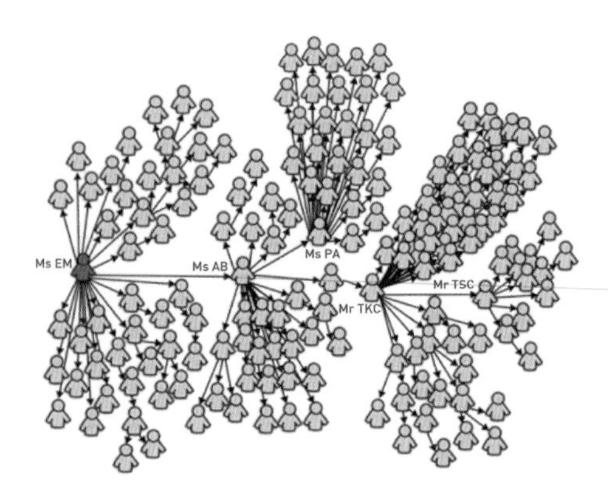

Figure 10.1 **Super-spreaders**
144 of the 206 SARS patients diagnosed in Singapore were traced to a chain of five individuals, including four *super-spreaders*. The most important of these was *Patient Zero*, the physician from Guangdong Province in China, who brought the disease to the Metropole Hotel.
After [349].

brought the virus to Hong Kong became an example of a *super-spreader*, an individual who is responsible for a disproportionate number of infections during an epidemic.

A network theorist will recognize super-spreaders as hubs, nodes with an exceptional number of links in the contact network on which a disease spreads. As hubs appear in many networks, super-spreaders have been documented in many infectious diseases, from smallpox to AIDS [350]. In this chapter we introduce a network-based approach to epidemic phenomena that allows us to understand and predict the true impact of these hubs. The resulting framework, that we call *network epidemics*, offers an analytical and numerical platform to quantify and forecast the spread of infectious diseases.

Infectious diseases account for 43% of the global burden of disease, as captured by the number of years of lost healthy life. They are called *contagious*, as they are transmitted by contact with an ill person or with their secretions. Cures and vaccines are rarely sufficient to stop an infectious disease – it is equally important to understand how the pathogen responsible for the disease spreads in the population, which in turn determines the way we administer the available cures or vaccines.

The diversity of phenomena regularly described as spreading processes on networks is staggering:

Biological
The spread of pathogens on their respective contact networks is the main subject of this chapter. Examples include airborne diseases like influenza, SARS or tuberculosis, transmitted when two individuals breathe the air in the same room; contagious diseases and parasites, transmitted when people touch each other; the Ebola virus, transmitted via contact with a patient's bodily fluids; HIV and other sexually transmitted diseases, passed on during sexual intercourse. Infectious diseases also include cancers carried by cancer-causing viruses, like HPV or EBV, or diseases carried by parasites, like bedbugs or malaria.

Digital
A computer virus is a self-reproducing program that can transmit a copy of itself from computer to computer. Its spreading pattern has many similarities to the spread of pathogens, but digital viruses also have many unique features, determined by the technology behind the specific virus. As mobile phones morphed into hand-held computers, lately we have also

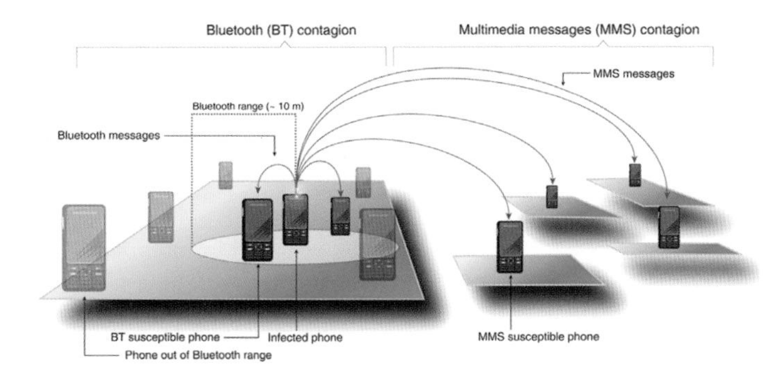

Figure 10.2 **Mobile Phone Viruses**

Smartphones, capable of sharing programs and data with each other, offer fertile ground for virus writers. Indeed, since 2004 hundreds of smartphone viruses have been identified, reaching a state of sophistication in a few years that took computer viruses about two decades to achieve [351]. Mobile viruses are transmitted using two main communication mechanisms [35]:

Bluetooth (BT) Viruses

A BT virus infects all phones found within BT range of the infected phone, which is about 10–30 meters. As physical proximity is essential for a BT connection, the transmission of a BT virus is determined by the owner's location and the underlying mobility network, connecting locations by individuals who travel between them (**Section 10.4**). Hence, BT viruses follow a spreading pattern similar to influenza.

Multimedia Messaging Services (MMS) Viruses

Viruses carried by MMS can infect all susceptible phones whose number is in the infected phone's phonebook. Hence, MMS viruses spread on the social network, following a long-range spreading pattern that is independent of the infected phone's physical location. Consequently, the spreading of MMS viruses is similar to the patterns characterizing computer viruses.

witnessed the appearance of mobile viruses and worms that infect smartphones (**Figure 10.2**).

Social

The role of the social and professional network in the spread and acceptance of innovations, knowledge, business practices, products, behavior, rumors and memes is a much-studied problem in social sciences, marketing and economics [352, 353]. Online environments, like Twitter, offer unprecedented ability to track such phenomena. Consequently, a staggering number of studies focus on social spreading, asking for example why some messages reach millions of individuals while others struggle to get noticed.

The examples discussed above involve diverse spreading agents, from biological to computer viruses, ideas and products; they spread on different types of network, from social to computer and professional networks; they are characterized by widely different time scales and follow different mechanisms of transmission (**Table 10.1**). Despite this diversity, as we show in this chapter, these spreading processes obey common patterns and can be described using the same network-based theoretical and modeling framework.

10.2 Epidemic Modeling

Epidemiology has developed a robust analytical and numerical framework to model the spread of pathogens. This framework relies on two fundamental hypotheses:

(i) **Compartmentalization**

Epidemic models classify each individual based on the stage of the disease affecting them. The simplest classification

Table 10.1 **Networks and Agents**

The spread of a pathogen, a meme or a computer virus is determined by the network on which the agent spreads and the transmission mechanism of the responsible agent. The table lists several much-studied spreading phenomena, together with the nature of the particular spreading agent and the network on which the agent spreads.

Phenomenon	Agent	Network
Venereal disease	Pathogens	Sexual network
Rumor spreading	Information, memes	Communication network
Diffusion of innovations	Ideas, knowledge	Communication network
Computer viruses	Malwares, digital viruses	Internet
Mobile phone virus	Mobile viruses	Social network/proximity network
Bedbugs	Parasitic insects	Hotel–traveler network
Malaria	Plasmodium	Mosquito–human network

assumes that an individual can be in one of three *states* or *compartments*.

- *Susceptible (S):* Healthy individuals who have not yet contacted the pathogen (**Figure 10.3**).
- *Infectious (I):* Contagious individuals who have contacted the pathogen and hence can infect others.
- *Recovered (R):* Individuals who have been infected before, but have recovered from the disease, hence are not infectious.

The modeling of some diseases requires additional states, like *immune* individuals, who cannot be infected, or *latent* individuals, who have been exposed to the disease but are not yet contagious.

Individuals can move between compartments. For example, at the beginning of a new influenza outbreak everyone is in the susceptible state. Once an individual comes into contact with an infected person, she can become infected. Eventually she will recover and develop immunity, losing her susceptibility to that particular strain of influenza.

(ii) **Homogenous Mixing**

The homogenous mixing hypothesis (also called *fully mixed* or *mass-action approximation*) assumes that each individual has the same chance of coming into contact with an infected individual. This hypothesis eliminates the need to know the precise contact network on which the disease spreads, replacing it with the assumption that anyone can infect anyone else.

In this section we introduce the epidemic modeling framework built on these two hypotheses. To be specific, we explore the dynamics of three frequently used epidemic models, the so-called SI, SIS and SIR models, that help us understand the basic building blocks of epidemic modeling.

10.2.1 Susceptible–Infected (SI) Model

Consider a disease that spreads in a population of N individuals. Denote by $S(t)$ the number of individuals who are susceptible (healthy) at time t and by $I(t)$ the number of individuals who have already been infected. At time $t = 0$ everyone is susceptible ($S(0) = N$) and no one is infected ($I(0) = 0$). Let us assume that a typical individual has $\langle k \rangle$ contacts and that the likelihood of the disease being transmitted from an infected to a susceptible individual in unit time is β. We ask the following: If a single individual becomes infected at time $t = 0$ (i.e., $I(0) = 1$), how many individuals will be infected at some later time t?

Within the homogenous mixing hypothesis, the probability that the infected person encounters a susceptible individual is $S(t)/N$. Therefore, the infected person comes into contact with $\langle k \rangle S(t)/N$ susceptible individuals in a unit time. Since $I(t)$ infected individuals are transmitting the pathogen, each at rate β, the average number of new infections $dI(t)$ during a time frame dt is

$$\beta \langle k \rangle \frac{S(t)I(t)}{N} dt.$$

Consequently, $I(t)$ changes at the rate

$$\frac{dI(t)}{dt} = \beta \langle k \rangle \frac{S(t)I(t)}{N}. \tag{10.1}$$

Throughout this chapter we will use the variables

$$s(t) = S(t)/N, \quad i(t) = I(t)/N \tag{10.2}$$

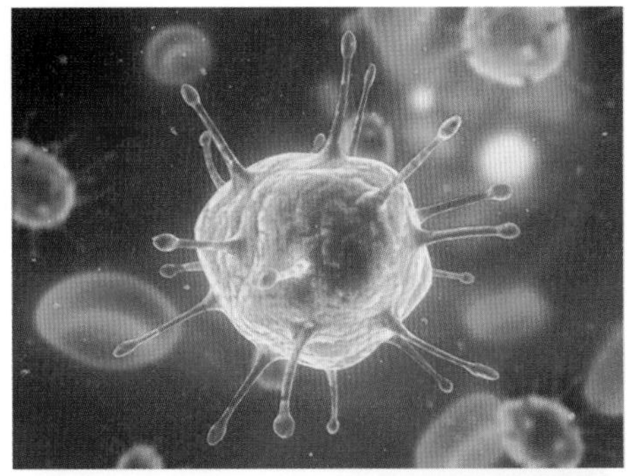

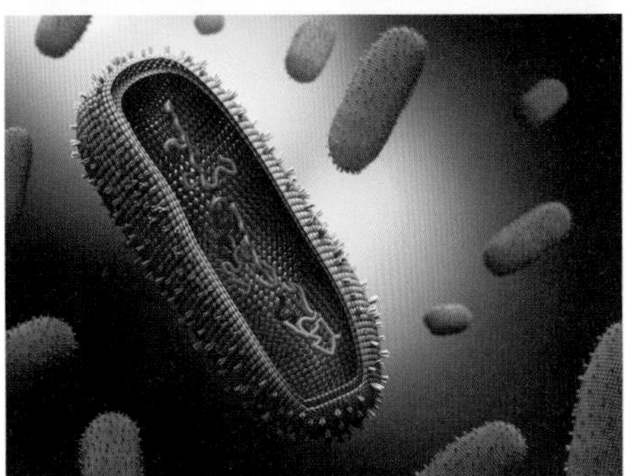

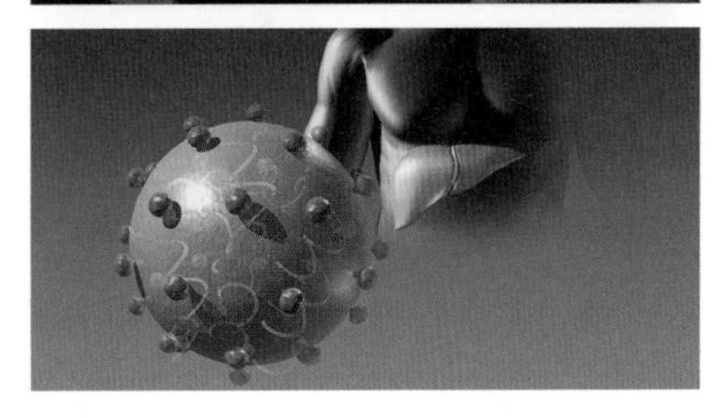

Figure 10.3 **Pathogens**
A *pathogen*, a word rooted in the Greek words "suffering, passion" (pathos) and "producer of" (genes), denotes an infectious agent or germ. A pathogen could be a disease-causing micro-organism, like a virus, a bacterium, a prion or a fungus. The figure shows several much-studied pathogens, like the HIV virus, responsible for AIDS, an influenza virus and the hepatitis C virus. After http://www.livescience.com/18107-hiv-therapeuticvaccines-promise.html and http://www.huffingtonpost.com/2014/01/13/deadly-viruses-beautiful-photos_n_4545309.html

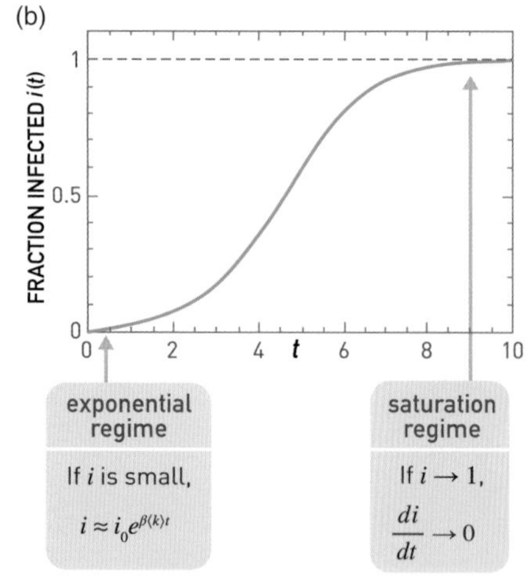

Figure 10.4 **The Susceptible–Infected (SI) Model**

(**a**) In the SI model an individual can be in one of two states: susceptible (healthy) or infected (sick). The model assumes that if a susceptible individual comes into contact with an infected individual, they become infected at a rate β. The arrow indicates that once an individual becomes infected they stay infected, and hence cannot recover.

(**b**) Time evolution of the fraction of infected individuals, as predicted by **(10.4)**. At early times the fraction of infected individuals grows exponentially. As eventually everyone becomes infected, at large times we have $i(\infty) = 1$.

to capture the fraction of the susceptible and of the infected population at time t. For simplicity we also drop the (t) variable from $i(t)$ and $s(t)$, rewriting **(10.1)** as (**Advanced topics 10.A**)

$$\frac{di}{dt} = \beta\langle k\rangle si = \beta\langle k\rangle i(1-i), \tag{10.3}$$

where the product $\beta\langle k\rangle$ is called the *transmission rate* or *transmissibility*. We solve **(10.3)** by writing

$$\frac{di}{i} + \frac{di}{(1-i)} = \beta\langle k\rangle dt.$$

Integrating both sides, we obtain

$$\ln i - \ln(1-i) + C = \beta\langle k\rangle t.$$

With the initial condition $i_0 = i(t=0)$, we get $C = i_0/(1-i_0)$, obtaining that the fraction of infected individuals increases in time as

$$i = \frac{i_0 e^{\beta\langle k\rangle t}}{1 - i_0 + i_0 e^{\beta\langle k\rangle t}}. \tag{10.4}$$

Equation **(10.4)** predicts that:

- At the beginning, the fraction of infected individuals increases exponentially (**Figure 10.4b**). Indeed, early on an infected individual encounters only susceptible individuals, hence the pathogen can easily spread.
- The *characteristic time* required to reach a $1/e$ fraction (about 36%) of all susceptible individuals is

$$\tau = \frac{1}{\beta\langle k\rangle}. \tag{10.5}$$

Hence, τ is the inverse of the speed with which the pathogen spreads through the population. Equation **(10.5)** predicts that increasing either the density of links $\langle k\rangle$ or β enhances the speed of the pathogen and reduces the characteristic time.

- With time, an infected individual encounters fewer and fewer susceptible individuals. Hence, the growth of i slows for large t (**Figure 10.4b**). The epidemic ends when everyone has been infected, that is when $i(t \to \infty) = 1$ and $s(t \to \infty) = 0$.

10.2.2 Susceptible–Infected–Susceptible (SIS) Model

Most pathogens are eventually defeated by the immune system or by treatment. To capture this fact we need to allow the infected individuals to recover, ceasing to spread the disease. With that we arrive at the so-called *SIS model*, which has the same two states as the SI model, susceptible and infected. The difference is that now infected individuals recover at a fixed rate μ, becoming susceptible again (**Figure 10.5a**). The equation describing the dynamics of this model is an extension of **(10.3)**,

$$\frac{di}{dt} = \beta\langle k\rangle i(1-i) - \mu i, \tag{10.6}$$

where μ is the *recovery rate* and the μi term captures the rate at which the population recovers from the disease. The solution of **(10.6)** provides the fraction of infected individuals as a function of time (**Figure 10.5b**)

$$i = \left(1 - \frac{\mu}{\beta\langle k\rangle}\right) \frac{Ce^{(\beta\langle k\rangle - \mu)t}}{1 + Ce^{(\beta\langle k\rangle - \mu)t}}, \tag{10.7}$$

where the initial condition $i_0 = i(t=0)$ gives $C = i_0/(1 - i_0 - \mu/\beta\langle k\rangle)$.

While in the SI model eventually everyone becomes infected, **(10.7)** predicts that in the SIS model the epidemic has two possible outcomes:

- **Endemic State** ($\mu < \beta\langle k\rangle$)
  For a low recovery rate the fraction of infected individuals, i, follows a logistic curve similar to that observed for the SI model. Yet, not everyone is infected, but i reaches a

(a)

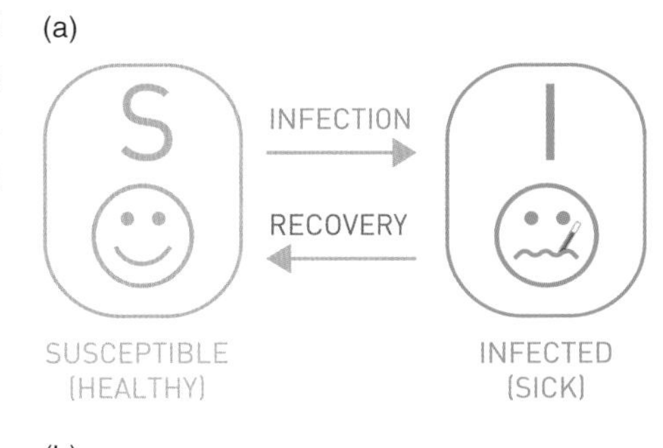

(b)

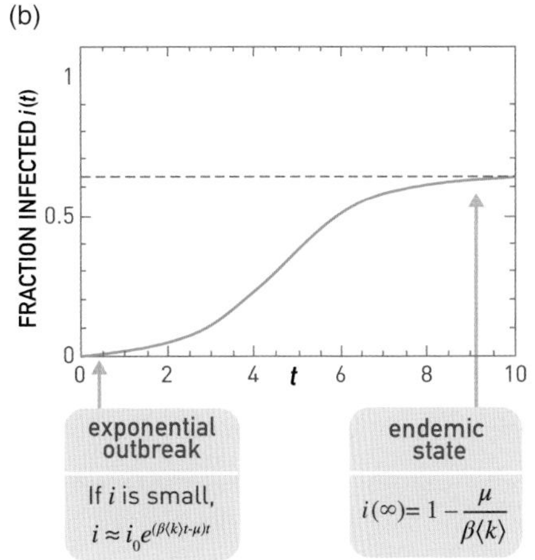

Figure 10.5 **The Susceptible–Infected–Susceptible (SIS) Model**

(a) The SIS model has the same states as the SI model: susceptible and infected. It differs from the SI model in that it allows recovery, that is infected individuals are cured, becoming susceptible again at rate μ.

(b) Time evolution of the fraction of infected individuals in the SIS model, as predicted by **(10.7)**. As recovery is possible, at large t the system reaches an *endemic state*, in which the fraction of infected individuals is constant, $i(\infty)$, given by **(10.8)**. Hence, in the endemic state only a finite fraction of individuals are infected. Note that for high recovery rate μ, the number of infected individuals decreases exponentially and the disease dies out.

constant $i(\infty) < 1$ value (**Figure 10.5b**). This means that at any moment only a finite fraction of the population is infected. In this stationary or *endemic state* the number of newly infected individuals equals the number of individuals who recover from the disease, hence the infected fraction of the population does not change with time. We can calculate $i(\infty)$ by setting $di/dt = 0$ in **(10.6)**, obtaining

$$i(\infty) = 1 - \frac{\mu}{\beta\langle k\rangle}. \qquad (10.8)$$

- **Disease-Free State** ($\mu > \beta\langle k\rangle$)
  For a sufficiently high recovery rate the exponent in **(10.7)** is negative. Therefore, *i decreases* exponentially with time, indicating that an initial infection will die out exponentially. This is because in this state the number of individuals cured per unit time exceeds the number of newly infected individuals. Therefore, with time the pathogen disappears from the population.

In other words, the SIS model predicts that some pathogens will persist in the population while others die out shortly. To understand what governs the difference between these two outcomes, we write the characteristic time of a pathogen as

$$\tau = \frac{1}{\mu(R_0 - 1)}, \qquad (10.9)$$

where

$$R_0 = \frac{\beta\langle k\rangle}{\mu} \qquad (10.10)$$

is the *basic reproductive number*. It represents the average number of susceptible individuals infected by an infected individual during its infectious period in a fully susceptible population. In other words, R_0 is the number of new infections each infected individual causes under ideal circumstances. The basic reproductive number is valuable for its predictive power:

Table 10.2 **The Basic Reproductive Number, R_0**

The reproductive number **(10.10)** provides the number of individuals an infectious individual infects if all its contacts are susceptible. For $R_0 < 1$ the pathogen naturally dies out, as the number of recovered individuals exceeds the number of new infections. If $R_0 > 1$ the pathogen will spread and persist in the population. The higher is R_0, the faster is the spreading process. The table lists R_0 for several well-known pathogens. After [354].

Disease	Transmission	R_0
Measles	Airborne	12–18
Pertussis	Airborne droplet	12–17
Diptheria	Saliva	6–7
Smallpox	Social contact	5–7
Polio	Fecal–oral route	5–7
Rubella	Airborne droplet	5–7
Mumps	Airborne droplet	4–7
HIV/AIDS	Sexual contact	2–5
SARS	Airborne droplet	2–5
Influenza (1918 strain)	Airborne droplet	2–3

- If $R_0 > 1$ then τ is positive, hence the epidemic is in the endemic state. Indeed, if each infected individual infects more than one healthy person, the pathogen is poised to spread and persist in the population. The higher is R_0, the faster is the spreading process.
- If $R_0 < 1$ then τ is negative and the epidemic dies out. Indeed, if each infected individual infects less than one additional person, the pathogen cannot persist in the population.

Consequently, the reproductive number is one of the first parameters epidemiologists estimate for a new pathogen, gauging the severity of the problem they face. For several well-studied pathogens, R_0 is listed in **Table 10.2**. The high R_0 of some of these pathogens underlies the dangers they pose: for example, each individual infected with measles causes over a dozen subsequent infections.

10.2.3 Susceptible–Infected–Recovered (SIR) Model

For many pathogens, like most strains of influenza, individuals develop immunity after they recover from the infection. Hence, instead of returning to the susceptible state, they are "removed" from the population. These recovered individuals do not count any longer from the perspective of the pathogen as they cannot be infected, nor can they infect others. The SIR model, whose properties are discussed in **Figure 10.6**, captures the dynamics of this process.

In summary, depending on the characteristics of a pathogen, we need different models to capture the dynamics of an epidemic outbreak. As shown in **Figure 10.7**, the predictions of the SI, SIS and SIR models agree with each other in the early stages of an epidemic: when the number of infected individuals is small, the disease spreads freely and the number of infected individuals increases exponentially. The outcomes are different for large times: in the SI model everyone becomes infected; the SIS model either reaches an endemic state, in which a finite fraction of individuals are

(a)

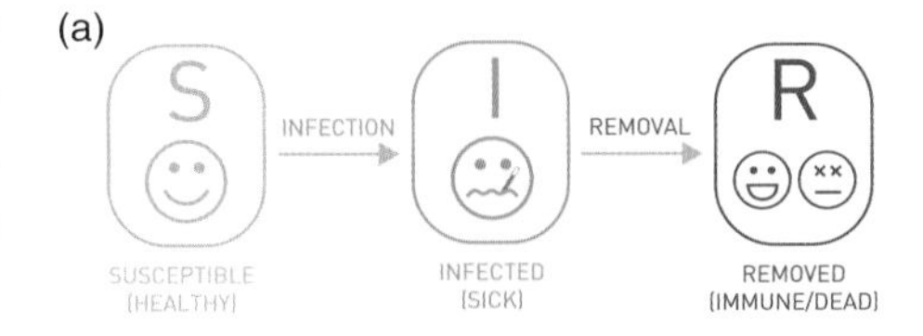

(b)

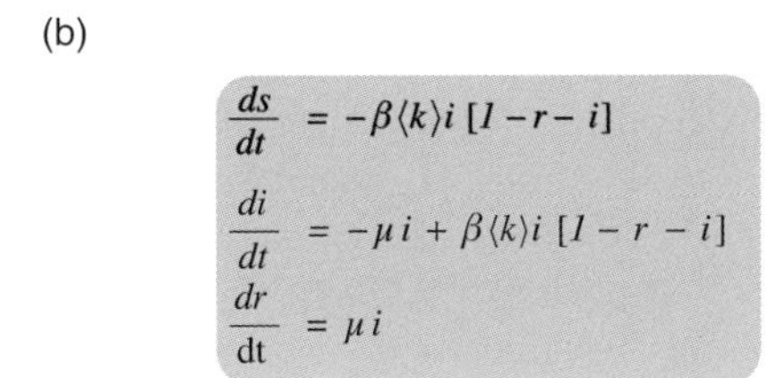

(c)

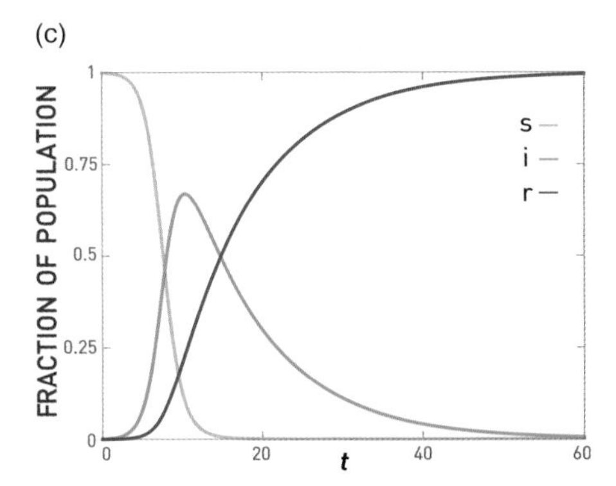

Figure 10.6 **The Susceptible–Infected–Recovered (SIR) Model**

(a) In contrast with the SIS model, in the SIR model recovered individuals enter a *recovered* state, meaning that they develop immunity rather than becoming susceptible again. Flu, SARS and the plague are diseases with this property, hence we must use the SIR model to describe their spread.

(b) The differential equations governing the time evolution of the fraction of individuals in the susceptible *s*, infected *i* and removed *r* state.

(c) The time-dependent behavior of *s*, *i* and *r* as predicted by the equations shown in (b). According to the model, all individuals transition from the susceptible (healthy) state to the infected (sick) state and then to the recovered (immune) state.

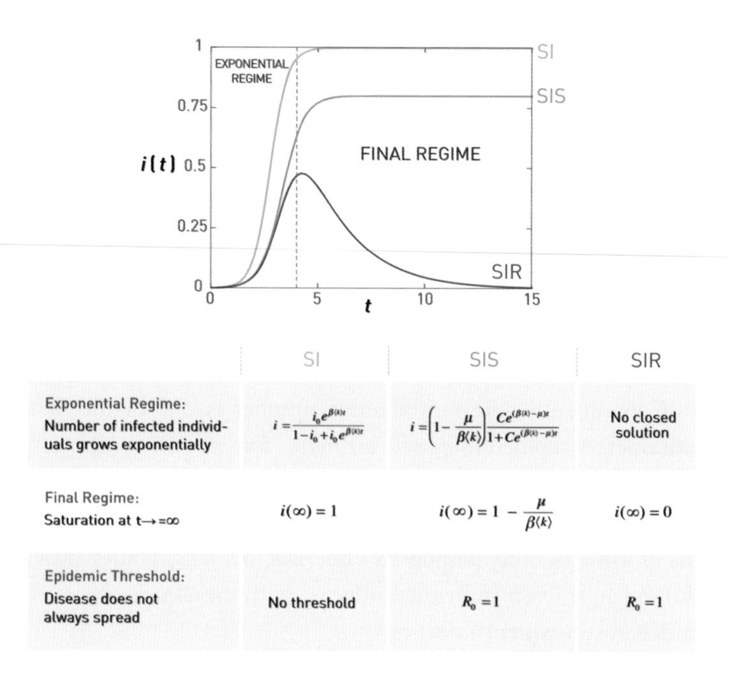

	SI	SIS	SIR
Exponential Regime: Number of infected individuals grows exponentially	$i = \frac{i_0 e^{\beta\langle k\rangle t}}{1 - i_0 + i_0 e^{\beta\langle k\rangle t}}$	$i = \left(1 - \frac{\mu}{\beta\langle k\rangle}\right)\frac{Ce^{(\beta\langle k\rangle-\mu)t}}{1+Ce^{(\beta\langle k\rangle-\mu)t}}$	No closed solution
Final Regime: Saturation at $t \to =\infty$	$i(\infty) = 1$	$i(\infty) = 1 - \frac{\mu}{\beta\langle k\rangle}$	$i(\infty) = 0$
Epidemic Threshold: Disease does not always spread	No threshold	$R_0 = 1$	$R_0 = 1$

Figure 10.7 **Comparing the SI, SIS and SIR Models**

The plot shows growth of the fraction of infected individuals, i, in the SI, SIS and SIR models. Two different regimes stand out:

Exponential Regime

The models predict an exponential growth in the number of infected individuals during the early stages of the epidemic. For the same β the SI model predicts the fastest growth (smallest τ, see **(10.5)**). For the SIS and SIR models the growth is slowed by recovery, resulting in a larger τ, as predicted by **(10.9)**. Note that for sufficiently high recovery rate μ the SIS and the SIR models predict a disease-free state, when the number of infected individuals *decays* exponentially with time.

Final Regime

The three models predict different long-term outcomes: in the SI model everyone becomes infected, $i(\infty) = 1$; in the SIS model a finite fraction of individuals are infected, $i(\infty) < 1$; in the SIR model all infected nodes recover, hence the number of infected individuals goes to zero, $i(\infty) = 0$.

The table summarizes the main properties of each model.

always infected, or the infection dies out; in the SIR model everyone recovers at the end. The reproductive number predicts the long-term fate of an epidemic: for $R_0 < 1$ the pathogen persists in the population, while for $R_0 > 1$ it dies out naturally.

The models discussed so far have ignored the fact that an individual comes into contact only with their network-based neighbors in the pertinent contact network. We assumed homogenous mixing instead, which means that an infected individual can infect any other individual. It also means that an infected individual typically infects only $\langle k\rangle$ other individuals, ignoring variations in node degree. To accurately predict the dynamics of an epidemic, we need to consider the precise role the contact network plays in epidemic phenomena.

10.3 Network Epidemics

The ease of air travel, allowing millions to cross continents on a daily basis, has dramatically accelerated the speed with which pathogens travel around the world. While in medieval times a virus took years to sweep a continent (**Figure 10.8**), today a

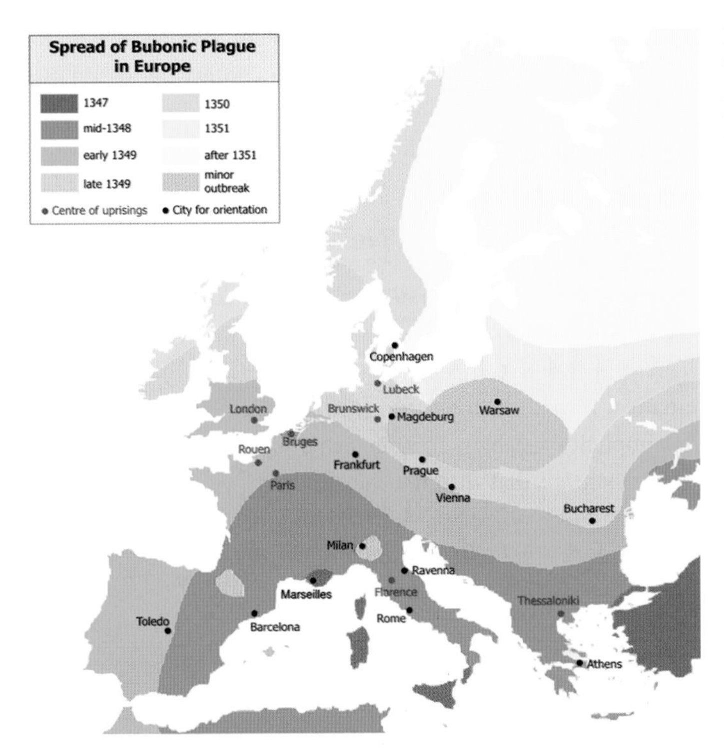

Figure 10.8 **The Great Plague**
The Black Death, one of the most devastating pandemics in human history, was an outbreak of bubonic plague caused by the bacterium *Yersinia pestis*. The figure shows the gradual advance of the disease throughout Europe, taking years to sweep the continent. It started in China and traveled along the Silk Road to reach Crimea in around 1346. From there, probably carried by Oriental rat fleas on the black rats that were regular passengers on merchant ships, it spread throughout the Mediterranean and Europe. Its slow spread reflected the slow travel speed of the era. The Black Death is estimated to have killed 30%–60% of Europe's population [355]. The resulting devastation caused a series of religious, social and economic upheavals, having a profound impact on the history of Europe.
After Roger Zenner, Wikipedia.

new virus can reach several continents in a matter of days. There is an acute need, therefore, to understand and predict the precise patterns that pathogens follow as they spread around the globe.

The epidemic models discussed in the previous section do not incorporate the structure of the contract network that facilitates the spread of a pathogen. Instead, they assume that any individual can come into contact with any other individual (homogenous mixing hypothesis) and that all individuals have comparable number of contacts, $\langle k \rangle$. Both assumptions are false: individuals can transmit a pathogen only to those they come into contact with, hence pathogens spread on a complex contact network. Furthermore, these contact networks are often scale-free, hence $\langle k \rangle$ is not sufficient to characterize their topology.

The failure of the basic hypotheses prompted a fundamental revision of the epidemic modeling framework. This change began with the work of Romualdo Pastor-Satorras and Alessandro Vespignani, who in 2001 extended the basic epidemic

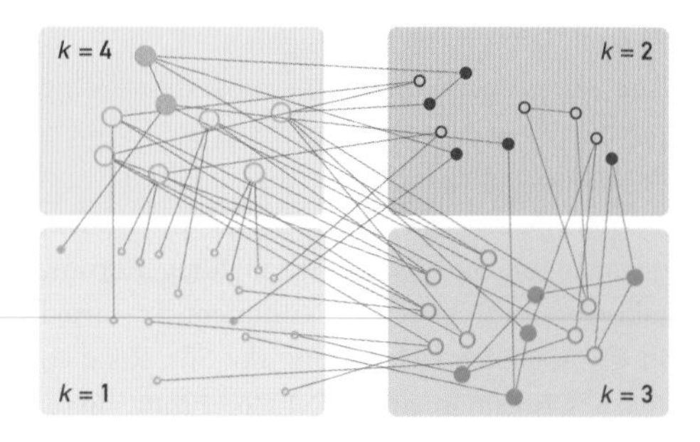

Figure 10.9 **Degree-Block Approximation**
The epidemic models discussed in **Section 10.2** grouped each node into compartments based on their state, placing them into susceptible, infected or recovered compartments. To account for the role of the network topology, the *degree-block approximation* adds an additional set of compartments, placing all *nodes that have the same degree into the same block.* In other words, we assume that nodes with the same degree behave similarly. This allows us to write a separate rate equation for each degree, as we did in **(10.13)**. The degree-block approximation does not eliminate the compartments based on the state of an individual: independent of degree, an individual can be susceptible to the disease (empty circles) or infected (full circles).

models to incorporate in a self-consistent fashion the topological characteristics of the underlying contact network [18]. In this section we introduce the formalism developed by them, familiarizing ourselves with *network epidemics.*

10.3.1 SI Model On A Network

If a pathogen spreads on a network, individuals with more links are more likely to be in contact with an infected individual, hence they are more likely to be infected. Therefore, the mathematical formalism must consider the degree of each node as an implicit variable. This is achieved by the *degree-block approximation* that distinguishes nodes based on their degree and assumes that nodes with the same degree are statistically equivalent (**Figure 10.9**). Therefore, we denote by

$$i_k = \frac{I_k}{N_k} \qquad (10.11)$$

the fraction of nodes with degree k that are infected among all N_k degree-k nodes in the network. The total fraction of infected nodes is the sum of all infected degree-k nodes

$$i = \sum_k p_k i_k. \qquad (10.12)$$

Given the different node degrees, we write the SI model for each degree k separately:

$$\frac{di_k}{dt} = \beta(1 - i_k)k\Theta_k. \qquad (10.13)$$

This equation has the same structure as **(10.3)**: the infection rate is proportional to β and the fraction of degree-k nodes that are not yet infected, which is $(1 - i_k)$. Yet, there are some key differences:

- The average degree $\langle k \rangle$ in **(10.3)** is replaced by each node's actual degree k.
- The density function Θ_k represents the fraction of infected neighbors of a susceptible node k. In the homogenous mixing assumption, Θ_k is simply the fraction of the infected nodes, i. In a network environment, however, the fraction of infected nodes in the vicinity of a node can depend on the node's degree k and time t.

- While **(10.3)** captures with a single equation the time-dependent behavior of the whole system, **(10.13)** represents a system of k_{max} coupled equations, one equation for each degree present in the network.

We start by exploring the early-time behavior of i_k, a choice driven by both theoretical interest and practical considerations. Indeed, developing vaccines, cures and other medical interventions for a new pathogen can take months to years. If we lack a cure, the only way to alter the course of an epidemic is to do so early, using quarantine, travel restrictions and transmission-slowing measures to halt its spread. To make the right decision about the nature, timing and magnitude of each intervention, we need an accurate estimate of the number of individuals infected in the early stages of the epidemic.

At the beginning of the epidemic i_k is small and the higher-order term in $\beta i_k k \Theta_k$ can be neglected. Hence, we can approximate **(10.13)** by

$$\frac{di_k}{dt} \approx \beta k \Theta_k. \tag{10.14}$$

As we show in **Advanced topics 10.B**, for a network lacking degree correlations the Θ_k function is independent of k, so using **(10.40)**, **(10.14)** becomes

$$\frac{di_k}{dt} \approx \beta k i_0 \frac{\langle k \rangle - 1}{\langle k \rangle} e^{t/\tau^{SI}}, \tag{10.15}$$

where τ^{SI} is the characteristic time for the spread of the pathogen

$$\tau^{SI} = \frac{\langle k \rangle}{\beta(\langle k^2 \rangle - \langle k \rangle)}. \tag{10.16}$$

Integrating **(10.15)** we obtain the fraction of infected nodes with degree k

$$i_k = i_0 \left(1 + \frac{k\langle k \rangle - 1}{\langle k^2 \rangle - \langle k \rangle} \left(e^{t/\tau^{SI}} - 1 \right) \right). \tag{10.17}$$

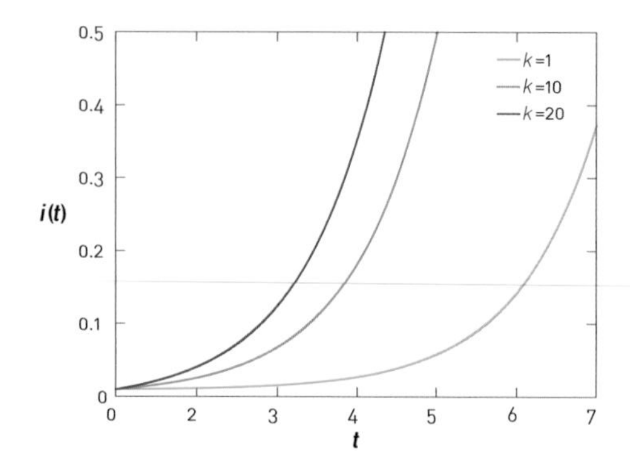

Figure 10.10 **Fraction of Infected Nodes in the SI Model**
Equation **(10.17)** predicts that a pathogen spreads with different speed on nodes with different degree. To be specific, we can write $i_k = g(t) + kf(t)$, indicating that at any time the fraction of high-degree nodes that are infected is higher than the fraction of low-degree nodes. The figure shows the fraction of infected nodes with degrees $k = 1$, 10 and 100 in an Erdős–Rényi network with average degree $\langle k \rangle = 2$. It shows that at $t = 3$ less than 3% of the $k = 1$ nodes are infected, in contrast with close to 20% of the $k = 10$ nodes and close to 30% of the $k = 20$ nodes. Consequently, at any time virtually all hubs are infected, but small-degree nodes tend to be disease free. Hence, the disease is maintained in the hubs, which in turn broadcast the disease to the rest of the network.

Equation **(10.17)** makes several important predictions:

- The higher the degree of a node, the more likely that it will become infected. Indeed, for any time t we can write **(10.17)** as $i_k = g(t) + kf(t)$, indicating that the group of nodes with higher degree has a higher fraction of infected nodes (**Figure 10.10**).
- According to **(10.12)**, the total fraction of infected nodes grows with time as

$$i = \int_0^{k_{\max}} i_k p_k dk = i_0 \left(1 + \frac{\langle k \rangle^2 - \langle k \rangle}{\langle k^2 \rangle - \langle k \rangle} \left(e^{t/\tau^{SI}} - 1 \right) \right). \qquad (10.18)$$

According to **(10.16)**, the characteristic time τ depends not only on $\langle k \rangle$, but also on the network's degree distribution through $\langle k^2 \rangle$. To fully understand the significance of the prediction **(10.16)**, let us derive τ^{SI} for different networks:

- **Random Network**
  For a random network $\langle k^2 \rangle = \langle k \rangle (\langle k \rangle + 1)$, obtaining

$$\tau_{ER}^{SI} = \frac{1}{\beta \langle k \rangle}, \qquad (10.19)$$

  recovering the result **(10.5)** for homogenous networks.
- **Scale-Free Network with $\gamma \geq 3$**
  If the contact network on which the disease spreads is scale-free with degree exponent $\gamma \geq 3$, both $\langle k \rangle$ and $\langle k^2 \rangle$ are finite. Consequently, τ^{SI} is also finite and the spreading dynamics is similar to the behavior predicted for a random network but with an altered τ^{SI}.
- **Scale-Free Networks with $\gamma \leq 3$**
  For $\gamma < 3$ in the $N \to \infty$ limit $\langle k^2 \rangle \to \infty$, hence **(10.16)** predicts $\tau^{SI} \to 0$. In other words, *the spread of a pathogen on a scale-free network is instantaneous*. This is perhaps the most unexpected prediction of network epidemics.

  The vanishing characteristic time reflects the important role hubs play in epidemic phenomena. Indeed, as illustrated in **Figure 10.10**, in a scale-free network the hubs are the first to be infected, as through the many links they have, they are very likely to be in contact with an infected node. Once a hub becomes infected, it "broadcasts" the disease to the rest of the network, turning into a super-spreader.

- **Inhomogenous Networks**
  A network does not need to be strictly scale-free for the impact of the degree heterogeneity to be detectable. Indeed, **(10.16)** predicts that as long as $\langle k^2 \rangle > \langle k \rangle (\langle k \rangle + 1)$, τ^{SI} is reduced. Hence, heterogenous networks enhance the speed of any pathogen.

In the SI model, with time, the pathogen reaches all individuals. Consequently, the degree heterogeneity affects only the characteristic time, which in turn determines the speed with which the pathogen sweeps through the population. To understand the full impact of the network topology, we need to explore the behavior of the SIS model on a network.

10.3.2 SIS Model and The Vanishing Epidemic Threshold

The continuum equation describing the dynamics of the SIS model on a network is a straightforward extension of the SI model discussed in **Section 10.2**,

$$\frac{di_k}{dt} = \beta(1 - i_k)k\Theta_k(t) - \mu i_k. \tag{10.20}$$

The difference between **(10.13)** and **(10.20)** is the presence of the recovery term $-\mu i_k$.This changes the characteristic time of the epidemic to (**Advanced topics 10.B**)

$$\tau^{SIS} = \frac{\langle k \rangle}{\beta \langle k^2 \rangle - \mu \langle k \rangle}. \tag{10.21}$$

For sufficiently large μ the characteristic time is negative, hence i_k decays exponentially. The condition for the decay depends not only on the recovery rate and $\langle k \rangle$, but also on the network heterogeneity, through $\langle k^2 \rangle$. To predict when a pathogen persists in the population, we define the *spreading rate*

$$\lambda = \frac{\beta}{\mu}, \tag{10.22}$$

which depends only on the biological characteristics of the pathogen, namely the transmission probability β and the recovery rate μ. The higher is λ, the more likely that the disease will spread. Yet, the number of infected individuals does not increase gradually with λ. Rather, the pathogen can spread only

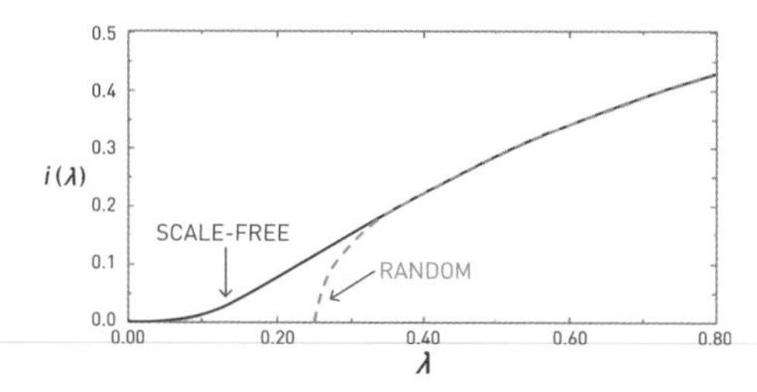

Figure 10.11 **Epidemic Threshold**
The fraction of infected individuals $i(\lambda) = i(t \to \infty)$ in the endemic state of the SIS model. The curves are for a random (green) and for a scale-free contact network (purple). The random network has a finite epidemic threshold λ_c, implying that a pathogen with a small spreading rate ($\lambda < \lambda_c$) must die out, i.e. $i(\lambda_c) = 0$. If, however, the spreading rate of the pathogen exceeds λ_c, the pathogen becomes endemic and a finite fraction of the population is infected at any time. For a scale-free network we have $\lambda_c = 0$, hence even viruses with a very small spreading rate λ can persist in the population.

if its spreading rate exceeds an *epidemic threshold* λ_c. Next we calculate λ_c for random and scale-free networks.

- **Random Network**
  If a pathogen spreads on a random network, we can use $\langle k^2 \rangle > \langle k \rangle (\langle k \rangle + 1)$ in **(10.21)**, obtaining that the pathogen persists in the population if

$$\tau^{SI}_{ER} = \frac{1}{\beta(\langle k \rangle + 1) - \mu} > 0. \tag{10.23}$$

Using **(10.22)** we obtain

$$\lambda > \frac{1}{\langle k \rangle + 1}, \tag{10.24}$$

obtaining the *epidemic threshold of a random network* as

$$\lambda_c = \frac{1}{\langle k \rangle + 1}, \tag{10.25}$$

As $\langle k \rangle$ is always finite, a random network always has a nonzero epidemic threshold (**Figure 10.11**), with key consequences:

- If the spreading rate λ exceeds the epidemic threshold λ_c, the pathogen will spread until it reaches an endemic state, where a finite fraction $i(\lambda)$ of the population is infected at any time.
- If $\lambda < \lambda_c$ the pathogen dies out, that is $i(\lambda) = 0$.
- Hence, the epidemic threshold allows us to decide if a pathogen can or cannot persist in a population. This transition from the absence to the presence of an epidemic outbreak by increasing the spreading rate λ is at the basis of most campaigns to stop a pathogen (**Section 10.6**).

- **Scale-Free Network**
  For a network with an arbitrary degree distribution we set $\tau^{SIS} > 0$ in **(10.21)**, obtaining the epidemic threshold as

$$\lambda_c = \frac{\langle k \rangle}{\langle k^2 \rangle}. \tag{10.26}$$

As for a scale-free network $\langle k^2 \rangle$ diverges in the $N \to \infty$ limit, for large networks the epidemic threshold is expected to vanish (**Figures 10.11 and 10.12**). This means that *even viruses that are hard to pass from individual to individual can spread successfully*, representing the second fundamental prediction of network epidemics (**Table 10.3**).

Table 10.3 **Epidemic Models on Networks**

The table shows the rate equation for the three basic epidemic models (SI, SIS, SIR) on a network with arbitrary $\langle k\rangle$ and $\langle k^2\rangle$, together with the corresponding characteristic τ and the epidemic threshold λ_c. For the SI model $\lambda_c = 0$, as in the absence of recovery ($\mu = 0$) a pathogen spreads until it reaches all susceptible individuals. The listed τ and λ_c are derived in **Advanced topics 10.B**.

Model	Continuum equation	τ	λ_c
SI	$\frac{di_k}{dt} = \beta[1 - i_k]k\Theta_k$	$\frac{\langle k\rangle}{\beta(\langle k^2\rangle - \langle k\rangle)}$	0
SIS	$\frac{di_k}{dt} = \beta[1 - i_k]k\Theta_k - \mu i_k$	$\frac{\langle k\rangle}{\beta(\langle k^2\rangle - \mu\langle k\rangle)}$	$\frac{\langle k\rangle}{\langle k^2\rangle}$
SIR	$\frac{di_k}{dt} = \beta s_k\Theta_k - \mu i_k$; $s_k = 1 - i_k - r_k$	$\frac{\langle k\rangle}{\beta\left(\langle k^2\rangle - (\mu+\beta)\langle k\rangle\right)}$	$\frac{1}{\frac{\langle k^2\rangle}{\langle k\rangle} - 1}$

$2 < \gamma < 3$: $\lambda_c = 0$; $\Theta(\lambda) \sim (k_{min}\lambda)^{(\gamma-2)/(3-\gamma)}$; $i(\lambda) \sim \lambda^{1/(3-\gamma)}$

$\gamma = 3$: $\lambda_c = 0$; $\Theta(\lambda) \approx \frac{e^{-1/k_{min}\lambda}}{\lambda k_{min}}(1 - e^{-1/k_{min}\lambda})^{-1}$; $i(\lambda) \sim 2e^{-1/k_{min}\lambda}$

$3 < \gamma < 4$: $\lambda_c > 0$; $i(\lambda) \sim \left(\lambda - \frac{\gamma - 3}{k_{min}(\gamma - 2)}\right)^{1/(\gamma-3)}$

$\gamma > 4$: $\lambda_c > 0$; $i(\lambda) \sim \lambda - \frac{\gamma - 3}{k_{min}(\gamma - 2)}$

Figure 10.12 **The Asymptotic Behavior of the SIS Model**

The fraction of individuals infected in the endemic state, $i(\lambda) = i(t \to \infty)$, depends on the structure of the underlying network and the disease parameters β and μ. The figure summarizes the key properties of the epidemic threshold λ, the density function $\Theta(\lambda)$ and $i(\lambda)$ for a scale-free network with degree exponent γ. The results indicate that only for $\gamma > 4$ does the epidemic on a scale-free network converge to the results of the traditional epidemic models. After [356].

The vanishing epidemic threshold is a direct consequence of the hubs. Indeed, a pathogen that fails to infect other nodes before the infected individual recovers will slowly disappear from the population (**Advanced topics 10.A**). In a random network all nodes have comparable degree, $k \approx \langle k\rangle$, hence if the spreading rate is under the epidemic threshold, the pathogen has no avenues to spread. In a scale-free network, however, even if a pathogen is only weakly infectious, if it infects a hub, the hub can pass it on to a large number of other nodes, allowing it to persist in the population.

In summary, the results of this section show that accounting for the network topology greatly alters the predictive power of the epidemic models. We derived two fundamental results:

- In a large scale-free network $\tau = 0$, which means that a virus can instantaneously reach most nodes.
- In a large scale-free network $\lambda_c = 0$, which means that even viruses with small spreading rate can persist in the population.

Both results are the consequence of hubs' ability to broadcast a pathogen to a large number of other nodes.

Note that these results are not limited to scale-free networks. Rather, **(10.26)** predicts that both τ and λ depend on $\langle k^2\rangle$, hence the effects discussed above will impact any network with high-degree heterogeneity. In other words, if $\langle k^2 \rangle$ is larger than the random expectation $\langle k \rangle(\langle k \rangle + 1)$, we will observe an enhanced spreading process, resulting in a smaller τ and λ than predicted by the traditional epidemic models. As this implies a faster spread of the pathogen than predicted by the traditional epidemic models, efforts to control an epidemic cannot ignore this difference.

The results of this section were based on the degree-block approximation, which treats the detailed time-dependent infection process in a mean-field fashion. Note, however, that this approximation, while simplifies the presentation, is not necessary. The underlying stochastic problem can be treated in its full mathematical complexity [357, 358, 359, 360]. Such calculations show that due to the fact that the hubs can be reinfected in the SIS model, the epidemic threshold vanishes even for $\gamma > 3$, in contrast with the finite threshold predicted by the mean-field approach (**Figure 10.12**). Hence, hubs play an even more important role than our earlier calculations indicate.

10.4 Contact Networks

Network epidemics predict that the speed with which a pathogen spreads depends on the degree distribution of the relevant contact network. Indeed, we found that $\langle k^2 \rangle$ affects both the characteristic time τ and the epidemic threshold λ_c. None of these findings are consequential if the network on which a pathogen spreads is random – in that case the predictions of network epidemics are indistinguishable from the predictions of the traditional epidemic models encountered in **Section 10.2**. In this section we inspect the structure of several contact networks encountered in epidemic phenomena, offering direct empirical evidence of the significance of the underlying degree heterogeneities.

10.4.1 Sexually Transmitted Diseases

The HIV virus, the pathogen responsible for AIDS, spreads mainly through sexual intercourse. Consequently, the relevant contact network captures who had a sexual relationship with whom. The structure of this sex web was first revealed by a study surveying the sexual habits of the Swedish population [361]. Through interviews and questionnaires, researchers collected information from 4,781 randomly chosen Swedes aged 18 to 74. The participants were not asked to reveal the identity of their sexual partners, but only to estimate the number of sexual partners they had had during their lifetime. Hence, the researchers could reconstruct the degree distribution of the sexual network [109], finding that it is well approximated by a power law (**Figure 10.13**). This was the first empirical evidence of the relevance of scale-free networks to the spread of pathogens. The finding was confirmed by data collected in the UK, USA and Africa [362].

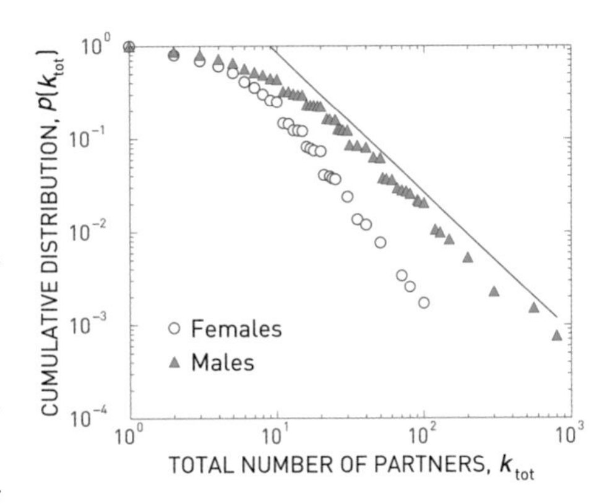

Figure 10.13 **The Sex Web**
Cumulative distribution of the total number of sexual partners k since sexual initiation for individuals interviewed in the 1996 study on sexual patterns in Sweden [361]. For women a power-law fit to the tail indicates $\gamma = 3.1 \pm 0.3$ for $k > 20$; for men $\gamma = 2.6 \pm 0.3$ in the range $20 < k < 400$. Note that for men the average number of partners is higher than for women. This difference may be rooted in social bias, prompting males to exaggerate and females to suppress the number of sexual partners they report.
After [109].

The scale-free nature of the sexual network indicates that most individuals have relatively few sexual partners. A few individuals, however, have had hundreds of sexual partners during their lifetime (**Box 10.1**). Consequently, the sexual network has a high $\langle k^2 \rangle$, which lowers both τ and λ.

10.4.2 Airborne Diseases

For airborne diseases, like influenza, SARS or H1N1, the contact network captures the set of individuals a person comes into physical proximity with.

The structure of this contact network is explored at two levels. First, the global travel network allows us to predict the worldwide spread of a pathogen, representing the input of several large-scale epidemic prediction tools (**Section 10.7**). Second, digital badges probe the local properties of the contact network, that is the number of individuals a person interacts with directly.

Global Travel Network

To predict the spread of pathogens, we must know how far infected individuals travel. Our understanding of individual travel patterns exploded with the use of mobile phones, that offer direct information about individual mobility [366, 367, 368, 369]. In the context of epidemic phenomena, the most studied mobility data comes from air travel, the mode of transportation that determines the speed with which a

Box 10.1 Sexual Hubs

Anecdotal evidence suggests that sexual hubs are real. Take for example Wilt Chamberlain, a Hall of Fame basketball player in the 1980s, who claimed to have had sex with a staggering 20,000 partners. "Yes, that's correct, twenty thousand different ladies," he wrote in his autobiography [363]. "At my age, that equals to having sex with 1.2 woman a day, every day, since I was fifteen years old." Within the AIDS literature the story of Geetan Dugas, a flight attendant with approximately 250 homosexual partners, is well documented [364]. He is often called *Patient Zero*, who, given his extensive travel, became a super-spreader of AIDS within the gay community. Hubs are observed even in high-school romantic networks (**Figure 10.14**).

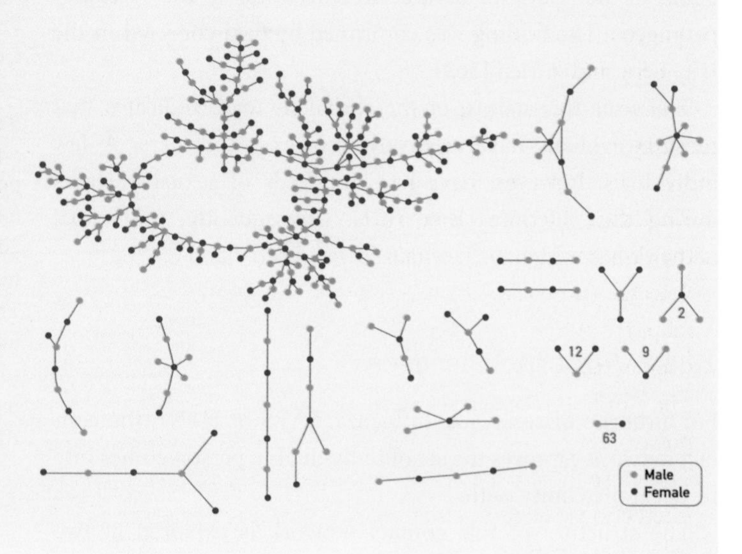

Figure 10.14 **Romantic Links in High School**
Romantic and sexual links between high-school students in midwestern United States. Each circle represents a student and the links represent romantic relationships during six months preceding the interview. The numbers indicate the frequency of each subgraph: there are 63 couples isolated from the rest of the network. After [365].

pathogen moves around the globe. Consequently the *air transportation network*, that connects airports with direct flights, plays a key role in modeling and predicting the spread of pathogens [370, 34, 371]. As **Figure 10.15** shows, this network is scale-free with degree exponent $\gamma = 1.8$. This low value is possible because there are multiple flights between two airports, hence the network is not simple. A similar power law distribution is detected for the link weights, indicating that the number of passengers traveling between two airports is typically low, but between some airports the traffic can be extraordinary. As we discuss in **Section 10.5**, these heterogeneities play a key role in the spread of specific pathogens.

Local Contact Patterns

Many airborne diseases spread thanks to face-to-face interactions [372, 373, 374, 375]. These interaction patterns can be monitored using radio-frequency identification devices (RFID) [373, 375], mobile phone-based sociometric badges [376] and other wireless technologies [377].

RFID are digital badges that detect the proximity of other individuals that wear a badge (**Online resource 10.1**). They have been deployed in various environments, capturing for example the interactions between more than 14,000 visitors to a science gallery over a three-month period or between 100 participants in a three-day conference [373]. An RFID-mapped network shown in **Figure 10.16** captures the interactions between high-school students and their teachers during a two-day period. Several findings stand out:

- RFID tags detect interactions only with individuals that wear the same badge and face each other, limiting the number of detected contacts. Consequently, the contact networks mapped out in these studies typically have an exponential degree distribution.
- The duration of each face-to-face interaction follows a power-law distribution over several orders of magnitude. Therefore most contacts are brief, but there are a few lasting interactions, documenting a bursty temporal pattern [378] with key consequences for the spread of pathogens (**Section 10.5**).
- The link weights, which capture the *cumulative time* two individuals have spent together, also follow a power-law distribution. Therefore, individuals spend most of their time with only a few others, again with important implications for spreading patterns (**Section 10.5**).
- For most airborne pathogens spatial proximity is sufficient for transmission. For example, standing next to an infected individual in the elevator may be sufficient to transmit SARS or H1N1, an interaction not recorded by an RFID tag.

In summary, RFID tags provide remarkably detailed temporal and spatial information about local contacts. To be useful these studies must be scaled up using, for example, mobile phone-based technologies [379].

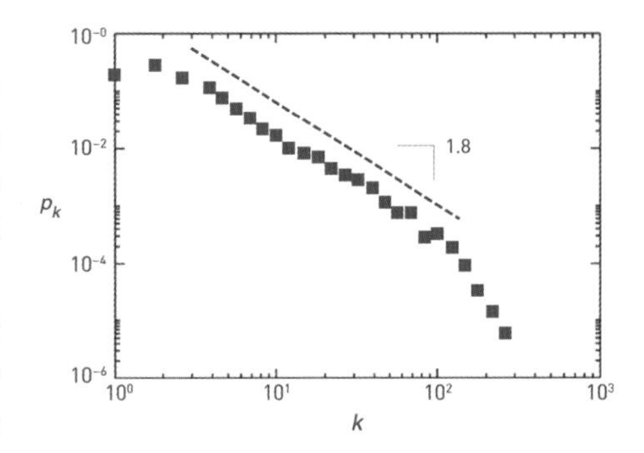

Figure 10.15 **Air Transportation Network**
The degree distribution of the air transportation network is well approximated by a power law with $\gamma = 1.8 \pm 0.2$. The map was built using the International Air Transport Association database, which contains the world list of airports and the direct flights between them in 2002. The resulting network is a weighted graph containing the $N = 3{,}100$ largest airports as nodes that are connected by $L = 17{,}182$ direct flights as links, together accounting for 99% of worldwide traffic.
After [370].

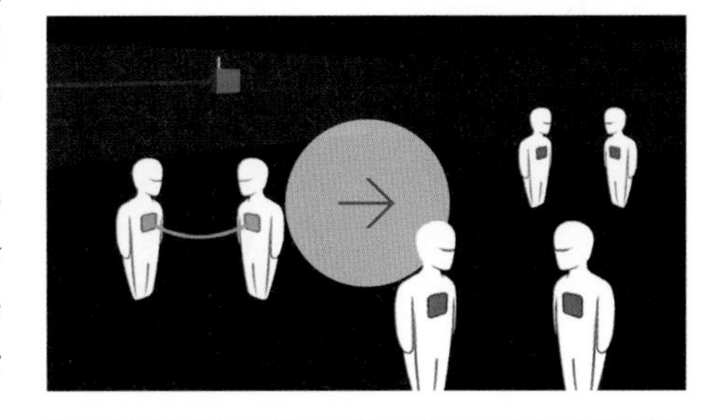

Online Resource 10.1
Detecting Networks via RFID
A video introducing RFID technology and its use in mapping social interactions.

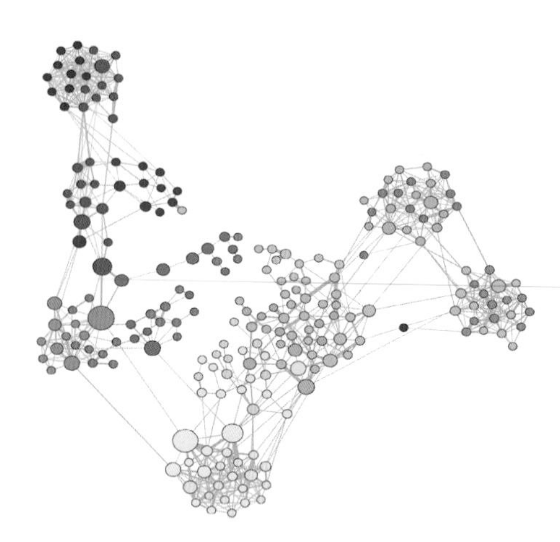

Figure 10.16 **Face-to-Face Interactions**
A face-to-face contact network mapped out using RFA tags, capturing interactions between 232 students and ten teachers across ten classes in a school [375]. The structure of the maps obtained by RFID tags depends on the context in which they are collected. For example, the school network shown here reveals the presence of clear communities. In contrast, a study capturing the interactions between individuals who visited a museum reveals an almost linear network [373]. Finally, a network of attendees at a small conference is rather dense, as most participants interact with most others [373]. After [375].

10.4.3 Location Networks

For many airborne pathogens the relevant contact network is the so-called *location network*, whose nodes are the locations that are connected by individuals who move regularly between them. Measurements combined with agent-based simulations indicate that the location network is fat tailed [380]: malls, airports, schools or supermarkets act as hubs, being linked to an exceptionally large number of smaller locations, like homes and offices. Therefore, once the pathogen infects a hub, the disease can rapidly reach many other locations.

10.4.4 Digital Viruses

The study of digital viruses, that infect computers and smartphones, represents an increasingly important application of epidemic phenomena. As we discuss next, the relevent contact networks are determined by the spreading mode of the respective digital pathogen.

Computer Viruses

Computer viruses display just as much diversity as biological viruses: depending on the nature of the virus and its spreading mechanism, the relevant contact network can differ dramatically. Many computer viruses spread as email attachments. Once a user opens the attachment, the virus infects the user's computer and mails a copy of itself to the email addresses found in the computer. Hence, the pertinent contact network is the email network, which, as we discussed in **Table 4.1**, is scale-free [113]. Other computer viruses exploit various communication protocols, spreading on networks that reflect the Internet's pattern of interconnectedness, which is again scale free (**Table 4.1**). Finally, some malware scan IP addresses, spreading on fully connected networks.

Mobile Phone Viruses

Mobile phone viruses spread via MMS and Bluetooth (**Figure 10.2**). An MMS virus sends a copy of itself to all phone numbers found in the phone's contact list. Therefore, MMS viruses exploit the social network behind mobile communications. As shown in **Table 4.1**, the mobile call network is scale-free with a high-degree exponent. Mobile viruses can also spread via Bluetooth, passing a copy of themselves to all susceptible phones with a BT connection in their physical proximity. As discussed above, this colocation network is also highly heterogenous [35].

In summary, in the past decade technological advances allowed us to map out the structure of several networks that support the spread of biological or digital viruses, from sexual to proximity-based contact networks (see also **Online resource 10.2**). Many of these, like the email network, the Internet or sexual networks, are scale-free. For others, like colocation networks, the degree distribution may not be fitted with a simple power law, yet show significant degree heterogeneity with high $\langle k^2 \rangle$. This means that the analytical results obtained in the previous section are of direct relevance to pathogens spreading on most networks. Consequently, the underlying heterogenous contact networks allow even weakly virulent viruses to easily spread in the population.

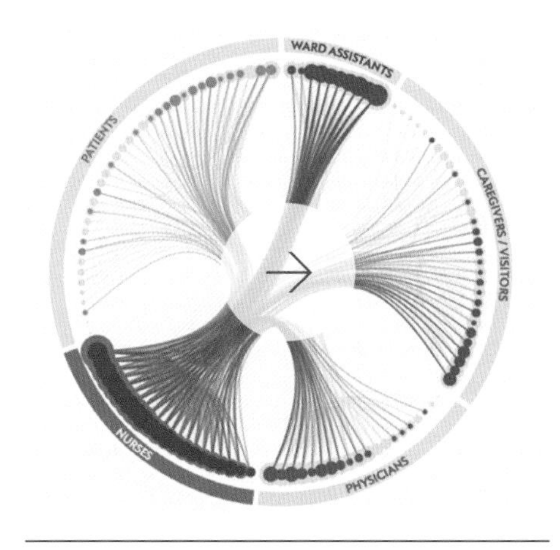

Online Resource 10.2

Hospital Outbreaks

Bacteria resistant to current antibiotics pose an important threat to global health. Such bacteria are particularly prevalent in hospitals and health care facilities. The Interactive Feature by *Scientific American* describes the tracking of bacterical outbreaks in hospitals.

10.5 Beyond the Degree Distribution

So far we have kept our models simple: we assumed that pathogens spread on an unweighted network uniquely defined by its degree distribution. Yet, real networks have a number of characteristics that are not captured by p_k alone, like degree correlations or community structure. Furthermore, the links are typically weighted and the interactions have a finite temporal duration. In this section we explore the impact of these properties on the spread of a pathogen.

10.5.1 Temporal Networks

Most interactions that we perceive as social links are brief and infrequent. As a pathogen can only be transmitted when there is an actual contact, an accurate modeling framework must also consider the timing and the duration of each interaction. Ignoring the timing of the interactions can lead to misleading conclusions [381, 382, 383]. For

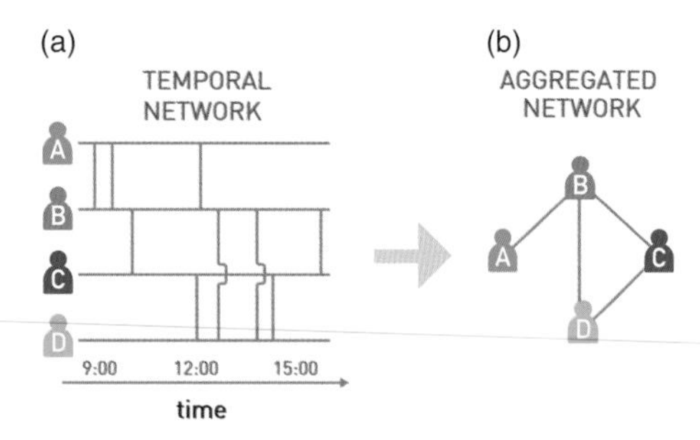

Figure 10.17 **Temporal Networks**
Most interactions in a network are not continuous, but have a finite duration. We must therefore view the underlying networks as *temporal networks*, an increasingly active research topic in network science.
(a) **Temporal Network**
The time line of the interactions between four individuals. Each vertical line marks the moment when two individuals come into contact with each other. If A is the first to be infected, the pathogen can spread from A to B and then to C, eventually reaching D. If, however, D is the first to be infected, the disease can reach C and B, but not A. This is because there is a temporal path from A to D.
(b) **Aggregated Network**
The network obtained by merging the temporal interactions shown in (a). If we only have access to this aggregated representation, the pathogen can reach all individuals, independent of its starting point.
After [382].

example, the static network of **Figure 10.17b** was obtained by aggregating the individual interactions shown in **Figure 10.17a**. On the aggregated network the infection has the same chance of spreading from D to A as from A to D. Yet, by inspecting the timing of each interaction, we realize that while an infection starting from A can infect D, an infection that starts at D cannot reach A. Therefore, to accurately predict an epidemic process we must consider the fact that pathogens spread on *temporal networks*, a topic of increasing interest in network science [382, 383, 384, 385]. By ignoring the temporality of these contact patterns, we typically overestimate the speed and the extent of an outbreak [384, 385].

10.5.2 Bursty Contact Patterns

The theoretical approaches discussed in **Sections 10.2 and 10.3** assume that the timing of the interactions between two connected nodes is random. This means that the inter-event times between consecutive contacts follow an exponential distribution, resulting in a random but uniform sequence of events (**Figure 10.18a–c**). The measurements indicate otherwise: the inter-event times in most social systems follow a power-law distribution [378, 386] (**Figure 10.18d–f**). This means that the sequence of contacts between two individuals is characterized by periods of frequent interactions, when multiple contacts follow each other within a relatively short time frame. Yet, the power law also implies that occasionally there are very long time gaps between two contacts. Therefore, the contact patterns have an uneven, "bursty" character in time (**Figure 10.18d,e**).

Bursty interactions are observed in a number of contact processes of relevance for epidemic phenomena, from email communications to call patterns and sexual contacts. Once present, burstiness alters the dynamics of the spreading process [385]. To be specific, power-law inter-event times increase the characteristic time τ, consequently the number of infected individuals decays slower than predicted by a random contact pattern. For example, if the time between consecutive emails follows a Poisson distribution, an email virus would decay following $i(t) \sim \exp(-t/\tau)$ with a decay time of $\tau \approx 1$ day. In

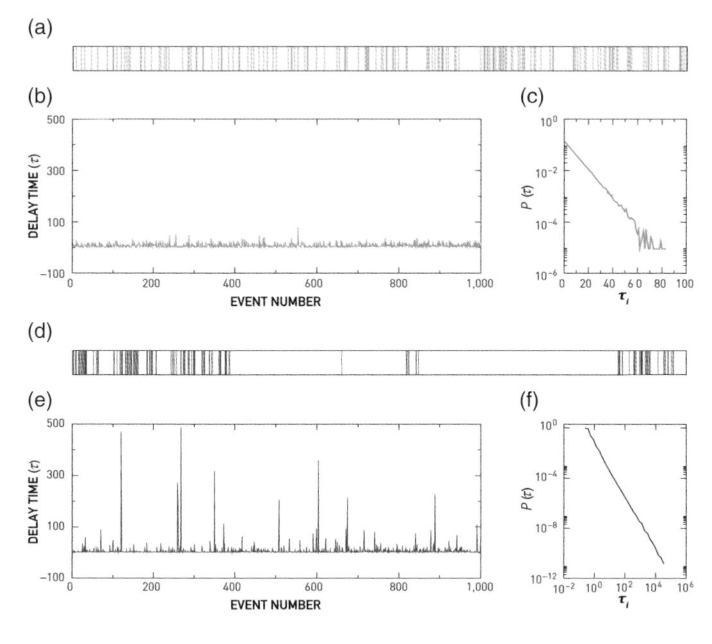

Figure 10.18 **Bursty Interactions**

(a) If the pattern of activity of an individual is random, the inter-event times follow a Poisson process, which assumes that at any moment an event takes place with the same probability q. The horizontal axis denotes time and each vertical line corresponds to an event whose timing is chosen at random. The observed inter-event times are comparable to each other and very long delays are rare.

(b) The absence of long delays is visible if we show the inter-event times τ for 1,000 consecutive random events. The height of each vertical line corresponds to the gaps seen in (a).

(c) The probability of finding exactly n events within a fixed time interval follows the Poisson distribution $P(n,q) \sim e^{-qt}(qt)^n/n!$, predicting that the inter-event time distribution follows $P(\tau_i) \sim e^{-q\tau_i}$, shown on a log-linear plot.

(d) The succession of events for a temporal pattern whose inter-event times follow a power law distribution. While most events follow each other closely, forming bursts of activity, there are a few exceptionally long inter-event times, corresponding to long gaps in the contact pattern. The time sequence is not as uniform as in (a), but has a bursty character.

(e) The waiting time τ_i of 1,000 consecutive events, where the mean event time is chosen to coincide with the mean event time of the Poisson process shown in (b). The large spikes correspond to exceptionally long delays.

(f) The delay time distribution $P(\tau_i) \sim \tau_i$ for the bursty process shown in (d) and (e).

After [378].

the real data, however, the decay time is $\tau \approx 21$ days, a much slower process, correctly predicted by the theory if we use power-law inter-event times [385].

10.5.3 Degree Correlations

As discussed in **Chapter 7**, many social networks are assortative, implying that high-degree nodes tend to connect to other high-degree nodes. Do these degree correlations affect the spread of a pathogen? The calculations indicate that degree correlations leave key aspects of network epidemics in place, but they alter the speed with which a pathogen spreads in a network:

- Degree correlations alter the epidemic threshold λ_c: assortative correlations decrease λ_c and dissasortative correlations increase it [387, 388].
- Despite the changes in λ_c, for the SIS model the epidemic threshold vanishes for a scale-free network with diverging second moment, whether the network is assortative, neutral or disassortative. Hence the fundamental results of **Section 10.3** are not affected by degree correlations.

- Given that hubs are the first to be infected in a network, assortativity accelerates the spread of a pathogen. In contrast, disassortativity slows the spreading process.
- Finally, in the SIR model assortative correlations were found to lower the prevalence but increase the average lifetime of an epidemic outbreak [389].

10.5.4 Link Weights and Communities

Throughout this chapter we have assumed that all link weights are equal, focusing our attention on pathogens spreading on an unweighted network. In reality, link weights vary considerably, a heterogeneity that plays an important role in spreading phenomena. Indeed, the more time an individual spends with an infected individual, the more likely that she too becomes infected.

In the same vein, previously we ignored the community structure of the network on which the pathogen spreads. Yet, the existence of communities (**Chapter 9**) lead to repeated interactions between the nodes within the same community, altering the spreading dynamics.

The mobile phone network allows us to explore the role of tie strengths and communities on spreading phenomena [115]. Let us assume that at $t = 0$ we provide a randomly selected individual with some key information. At each time step this "infected" individual i passes the information to her contact j with probability $p_{ij} \sim \beta w_{ij}$, where β is the spreading probability and w_{ij} is the strength of the ties captured by the number of minutes i and j have spent with each other on the phone. Indeed, the more time two individuals talk, the higher is the chance that they will pass on the information. To understand the role of the link weights in the spreading process, we also consider the situation when the spreading takes place on a *control network* that has the same wiring diagram but all tie strengths set equal to $w = \langle w_{ij} \rangle$.

As **Figure 10.19a** illustrates, information travels significantly faster on the control network. The reduced speed observed in the real system indicates that the information is trapped within communities. Indeed, as we discussed in **Chapter 9**, strong ties tend to be within communities while weak ties are between them [24]. Therefore, once the information reaches a member of a community, it can rapidly reach all other members of the same community, given the strong ties between them. Yet, as the ties between the communities are weak, the information has difficulty escaping the community. Consequently, the rapid invasion of the community is followed by long intervals during which the infection is trapped within a community. When all link weights are equal (control), the bridges between communities are strengthened and the trapping vanishes.

The difference between the real and the control spreading process is illustrated in **Figure 10.19b,c**, that shows the spreading pattern in a small neighborhood of the mobile call network. In the control simulation the information tends to follow the shortest path. When the link weights are taken into account, information flows along a longer backbone with strong ties. For example, the information rarely reaches the lower half of the network in **Figure 10.20b**, a region always reached in the control simulation shown in (c).

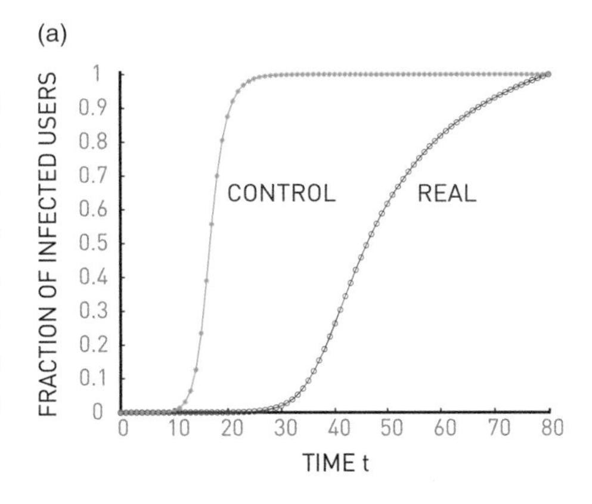

10.5.5 Complex Contagion

Communities have multiple consequences of spreading, from inducing global cascades [390, 391] to altering the activity of individuals [392].

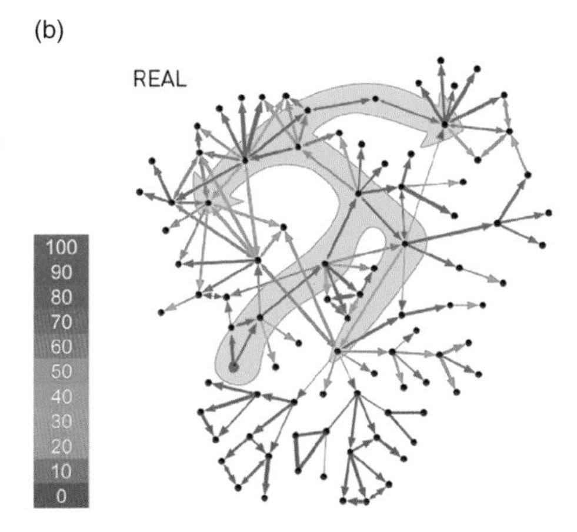

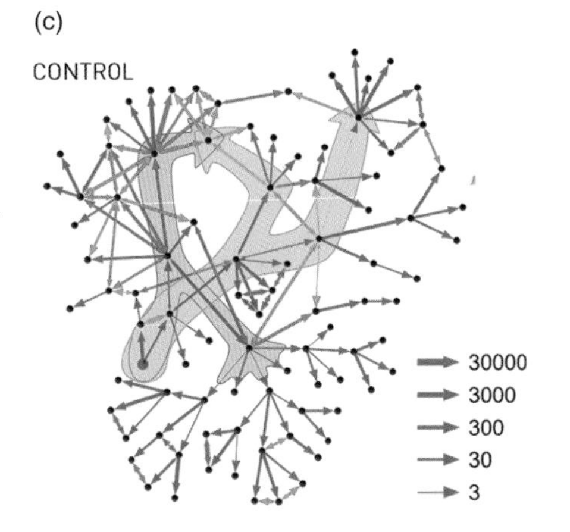

Figure 10.19 **Information Diffusion in Mobile Phone Networks**
The spread of information on a weighted mobile call graph, where the probability that a node passes information to one of its neighbors is proportional to the strength of the tie between them. The tie strength is the number of minutes two individuals talk on the phone.
(a) The fraction of infected nodes as a function of time. The blue circles capture the spread on the network with the real tie strengths; the green symbols represent the control case, when all tie strengths are equal.
(b) Spreading in a small network neighborhood, following the real link weights. The information is released from the red node, the arrow weight indicating the tie strength. The simulation was repeated 1,000 times; the size of the arrowheads is proportional to the number of times the information was passed along the corresponding direction, and the color indicates the total number of transmissions along that link. The background contours highlight the difference in the direction the information follows in the real and the control simulations.
(c) Same as (b), but we assume that each link has the same weight $w = \langle w_{ij} \rangle$ (control).
After [345].

The diffusion of memes, representing ideas or behavior that spread from individual to individual, further highlights the important role of communities [393]. Meme diffusion has attracted considerable attention from marketing [352, 394] to network science [395, 396], communications [397] and social media [398, 399, 400]. Pathogens and memes can follow different spreading patterns, prompting us to systematically distinguish simple from complex contagion [393, 401].

Simple contagion is the process we have explored so far: it is sufficient to come into contact with an infected individual to be infected. The spread of memes, products and behavior is often described by *complex contagion*, capturing the fact that most individuals do not adopt a new meme, product or behavioral pattern at the first contact. Rather, adoption requires reinforcement [402], that is repeated contact with several individuals who have already adopted. For example, the higher is the fraction of a person's friends that have a mobile phone, the more likely that she will also buy one.

In simple contagion, communities trap information or a pathogen, slowing the spread (**Figure 10.19a**). The effect is reversed in complex contagion: because communities have redundant ties, they offer social reinforcement, exposing an individual to multiple examples of adoption. Hence communities can incubate a meme, a product or a behavioral pattern, enhancing its adoption.

The difference between simple and complex contagion is well captured by Twitter data. Tweets, or short messages, are often labeled with *hashtags*, which are keywords acting as memes. Twitter users can follow other users, receiving their messages; they can forward tweets to their own followers (*re tweet*) or mention others in tweets. The measurements indicate that most hashtags are trapped in specific communities, a signature of complex contagion [393]. A high concentration of a meme within a certain community is evidence of reinforcement. In contrast, viral memes spread across communities, following a pattern similar to that encountered in biological pathogens. In general, the more communities a meme reaches, the more viral it is (**Figure 10.20**).

In summary, several network characteristics can affect the spread of a pathogen in a network, from degree correlations to link weights and the bursty nature of the contact pattern. As

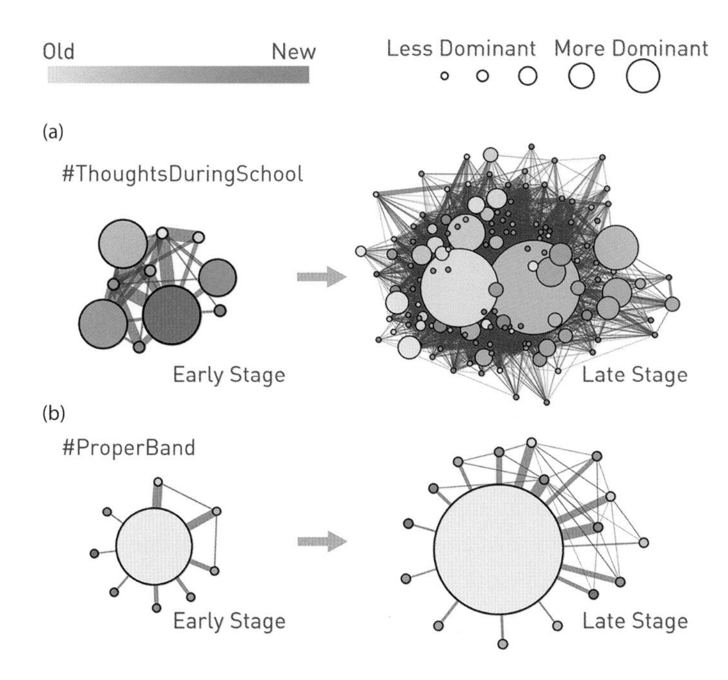

Figure 10.20 **Simple vs. Complex Contagion**
The community structure of the Twitter follower network. Each circle corresponds to a community and its size is proportional to the number of tweets produced by the respective community. The color of a community represents the time when the studied hashtag (meme) is first used in the community. Lighter colors denote the first communities to use a hashtag, darker colors denote the last communities to adopt it.
(a) **Simple Contagion**
The evolution of the viral meme captured by the #Thoughts During School hashtag from its early stage (30 tweets, left) to its late stage (200 tweets, right). The meme jumps easily between communities, infecting many of them, following a contagion pattern encountered in the case of biological pathogens.
(b) **Complex Contagion**
The evolution of a non-viral meme captured by the #ProperBand hashtag from the early stage (left) to the final stage (65 tweets, right). The tweet is trapped in a few communities, having difficulty escaping them. This is a signature of reinforcement, an indication that the meme follows complex contagion.
After [393].

we have discussed in this section, some network characteristics slow a pathogen, others aid their spread. These effects must therefore be accounted for if we wish to predict the spread of a real pathogen. While these patterns are of obvious relevance for infectious diseases, they also influence the spread of such non-infectious diseases as obesity (**Box 10.2**).

10.6 Immunization

Immunization strategies specify how vaccines, treatments or drugs are distributed in the population. Ideally, should a treatment or vaccine exist, it should be given to every infected individual or those at risk of contracting the pathogen. Yet, often cost considerations, the difficulty of reaching all individuals at risk, and real or perceived side-effects of the treatment prohibit full coverage. Given these constraints, immunization strategies aim to minimize the threat of a pandemic by most effectively distributing the available vaccines or treatments.

Immunization strategies are guided by an important prediction of the traditional epidemic models: if a pathogen's spreading rate λ is reduced under its critical threshold λ_c, the virus naturally dies out (**Figure 10.11**). Yet, the epidemic threshold

Box 10.2 Do Our Friends Make Us Fat?

Infectious diseases, like influenza, SARS or AIDS, spread through the transmission of a pathogen. But could the social network aid the spread of non-infectious diseases as well? Recent measurements indicate that it does, offering evidence that social networks can impact the spread of obesity, happiness and behavioral patterns, like giving up smoking [403, 404].

Obesity is diagnosed through an individual's body-mass index (BMI), which is determined by numerous factors, from genetics to diet and exercise. The measurements show that our friends also play an important role. The analysis of the social network of 5,209 men and women has found that if one of our friends is obese, the risk that we too gain weight in the next two to four years increases by 57% [403]. The risk triples if our best friend is overweight: in this case, our chances of weight gain jumps by 171% (**Figure 10.21**). For all practical purposes, obesity appears to be just as contagious as influenza or AIDS, despite the fact that there is no "obesity pathogen" that transmits it (**Online resource 10.3**).

Online Resource 10.3
Spreading in Social Networks
"If your friends are obese, your risk of obesity is 45 percent higher. . . . If your friend's friends are obese, your risk of obesity is 25 percent higher. . . . If your friend's friend's friend, someone you probably don't even know, is obese, your risk of obesity is 10 percent higher. It's only when you get to your friend's friend's friend's friends that there's no longer a relationship between that person's body size and your own body size."
Watch Nicholas Christakis explaining the spread of health patterns in social networks.

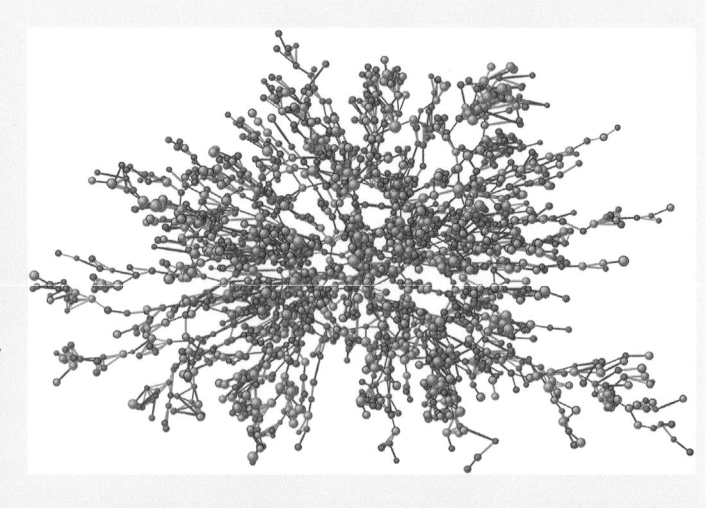

Figure 10.21 **The Web of Obesity**
The largest connected component of the social network capturing the friendship ties between 2,200 individuals enrolled in the Framingham Heart Study. Each node represents an individual; nodes with blue borders are men, those with red borders are women. The size of each node is proportional to the person's BMI, yellow nodes denoting obese individuals (BMI $\geq$ 30). Purple links are friendship or marital ties and orange links are family ties (e.g., siblings). Clusters of obese and non-obese individuals are visible in the network. The analysis indicates that these clusters cannot be attributed to homophily, i.e. the fact that individuals of similar body size may befriend each other. They document instead a complex contagion process, capturing the "spread" of obesity along the links of the social network. After [393].

vanishes in scale-free networks, questioning the effectiveness of this strategy. Indeed, if the epidemic threshold vanishes, immunization strategies cannot move λ under λ_c. In this section we discuss how to use our understanding of the network topology to design effective network-based immunization strategies that counter the impact of the vanishing epidemic threshold.

10.6.1 Random Immunization

The main purpose of immunization is to protect the immunized individual from an infection. Equally important, however, is its secondary role: immunization reduces the speed with which the pathogen spreads in a population. To illustrate this effect, consider the situation when a randomly selected fraction g of individuals are immunized in a population [355].

Let us assume that the pathogen follows the SIS model **(10.3)**. The immunized nodes are invisible to the pathogen, and only the remaining $(1-g)$ fraction of the nodes can contact and spread the disease. Consequently, the effective degree of each susceptible node changes from $\langle k \rangle$ to $\langle k \rangle(1-g)$, which decreases the spreading rate of the pathogen from $\lambda = \beta/\mu$ to $\lambda' = \lambda(1-g)$. Next we explore the consequences of this reduction in both random and scale-free contact networks.

- **Random Networks**

  If the pathogen spreads on a random network, for a sufficiently high g the spreading rate λ' could fall below the epidemic threshold **(10.25)**. The immunization rate g_c necessary to achieve this is calculated by setting

$$\frac{(1-g_c)\beta}{\mu} = \frac{1}{\langle k \rangle + 1},$$

  obtaining

$$g_c = 1 - \frac{\mu}{\beta}\frac{1}{\langle k \rangle + 1}. \qquad (10.27)$$

  Consequently, if vaccination increases the fraction of immunized individuals above g_c, it pushes the spreading rate under the epidemic threshold λ_c. In this case τ becomes negative and the pathogen dies out naturally. This explains why health officials encourage a high fraction of the population to take the influenza vaccine: the vaccine protects not only the individual, but also the rest of the population by decreasing the pathogen's spreading rate. Similarly, a condom not only protects the individual who uses it from contacting the HIV virus, but also decreases the rate at which AIDS spreads in the sexual network. Hence, for random networks a sufficiently high immunization rate can eliminate the pathogen from the population.

- **Heterogenous Networks**
If the pathogen spreads on a network with high $\langle k^2 \rangle$, and random immunization changes λ to $\lambda(1-g)$, we can use **(10.26)** to determine the critical immunization g_c:

$$\frac{\beta}{\mu}(1-g_c) = \frac{\langle k \rangle}{\langle k^2 \rangle} \tag{10.28}$$

obtaining

$$g_c = 1 - \frac{\mu}{\beta}\frac{\langle k \rangle}{\langle k^2 \rangle}. \tag{10.29}$$

For a random network **(10.29)** reduces to **(10.27)**. For a scale-free network with $\gamma < 3$ we have $\langle k^2 \rangle \to \infty$, hence **(10.29)** predicts $g_c \to 1$. In other words, if the contact network has a high $\langle k \rangle$, *we need to immunize virtually all nodes to stop the epidemic.* This prediction is consistent with the finding that for many diseases we must immunize 80%–100% of the population to eradicate the pathogen. For example, measles requires 95% of the population to be immunized [355]; for digital viruses the strategies relying on random immunization call for close to 100% of computers to install the appropriate antivirus software.

To illustrate the role degree heterogeneity plays in immunization, let us consider a digital virus spreading on the email network. If we make the email network random and undirected, we have $\langle k \rangle = 3.26$. Using $\lambda = 1$ in **(10.27)**, we obtain $g_c = 0.76$. In other words, to eradicate the virus we need to convince 76% of computer users to update their antivirus software. Yet, the email network is scale free with $\langle k^2 \rangle = 1,271$ (undirected version), hence **(10.27)** does not apply. In this case **(10.29)** predicts $g_c = 0.997$ for $\lambda = 1$, meaning that more than 99.7% of users must install the software to halt the email virus. It is virtually impossible to achieve this level of compliance – many users simply ignore all warnings. This is the reason why email viruses linger for years and disappear only after the operating systems that support them are phased out.

10.6.2 Vaccination Strategies In Scale-Free Networks

The ineffectiveness of random immunization is rooted in the vanishing epidemic threshold (**Box 10.3**). Consequently, to successfully eradicate a pathogen in heterogenous networks, we must find ways to increase the epidemic threshold. This requires us to reduce the variance, $\langle k^2 \rangle$, of the underlying contact network.

The hubs are responsible for the large variance of heterogenous networks. Therefore, if we immunize the hubs (i.e., all nodes whose degree exceeds some preselected $k'_{\max}$), we decrease the variance and increase the epidemic threshold according to **(10.26)** [405, 406]. Indeed, if nodes with degrees $k > k'_{\max}$ are absent, the epidemic threshold changes to (**Advanced topics 10.C**)

$$\lambda'_c \approx \frac{\gamma - 2}{3 - \gamma} \frac{k_{\min}^{2-\gamma}}{(k'_{\max})^{\gamma-3}}. \tag{10.30}$$

Therefore, for $\gamma < 3$, the more hubs we cure (i.e., the smaller is $k'_{\max}$), the larger will be the epidemic threshold (**Figure 10.22**). By immunizing a sufficient fraction of the hubs we can drop λ_c below $\lambda = \beta/\mu$, which characterizes the pathogen. This procedure is equivalent to altering the underlying network: by immunizing the hubs, we are fragmenting the contact network, making it more difficult for the pathogen to reach the nodes in other components (**Figure 10.23**).

Hub immunization represents a perspective change in immunization protocols: instead of trying to decrease the

Box 10.3 How to Halt An Epidemic?

Health safety officials rely on several interventions to control or delay an epidemic outbreak. Some of the most common interventions include the following.

Transmission-Reducing Interventions
Face masks, gloves and hand washing reduce the transmission rate of airborne or contact-based pathogens. Similarly, condoms reduce the transmission rate of sexually transmitted pathogens.

Contact-Reducing Interventions
For diseases with severe health consequences, officials can quarantine patients, close schools and limit access to frequently visited public spaces, like movie theaters and malls. These make the network sparser by reducing the number of contacts between individuals, hence decreasing the transmission rate.

Vaccinations
Vaccinations permanently remove the vaccinated nodes from the network, as they cannot be infected nor can they spread the disease. Vaccinations also reduce the spreading rate, enhancing the likelihood that the pathogen will die out.

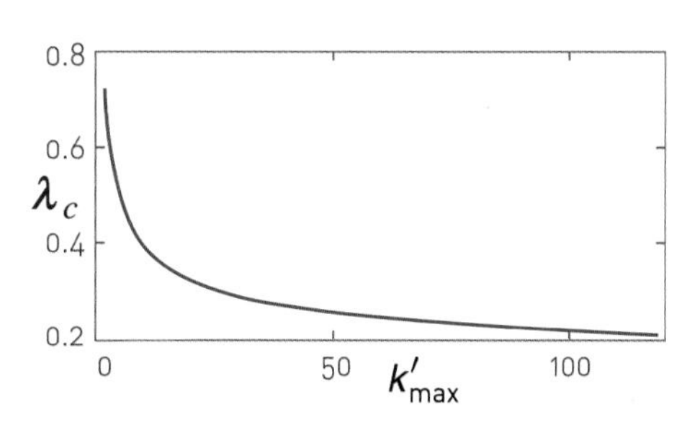

Figure 10.22 **Immunizing the Hubs**
In heterogenous networks a virus can be eradicated by increasing the epidemic threshold through hub immunization. The figure shows the expected epidemic threshold if we immunize all nodes with degree larger than $k'_{\max}$. The more hubs are immunized (i.e., the smaller is $k'_{\max}$, the larger is λ_c, increasing the chance that the disease dies out. Immunizing the hubs changes the network on which the disease spreads, making the hubs invisible to the pathogen (**Figure 10.23**).

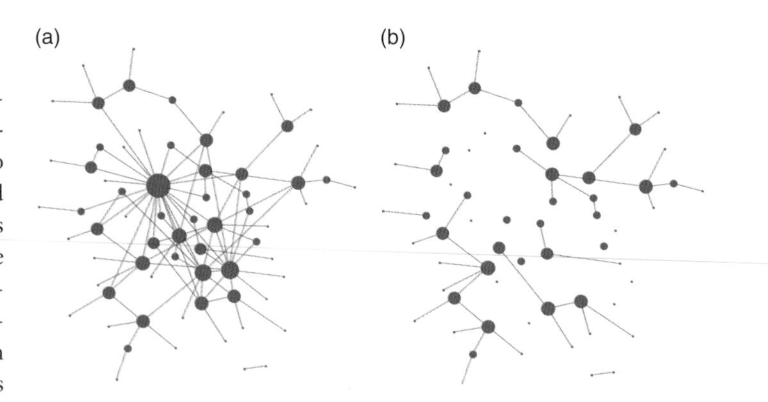

Figure 10.23 **Robustness and Immunization**
Scale-free networks show a remarkable resilience to random node and link failures (**Chapter 8**). At the same time, they are vulnerable to attacks: if we remove their most connected nodes, scale-free networks break apart. This phenomenon has many similarities to the immunization problem: random immunization is unable to eradicate a disease, but selective immunization, that targets the hubs, can restore a finite critical threshold, helping us eradicate the disease. The analogy is not accidental: the robustness and the immunization problem can both be linked to the diverging $\langle k^2 \rangle$. Indeed, the vanishing epidemic threshold is equivalent to finding that the percolation threshold under random node removal problem converges to one (**Advanced topics 10.D**). Similarly, the re-emergence of the epidemic threshold under hub immunization is equivalent to the small percolation threshold characterizing a scale-free network under attack. Therefore, the attack and targeted immunization problems represent two sides of the same coin. To illustrate the equivalence between attacks and targeted immunization, consider the network shown in (**a**). An attack that removes its five largest hubs breaks the network into many isolated islands, as shown in (**b**). Targeted immunization plays the same role: by making the hubs immune to the disease, the network on which the pathogen spreads becomes the fragmented network in (**b**). As the immunized network is broken into small islands, the pathogen will be stuck in one of the small clusters, unable to infect the nodes in the other clusters.

spreading rate using random immunization, we must alter the topology of the contact network, which in turn increases λ_c above the biologically determined $\lambda = \beta/\mu$.

The problem with a hub-based immunization strategy is that for most epidemic processes we lack a detailed map of the contact network. Indeed, we do not know the number of sexual partners each individual has in a population, nor can we accurately identify the super-spreaders during an influenza outbreak. In other words, it is difficult to identify the hubs. Yet, we can still exploit the network topology to design more efficient immunization strategies. To do so, we rely on the friendship paradox, the fact that on average the neighbors of a node have higher degree than the node itself (**Box 7.1**). Therefore, by immunizing the acquaintances of a randomly selected individual, we target the hubs without having to know precisely which individuals are hubs. The procedure consists of the following steps [407]:

1. Choose randomly a fraction p of nodes, like we do during random immunization. Call these nodes Group 0.
2. Select randomly a link for each node in Group 0. We call Group 1 the set of nodes to which these links connect. For example, we ask each individual from Group 0 to nominate one of their acquaintances with whom they engaged in an activity that could have resulted in the transmission of the pathogen. In the case of HIV, ask them to name a sexual partner.
3. Immunize the Group 1 individuals.

This strategy requires no information about the global structure of the network. Yet, according to (**7.3**), the probability that a node with k links belongs to Group 1 is proportional to kp_k. Consequently, the Group 1 individuals have higher average degree than the Group 0 individuals. The implications of this bias are illustrated in **Figure 10.24**, which shows the critical threshold required to eradicate a pathogen for a scale-free network with degree exponent γ. The figure offers several key insights:

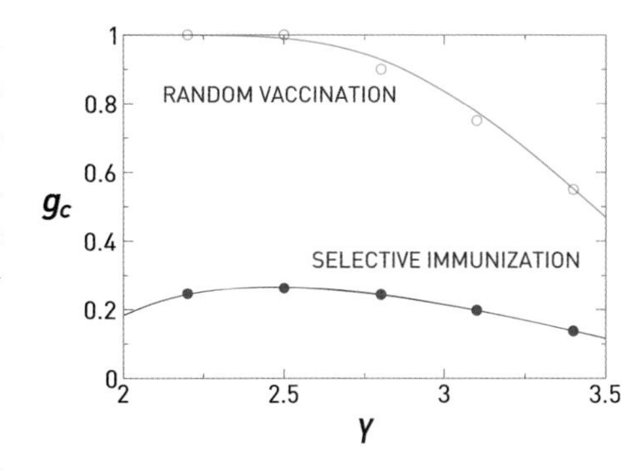

Figure 10.24 **Selective Immunization of Scale-Free Networks**

The critical immunization threshold g_c as a function of the degree exponent γ of the contact network on which the pathogen spreads following the SIS model. The curves correspond to two immunization strategies: *random immunization* (green) and *selective immunization* (purple), that immunize a first neighbor of a randomly selected node. The continuous lines represent the analytical results while the symbols represent simulation data for $N = 10^6$ and $m = 1$. As the population has a finite size, we have $g_c < 1$ for random immunization even for $\gamma < 3$. Redrawn after [407].

1. Random **Immunization**
   The top curve shows g_c for random immunization. For heterogenous networks (small γ) we find that $g_c \approx 1$, indicating that we must immunize all nodes to eradicate the disease. As γ approaches 3, the network develops a finite epidemic threshold and g_c drops. Hence, for large γ, immunizing a sufficiently high fraction of the population can eradicate the pathogen.
2. **Selective Immunization**
   For the biased strategy g_c is systematically under 30%. Therefore, by immunizing a randomly chosen neighbor of 30% of the nodes, we could eradicate the disease. The efficiency of this strategy depends only weakly on γ. Selective immunization is more efficient than random immunization even for high γ, when hubs are less prominent.

In summary, if we have the resources to immunize everyone at risk of contacting a pathogen, we should do that – this was the strategy of the eradication campaigns (**Box 10.4**). When extensive immunization is not feasible, we need to employ various immunization strategies to maximize the impact of our resources. The effectiveness of each strategy depends on the structure of the contact network on which the pathogen spreads. In general, random immunization is inefficient for pathogens that spread on heterogenous networks: for random immunization to succeed we need to reach and immunize close to 100% of the susceptable nodes, which is impossible in most circumstances. In contrast, strategies that immunize the hubs have high effectiveness. Selective immunization, that immunizes the neighbors of randomly selected nodes, can significantly enhance effectiveness, without requiring an accurate map of the contact network. This strategy is efficient for both random and heterogenous networks.

Box 10.4 Can Pathogens be Eradicated?

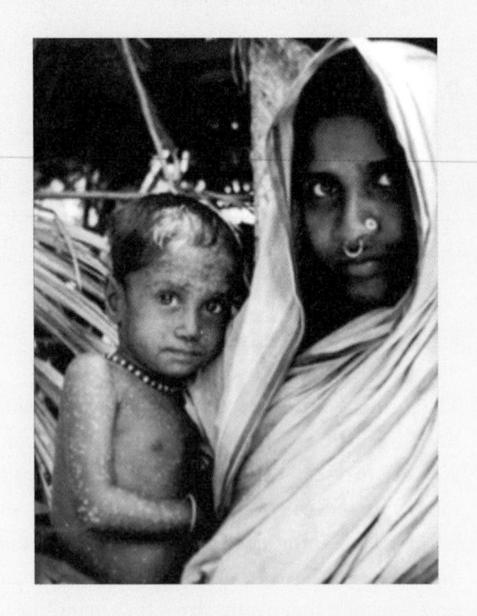

Figure 10.25 **Eradicating Smallpox**
Rahima Banu, the last smallpox-infected patient in Bangladesh in 1976. After [408].

At the end of the 1960s smallpox was still widespread in Africa and Asia. Before 1967 the smallpox eradication strategy relied on mass vaccination, a strategy that was ineffective in densely populated areas. Health officials eventually developed network-based protocols to stop the transmission: they set out to find and treat anyone who had been in contact with an infected individual. This strategy allowed smallpox to become the first disease to be officially eradicated (**Figure 10.25**).

Eradication is the complete elimination of a pathogen from the population. To select an infectious disease for eradication, health officials must make sure that the targeted pathogen does not have a non-human reservoir, so human vaccination can truly eradicate it. There is also a need for an efficient and practical vaccine or drug to interrupt its transmission. So far eradication campaigns have had mixed success: smallpox and rinderpest were successfully eradicated, but programs targeting hookworm, malaria and yellow fever have failed.

10.7 Epidemic Prediction

During much of its history humanity has been helpless when faced with a pandemic. Lacking drugs and vaccines, infectious diseases repeatedly swept through continents, decimating the world's population. The first vaccine was tested only in 1796, and the systematic development of vaccines and cures against new pathogens became possible only in the 1990s. Despite spectacular medical advances, we have effective vaccines only against a small number of pathogens. Consequently, transmission-reducing and quarantine-based measures remain the main tools of health professionals in combatting new pathogens. For the combination of vaccines, treatments and quarantine-based measures to be effective, we need to predict when and where the pathogen will emerge next, allowing local health officials to best deploy their resources.

The real-time prediction of an epidemic outbreak is a very recent development. The ground was set by the development of the epidemic modeling framework in the 1980s [409] and by the 2003 SARS epidemic, which resulted in worldwide reporting guidelines about ongoing outbreaks. The subsequent systematic availability of data pertaining to a pandemic [349] offered real-time input to modeling efforts. The 2009 H1N1 outbreak was the first beneficiary of these developments, becoming the first pandemic whose spread was predicted in real time.

The emergence of any new pathogen raises several key questions:

- Where did the pathogen originate?
- Where do we expect new cases?
- When will the epidemic arrive at various densely populated areas?
- How many infections are to be expected?
- What can we do to slow its spread?
- How can we eradicate it?

Today these questions are addressed using powerful epidemic simulators that consider as input demographic, mobility-related (**Online resource 10.4**) and epidemiological data [410, 411, 412]. The algorithms behind these tools range from stochastic meta-population models [413, 414, 415] to agent-based computer simulations that capture the behavior and interactions of millions of individuals [416]. In this section we summarize the capabilities of these tools, highlighting the role of network science in these developments.

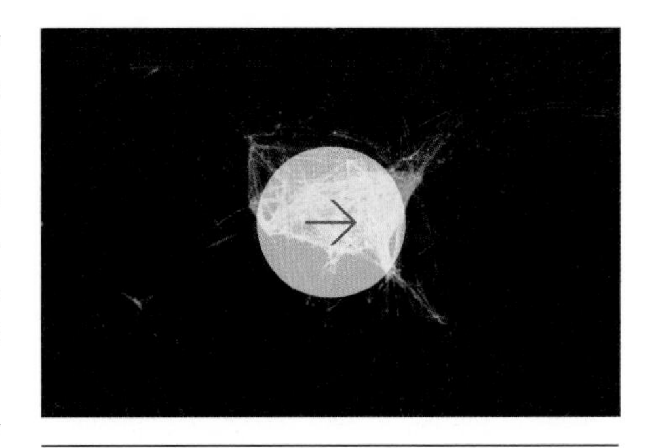

Online Resource 10.4

North American Flight Patterns

Real-time flights across North America, relying on data released by the Federal Aviation Administration. This global transportation network is responsible for the spread of pathogens across continents. Consequently, flight schedules represent the input for epidemic forecasts. While this video, produced by Aaron Koblin, could easily be seen as a purely scientific illustration, it is also viewed as digital art by the art community. Indeed, the video is now in the Media Art collection of the Museum of Modern Art (MoMA) in New York.

10.7.1 Real-Time Forecast

Epidemic forecasts aim to foresee the real-time spread of a pathogen, predicting the number of infected individuals expected each week in each major city [416, 417]. The first successful real-time pandemic forecast based on network science relied on the Global Epidemic and Mobility (GLEAM) computational model [417] (**Figure 10.26**, **Online resource 10.5**), a stochastic framework that uses as input high-resolution data on worldwide human demography and mobility. GLEAM employs a network-based computational model:

- GLEAM maps each geographic location into the nodes of a network.

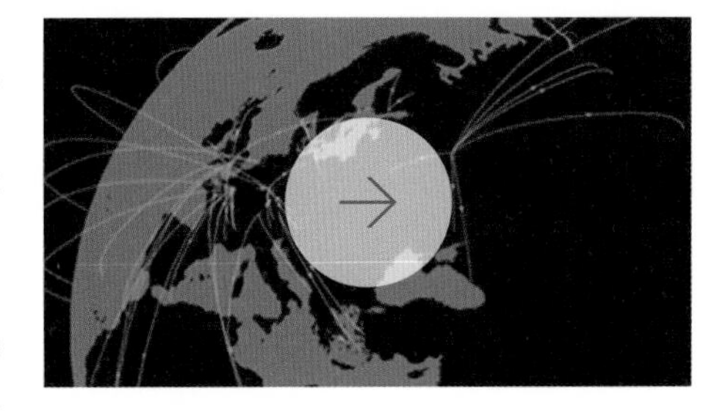

Online Resource 10.5

Gleam

A video describing the GLEAM software package for epidemic prediction.

(a)

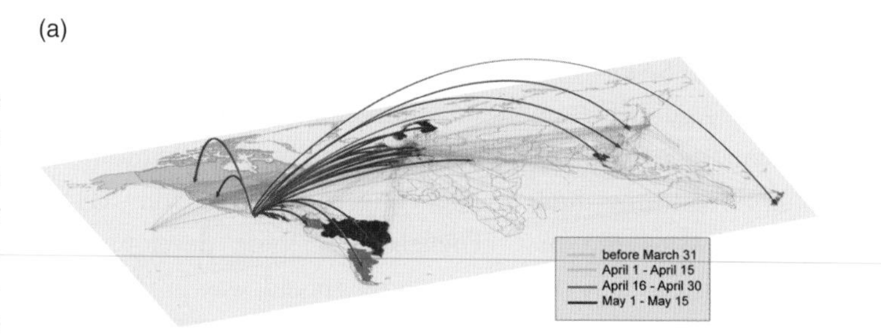

(b)

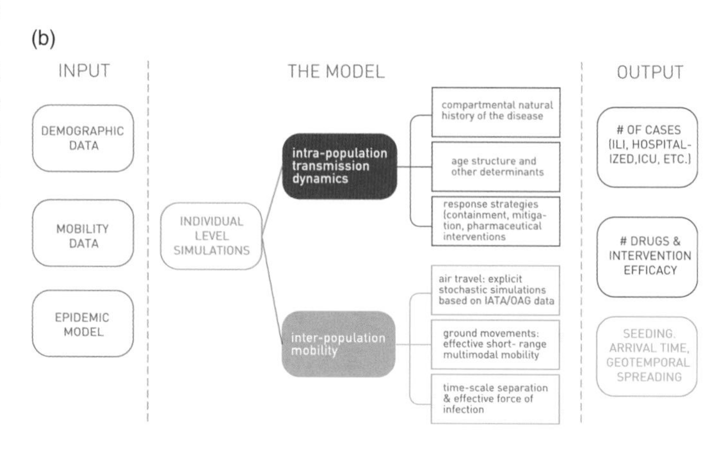

Figure 10.26 **Modeling the 2009 H1N1 Pandemic**
(**a**) The spread of the H1N1 virus during the early stage of the 2009 outbreak. The arrows represent the arrival of the first infections in previously unaffected countries. The color code indicates the time of the virus's arrival.
(**b**) The flowchart of the GLEAM computational model, used to predict the real-time spread of pathogens like H1N1 or Ebola. The left column (Input) represents the input databases, capturing demographic, mobility and epidemiological information. The center column (Model) describes the network-based dynamic processes that are modeled at each time step. The right column (Output) offers examples of quantities the model can predict. After [418].

- Transport between these nodes, representing the links, is provided by global transportation data, like airline schedules (**Online resource 10.4**).
- GLEAM estimates the epidemic parameters, like the transmission rate or reproduction number, using a network-based approach: it relies on chronological data that captures the worldwide spread of the pandemic, rather than medical reports [33].

GLEAM then implements the network-based epidemic framework described in **Section 10.3**, generating a large number of potential outcomes of the pathogen's global progression for the coming months. For H1N1 the predictions were compared with data collected from surveillance and virologic sources in 48 countries during the full course of the pandemic [417], resulting in several key findings:

- **Peak Time**

  Peak time corresponds to the week when most individuals are infected in a particular country. Predicting the peak time helps health officials decide the timing and quantity of the vaccines or treatments they distribute. The peak time depends on the arrival time of the first infection and the demographic and mobility characteristics of each country. The observed peak time fell within the prediction interval for 87% of countries (**Figure 10.27**). In the remaining cases the difference between the real and the predicted peak was at most two weeks.

- **Early Peak**
  GLEAM predicted that the H1N1 epidemic would peak in November, rather than in January or February, the typical peak time of influenza-like viruses. This unexpected prediction turned out to be correct, confirming the model's predictive power. The early peak time was a consequence of the fact that H1N1 originated in Mexico rather than South Asia (where many flu viruses come from), hence it took the virus less time to arrive in the northern hemisphere.
- **The Impact of Vaccination**
  Several countries implemented vaccination campaigns to accelerate the decline of the pandemic. The simulations indicated that these mass vaccination campaigns had only negligible impact on the course of the epidemic. The reason is that the timing of these campaigns was guided by the expectation of a January peak time, prompting the deployment of the vaccines after the November 2009 peak [419], too late to have a strong effect.

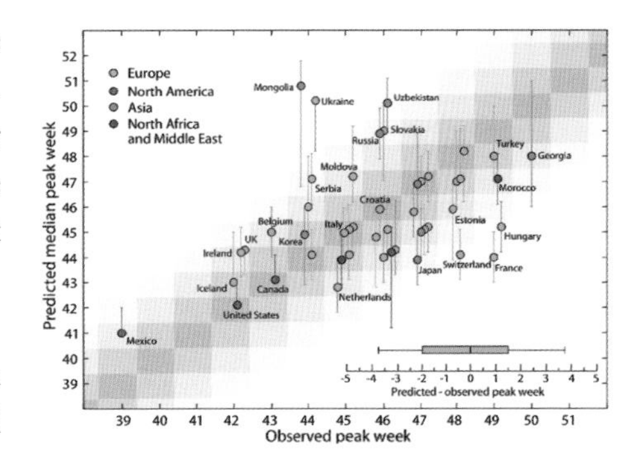

Figure 10.27 **Activity Peaks for H1N1**
The predicted and observed peak times for the H1N1 virus in several countries. The peak time corresponds to the week when most individuals are infected by the pathogen, and is measured in weeks after the beginning of the epidemic. The model predictions were obtained by analyzing 2,000 stochastic realizations of the outbreak, generating the error bars in the figure. After [418].

10.7.2 "What if" Analysis

By incorporating the time and nature of each containment and mitigation procedure, simulations can estimate the efficiency of specific contingency plans [410, 411, 412, 414, 420]. Next we discuss the impact of two such interventions.

- **Travel Restrictions**
  Given the important role air travel plays in the spread of a pathogen, faced with a dangerous pandemic like an Ebola outbreak (**Figure 10.28**), the first instinct is to restrict travel. Yet, in a world where key resources travel by air, a travel ban can lead to economic collapse as often illustrated by fictional takes on pandemics (**Box 10.5**). Therefore, before resorting to a travel ban we must make sure that travel restrictions have beneficial effects on the pandemic. For this we must

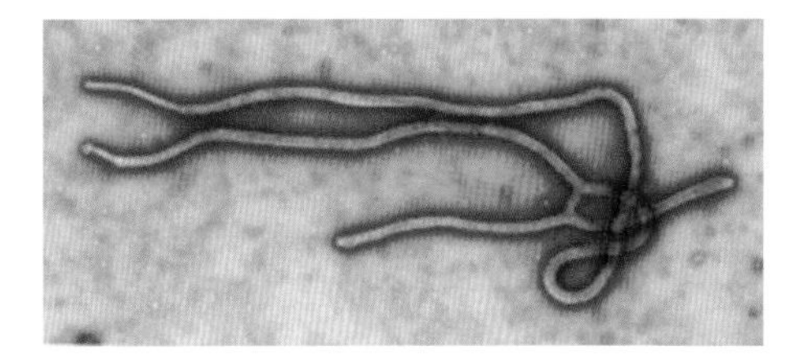

Figure 10.28 **The Deadliest Outbreak**
With a fatality rate in the vicinity of 80%, the Ebola virus is one of the deadliest viruses known to humans. Its first known incidence was in 1976 in Zaire, killing 280 of the 312 infected individuals by hemorrhagic fever, a combination of high fever and a bleeding disorder. The virus can be transmitted by contact with the blood or secretions of an infected individual.

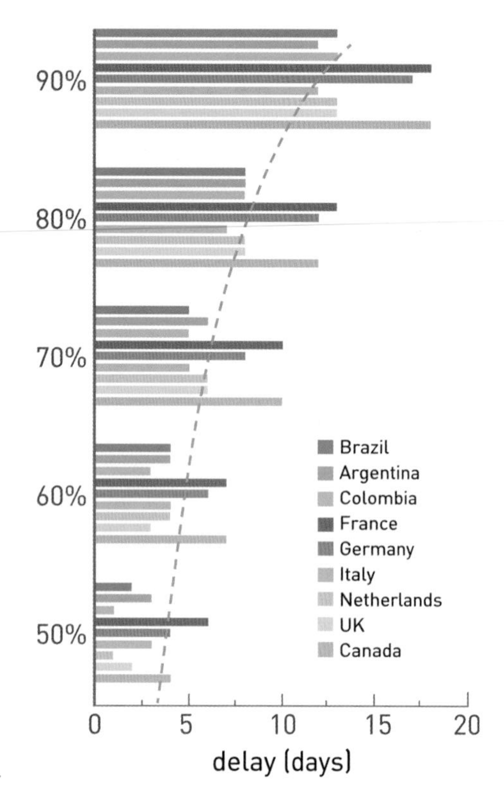

Figure 10.29 **The Impact of Travel Reduction**
The impact of travel reduction on the arrival time of the H1N1 virus from Mexico in various countries, compared with the reference scenario of no travel reduction. The percentages on the vertical axis show the degree of travel reduction implemented around the world. The largest delay is less than 20 days, observed for a 90% travel restriction.
After [414].

realize that awareness of a viral outbreak results in self-imposed travel reductions. For example, there was a 40% decline in travel to and from Mexico in May 2009, during the H1N1 outbreak, as individuals canceled non-necessary business and leisure activities in the infected region. The modeling indicates [417, 418] that this 40% reduction delayed the arrival of the first infection by less than 3 days in various countries around the world. Furthermore, even if travel dropped 90%, the peak time is delayed by less than 20 days (**Figure 10.29**).

Most important, travel restrictions do not decrease the number of infected individuals. They only delay the outbreak, offering local authorities more time to prepare for the pandemic. Hence, travel restrictions are effective only if the delay caused by them increases local vaccination levels or helps the deployment of cures.

- **Antiviral Treatment**
During the 2009 H1N1 pandemic Canada, Germany, Hong Kong, Japan, the UK and the USA distributed antiviral drugs to mitigate the impact of the disease [421]. This prompted modelers to ask what would have been the impact if all countries that had drug stock piles distributed them to their population [422]. The simulations indicate that peak times would have been delayed by about 3 to 4 weeks, offering time to immunize a larger fraction of the population before the pandemic reached its peak.

10.7.3 Effective Distance

Before cars and airplanes, pathogens traveled on foot or at most with the speed of a horse. Hence a pandemic like the Black Death in Europe moved slowly from village to village (**Figure 10.8**), following a diffusive process described by simple reaction–diffusion models [423, 424]. As the next infection always emerged in the geographic proximity of the previous infections, there was a strong correlation between the time of the outbreak and the physical distance from the origin of the outbreak.

Today, with airline travel, physical distance has lost its relevance for epidemic phenomena. A pathogen that emerges

Box 10.5 A Night At The Movies

For a fictionalized but plausible depiction of a major pandemic, watch *Contagion*, the 2011 medical thriller directed by Steven Soderbergh, featuring Marion Cotillard, Bryan Cranston, Matt Damon, Laurence Fishburne, Jude Law, Gwyneth Paltrow, Kate Winslet and Jennifer Ehle. The movie follows the desperate attempts of public health officials to stop a virus and the ensuing panic from sweeping the globe, hence addressing the impact of both biological and social contagion. The 1995 medical disaster film *Outbreak* (**Figure 10.30**), directed by Wolfgang Petersen, starring Dustin Hoffman, Rene Russo and Morgan Freeman, focuses on a deadly Ebola-like virus that starts from a small village in Zaire and reaches the United States. Both movies illustrate the difficult choices civilian and military agencies must take to contain the spread of a deadly pathogen.

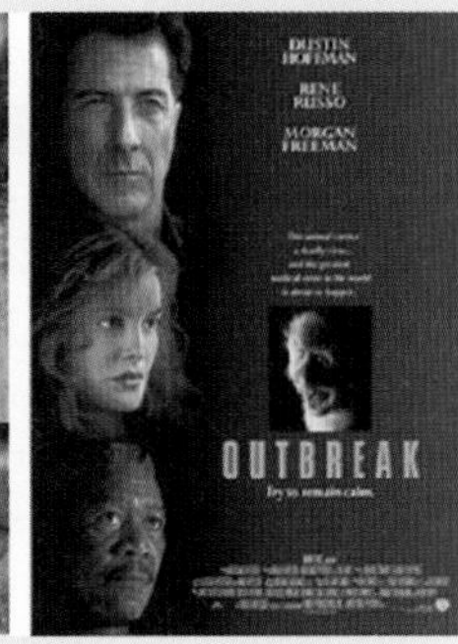

Figure 10.30 **Outbreak: Fiction and Truth**
The theatrical release posters of two pandemic-related movies, *Contagion* and *Outbreak*.

in Manhattan can just as easily travel to London as to Garrison, NY, a village an hour's drive from Manhattan. This prompts us to ask: Is there a better space to view the spread of an epidemic than the physical space? Such a space does exist if we replace the conventional geographic distance with an effective distance derived from the mobility network [425]. The nodes of the mobility network are cities and the links represent the amount of travel between them. Each link is directed and weighted, characterized by a flux fraction $0 \leq p_{ij} \leq 1$ that represents the fraction of travelers that leave node i and arrive at node j. The values of p_{ij} can be extracted from airline schedules, having $p_{ij} > 0$ only if there is direct travel from i to j.

Given the multiple routes a person can take between any two cities, a pathogen can follow multiple paths on the mobility network. Yet, its spread is dominated by the most probable trajectories predicted by the mobility matrix p_{ij}. This allows us to define the *effective distance* d_{ij} between two connected locations i and j, as

$$d_{ij} = (1 - \ln p_{ij}) \geq 0. \tag{10.31}$$

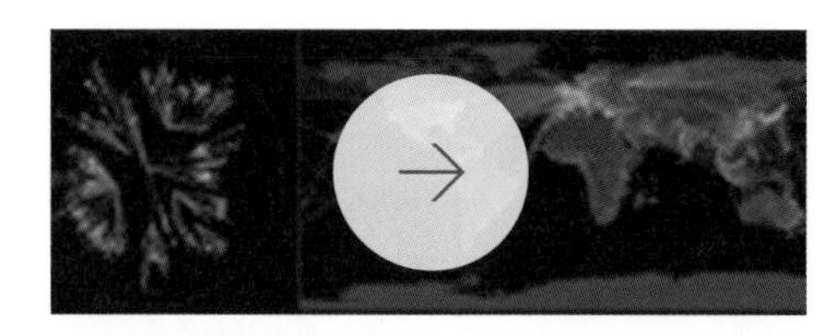

Online Resource 10.6

The Speed of a Pandemic

The spread of a pathogen, as predicted by GLEAM, from three initial outbreak locations. While the geographic spreading pattern is difficult to interpret, in the effective distance representation the pandemic follows a regular radial pattern (**Figure 10.31**).

The observed spreading patterns prompt us to ask: What is the speed of a typical pathogen as it spreads around the globe? The speed depends on three key parameters:

1. The basic reproduction number R_0, which is in the vicinity of 2 for influenza-type viruses (**Table 10.2**).
2. The recovery rate, which is approximately 3 days for influenza.
3. The mobility rate, which represents the total fraction of the population that travels during a day. This parameter is in the range of 0.01–0.001.

Running GLEAM (**Figure 10.26**) with these parameters, we can compute the correlation between the arrival time and the geographic distance to the source of the epidemic, obtaining a speed of about 250–300 km/day. Therefore, an influenza virus moves through a continent with the speed of a sports car or of a small airplane [425].

If p_{ij} is small, implying that only a small fraction of individuals that leave from i travel to j, then the effective distance between i and j is large. Note that $d_{ij} \neq d_{ji}$: for a small village i located near a metropolis j we expect d_{ij} to be small, as many travelers from i go to j. Yet, d_{ji} is large as only a small fraction of travelers leaving the metropolis head to the small village. The logarithm in **(10.31)** accounts for the fact that effective distances are additive, whereas probabilities along multi-step paths are multiplicative.

As **Figure 10.31** indicates (see also **Online resource 10.6**), if we use **(10.31)** to represent the distance of each city from the source of an epidemic, the pathogen follows circular wave fronts. This is in contrast to the complex spreading pattern we observe if we view the pandemic in the geographical space. Furthermore, while the arrival time of H1N1 appears to be random if plotted as a function of the physical distance, it correlates strongly with the effective distance (**Figure 10.32**). We can therefore use the effective distance to determine the speed of a pathogen (**Online resource 10.6**).

A surprising but welcome aspect of epidemic forecasts is that the predictions of different models are rather similar, despite the fact that they use different mobility data (airline schedules [370] or dollar bill movement [369]) and different assumptions about the epidemic parameters (recovery rate, transmission rate, etc.). The effective distance helps us understand why the various model predictions converge. Indeed, we can write the arrival time of a pathogen at location a as [425]

$$T_a = \frac{d_{\text{eff}}(P)}{V_{\text{eff}}(\beta, R_0, \gamma, \varepsilon)}. \quad (10.32)$$

Therefore, the arrival time is the ratio of the effective distance d_{eff} and an effective speed V_{eff}. The effective speed is determined only by the epidemiological parameters of the pathogen, whereas the effective distance d_{eff} depends only on the topology of the mobility network encoded by p_{ij}. When confronted with a new outbreak, the pathogen-specific epidemiological parameters are unknown in the beginning. However, **(10.32)** predicts that the *relative arrival times are independent of the epidemiological parameters.* For example, for an

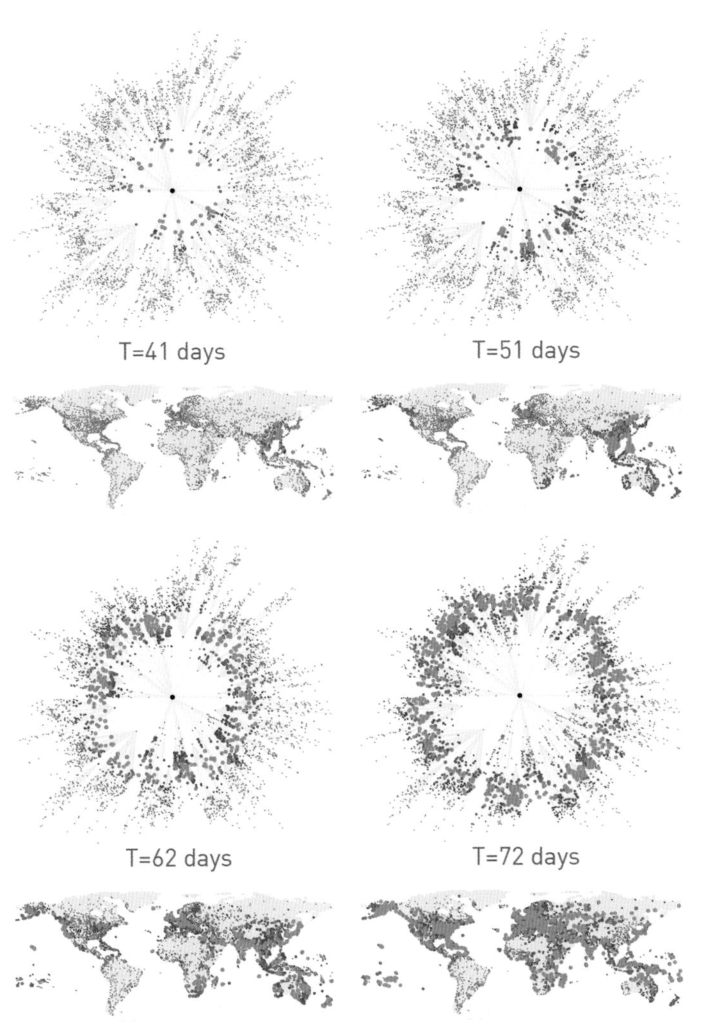

Figure 10.31 **Effective Distance**
The spread of a pandemic with an initial outbreak in Hong Kong. Regions with a large number of infections are shown as red nodes. Each panel compares the state of the system in the conventional geographic representation (bottom) with the effective distance representation (top). The complex spatial pattern observed in the geographic representation becomes a circular wave that moves outwards at constant speed in the effective distance representation (see also **Online resource 10.6**). After [425].

outbreak that starts at node i, the ratio of the arrival times to nodes j and l is

$$\frac{T_a(j/i)}{T_a(l/i)} = \frac{d_{\text{eff}}(j/i)}{d_{\text{eff}}(l/i)},$$

that is the ratio depends only on the effective distances. Therefore, the relative arrival times of the disease depend only on the topology of the mobility network. As the mobility patterns around the world are unique and model independent, the predictions of different models converge, independent of the choice of epidemiological parameters.

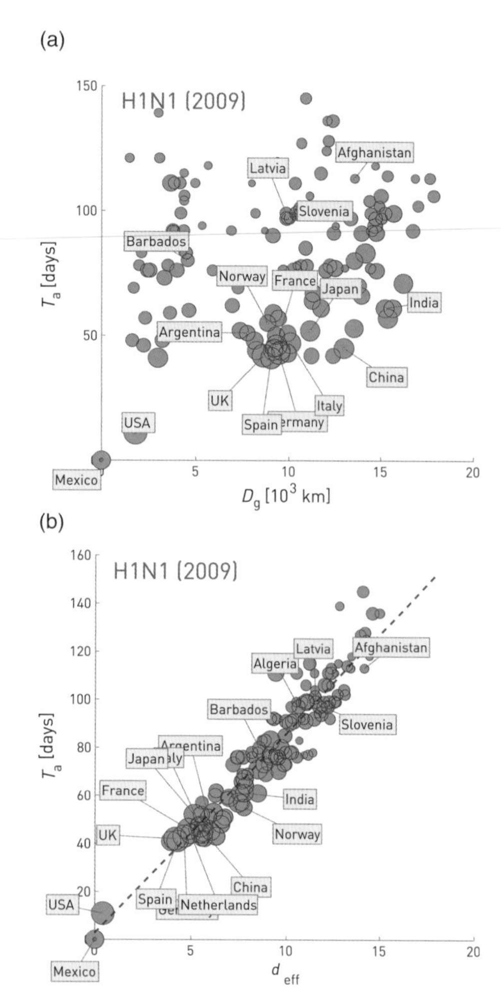

Figure 10.32 **Effective Distance and Arrival Time**

(**a**) Geographic Distance

Arrival times vs. geographic distance from its source (Mexico) for the 2009 H1N1 pandemic. Each circle represents one of the 140 affected countries and the symbol size indicates the total traffic in each country. Arrival times are the date of the first confirmed case in a given country after the beginning of the outbreak on March 17, 2009. In this representation the arrival time and the geographic distance are largely independent of each other ($R_0 = 0.0394$).

(**b**) Effective Distance

Epidemic arrival time T_a vs. effective distance d_{eff} for H1N1, demonstrating the strong correlations between the effective distance **(10.31)** and the arrival time.

After [425].

In summary, joint advances in data collection and network epidemics have offered the capability to predict the real-time spread of a pathogen. The developed models can help design response and mitigation scenarios, train health and emergency personnel, can be used to explore the impact of various interventions, from quarantine to travel restrictions, to optimize the deployment of treatments and vaccines, and identify the source of a pandemic (**Box 10.6**).

Interestingly, the recent success of epidemic forecasts is not due to the improved understanding of the underlying biology of infectious pathogens. It can be attributed instead to the lucky situation that when it comes to the spreading of a pathogen, the epidemic parameters are of secondary importance. The most important factor is the structure of the mobility network. That, however, can be accurately estimated from travel schedules, allowing us to turn human mobility patterns into accurate predictions about the course of a pandemic.

10.8 Summary

Most networks facilitate transfer along their links: transfer of trust, knowledge, habits or information (social networks), electricity (power grid), money (financial networks), goods (trade networks). To understand these phenomena, we must understand how the network topology affects these dynamical processes. In this chapter we focused on the spread of pathogens along the links of the network, the area where our understanding of the interplay between dynamical phenomena and network topology is the most advanced. We showed that the network topology has a drastic impact on the dynamics of the spreading process, offering distinct predictions for spreading on random and scale-free networks. This finding laid the ground for addressing a wider class of problems: the need to systematically understand the impact networks have on various dynamical processes [436], an increasingly active chapter of network science [437, 438].

Modeling the spread of pathogens also represents an important practical application of network science. The advances in this area were rather spectacular, giving birth to accurate epidemic forecasts, something that was only a dream

Box 10.6 Identifying The Source Of A Pandemic

Identifying the source of an epidemic is an important component of epidemic control. The source could be the first individual in a contact network, or the city where the pathogen first emerged in the mobility network. The mathematical formulation of the problem [427] inspired a burst of research on the subject [428–435].

The difficulty in finding the source is rooted in the stochastic nature of the infection process: different initial conditions can lead to similar infection patterns at the observation time. The approach we take depends on the information we have about the epidemic:

- In the simplest case, at a given moment t we know the nodes that have been infected and the network on which the pathogen spreads. The task is to find the source i [427] (**Figure 10.33**).
- If we also have the time of infection for each node, we can reconstruct the dynamics of the epidemic, significantly enhancing our ability to detect the source.
- The best strategy is to monitor the hubs, as they have the earliest and the most accurate information about a breakout. For example, for a pathogen spreading on a scale-free network, monitoring the state of 18% of the highest-degree nodes can offer a 90% success rate in detecting the source. In contrast, to achieve the same level of accuracy we need to monitor 41% of the nodes if we select randomly the nodes we monitor [429].
- In the effective distance representation (**Figure 10.31**), the infection follows a circular pattern only if we use the right outbreak location. Otherwise, the observed pattern is asymmetric. Therefore, we can detect the source by finding the location (node) from which the outbreak pattern shows the highest radial symmetry [425].

Figure 10.33 **Epidemic Sources**
Finding the source of an epidemic is like finding the source of a water ripple. As pathogens do not spread in a uniform medium, the challenge is to identify the appropriate "ripples" in the mobility network.

a decade earlier. Two advances have made this possible. The first is the emergence of a robust theoretical framework to describe network-based epidemics. The second is access to accurate real-time data on human travel and demographics, allowing us to reconstruct the mobility network that is responsible for the global spread of a pathogen. As we have seen in **Section 10.7**, the biological parameters and the network contributions to the accuracy of the observed predictive power are decoupled. Consequently, an accurate forecast requires primarily an accurate knowledge of the mobility network.

The analytical framework of network epidemics has offered a number of unexpected results, the most important being the vanishing characteristic spreading time and epidemic threshold in heterogenous networks. As most contact networks encountered in epidemic processes have a broad degree distribution, these results are of immediate and lasting theoretical and practical interest.

Equally important are the insights network epidemiology offers for immunization strategies. As we showed in **Section 10.6**, while random immunization can successfully eradicate a virus that spreads on a random network, this strategy is sub-optimal in a scale-free network. As most contact networks are heterogenous, this is a rather depressing conclusion. Yet, we showed that selective immunization strategies can restore the epidemic threshold and suppress the prevalence of a pathogen. Selective immuniztion succeeds by systematically altering the topology of the network on which a pathogen spreads.

Box 10.7 At A Glance: Network Epidemics

Infection rate: β
Recovery rate: μ
Spreading rate: $\lambda = \frac{\beta}{\mu}$
Reproductive number: $R_0 = \frac{\beta\langle k\rangle}{\mu}$

SI Model

$$i(t) = \frac{i_0 e^{\beta\langle k\rangle t}}{1 - i_0 + i_0 e^{\beta\langle k\rangle t}}$$

SIS Model

$$i(t) = \left(1 - \frac{\mu}{\beta\langle k\rangle}\right)\frac{Ce^{(\beta\langle k\rangle - \mu)t}}{1 + Ce^{(\beta\langle k\rangle - \mu)t}}$$

Characteristic Time

$$\text{SI: } \tau = \frac{\langle k\rangle}{\beta(\langle k^2\rangle - \langle k\rangle)}.$$

$$\text{SIS: } \tau = \frac{\langle k\rangle}{\beta(\langle k^2\rangle - \mu\langle k\rangle)}.$$

$$\text{SIR: } \tau = \frac{\langle k\rangle}{\beta\langle k^2\rangle - (\mu + \beta)\langle k\rangle}.$$

Epidemic Threshold

$$\text{SIS: } \lambda_c = \frac{\langle k\rangle}{\langle k^2\rangle}$$

$$\text{SIR: } \lambda_c = \frac{1}{\frac{\langle k^2\rangle}{\langle k\rangle} - 1}$$

Immunization Threshold (SIS):

$$g_c = 1 - \frac{\mu}{\beta}\frac{\langle k\rangle}{\langle k^2\rangle}$$

10.9 Homework

10.9.1 Epidemics on Networks

Calculate the characteristic time τ and the epidemic threshold λ_c of the SI, SIS and SIR models for networks with:

(a) Exponential degree distribution.
(b) Stretched exponential degree distribution.
(c) Delta distribution (all nodes have the same degree).

Assume that the networks are uncorrelated and infinite. Refer to **Table 4.2** for the functional form of the distribution and the corresponding first and second moments.

Box 10.8 Historical Note: Network Epidemics

Epidemic phenomena became a central topic in network science after Romualdo Pastor-Satorras and Alessandro Vespignani introduced the continuum theory that can account for the properties of the underlying contact network (**Figure 10.34**). They also discovered the dependence of the epidemic threshold and characteristic time on the second moment of the degree distribution, a central result of network epidemics. Subsequently, Vespignani and his research group developed GLEAM, a computational framework that offers real-time predictions for the spread of a pathogen.

Figure 10.34 **Romualdo Pastor-Satorras and Alessandro Vespignani**
Physicists by training, Pastor-Satorras was a postdoctoral associate with Vespignani at ICTP in Trieste when they discovered the impact of the scale-free property on the epidemic threshold. Subsequently, both researchers made major contributions to network science, from the discovery of degree correlations (**Chapter 7**) to our understanding of weighted networks.

10.9.2 Random Obesity in Social Networks

Consider a social network with degree distribution p_k, where 50% of the nodes are obese. Make the assumption that obese nodes are distributed randomly within the network.

(a) If the network has degree correlation, encoded in the join probability e'_{kk}, what is the probability P (øo) that a non-obese (ø) individual is friends with an obese individual (o)? And what is the probability P (oo) that two obese individuals are friends?
(b) Assume that the network is uncorrelated. How many second neighbors of a degree-k node are obese?

Calculate the same quantities (a) and (b) if the percentage of obese nodes increases to 70%.

10.9.3 Immunization

Choose four networks from **Table 4.1** (assume that directed networks behave like undirected and uncorrelated networks with $p_k = p_{k_{in}}$) and consider an epidemic process spreading on them. Remember: not only pathogens, but also ideas or opinions can spread on a network! Determine for each network the critical fraction g_c necessary to stop the epidemic if we randomly immunize a fraction g_c of the nodes. How would the epidemic threshold λ_c change if all nodes with degree higher than 1000 were immunized?

10.9.4 Epidemic on Bipartite Networks

Consider a bipartite network with two types of node, which we indicate as male (M) and female (F). On this network a pathogen can be transmitted only from a node of one set to a node of the other set. Assume that the rate of transmission from an M node to an F node, $\beta_{M \to F}$, is different from the rate of transmission from an F node to an M node, $\beta_{F \to M}$. Write the equations of the corresponding SI model, assuming the degree-block approximation and that the network is uncorrelated.

10.10 Advanced Topics 10.A Microscopic Models of Epidemic Processes

In **Sections 10.2 and 10.3** we relied on the continuum approach to describe epidemic phenomena. In this section we show that the key results can be derived using microscopic models and probability-based reasoning. These arguments help us understand the origin of the continuum approach and improve our understanding of epidemic phenomena.

10.10.1 Deriving The Epidemic Equation

We start by deriving the continuum model **(10.3)** from the microscopic processes that describe the interactions between two individuals [437]. Consider a susceptible individual in contact with an infected individual, so that the susceptible individual becomes infected with probability βdt during the time interval dt. The probability that the susceptible individual is *not* infected in the dt interval is $(1-\beta dt)$. If the susceptible individual i has degree k_i, each of its k_i links could in principle infect it. Therefore, the probability that it avoids infection is $(1-\beta dt)^{k_i}$. Finally, the total probability that node i becomes infected in time dt is $1-(1-\beta dt)^{k_i}$, or one minus the total probability that it is not infected. Assuming $\beta dt \ll 1$, at leading order the probability that a susceptible individual becomes infected is

$$1-(1-\beta dt)^{k_i} \approx \beta k_i dt. \tag{10.33}$$

In a random network all nodes have approximately $\langle k \rangle$ neighbors. Replacing k_i with $\langle k \rangle$ in **(10.33)** we obtain the first term of the continuum equation **(10.3)**. If we do not replace k_i with $\langle k \rangle$ we obtain the first term of **(10.13)**, capturing the spread of a pathogen in a heterogenous network.

10.10.2 Epidemic Threshold and Network Topology

A key result of **Section 10.3** connects the network topology to the epidemic threshold λ_c, a result derived using the continuum theory. We can arrive at the same result using a mechanistic argument that illustrates the connection between the epidemic threshold and the network topology.

Consider a pathogen that is transmitted with probability β in unit time. Therefore, in unit time an infected node with degree k will infect βk neighbors. If each infected node recovers at rate μ, then the characteristic time that a node stays infected is $1/\mu$. The pathogen can persist in the population only if during this $1/\mu$ time interval the infected node infects at least one other node. Otherwise, the pathogen gradually dies out.

In other words, if $\beta k/\mu < 1$, then our degree-k node recovers before it could infect other nodes. If we consider a random network, where most nodes have comparable degrees, $k \sim \langle k \rangle$, the condition $\beta k/\mu = 1$ allows us to calculate the epidemic threshold. Using $\lambda = \beta/\mu$ we obtain $\lambda_c = 1/\langle k \rangle$, which is the high-$k$ limit of the result **(10.25)** derived for random networks. It tells us that the ability of a pathogen to spread is determined by the interplay between the epidemiological characteristics of the pathogen (β and μ) and the network topology ($\langle k \rangle$).

In a scale-free network nodes have widely different degrees. Therefore, while the network's average degree may satisfy $\beta\langle k\rangle/\mu < 1$, suggesting that the virus will die out, for all nodes with $k > \langle k\rangle$ we have $\beta\langle k\rangle/\mu > 1$. If such a high-degree node is infected, even if the spreading rate λ is under the threshold $1/\langle k \rangle$, the disease can spread, persisting in the hubs. This is the reason why the epidemic threshold vanishes in networks with high $\langle k^2\rangle$.

10.11 Advanced Topics 10.B Analytical Solution of the SI, SIS and SIR Models

In this section we solve the SI, SIS and SIR models on a network, deriving the results summarized in **Table 10.3**, namely the characteristic spreading time τ and the epidemic threshold λ_c for each model.

10.11.1 The Density Function

The density function Θ_k provides the fraction of infected nodes in the neighborhood of a susceptible node with degree k. As discussed in **Section 10.3**, to calculate i_k we must first determine Θ_k. If a network lacks degree correlations, the probability that a link points from a node with degree k to a node with degree k' is independent of k. Hence, the probability that a randomly chosen link points to a node with degree k' is the excess degree **(7.3)**,

$$\frac{k' p_{k'}}{\sum_k k p_k} = \frac{k' p_{k'}}{\langle k\rangle}.$$

At least one link of each infected node is connected to another infected node, the one that transmitted the infection. Therefore, the number of links available for future transmission is $(k' - 1)$, allowing us to write

$$\Theta_k = \frac{\sum_{k'} (k' - 1) p_{k'} i_{k'}}{\langle k\rangle} = \Theta. \qquad (10.34)$$

In other words, in the absence of degree correlations Θ_k is independent of k. Differentiating **(10.34)** we obtain

$$\frac{d\Theta}{dt} = \sum_k \frac{(k-1)p_k}{\langle k \rangle} \frac{di_k}{dt}. \tag{10.35}$$

To make further progress, we need to consider the specific model the pathogen follows.

10.11.2 SI Model

Using **(10.13)** and **(10.35)**, we obtain

$$\frac{d\Theta}{dt} = \beta \sum_k \frac{(k^2-k)p_k}{\langle k \rangle}[1-i_k]\Theta. \tag{10.36}$$

To predict the early behavior of epidemics, we consider the fact that for small t the fraction of infected individuals is much smaller than one. Therefore, we can neglect the second-order terms in **(10.36)**, obtaining

$$\frac{d\Theta}{dt} = \beta \left(\frac{\langle k^2 \rangle}{\langle k \rangle} - 1 \right) \Theta. \tag{10.37}$$

This has the solution

$$\Theta(t) = Ce^{t/\tau} \tag{10.38}$$

where

$$\tau = \frac{\langle k \rangle}{\beta(\langle k^2 \rangle - \langle k \rangle)}. \tag{10.39}$$

Using the initial condition

$$\Theta(t=0) = C = i_0 \frac{\langle k \rangle - 1}{\langle k \rangle},$$

which means that initially a fraction i_0 of nodes are infected uniformly (hence $i_k(t=0) = i_0$ for all k), we obtain the time-dependent Θ as

$$\Theta(t) = i_0 \frac{\langle k \rangle - 1}{\langle k \rangle} e^{t/\tau}. \tag{10.40}$$

We insert this into **(10.13)** to arrive at **(10.15)**.

10.11.3 SIR Model

In the SIR model the density of infected nodes follows

$$\frac{di_k}{dt} = \beta(1 - i_k - r_k)k\Theta - \mu i_k, \tag{10.41}$$

where r_k is the fraction of recovered nodes with degree k. Keeping only the first-order terms (which means that we ignore i_k and r_k in the parentheses above, as for small t they are much smaller than one), we obtain

$$\frac{di_k}{dt} = \beta k\Theta - \mu i_k. \tag{10.42}$$

Multiplying this equation by $(k-1)p_k/\langle k\rangle$ and summing over k, we have

$$\frac{d\Theta}{dt} = \left(\beta\frac{\langle k^2\rangle - \langle k\rangle}{\langle k\rangle} - \mu\right)\Theta. \tag{10.43}$$

The solution of **(10.43)** is

$$\Theta(t) = Ce^{t/\tau}, \tag{10.44}$$

where the characteristic time for the SIR model is

$$\tau = \frac{\langle k\rangle}{\beta\langle k^2\rangle - \langle k\rangle(\beta + \mu)}. \tag{10.45}$$

A global outbreak is possible only if $\tau > 0$, that is when the number of infected nodes grows exponentially with time. This yields the condition for a global outbreak as

$$\lambda = \frac{\beta}{\mu} > \frac{\langle k\rangle}{\langle k^2\rangle - \langle k\rangle}, \tag{10.46}$$

allowing us to write the epidemic threshold for the SIR model as (**Table 10.3**)

$$\lambda_c = \frac{1}{\frac{\langle k^2\rangle}{\langle k\rangle} - 1}. \tag{10.47}$$

10.11.4 SIS Model

In the SIS model the density of infected nodes is given by **(10.18)**

$$\frac{di_k}{dt} = \beta(1 - i_k)k\Theta - \mu i_k. \tag{10.48}$$

There is a small but important difference in the density function of the SIS model. For the SI and the SIR models, if a node is infected, then at least one of its neighbors must also be infected or recovered, hence at most $(k - 1)$ of its neighbors are susceptible, the origin of the (-1) term in the parentheses of **(10.34)**. However, in the SIS model the previously infected neighbor can become susceptible again, therefore all k links of a node can be available to spread the disease. Hence, we modify the definition **(10.34)** to obtain

$$\Theta_k = \frac{\sum_{k'} k' p_{k'} i_{k'}}{\langle k \rangle} = \Theta. \tag{10.49}$$

Again keeping only the first-order terms we obtain

$$\frac{di_k}{dt} = \beta k\Theta - \mu i_k. \tag{10.50}$$

Multiplying the equation by $(k - 1)p_k/\langle k \rangle$ and summing over k, we have

$$\frac{d\Theta}{dt} = \left(\beta\frac{\langle k^2 \rangle}{\langle k \rangle} - \mu\right)\Theta. \tag{10.51}$$

This again has the solution

$$\Theta(t) = Ce^{t/\tau}, \tag{10.52}$$

where the characteristic time of the SIS model is

$$\tau = \frac{\langle k \rangle}{\beta\langle k^2 \rangle - \langle k \rangle\mu}. \tag{10.53}$$

A global outbreak is possible if $\tau > 0$, which yields the condition for a global outbreak as

$$\lambda = \frac{\beta}{\mu} > \frac{\langle k \rangle}{\langle k^2 \rangle}, \tag{10.54}$$

and the epidemic threshold for the SIS model as (**Table 10.3**)

$$\lambda_c = \frac{\langle k \rangle}{\langle k^2 \rangle}. \tag{10.55}$$

10.12 Advanced Topics 10.C Targeted Immunization

In this section we derive the epidemic threshold for the SIS and SIR models on scale-free networks under hub immunization. We start with an uncorrelated network with power-law degree distribution $p_k = c \cdot k^{-\gamma}$, where $c \approx (\gamma - 1)/k_{min}^{-\gamma+1}$ and $k \geq k_{min}$. In **Section 10.16** we obtained, for the critical spreading rate,

$$\lambda_c = \frac{\langle k \rangle}{\langle k^2 \rangle} = \frac{1}{k} \qquad \text{(SIS model)}$$

and

$$\lambda_c = \frac{1}{\frac{\langle k^2 \rangle}{\langle k \rangle} - 1} = \frac{1}{k - 1} \qquad \text{(SIR model).}$$

Under hub immunization we immunize all nodes whose degree is larger than k_0. From the perspective of the epidemic this is equivalent to removing the high-degree nodes from the network. Therefore, to calculate the new critical spreading rate, we need to determine the average degree $\langle k' \rangle$ and the second moment $\langle k'^2 \rangle$ after the hubs have been removed. This problem was addressed in **Advanced topics 8.F**, where we studied the robustness of a network under attack. We have seen that hub removal has two effects:

(1) The maximum degree of the network changes to k_0.
(2) The links connected to the removed hubs are also removed, as if we randomly remove a fraction

$$\tilde{f} = \left(\frac{k_0}{k_{min}} \right)^{-\gamma+2} \qquad (10.56)$$

of links.

The degree distribution of the resulting network is

$$p'_{k'} = \sum_{k=k_{min}}^{k_0} \binom{k}{k'} \tilde{f}^{k-k'} (1 - \tilde{f})^{k'} p_k.$$

According to **(8.39)** and **(8.40)**, this yields

$$\langle k' \rangle = (1 - \tilde{f}) \langle k \rangle,$$

$$\langle k'^2 \rangle = (1-\tilde{f})^2 \langle k^2 \rangle + \tilde{f}(1-\tilde{f})\langle k \rangle,$$

where $\langle k \rangle$ is the average and $\langle k^2 \rangle$ is the second moment of the degree distribution before the link removal, but with maximum degree k_0. For the SIS model this means

$$\lambda'_c = \frac{(1-\tilde{f})\langle k \rangle}{(1-\tilde{f})^2 \langle k^2 \rangle + \tilde{f}(1-\tilde{f})\langle k \rangle} = \frac{1}{(1-\tilde{f})\kappa + \tilde{f}}, \quad (10.57)$$

where, according to equation **(8.47)**, for $2 > \gamma > 3$

$$\kappa = \frac{\gamma - 2}{3 - \gamma} k_0^{3-\gamma} k_{\min}^{\gamma-2}. \quad (10.58)$$

Combining **(10.56)**, **(10.57)** and **(10.58)**, we obtain

$$\lambda'_c = \left[\frac{\gamma-2}{3-\gamma} k_0^{3-\gamma} k_{\min}^{\gamma-2} - \frac{\gamma-2}{3-\gamma} k_0^{5-2\gamma} k_{\min}^{2\gamma-4} + k_0^{2-\gamma} k_{\min}^{\gamma-2} \right]^{-1}. \quad (10.59)$$

For the SIR model a similar calculation yields

$$\lambda'_c = \left[\frac{\gamma-2}{3-\gamma} k_0^{3-\gamma} k_{\min}^{\gamma-2} - \frac{\gamma-2}{3-\gamma} k_0^{5-2\gamma} k_{\min}^{2\gamma-4} + k_0^{2-\gamma} k_{\min}^{\gamma-2} - 1 \right]^{-1}. \quad (10.60)$$

For both the SIR and SIS models, if $k_0 \gg k_{\min}$ we have

$$\lambda'_c \approx \frac{3-\gamma}{\gamma-2} k_0^{\gamma-3} k_{\min}^{2-\gamma}. \quad (10.61)$$

10.13 Advanced Topics 10.D The SIR Model and Bond Percolation

The SIR model is a dynamical model that captures the time-dependent spread of an infection in a network. Yet, it can be mapped into a static bond percolation problem [280, 439, 440, 441]. This mapping offers analytical tools that help us predict the model's behavior.

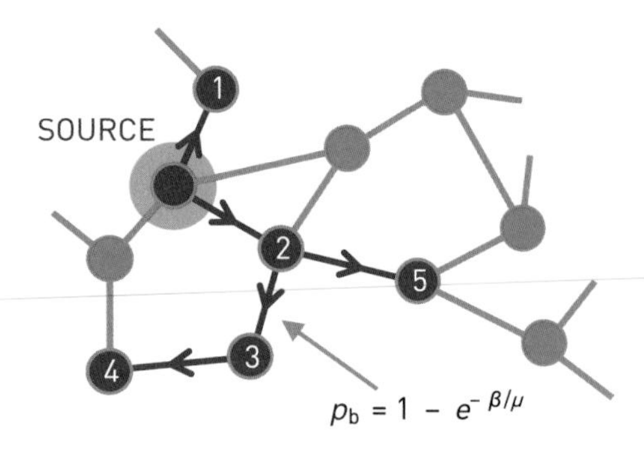

Figure 10.35 **Mapping Epidemics into Percolation**

Consider the contact network on which the epidemic spreads. To map the spreading process into percolation, we leave in place each link with probability $p_b = 1 - e^{-\beta/\mu}$, a probability determined by the biological characteristics of the pathogen. Therefore links are removed with probability $e^{-\beta/\mu}$. The cluster size distribution of the remaining network can be mapped exactly into the outbreak size. For large β/μ we will likely have a giant component, indicating that we could face a global outbreak. β/μ corresponds to a virus that has difficulty spreading and we end up with numerous small clusters, indicating that the pathogen will likely die out.

Consider an epidemic process on a network, so that each infected node transmits a pathogen to each of its neighbors with rate β, and recovers after a recovery time $\tau = 1/\mu$. We view the infection as a Poisson process, consisting of a series of random contacts with average inter-event time $\beta\tau$. Therefore, the probability that an infected node does not transmit the pathogen to susceptible neighbors decreases exponentially in time, or $e^{-\beta\tau}$. The infected node stays infected until it recovers in $\tau = 1/\mu$ time. Therefore, the overall probability that the pathogen is passed on is $1 - e^{-\beta\tau}$.

This process is equivalent to bond percolation on the same network, where each directed link is occupied with probability $p_b = 1 - e^{-\beta\tau}$ (**Figure 10.35**). If β and τ are the same for each node, the network can be considered undirected. Although this mapping loses the temporal dynamics of the epidemic process, it has several advantages:

- The total fraction of infected nodes in the endemic state maps into the size of the giant component of the percolation problem.
- The probability that a pathogen dies out before reaching the endemic state equals the fraction of the nodes in a randomly selected finite component in the percolation problem.
- We can determine the epidemic threshold by exploiting the known properties of bond percolation. Consider the average number of links outgoing from a node that can be reached by a link. This allows us to retrace the course of the epidemic: if an infected individual infects on average at least one other individual, then the epidemic can reach an endemic state. Since a node can be reached by one of its k links, the probability of being reached is $kp_k/N\langle k\rangle$. The probability of each of its $k-1$ outgoing links infecting its neighbor is p_b.

Since the network is randomly connected, as long as the epidemic has not spread yet, the average number of neighbors infected by the selected node is

$$\langle R_i \rangle = p_b \sum \frac{p_k k(k-1)}{\langle k \rangle}.$$

An endemic state can be reached only if $\langle R_i \rangle > 1$, obtaining the condition for the epidemic as [16, 239]

$$\left(\frac{\langle k^2\rangle}{\langle k\rangle}-1\right)>\frac{1}{p_b}. \tag{10.62}$$

Equation **(10.62)** agrees with the result **(10.46)** derived earlier from the dynamical models: scale-free networks with $\gamma \leq 3$ have a divergent second moment, hence such networks undergo a percolation transition even at $p_b \to 0$. That is, a virus can spread on this network regardless of how small the infection probability β is or how small the recovery time τ is.

References

[1] A.-L. Barabási. Invasion Percolation and Global Optimization. *Physical Review Letters*, 76:3750, 1996.

[2] B. Bollobás. *Random Graphs*. Cambridge: Cambridge University Press, 2001.

[3] P. Erdős and A. Rényi. On random graphs, I. *Publicationes Mathematicae (Debrecen)*, 6:290–297, 1959.

[4] S. Kaufmann. *Origins of Order: Self-Organization and Selection in Evolution*. Oxford: Oxford University Press, 1993.

[5] A.-L. Barabási. Dynamics of Random Networks: Connectivity and First Order Phase Transitions. http://arxiv.org/abs/cond-mat/9511052, 1995.

[6] I. de Sola Pool and M. Kochen. Contacts and influence. *Social Networks*, 1:5–51, 1978.

[7] S. Milgram. The Small World Problem. *Psychology Today*, 2: 60–67, 1967.

[8] D. J. Watts and S. H. Strogatz. Collective dynamics of 'small-world' networks. *Nature*, 393: 409–410, 1998.

[9] H. Jeong, R. Albert and A. L. Barabási. Internet: Diameter of the worldwide web. *Nature*, 401:130–131, 1999.

[10] A.-L. Barabási and R. Albert. Emergence of scaling in random networks. *Science*, 286:509–512, 1999.

[11] A.-L. Barabási, H. Jeong, R. Albert. Mean-field theory for scale-free random networks. *Physica A*, 272:173–187, 1999.

[12] R. Albert, H. Jeong and A.-L. Barabási. Error and attack tolerance of complex networks. *Nature*, 406: 378–482, 2000.

[13] S. N. Dorogovtsev, J. F.F. Mendes, and A. N. Samukhin. Structure of growing networks with preferential linking, *Physical Review Letters*, 85: 4633, 2000.

[14] P. L. Krapivsky and S. Redner. Statistics of changes in lead node in connectivity-driven networks, *Physical Review Letters*, 89:258703, 2002.

[15] B. Bollobás, O. Riordan, J. Spencer, and G. Tusnády. The degree sequence of a scale-free random graph process. *Random Structures and Algorithms*, 18:279–290, 2001.

[16] R. Cohen, K. Erez, D. ben-Avraham and S. Havlin. Resilience of the Internet to random breakdowns. *Physical Review Letters*, 85:4626, 2000.

[17] B. Bollobás and O. Riordan. Robustness and Vulnerability of Scale-Free Random Graphs. *Internet Mathematics*, 1, 2003.

[18] R. Pastor-Satorras and A. Vespignani. Epidemic spreading in scale-free networks. *Physical Review Letters*, 86:3200–3203, 2001.

[19] R. Albert and A.-L. Barabási. Statistical Mechanics of Complex Networks. *Reviews of Modern Physics*, 74: 47, 2002.

[20] H. Jeong, B. Tombor, R. Albert, Z. N. Oltvai, and A.-L. Barabási. The large-scale organization of metabolic networks. *Nature*, 407:651–655, 2000.

[21] H. Jeong, S.P. Mason, A.-L. Barabási, and Z. N. Oltvai. Lethality and centrality in protein networks. *Nature*, 411:41–42, 2001.

[22] A.-L. Barabási, and Z. N. Oltvai. Network biology: understanding the cell's functional organization. *Nature Reviews Genetics*, 5:101–113, 2004.

[23] J. Richards and R. Hobbs. *Mark Lombardi: Global Networks*. New York: Independent Curators International, 2003.

[24] M. S. Granovetter. The strength of weak ties. *American Journal of Sociology*, 78: 1360, 1973.

[25] S. W. Oh et.al. A mesoscale connectome of the mouse brain. *Nature*, 508: 207–214, 2014.

[26] International Human Genome Sequencing Consortium. Initial sequencing and analysis of the human genome. *Nature*, 409: 6822, 2001.

[27] J. C. Venter et al. The Sequence of the Human Genome. *Science*, 291: 1304, 2001.

[28] A. L. Hopkins, Network Pharmacology. Nature *Biotechnology*, 25: 1110–1111, 2007.

[29] N. Gulbahce, A.-L. Barabási, and J. Loscalzo. Network medicine: A network-based approach to human disease. *Nature Reviews Genetics*, 12: 56, 2011.

[30] C. Wilson. Searching for Saddam: A five-part series on how the US military used social networking to capture the Iraqi dictator. 2010. www. slate.com/id/2245228/.

[31] J. Arquilla and D. Ronfeldt. *Networks and Netwars: The Future of Terror, Crime, and Militancy*. Santa Monica, CA, RAND, 2001.

[32] A. L. Barabási, Scientists must spearhead ethical use of big data. Politico.com, September 30, 2013.

[33] D. Balcan, H. Hu, B. Goncalves, P. Bajardi, C. Poletto, J. J. Ramasco, D. Paolotti, N. Perra, M. Tizzoni, W. Van den Broeck, V. Colizza, and A. Vespignani. Seasonal transmission potential and activity peaks of the new influenza A(H1N1): a Monte Carlo likelihood analysis based on human mobility. *BMC Medicine*, 7: 45, 2009.

[34] L. Hufnagel, D. Brockmann, and T. Geisel. Forecast and control of epidemics in a globalized world. *PNAS*, 101: 15124, 2004.

[35] P. Wang, M. Gonzalez, C. A. Hidalgo, and A.-L. Barabási. Understanding the spreading patterns of mobile phone viruses. *Science*, 324: 1071, 2009.

[36] O. Sporns, G. Tononi, and R. Kötter. The Human Connectome: A Structural Description of the Human Brain. *PLoS Computional Biology*, 1: 4, 2005.

[37] L. Wu, B. N. Waber, S. Aral, E. Brynjolfsson, and A. Pentland. *Mining Face-to-Face Interaction Networks using Sociometric Badges: Predicting Productivity in an IT Configuration Task*. Proceedings of the International Conference on Information Systems, Paris, France, December 14-17, 2008.

[38] E. N. Lorenz. Deterministic Nonperiodic Flow. Journal of the Atmospheric Sciences, 20: 130, 1963.

[39] K. G. Wilson. The renormalization group: Critical phenomena and the Kondo problem. *Reviews of Modern Physics*, 47: 773, 1975.

[40] S. F. Edwards and P. W. Anderson. Theory of Spin Glasses. *Journal of Physics, F* 5: 965, 1975.

[41] B. B. Mandelbrot. *The Fractal Geometry of Nature*. New York: W.H. Freeman and Company. 1982.

[42] T. Witten, Jr. and L. M. Sander. Diffusion-Limited Aggregation, a Kinetic Critical Phenomenon. *Physical Review Letters*, 47: 1400, 1981.

[43] J. J. Hopfield. Neural networks and physical systems with emergent collective computational abilities. *PNAS*, 79: 2554, 1982.

[44] P. Bak, C. Tang, and K. Wiesenfeld. Self-organized criticality: an explanation of $1/f$ noise. *Physical Review Letters*, 59: 4, 1987.

[45] M. E. J. Newman. The structure and function of complex networks. *SIAM Review*. 45: 167, 2003.

[46] S. Chandrasekhar. Stochastic Problems in Physics and Astronomy. *Reviews Modern Physics*, 15: 1, 1943.

[47] M. Girvan and M. E. J. Newman. Community structure in social and biological networks. *PNAS*, 99: 7821, 2002.

[48] National Research Council. *Network Science*. Washington, DC: The National Academies Press, 2005.

[49] National Research Council. *Strategy for an Army Center for Network Science, Technology, and Experimentation*. Washington, DC: The National Academies Press, 2007.

[50] A.-L. Barabási. *Linked: The New Science of Networks*. New York: Perseus, 2002.

[51] M. Buchanan. *Nexus: Small Worlds and the Groundbreaking Science of Networks*. Norton, 2003.

[52] D. Watts. *Six Degrees: The Science of a Connected Age*. New York: Norton, 2004.

[53] N. Christakis and J. Fowler. *Connected: The Surprising Power of Our Social Networks and How They Shape Our Lives*. New York: Back Bay Books, 2011.

[54] M. Schich, R. Malina, and I. Meirelles (eds). Arts, Humanities, and Complex Networks [Kindle Edition], 2012.

[55] K.-I. Goh, M. E. Cusick, D. Valle, B. Childs, M. Vidal, and A.-L. Barabási. The human disease network. *PNAS*, 104:8685–8690, 2007.

[56] H.U. Obrist. *Mapping it out: An alternative atlas of contemporary cartographies*. London: Thames and Hudson, 2014.

[57] I. Meirelles. *Design for Information*. Beverley, MA: Rockport, 2013.

[58] K. Börner. *Atlas of Science: Visualizing What We Know*. Boston, MA: MIT Press, 2010.

[59] L. B. Larsen. *Networks: Documents of Contemporary Art*. Boston, MA: MIT Press. 2014.

[60] L. Euler, Solutio Problemat is ad Geometriam Situs Pertinentis. *Commentarii Academiae Scientiarum Imperialis Petropolitanae* 8:128–140, 1741.

[61] G. Alexanderson. Euler and Königsberg's bridges: a historical view. *Bulletin of the American Mathematical Society* 43: 567, 2006.

[62] G. Gilder. *Metcalfe's law and legacy*. Forbes ASAP, 1993.

[63] B. Briscoe, A. Odlyzko, and B. Tilly. Metcalfe's law is wrong. *IEEE Spectrum*, 43:34–39, 2006.

[64] Y.-Y. Ahn, S. E. Ahnert, J. P. Bagrow and A.-L. Barabási. Flavor network and the principles of food pairing, *Scientific Reports*, 196, 2011.

[65] A. Barrat, M. Barthélemy, R. Pastor-Satorras, and A. Vespignani. The architecture of complex weighted networks. *PNAS*, 101:3747–3752, 2004.

[66] J. P. Onnela, J. Saramäki, J. Kertész, and K. Kaski. Intensity and coherence of motifs in weighted complex networks. *Physical Review E*, 71:065103, 2005.

[67] B. Zhang and S. Horvath. A general framework for weighted gene coexpression network analysis. *Statistical Applications in Genetics and Molecular Biology*, 4:17, 2005.

[68] P. Holme, S. M. Park, J. B. Kim, and C. R. Edling. Korean university life in a network perspective: Dynamics of a large affiliation network. *Physica A*, 373:821–830, 2007.

[69] R. D. Luce and A. D. Perry. A method of matrix analysis of group structure. *Psychometrika*, 14:95–116, 1949.

[70] S. Wasserman and K Faust. *Social Network Analysis: Methods and Applications*. Cambridge: Cambridge University Press, 1994.

[71] B. Bollobás and O. M. Riordan. Mathematical results on scale-free random graphs. In Bornholdt S., Schuster H. G., *Handbook of Graphs and Networks: From the Genome to the Internet*. Wiley-VCH Verlag, 2003.

[72] P. Erdős and A. Rényi. On the evolution of random graphs. *Publications of the Mathematical Institute of the Hungarian Academy of Sciences*, 5:17–61, 1960.

[73] P. Erdős and A. Rényi. On the evolution of random graphs. *Bulletin of the International Statistical Institute*, 38:343–347, 1961.

[74] P. Erdős and A. Rényi. On the Strength of Connectedness of a Random Graph, *Acta Mathematica Academiae Scientiarum Hungaricae*, 12: 261–267, 1961.

[75] P. Erdős and A. Rényi. Asymmetric graphs. *Acta Mathematica Academiae Scientiarum Hungaricae*, 14:295–315, 1963.

[76] P. Erdős and A. Rényi. On random matrices. *Publications of the Mathematical Institute of the Hungarian Academy of Sciences*, 8:455–461, 1966.

[77] P. Erdős and A. Rényi. On the existence of a factor of degree one of a connected random graph. *Acta Mathematica Academiae Scientiarum Hungaricae*, 17:359–368, 1966.

[78] P. Erdős and A. Rényi. On random matrices II. *Studia Scientiarum Mathematicarum Hungarica*, 13:459–464, 1968.

[79] E. N. Gilbert. Random graphs. *The Annals of Mathematical Statistics*, 30:1141–1144, 1959.

[80] R. Solomonoff and A. Rapoport. *Connectivity of random nets. Bulletin of Mathematical Biology*, 13:107–117, 1951.

[81] P. Hoffman. *The Man Who Loved Only Numbers: The Story of Paul Erdős and the Search for Mathematical Truth*. New York: Hyperion Books, 1998.

[82] B. Schechter. *My Brain is Open: The Mathematical Journeys of Paul Erdős*. New York: Simon and Schuster, 1998.

[83] G. P. Csicsery. N is a Number: A Portrait of Paul Erdős, 1993.

[84] L. C. Freeman and C. R. Thompson. Estimating Acquaintanceship. In Kochen, M. (ed.), *The Small World*, Norwood, NJ: Ablex, 1989.

[85] H. Rosenthal. Acquaintances and contacts of Franklin Roosevelt. Unpublished thesis. Massachusetts Institute of Technology, 1960.

[86] L. Backstrom, P. Boldi, M. Rosa, J. Ugander, and S. Vigna. *Four degrees of separation. In ACM Web Science 2012: Conference Proceedings*, p. 45–54. ACM Press, 2012.

[87] S. Lawrence and C.L. Giles. Accessibility of information on the Web. *Nature*, 400:107, 1999.

[88] A. Broder, R. Kumar, F. Maghoul, P. Raghavan, S. Rajagopalan, R. Stata, A. Tomkins, and J. Wiener. Graph structure in the web. *Computer Networks*, 33:309–320, 2000.

[89] J. Travers and S. Milgram. An Experimental Study of the Small World Problem. *Sociometry*, 32:425–443, 1969.

[90] F. Karinthy, "Láncszemek." In R. T. Kiadasa (ed.) *Minden másképpen van*. Budapest: Atheneum Irodai es Nyomdai, 1929, pp. 85–90. English translation available in [91].

[91] M. Newman, A.-L. Barabási, and D. J. Watts. *The Structure and Dynamics of Networks*. Princeton, NJ: Princeton University Press, 2006.

[92] J. Guare. *Six Degrees of Separation*. New York: Dramatist Play Service, 1992.

[93] T. S. Kuhn. *The Structure of Scientific Revolutions*. Chicago: University of Chicago Press, 1962.

[94] M. Newman. *Networks: An Introduction*. Oxford: Oxford University Press, 2010.

[95] K. Christensen, R. Donangelo, B. Koiller, and K. Sneppen. Evolution of Random Networks. *Physical Review Letters*, 81:2380–2383, 1998.

[96] H. E. Stanley. *Introduction to Phase Transitions and Critical Phenomena*. Oxford: Oxford University Press, 1987.

[97] D. Fernholz and V. Ramachandran. The diameter of sparse random graphs. *Random Structures and Algorithms*, 31:482–516, 2007.

[98] V. Pareto. *Cours d'Économie Politique: Nouvelle édition*. Geneva: Librairie Droz, 299–345, 1964.

[99] M. Faloutsos, P. Faloutsos, and C. Faloutsos. On power-law relationships of the internet topology. Proceedings of SIGCOMM. *Computer Communication Review*, 29: 251–262, 1999.

[100] R. Pastor-Satorras and A. Vespignani. *Evolution and Structure of the Internet: A Statistical Physics Approach*. Cambridge: Cambridge University Press, 2004.

[101] D. J. De Solla Price. Networks of Scientific Papers. *Science* 149: 510–515, 1965.

[102] S. Redner. How Popular is Your Paper? An Empirical Study of the Citation Distribution. *European Physics Journal*, B 4: 131, 1998.

[103] R. Kumar, P. Raghavan, S. Rajalopagan, and A. Tomkins. *Extracting Large-Scale Knowledge Bases from the Web*. Proceedings of the 25thVLDBConference, Edinburgh, Scotland, pp. 639–650, 1999.

[104] A. Wagner, A. and D.A. Fell. The small world inside large metabolic networks. *Proceedings of the Royal Society London, Series B*, 268: 1803–1810, 2001.

[105] W. Aiello, F. Chung, and L.A. Lu. Random graph model for massive graphs, *Proceedings of the 32nd ACM Symposium on Theoretical Computing*, 2000.

[106] A. Wagner. How the global structure of protein interaction networks evolves. *Proceedings of the Royal Society London, Series B*, 270: 457–466, 2003.

[107] M. E. J. Newman. The structure of scientific collaboration networks. *Proceedings of the National Academy of Science*, 98: 404–409, 2001.

[108] A.-L. Barabási, H. Jeong, E. Ravasz, Z. Néda, A. Schubert, and T. Vicsek. Evolution of the social network of scientific collaborations. *Physica A* 311: 590–614, 2002.

[109] F. Liljeros, C.R. Edling, L.A.N. Amaral, H.E. Stanley, and Y. Aberg. The Web of Human Sexual Contacts. *Nature* 411: 907–908, 2001.

[110] R. Ferrer i Cancho and R.V. Solé. The small world of human language. *Proceedings of the Royal Society London, Series B*, 268: 2261–2265, 2001.

[111] R. Ferrer i Cancho, C. Janssen, and R.V. Solé. Topology of technology graphs: Small world patterns in electronic circuits. *Physical Review E*, 64: 046119, 2001.

[112] S. Valverde and R.V. Solé. Hierarchical Small Worlds in Software Architecture. arXiv:cond-mat/0307278, 2003.

[113] H. Ebel, L.-I. Mielsch, and S. Bornholdt. Scale-free topology of e-mail networks. *Physical Review E*, 66: 035103(R), 2002.

[114] J.P.K. Doye. Network Topology of a Potential Energy Landscape: A Static Scale-Free Network. *Physical Review Letters*, 88: 238701, 2002.

[115] J.-P. Onnela, J. Saramaki, J. Hyvonen, G. Szabó, D. Lazer, K. Kaski, J. Kertesz, and A.-L. Barabási. Structure and tie strengths in mobile communication networks. *Proceedings of the National Academy of Sciences, USA,* 104: 7332–7336 (2007).

[116] H. Kwak, C. Lee, H. Park, and S. Moon. What is Twitter, a social network or a news media? Proceedings of the 19th International Conference on World Wide Web, 591-600, 2010.

[117] M. Cha, H. Haddadi, F. Benevenuto and K. P. Gummadi. Measuring user influence in Twitter: The million follower fallacy. Proceedings of International AAAI Conference on Weblogs and Social, 2010.

[118] J. Ugander, B. Karrer, L. Backstrom, and C. Marlow. The Anatomy of the Facebook Social Graph. ArXiv:1111.4503, 2011.

[119] L.A.N. Amaral, A. Scala, M. Barthelemy and H.E. Stanley. Classes of small-world networks. *Proceedings of the National Academy of Sciences, USA*, 97:11149–11152, 2000.

[120] R. Cohen and S. Havlin. Scale free networks are ultrasmall. *Physical Review Letters*, 90, 058701, 2003.

[121] B. Bollobás and O. Riordan. The Diameter of a Scale-Free Random Graph. *Combinatorica*, 24: 5–34, 2004.

[122] R. Cohen and S. Havlin. *Complex Networks - Structure, Robustness and Function*. Cambridge: Cambridge University Press, 2010.

[123] K.-I. Goh, B. Kahng, and D. Kim. Universal behavior of load distribution in scale-free networks. *Physical Review Letters*, 87: 278701, 2001.

[124] P.S. Dodds, R. Muhamad and D.J. Watts. An experimental study to search in global social networks. *Science* 301: 827–829, 2003.

[125] P. Erdős and T. Gallai. Graphs with given degrees of vertices. *Matematikai Lapok*, 11:264–274, 1960.

[126] C.I. Del Genio, H. Kim, Z. Toroczkai, and K.E. Bassler. Efficient and exact sampling of simple graphs with given arbitrary degree sequence. *PLoS ONE*, 5: e10012, 04 2010.

[127] V. Havel. A remark on the existence of finite graphs. *Časopis pro pěstováni matematiky a fysiky*, 80:477–480, 1955.

[128] S. Hakimi. On the realizability of a set of integers as degrees of the vertices of a graph. SIAM *Journal of Applied Mathematics*, 10: 496–506, 1962.

[129] I. Charo Del Genio, G. Thilo, and K.E. Bassler. All scale-free networks are sparse. *Physical Review Letters*, 107:178701, 10 2011.

[130] B. Bollobás. A probabilistic proof of an asymptotic formula for the number of labelled regular graphs. *European Journal of Combinatorics*, 1: 311–316, 1980.

[131] M. Molloy and B. A. Reed. Critical Point for Random Graphs with a Given Degree Sequence. *Random Structures and Algorithms*, 6: 161–180, 1995.

[132] S. Maslov and K. Sneppen. Specificity and stability in topology of protein networks. *Science*, 296:910–913, 2002.

[133] G. Caldarelli, I. A. Capocci, P. De Los Rios, and M.A. Muñoz. Scale-Free Networks from Varying Vertex Intrinsic Fitness. *Physical Review Letters*, 89: 258702, 2002.

[134] B. Söderberg. General formalism for inhomogeneous random graphs. *Physical Review E*, 66: 066121, 2002.

[135] M. Boguñá and R. Pastor-Satorras. Class of correlated random networks with hidden variables. *Physical Review E*, 68: 036112, 2003.

[136] A. Clauset, C.R. Shalizi, and M.E.J. Newman. Power-law distributions in empirical data. *SIAM Review* S1: 661–703, 2009.

[137] S. Redner. Citation statistics from 110 years of physical review. *Physics Today*, 58:49, 2005.

[138] F. Eggenberger and G. Pólya. Über die Statistik Verketteter Vorgänge. *Zeitschrift für Angewandte Mathematik und Mechanik*, 3:279–289, 1923.

[139] G.U. Yule. A mathematical theory of evolution, based on the conclusions of Dr. J. C. Willis. *Philosophical Transactions of the Royal Society of London. Series B*, 213:21–87, 1925.

[140] R. Gibrat. *Les Inégalités économiques*. Paris Librairie du Recueil Sirey, 1931.

[141] G. K. Zipf. *Human behavior and the principle of least resort*. Oxford: Addison-Wesley Press, 1949.

[142] H. A. Simon. On a class of skew distribution functions. *Biometrika*, 42:425–440, 1955.

[143] D. De Solla Price. A general theory of bibliometric and other cumulative advantage processes. *Journal of the American Society for Information Science*, 27:292–306, 1976.

[144] R. K. Merton. The Matthew effect in science. *Science*, 159:56–63, 1968.

[145] P.L. Krapivsky, S. Redner, and F. Leyvraz. Connectivity of growing random networks. *Physical Review Letters*, 85:4629–4632, 2000.

[146] H. Jeong, Z. Néda. A.-L. Barabási. Measuring preferential attachment in evolving networks. *Europhysics Letters*, 61:567–572, 2003.

[147] M.E.J. Newman. Clustering and preferential attachment in growing networks. *Physical Review E*, 64:025102, 2001.

[148] S.N. Dorogovtsev and J.F.F. Mendes. *Evolution of networks*. Oxford: Clarendon Press, 2002.

[149] J.M. Kleinberg, R. Kumar, P. Raghavan, S. Rajagopalan, and A. Tomkins. The Web as a graph: measurements, models and methods. Proceedings of the International Conference on Combinatorics and Computing, 1999.

[150] R. Kumar, P. Raghavan, S. Rajalopagan, D. Divakumar, A.S. Tomkins, and E. Upfal. The Web as a graph. Proceedings of the 19th Symposium on principles of database systems, 2000.

[151] R. Pastor-Satorras, E. Smith, and R. Sole. Evolving protein minteraction networks through gene duplication. *Journal of Theoretical Biology*, 222:199–210, 2003.

[152] A. Vazquez, A. Flammini, A. Maritan, and A. Vespignani. Modeling of protein interaction networks. *ComPlexUs* 1:38–44, 2003.

[153] G.S. Becker. *The economic approach to Human Behavior*. Chicago: Chicago University Press, 1976.

[154] A. Fabrikant, E. Koutsoupias, and C. Papadimitriou. *Heuristically optimized trade-offs: a new paradigm for power laws in the internet*. In Proceedings of the 29th International Colloquium on Automata, Languages, and Programming (ICALP), Malaga, Spain, July 2002, pp. 110–122.

[155] R.M. D'Souza, C. Borgs, J.T. Chayes, N. Berger, and R.D. Kleinberg. Emergence of tempered preferential attachment from optimization. *PNAS* 104, 6112–6117, 2007.

[156] F. Papadopoulos, M. Kitsak, M. Angeles Serrano, M. Boguna, and D. Krioukov. Popularity versus similarity in growing networks. *Nature*, 489: 537, 2012.

[157] A.-L. Barabási. Network science: luck or reason. *Nature*, 489: 1–2, 2012.

[158] B. Mandelbrot. An Informational Theory of the Statistical Structure of Languages. In W. Jackson (ed.), *Communication Theory*, pp. 486–502. Woburn, MA: Butterworth, 1953.

[159] B. Mandelbrot. A note on a class of skew distribution function: analysis and critique of a Paper by H.A. Simon. *Information and Control*, 2: 90–99, 1959.

[160] H.A. Simon. Some Further Notes on a Class of Skew Distribution Functions. *Information and Control* 3: 80–88, 1960.

[161] B. Mandelbrot. Final Note on a Class of Skew Distribution Functions: Analysis and Critique of a Model due to H.A. Simon. *Information and Control*, 4: 198–216, 1961.

[162] H.A. Simon. Reply to final note. *Information and Control*, 4: 217–223, 1961.

[163] B. Mandelbrot. Post scriptum to final note. *Information and Control*, 4: 300–304, 1961.

[164] H.A. Simon. Reply to Dr. Mandelbrot's Post Scriptum. *Information and Control*, 4: 305–308, 1961.

[165] K. Klemm and V.M. Eguluz. Growing scale-free networks with small-world behavior. *Physical Review E*, 65:057102, 2002.

[166] G. Bianconi and A.-L. Barabási. Competition and multiscaling in evolving networks. *Europhysics Letters*, 54: 436–442, 2001.

[167] A.-L. Barabási, R. Albert, H. Jeong, and G. Bianconi. Power-law distribution of the world wide web. *Science*, 287: 2115, 2000.

[168] C. Godreche and J. M. Luck. On leaders and condensates in a growing network. *Journal of Statistical Mechanics*, P07031, 2010.

[169] J. H. Fowler, C. T. Dawes, and N. A. Christakis. Model of Genetic Variation in Human Social Networks. *PNAS*, 106: 1720–1724, 2009.

[170] M. O. Jackson. Genetic influences on social network characteristics. *PNAS*, 106:1687–1688, 2009.

[171] S.A. Burt. Genes and popularity: Evidence of an evocative gene environment correlation. *Psychological Science*, 19:112–113, 2008.

[172] J. S. Kong, N. Sarshar, and V. P. Roychowdhury. Experience versus talent shapes the structure of the Web. *PNAS*, 105:13724–9, 2008.

[173] A.-L. Barabási, C. Song, and D. Wang. Handful of papers dominates citation. *Nature*, 491:40, 2012.

[174] D. Wang, C. Song, and A.-L. Barabási. Quantifying Long term scientific impact. *Science*, 342:127–131, 2013.

[175] M. Medo, G. Cimini, and S. Gualdi. Temporal effects in the growth of networks. *Physical Review Letters*, 107:238701, 2011.

[176] G. Bianconi and A.-L. Barabási. Bose-Einstein condensation in complex networks. *Physical Review Letters*, 86: 5632–5635, 2001.

[177] C. Borgs, J. Chayes, C. Daskalakis, and S. Roch. *First to market is not everything: analysis of preferential attachment with fitness*. STOC'07, San Diego, California, 2007.

[178] C. Godreche, H. Grandclaude, and J.M. Luck. Finite-time fluctuations in the degree statistics of growing networks. *Journal of Statistical Physics*, 137:1117–1146, 2009.

[179] Y.-H. Eom and S. Fortunato. Characterizing and Modeling Citation Dynamics. *PLoS ONE*, 6: e24926, 2011.

[180] R. Albert, and A.-L. Barabási. Topology of evolving networks: local events and universality. *Physical Review Letters*, 85:5234–5237, 2000.

[181] G. Goshal, L. Chi, and A.-L Barabási. Uncovering the role of elementary processes in network evolution. *Scientific Reports*, 3:1–8, 2013.

[182] J.H. Schön, Ch. Kloc, R.C. Haddon, and B. Batlogg. A superconducting field-effect switch. *Science*, 288: 656–8. 2000.

[183] D. Agin. *Junk Science: An Overdue Indictment of Government, Industry, and Faith Groups That Twist Science for Their Own Gain*. New York: Macmillan, 2007.

[184] S. Saavedra, F. Reed-Tsochas, and B. Uzzi. Asymmetric disassembly and robustness in declining networks. *PNAS*, 105:16466–16471, 2008.

[185] F. Chung and L. Lu. Coupling on-line and off-line analyses for random power-law graphs. *Internet Mathematics*, 1: 409–461, 2004.

[186] C. Cooper, A. Frieze, and J. Vera. Random deletion in a scalefree random graph process. *Internet Mathematics*, 1, 463–483, 2004.

[187] S. N. Dorogovtsev and J. Mendes. Scaling behavior of developing and decaying networks. *Europhysics Letters*, 52: 33–39, 2000.

[188] C. Moore, G. Ghoshal, and M. E. J. Newman. Exact solutions for models of evolving networks with addition and deletion of nodes. *Physical Review E*, 74: 036121, 2006.

[189] H. Bauke, C. Moore, J. Rouquier, and D. Sherrington. Topological phase transition in a network model with preferential attachment and node removal. *The European Physical Journal B*, 83: 519–524, 2011.
[190] M. Pascual and J. Dunne, (eds). *Ecological Networks: Linking Structure to Dynamics in Food Webs.* Oxford: Oxford University Press, 2005.
[191] R. Sole and J. Bascompte. *Self-Organization in Complex Ecosystems.* Princeton: Princeton University Press, 2006.
[192] U. T. Srinivasan, J. A. Dunne, J. Harte, and N. D. Martinez. Response of complex food webs to realistic extinction sequencesm. *Ecology*, 88:671–682, 2007.
[193] J. Leskovec, J. Kleinberg, and C. Faloutsos, Graph evolution: Densification and shrinking diameters. *ACM Transactions on Knowledge Discovery from Data*, 1:1, 2007.
[194] S. Dorogovtsev and J. Mendes. Effect of the accelerating growth of communications networks on their structure. *Physical Review E*, 63: 025101(R), 2001.
[195] M. J. Gagen and J. S. Mattick. Accelerating, hyperaccelerating, and decelerating networks. *Physical Review E*, 72: 016123, 2005.
[196] C. Cooper and P. Prałat. Scale-free graphs of increasing degree. *Random Structures & Algorithms*, 38: 396–421, 2011.
[197] N. Deo and A. Cami. Preferential deletion in dynamic models of web-like networks. *Information Processing Letters*, 102: 156–162, 2007.
[198] S.N. Dorogovtsev and J.F.F. Mendes. Evolution of networks with aging of sites. *Physical Review E*, 62:1842, 2000.
[199] K. Klemm and V. M. Eguiluz. Highly clustered scale free networks, *Physical Review E*, 65: 036123, 2002.
[200] X. Zhu, R. Wang, and J.-Y. Zhu. The effect of aging on network structure. *Physical Review E*, 68: 056121, 2003.
[201] P. Uetz, L. Giot, G. Cagney, T. A. Mansfield,R.S. Judson, J.R. Knight, D. Lockshon, V. Narayan, M. Srinivasan, P. Pochart, A. Qureshi-Emili, Y. Li, B. Godwin, D. Conover, T. Kalbfleisch, G. Vijayadamodar, M. Yang, M. Johnston, S. Fields, J. M. Rothberg. A comprehensive analysis of protein- protein interactions in Saccharomyces cerevisiae. *Nature*, 403: 623–627, 2000.
[202] I. Xenarios, D. W. Rice, L. Salwinski, M. K. Baron, E. M. Marcotte, D. Eisenberg, DIP: the database of interacting proteins *Nucleic Acids Research*, 28: 289–29, 2000.
[203] R. Pastor-Satorras, A. Vázquez, and A. Vespignani, Dynamical and correlation properties of the Internet. *Physical Review Letters*, 87: 258701, 2001.
[204] A. Vazquez, R. Pastor-Satorras, and A. Vespignani. Large-scale topological and dynamical properties of Internet. *Physical Review E*, 65: 066130, 2002.
[205] S.L. Feld. Why your friends have more friends than you do. *American Journal of Sociology*, 96: 1464–1477, 1991.
[206] E.W. Zuckerman and J.T. Jost. What makes you think you're so popular? Self evaluation maintenance and the subjective side of the "friendship paradox". *Social Psychology Quarterly*, 64: 207–223, 2001.
[207] M. E. J. Newman, Assortative mixing in networks. *Physical Review Letters*, 89: 208701, 2002.
[208] M. E. J. Newman, Mixing patterns in networks. *Physical Review E*, 67: 026126, 2003.
[209] S. Maslov, K. Sneppen, and A. Zaliznyak. Detection of topological pattern in complex networks: Correlation profile of the Internet. *Physica A*, 333: 529–540, 2004.
[210] M. Boguna, R. Pastor-Satorras, and A. Vespignani. Cut-offs and finite size effects in scale-free networks. *European Physics Journal, B*, 38: 205, 2004.
[211] M. E. J. Newman and Juyong Park. Why social networks are different from other types of networks. *Physical Review E*, 68: 036122, 2003.
[212] M. McPherson, L. Smith-Lovin, and J. M. Cook. Birds of a feather: homophily in social networks. *Annual Review of Sociology*, 27:415–444, 2001.
[213] J. G. Foster, D. V. Foster, P. Grassberger, and M. Paczuski. Edge direction and the structure of networks. *PNAS*, 107: 10815, 2010.

[214] A. Barrat and R. Pastor-Satorras, Rate equation approach for correlations in growing network models. *Physical Review E*, 71: 036127, 2005.
[215] S. N. Dorogovtsev and J. F. F. Mendes. Evolution of networks. *Advanced Physics*, 51: 1079, 2002.
[216] J. Berg and M. Lässig. Correlated random networks. *Physical Review Letters*, 89: 228701, 2002.
[217] R. Xulvi-Brunet and I. M. Sokolov. Reshuffling scale-free networks: From random to assortative. *Physical Review E*, 70: 066102, 2004.
[218] R. Xulvi-Brunet and I. M. Sokolov. Changing correlations in networks: assortativity and dissortativity. *Acta Physica Polonica B*, 36: 1431, 2005.
[219] J. Menche, A. Valleriani, and R. Lipowsky. Asymptotic properties of degree-correlated scale-free networks. *Physical Review E*, 81: 046103, 2010.
[220] V. M. Eguíluz and K. Klemm. Epidemic threshold in structured scale-free networks. *Physical Review Letters*, 89:108701, 2002.
[221] M. Boguñá and R. Pastor-Satorras. Epidemic spreading in correlated complex networks. *Physical Review Letters*, 66: 047104, 2002.
[222] M. Boguñá, R. Pastor-Satorras, and A. Vespignani. Absence of epidemic threshold in scale-free networks with degree correlations. *Physical Review Letters*, 90: 028701, 2003.
[223] A. Vázquez and Y. Moreno. Resilience to damage of graphs with degree correlations. *Physical Review E*, 67: 015101R, 2003.
[224] S.J. Wang, A.C. Wu, Z.X. Wu, X.J. Xu, and Y.H. Wang. Response of degree-correlated scale-free networks to stimuli. *Physical Review E*, 75: 046113, 2007.
[225] F. Sorrentino, M. Di Bernardo, G. Cuellar, and S. Boccaletti. Synchronization in weighted scale-free networks with degree–degree correlation *Physica D*, 224: 123, 2006.
[226] M. Di Bernardo, F. Garofalo, and F. Sorrentino. Effects of degree correlation on the synchronization of networks of oscillators. *International Journal of Bifurcation and Chaos in Applied Sciences and Engineering*, 17: 3499, 2007.
[227] A. Vazquez and M. Weigt. Computational complexity arising from degree correlations in networks. *Physical Review E*, 67: 027101, 2003.
[228] M. Posfai, Y Y. Liu, J-J Slotine, and A.-L. Barabási. Effect of correlations on network controllability. *Scientific Reports*, 3: 1067, 2013.
[229] M. Weigt and A. K. Hartmann. The number of guards needed by a museum: A phase transition in vertex covering of random graphs. *Physical Review Letters*, 84: 6118, 2000.
[230] L. Adamic and N. Glance. The political blogosphere and the 2004 U.S. election: Divided they blog (2005).
[231] J. Park and M. E. J. Newman. The origin of degree correlations in the Internet and other networks. *Physical Review E*, 66: 026112, 2003.
[232] F. Chung and L. Lu. Connected components in random graphs with given expected degree sequences. *Annals of Combinatorics*, 6: 125, 2002.
[233] Z. Burda and Z. Krzywicki. Uncorrelated random networks. *Physical Review E*, 67: 046118, 2003.
[234] R. Albert, H. Jeong, and A.-L. Barabási. Attack and error tolerance of complex networks. *Nature*, 406: 378, 2000.
[235] D. Stauffer and A. Aharony. *Introduction to Percolation Theory*. London: Taylor and Francis. 1994.
[236] A. Bunde and S. Havlin. *Fractals and Disordered Systems*. Berlin: Springer-Verlag, 1996.
[237] B. Bollobás and O. Riordan. *Percolation*. Cambridge: Cambridge University Press. 2006.
[238] S. Broadbent and J. Hammersley. Percolation processes I. Crystals and mazes. *Proceedings of the Cambridge Philosophical Society*, 53: 629, 1957.
[239] D. S. Callaway, M. E. J. Newman, S. H. Strogatz. and D. J. Watts. Network robustness and fragility: Percolation on random graphs. *Physical Review Letters*, 85: 5468–5471, 2000.

[240] R. Cohen, K. Erez, D. ben-Avraham, and S. Havlin. Breakdown of the Internet under intentional attack. *Physical Review Letters*, 86: 3682, 2001.

[241] P. Baran Introduction to Distributed Communications Networks. Rand Corporation Memorandum, RM-3420-PR, 1964.

[242] D.N. Kosterev, C.W. Taylor, and W.A. Mittlestadt. Model Validation of the August 10, 1996 WSCC System Outage. *IEEE Transactions on Power Systems* 14: 967–979, 1999.

[243] C. Labovitz, A. Ahuja and F. Jahasian. *Experimental Study of Internet Stability and Wide-Area Backbone Failures. Proceedings of IEEE FTCS.* Madison, WI, 1999.

[244] A. G. Haldane and R. M. May. Systemic risk in banking ecosystems. *Nature*, 469: 351–355, 2011.

[245] T. Roukny, H. Bersini, H. Pirotte, G. Caldarelli, and S. Battiston. Default Cascades in Complex Networks: Topology and Systemic Risk. *Scientific Reports*, 3: 2759, 2013.

[246] G. Tedeschi, A. Mazloumian, M. Gallegati, and D. Helbing. Bankruptcy cascades in interbank markets. *PLoS One*, 7: e52749, 2012.

[247] D. Helbing. Globally networked risks and how to respond. *Nature*, 497: 51–59, 2013.

[248] I. Dobson, B. A. Carreras, V. E. Lynch, and D. E. Newman. Complex systems analysis of series of blackouts: Cascading failure, critical points, and self-organization. *CHAOS*, 17: 026103, 2007.

[249] E. Bakshy, J. M. Hofman, W. A. Mason, and D. J. Watts. *Everyone's an influencer: quantifying influence on Twitter. Proceedings of the fourth ACM international conference on Web search and data mining (WSDM '11).* New York: ACM, pp. 65–74, 2011.

[250] Y. Y. Kagan, Accuracy of modern global earthquake catalogs. *Physics of the Earth and Planetary Interiors*, 135: 173, 2003.

[251] M. Nagarajan, H. Purohit, and A. P. Sheth. *A Qualitative Examination of Topical Tweet and Retweet Practices.* ICWSM, 295–298, 2010.

[252] P. Fleurquin, J.J. Ramasco, and V.M. Eguiluz. Systemic delay propagation in the US airport network. *Scientific Reports*, 3: 1159, 2013.

[253] B. K. Ellis, J. A. Stanford, D. Goodman, C. P. Stafford, D.L. Gustafson, D. A. Beauchamp, D. W. Chess, J. A. Craft, M. A. Deleray, and B. S. Hansen. Long-term effects of a trophic cascade in a large lake ecosystem. *PNAS*, 108: 1070, 2011.

[254] V. R. Sole, M. M. Jose. Complexity and fragility in ecological networks. *Proceedings of the Royal Society London, Series B*, 268: 2039, 2001.

[255] F. Jordán, I. Scheuring and G. Vida. Species Positions and Extinction Dynamics in Simple Food Webs. *Journal of Theoretical Biology*, 215: 441–448, 2002.

[256] S.L. Pimm and P. Raven. Biodiversity: Extinction by numbers. *Nature*, 403: 843, 2000.

[257] World Economic Forum, *Building Resilience in Supply Chains.* Geneva: World Economic Forum, 2013.

[258] Joint Economic Committee of US Congress. Your flight has been delayed again: Flight delays cost passengers, airlines and the U.S. economy billions. Available at http://www.jec.senate.gov, May 22. 2008.

[259] I. Dobson, A. Carreras, and D.E. Newman. A loading dependent model of probabilistic cascading failure. *Probability in the Engineering and Informational Sciences*, 19: 15, 2005.

[260] D.J. Watts. A simple model of global cascades on random networks. *PNAS*, 99: 5766, 2002.

[261] K.-I. Goh, D.-S. Lee, B. Kahng, and D. Kim. Sandpile on scale-free networks. *Physical Review Letters*, 91: 148701, 2003.

[262] D.-S. Lee, K.-I. Goh, B. Kahng, and D. Kim. Sandpile avalanche dynamics on scale-free networks. *Physica A*, 338: 84, 2004.

[263] M. Ding and W. Yang. Distribution of the first return time in fractional Brownian motion and its application to the study of onoff intermittency. *Physical Review E*, 52: 207–213, 1995.

[264] A. E. Motter and Y.-C. Lai. Cascade-based attacks on complex networks. *Physical Review E*, 66: 065102, 2002.

[265] Z. Kong and E. M. Yeh. Resilience to Degree-Dependent and Cascading Node Failures in Random Geometric Networks. *IEEE Transactions on Information Theory*, 56: 5533, 2010.

[266] G. Paul, S. Sreenivas, and H. E. Stanley. Resilience of complex networks to random breakdown. *Physical Review E*, 72: 056130, 2005.

[267] G. Paul, T. Tanizawa, S. Havlin, and H. E. Stanley. Optimization of robustness of complex networks. *European Physical Journal B*, 38: 187–191, 2004.

[268] A.X.C.N. Valente, A. Sarkar, and H. A. Stone. Two-peak and three- peak optimal complex networks. *Physical Review Letters*, 92: 118702, 2004.

[269] T. Tanizawa, G. Paul, R. Cohen, S. Havlin, and H. E. Stanley. Optimization of network robustness to waves of targeted and random attacks. *Physical Review E*, 71: 047101, 2005.

[270] A.E. Motter, Cascade control and defense in complex networks. *Physical Review Letters*, 93: 098701, 2004.

[271] A. Motter, N. Gulbahce, E. Almaas, and A.-L. Barabási. Predicting synthetic rescues in metabolic networks. *Molecular Systems Biology*, 4: 1–10, 2008.

[272] R.V. Sole, M. Rosas-Casals, B. Corominas-Murtra, and S. Valverde. Robustness of the European power grids under intentional attack. *Physical Review E*, 77: 026102, 2008.

[273] R. Albert, I. Albert, and G.L. Nakarado. Structural Vulnerability of the North American Power Grid. *Physical Review E*, 69: 025103 R, 2004.

[274] C.M. Schneider, N. Yazdani, N.A.M. Araújo, S. Havlin, and H.J. Herrmann. Towards designing robust coupled networks. *Scientific Reports*, 3: 1969, 2013.

[275] C.M. Song, S. Havlin, and H.A. Makse. Self-similarity of complex networks. *Nature*, 433: 392, 2005.

[276] S.V. Buldyrev, R. Parshani, G. Paul, H.E. Stanley, and S. Havlin. Catastrophic cascade of failures in interdependent networks. *Nature*, 464: 08932, 2010.

[277] R. Cohen, D. ben-Avraham, and S. Havlin. Percolation critical exponents in scale-free networks. *Physical Review E*, 66: 036113, 2002.

[278] S. N. Dorogovtsev, J. F. F. Mendes, and A. N. Samukhin. Anomalous percolation properties of growing networks. *Physical Review E*, 64: 066110, 2001.

[279] M. E. J. Newman, S. H. Strogatz, and D. J. Watts. Random graphs with arbitrary degree distributions and their applications. *Physical Review E*, 64: 026118, 2001.

[280] R. Cohen and S. Havlin. *Complex Networks: Structure, Robustness and Function*. Cambridge: Cambridge University Press. 2010.

[281] B. Droitcour. Young Incorporated Artists. Art in America, 92-97, April 2014.

[282] V. D. Blondel, J.-L. Guillaume, R. Lambiotte, and E. Lefebvre. Fast unfolding of communities in large networks. *Journal of Statistical Mechanics:* P10008, 2008.

[283] G.C. Homans. *The Human Groups*. New York: Harcourt, Brace & Co, 1950.

[284] S.A. Rice. The identification of blocs in small political bodies. *American Political Science Review*, 21:619–627, 1927.

[285] R.S. Weiss and E. Jacobson. A method for the analysis of the structure of complex organizations. *American Sociology Review*, 20:661–668, 1955.

[286] W.W. Zachary. An information flow model for conflict and fission in small groups. *Journal of Anthropological Research*, 33:452–473, 1977.

[287] L. Donetti and M.A. Muñoz. Detecting network communities: a new systematic and efficient algorithm. *Journal of Statistical Mechanics*, P10012, 2004.

[288] L.H. Hartwell, J.J. Hopfield, and A.W. Murray. From molecular to modular cell biology. *Nature*, 402:C47–C52, 1999.

[289] E. Ravasz, A. L. Somera, D. A. Mongru, Z. N. Oltvai, and A.-L. Barabási. Hierarchical organization of modularity in metabolic networks. *Science*, 297:1551–1555, 2002.

[290] J. Menche, A. Sharma, M. Kitsak, S. Ghiassian, M. Vidal, J. Loscalzo, A.-L. Barabási. Oncovering disease-disease relationships through the human interactome. 2014.

[291] G. W. Flake, S. Lawrence, and C.L. Giles. Efficient identification of web communities. Proceedings of the sixth ACM SIGKDD international conference on Knowledge discovery and data mining, 150-160, 2000.

[292] F. Radicchi, C. Castellano, F. Cecconi, V. Loreto, and D. Parisi. Defining and identifying communities in networks. *PNAS*, 101:2658–2663, 2004.

[293] A.B. Kahng, J. Lienig, I.L. Markov, and J. Hu. *VLSI Physical Design: From Graph Partitioning to Timing Closure.* Berlin: Springer-Verlag, 2011.

[294] B.W. Kernighan and S. Lin. An efficient heuristic procedure for partitioning graphs. *Bell Systems Technical Journal*, 49:291–307, 1970.

[295] G.E. Andrews. *The Theory of Partitions.* Boston: Addison-Wesley, 1976.

[296] L. Lovász. *Combinatorial Problems and Exercises.* Amsterdam: North-Holland, 1993.

[297] G. Pólya and G. Szegő. *Problems and Theorems in Analysis I.* Berlin: Springer-Verlag, 1998.

[298] V. H. Moll. *Numbers and Functions: From a classical-experimental mathematician's point of view. American Mathematical Society*, 2012.

[299] M.E.J. Newman and M. Girvan. Finding and evaluating community structure in networks. *Physical Review E*, 69:026113, 2004.

[300] M.E.J. Newman. A measure of betweenness centrality based on random walks. *Social Networks*, 27:39–54, 2005.

[301] U. Brandes. A faster algorithm for betweenness centrality. *Journal of Mathetical Sociology*, 25:163–177, 2001.

[302] T. Zhou, J.-G. Liu, and B.-H. Wang. Notes on the calculation of node betweenness. *Chinese Physics Letters*, 23:2327–2329, 2006.

[303] E. Ravasz and A.-L. Barabasi. Hierarchical organization in complex networks. *Physical Review E*, 67:026112, 2003.

[304] S. N. Dorogovtsev, A. V. Goltsev, and J. F. F. Mendes. Pseudofractal scale-free web. *Physical Review E*, 65:066122, 2002.

[305] E. Mones, L. Vicsek, and T. Vicsek. Hierarchy Measure for Complex Networks. *PLoS ONE*, 7:e33799, 2012.

[306] A. Clauset, C. Moore, and M. E. J. Newman. Hierarchical structure and the prediction of missing links in networks. *Nature*, 453:98–101, 2008.

[307] L. Danon, A. Díaz-Guilera, J. Duch, and A. Arenas. Comparing community structure identification. *Journal of Statistical Mechanics*, P09008, 2005.

[308] S. Fortunato and M. Barthélemy. Resolution limit in community detection. *PNAS*, 104:36–41, 2007.

[309] M.E.J. Newman. Fast algorithm for detecting community structure in networks. *Physical Review E*, 69:066133, 2004.

[310] A. Clauset, M.E.J. Newman, and C. Moore. Finding community structure in very large networks. *Physical Review E*, 70:066111, 2004.

[311] G. Palla, I. Derényi, I. Farkas, and T. Vicsek. Uncovering the overlapping community structure of complex networks in nature and society. *Nature*, 435:814, 2005.

[312] R. Guimerà, L. Danon, A. Díaz-Guilera, F. Giralt, and A. Arenas. Self- similar community structure in a network of human interactions. *Physical Review E*, 68:065103, 2003.

[313] J. Ruan and W. Zhang. Identifying network communities with a high resolution. *Physical Review E* 77: 016104, 2008.

[314] B. H. Good, Y.-A. de Montjoye, and A. Clauset. The performance of modularity maximization in practical contexts. *Physical Review E*, 81:046106, 2010.

[315] R. Guimerá, M. Sales-Pardo, and L.A.N. Amaral. Modularity from fluctuations in random graphs and complex networks. *Physical Review E*, 70:025101, 2004.

[316] J. Reichardt and S. Bornholdt. Partitioning and modularity of graphs with arbitrary degree distribution. *Physical Review E*, 76:015102, 2007.

[317] J. Reichardt and S. Bornholdt. When are networks truly modular? *Physica D*, 224:20–26, 2006.

[318] M. Rosvall and C.T. Bergstrom. Maps of random walks on complex networks reveal community structure. *PNAS*, 105:1118, 2008.

[319] M. Rosvall, D. Axelsson, and C.T. Bergstrom. The map equation. *European Physics Journal Special Topics*, 178:13, 2009.

[320] M. Rosvall and C.T. Bergstrom. Mapping change in large networks. *PLoS ONE*, 5:e8694, 2010.

[321] A. Perey. Oksapmin Society and World View. *Dissertation for Degree of Doctor of Philosophy*. Columbia University, 1973.

[322] I. Derényi, G. Palla, and T. Vicsek. Clique percolation in random networks. *Physical Review Letters*, 94:160202, 2005.

[323] J.M. Kumpula, M. Kivelä, K. Kaski, and J. Saramäki. A sequential algorithm for fast clique percolation. *Physical Review E*, 78:026109, 2008.

[324] G. Palla, A.-L. Barabási, and T. Vicsek. Quantifying social group evolution. *Nature*, 446:664–667, 2007.

[325] Y.-Y. Ahn, J. P. Bagrow, and S. Lehmann. Link communities reveal multiscale complexity in networks. *Nature*, 466:761–764, 2010.

[326] T.S. Evans and R. Lambiotte. Line graphs, link partitions, and overlapping communities. *Physical Review E*, 80:016105, 2009.

[327] M. Chen, K. Kuzmin, and B.K. Szymanski. Community Detection via Maximization of Modularity and Its Variants. *IEEE Transactions on Computational Social Systems*, 1:46–65, 2014.

[328] I. Farkas, D. Ábel, G. Palla, and T. Vicsek. Weighted network modules. *New Journal of Physics*, 9:180, 2007.

[329] S. Lehmann, M. Schwartz, and L.K. Hansen. Biclique communities. *Physical Review E*, 78:016108, 2008.

[330] N. Du, B. Wang, B. Wu, and Y. Wang. *Overlapping community detection in bipartite networks*. IEEE/WIC/ACM International Conference on Web Intelligence and Intelligent Agent Technology, IEEE Computer Society, Los Alamitos, CA, pp. 176–179, 2008.

[331] A. Condon and R.M. Karp. Algorithms for graph partitioning on the planted partition model. *Random Structures and Algorithms*, 18:116–140, 2001.

[332] A. Lancichinetti, S. Fortunato, and F. Radicchi. Benchmark graphs for testing community detection algorithms. *Physical Review E*, 78:046110, 2008.

[333] A. Lancichinetti, S. Fortunato, and J. Kertész. Detecting the overlapping and hierarchical community structure of complex networks. *New Journal of Physics*, 11:033015, 2009.

[334] S. Fortunato. Community detection in graphs. *Physics Reports*, 486:75–174, 2010.

[335] D. Hric, R.K. Darst, and S. Fortunato. Community detection in networks: structural clusters versus ground truth. *Physical Review E*, 90:062805, 2014.

[336] A. Maritan, F. Colaiori, A. Flammini, M. Cieplak, and J.R. Banavar. Universality Classes of Optimal Channel Networks. *Science*, 272:984–986, 1996.

[337] L.C. Freeman. A set of measures of centrality based upon betweenness. *Sociometry*, 40:35–41, 1977.

[338] J. Hopcroft, O. Khan, B. Kulis, and B. Selman. Tracking evolving communities in large linked networks. *PNAS*, 101:5249–5253, 2004.

[339] S. Asur, S. Parthasarathy, and D. Ucar. *An event-based framework for characterizing the evolutionary behavior of interaction graphs*. KDD '07: Proceedings of the 13th ACM SIGKDD International Conference on Knowledge Discovery and Data Mining, New York: ACM, pp. 913–921, 2007.

[340] D.J. Fenn, M.A. Porter, M. McDonald, S. Williams, N.F. Johnson, and N.S. Jones. Dynamic communities in multichannel data: An application to the foreign exchange market during the 2007–2008 credit crisis. *Chaos*, 19:033119, 2009.

[341] D. Chakrabarti, R. Kumar, and A. Tomkins. *Evolutionary clustering*, in: KDD '06: Proceedings of the 12th ACM SIGKDD International Conference on Knowledge Discovery and Data Mining, New York: ACM, pp. 554–560, 2006.

[342] Y. Chi, X. Song, D. Zhou, K. Hino, and B.L. Tseng. *Evolutionary spectral clustering by incorporating temporal smoothness*. KDD '07: Proceedings of the 13th ACM SIGKDD International Conference on Knowledge Discovery and Data Mining, New York: ACM, pp. 153–162, 2007.

[343] Y.-R. Lin, Y. Chi, S. Zhu, H. Sundaram, and B.L. Tseng. *Facetnet: a framework for analyzing communities and their evolutions in dynamic networks.* in: WWW '08: Proceedings of the 17th International Conference on the World Wide Web, New York: ACM, pp. 685–694, 2008.

[344] L. Backstrom, D. Huttenlocher, J. Kleinberg, and X. Lan. *Group formation in large social networks: membership, growth, and evolution.* KDD '06: Proceedings of the 12th ACM SIGKDD International Conference on Knowledge Discovery and Data Mining, New York: ACM, pp. 44–54, 2006.

[345] B. Krishnamurthy and J. Wang. On network-aware clustering of web clients. *SIGCOMM Computer Communication Review*, 30:97–110, 2000.

[346] K.P. Reddy, M. Kitsuregawa, P. Sreekanth, and S.S. Rao. *A graph based approach to extract a neighborhood customer community for collaborative filtering.* DNIS '02: Proceedings of the Second International Workshop on Databases in Networked Information Systems, London: Springer-Verlag, pp. 188–200, 2002.

[347] R. Agrawal and H.V. Jagadish. Algorithms for searching massive graphs. *Knowledge and Data Engineering*, 6:225–238, 1994.

[348] A.Y. Wu, M. Garland, and J. Han. *Mining scale-free networks using geodesic clustering.* KDD '04: Proceedings of the Tenth ACM SIGKDD International Conference on Knowledge Discovery and Data Mining, New York: ACM Press, 2004, pp. 719–724, 2004.

[349] D. Normile. The Metropole, Superspreaders and Other Mysteries. *Science*, 339:1272–1273, 2013.

[350] J.O. Lloyd-Smith, S.J. Schreiber, P.E. Kopp, and W.M. Getz. Superspreading and the effect of individual variation on disease emergence. *Nature*, 438:355–359, 2005.

[351] M. Hypponen. Malware Goes Mobile. *Scientific American*, 295:70, 2006.

[352] E.M. Rogers. *Diffusion of Innovations.* New York: Free Press, 2003.

[353] T.W. Valente. *Network models of the diffusion of innovations.* Hampton Press, Cresskill, NJ, 1995.

[354] The CDC and the World Health Organization. History and Epidemiology of Global Smallpox Eradication From the training course titled "Smallpox: Disease, Prevention, and Intervention". Slides 16–17.

[355] R.M. Anderson and R.M. May. *Infectious Diseases of Humans: Dynamics and Control.* Oxford: Oxford University Press, 1992.

[356] R. Pastor-Satorras and A. Vespignani. Epidemic dynamics and endemic states in complex networks. *Physical Review E*, 63:066117, 2001.

[357] Y. Wang, D. Chakrabarti, C, Wang, and C. Faloutsos. Epidemic spreading in real networks: an eigenvalue viewpoint. Proceedings of 22nd International Symposium on Reliable Distributed Systems, pg. 25–34, 2003.

[358] R. Durrett. Some features of the spread of epidemics and information on a random graph. *PNAS*, 107:4491–4498, 2010.

[359] S. Chatterjee and R. Durrett. Contact processes on random graphs with power law degree distributions have critical value 0. *Annals of Probability*, 37:2332–2356, 2009.

[360] C Castellano, and R Pastor-Satorras. Thresholds for epidemic spreading in networks. *Physical Review Letters*, 105:218701, 2010.

[361] B. Lewin. (ed.), Sex i Sverige. *Om sexuallivet i Sverige 1996 [Sex in Sweden. On the Sexual Life in Sweden 1996].* Stockholm: National Institute of Public Health, 1998.

[362] A. Schneeberger, C. H. Mercer, S. A. Gregson, N. M. Ferguson, C. A. Nyamukapa, R. M. Anderson, A. M. Johnson, and G. P. Garnett. Scale-free networks and sexually transmitted diseases: a description of observed patterns of sexual contacts in Britain and Zimbabwe. *Sexually Transmitted Diseases*, 31:380–387, 2004.

[363] W. Chamberlain. *A View from Above.* New York: Villard Books, 1991.

[364] R. Shilts. *And the Band Played On.* New York: St. Martin's Press, 2000.

[365] P. S. Bearman, J. Moody, and K. Stovel. Chains of affection: the structure of adolescent romantic and sexual networks. *American Journal of Sociology*, 110:44–91, 2004.

[366] M. C. González, C. A. Hidalgo, and A.-L. Barabási. Understanding individual human mobility patterns. *Nature*, 453:779–782, 2008.

[367] C. Song, Z. Qu, N. Blumm, and A.-L. Barabási. Limits of Predictability in Human Mobility. *Science*, 327:1018–1021, 2010.

[368] F. Simini, M. González, A. Maritan, and A.-L. Barabási. A universal model for mobility and migration patterns. *Nature*, 484:96–100, 2012.

[369] D. Brockmann, L. Hufnagel, and T. Geisel. The scaling laws of human travel. *Nature*, 439:462–465, 2006.

[370] V. Colizza, A. Barrat, M. Barthelemy, and A. Vespignani. The role of the airline transportation network in the prediction and predictability of global epidemics. *PNAS*, 103:2015, 2006.

[371] R. Guimerà, S. Mossa, A. Turtschi, and L. A. N. Amaral. The worldwide air transportation network: Anomalous centrality, community structure, and cities' global roles. *PNAS*, 102:7794, 2005.

[372] C. Cattuto, et al. Dynamics of Person-to-Person Interactions from Distributed RFID Sensor Networks. *PLoS ONE*, 5:e11596, 2010.

[373] L. Isella, C. Cattuto, W. Van den Broeck, J. Stehle, A. Barrat, and J.-F. Pinton. What's in a crowd? Analysis of face-to-face behavioral networks. *Journal of Theoretical Biology*, 271:166–180, 2011.

[374] K. Zhao, J. Stehle, G. Bianconi, and A. Barrat. Social network dynamics of face-to-face interactions. *Physical Review E*, 83:056109, 2011.

[375] J. Stehlé, N. Voirin, A. Barrat, C Cattuto, L. Isella, J-F. Pinton, M. Quaggiotto, W. Van den Broeck, C. Régis, B. Lina, and P. Vanhems. High-resolution measurements of face-to-face contact patterns in a primary school. *PLoS ONE*, 6:e23176, 2011.

[376] B.N. Waber, D. Olguin, T. Kim, and A. Pentland. *Understanding Organizational Behavior with Wearable Sensing Technology*. Academy of Management Annual Meeting. Anaheim, CA. August, 2008.

[377] M. Salathé, M. Kazandjievab, J.W. Leeb, P. Levisb, M.W. Feldmana, and J.H. Jones. A high-resolution human contact network for infectious disease transmission. *PNAS*, 107:22020–22025, 2010.

[378] A.-L. Barabási. The origin of bursts and heavy tails in human dynamics. *Nature*, 435:207–11, 2005.

[379] V. Sekara, and S. Lehmann. Application of network properties and signal strength to identify face-to-face links in an electronic dataset. Proceedings of CoRR, 2014.

[380] S. Eubank, H. Guclu, V.S.A. Kumar, M.V. Marathe, A. Srinivasan, Z. Toroczkai, and N. Wang. Modelling disease outbreaks in realistic urban social networks. *Nature*, 429:180–184, 2004.

[381] M. Morris, and M. Kretzschmar. Concurrent partnerships and transmission dynamics in networks. *Social Networks*, 17:299–318, 1995.

[382] N. Masuda and P. Holme. Predicting and controlling infectious diseases epidemics using temporal networks. *F1000 Prime Reports*, 5:6, 2013.

[383] P. Holme, and J. Saramäki. Temporal networks. *Physics Reports*, 519:97–125, 2012.

[384] M. Karsai, M. Kivelä, R. K. Pan, K. Kaski, J. Kertész, A.-L. Barabási, and J. Saramäki. Small but slow world: how network topology and burstiness slow down spreading. *Physical Review E*, 83:025102(R), 2011.

[385] A. Vazquez, B. Rácz, A. Lukács, and A.-L. Barabási. Impact of non-Poissonian activity patterns on spreading processes. *Physical Review Letters*, 98:158702, 2007.

[386] A. Vázquez, J.G. Oliveira, Z. Dezsö, K.-I. Goh, I. Kondor, and A.-L. Barabási. Modeling bursts and heavy tails in human dynamics. *Physical Review E*, 73:036127, 2006.

[387] A.V. Goltsev, S.N. Dorogovtsev, and J.F.F. Mendes. Percolation on correlated networks. *Physical Review E.*, 78:051105, 2008.

[388] P. Van Mieghem, H. Wang, X. Ge, S. Tang and F. A. Kuipers. Influence of assortativity and degree-preserving rewiring on the spectra of networks. *The European Physical Journal B*, 76:643, 2010.

[389] Y. Moreno, J. B. Gómez, and A.F. Pacheco. Epidemic incidence in correlated complex networks. *Physical Review E*, 68:035103, 2003.

[390] A. Galstyan, and P. Cohen. Cascading dynamics in modular networks. *Physical Review E*, 75:036109, 2007.

[391] J. P. Gleeson. Cascades on correlated and modular random networks. *Physical Review E*, 77:046117, 2008.

[392] P. A. Grabowicz, J. J. Ramasco, E. Moro, J. M. Pujol, and V. M. Eguiluz. Social features of online networks: The strength of intermediary ties in online social media. *PLOS ONE*, 7:e29358, 2012.

[393] L. Weng, F. Menczer and Y.-Y. Ahn. Virality Prediction and Community Structure in Social Networks. *Scientific Reports*, 3:2522, 2013.

[394] S. Aral, and D. Walker. Creating social contagion through viral product design: A randomized trial of peer influence in networks. *Management Science*, 57:1623–1639, 2011.

[395] J. Leskovec, L. Adamic, and B. Huberman. The dynamics of viral marketing. *ACM Transactions on the Web*, 1, 2007.

[396] L. Weng, A Flammini, A. Vespignani, and F. Menczer. Competition among memes in a world with limited attention. *Scientific Reports*, 2:335, 2012.

[397] J. Berger, and K. L. Milkman. What makes online content viral? *Journal of Marketing Research*, 49:192–205, 2009.

[398] S. Jamali, and H. Rangwala. Digging digg: Comment mining, popularity prediction and social network analysis. Proceedings of the International Conference on Web Information Systems and Mining (WISM), 32–38, 2009.

[399] G. Szabó and, B. A. Huberman. Predicting the popularity of online content. *Communications of the ACM*, 53:80–88, 2010.

[400] B. Suh, L. Hong, P. Pirolli, and E. H. Chi. Want to be retweeted? Large scale analytics on factors impacting retweet in Twitter network. Proceedings of IEEE International Conference on Social Computing, pp. 177–184, 2010.

[401] D. Centola. The spread of behavior in an online social network experiment. Science, 329:1194–1197, 2010.

[402] M. Granovetter. Threshold Models of Collective Behavior. *American Journal of Sociology*, 83:1420–1443, 1978.

[403] N.A. Christakis, and J.H. Fowler. The Spread of Obesity in a Large Social Network Over 32 Years. *New England Journal of Medicine*, 35:370–379, 2007.

[404] N. A. Christakis and J. H. Fowler. The collective dynamics of smoking in a large social network. *New England Journal of Medicine*, 358:2249–2258, 2008.

[405] Z. Dezső and A-L. Barabási. Halting viruses in scale-free networks. *Physical Review E*, 65:055103, 2002.

[406] R. Pastor-Satorras and A. Vespignani. Immunization of complex networks. *Physical Review E*, 65:036104, 2002.

[407] R. Cohen, S. Havlin, and D. ben-Avraham. Efficient Immunization Strategies for Computer Networks and Populations. *Physical Review Letters*, 91:247901, 2003.

[408] F. Fenner et al. *Smallpox and its Eradication*. WHO, Geneva, 1988. http://www.who.int/features/2010/smallpox/en/

[409] L. A. Rvachev, and I. M. Longini Jr. A mathematical model for the global spread of influenza. *Mathematical Biosciences*, 75:3–22, 1985.

[410] A. Flahault, E. Vergu, L. Coudeville, and R. Grais. Strategies for containing a global influenza pandemic. *Vaccine*, 24:6751–6755, 2006.

[411] I. M. Longini Jr, M. E. Halloran, A. Nizam, and Y. Yang. Containing pandemic influenza with antiviral agents. *American Journal of Epidemiology*, 159:623–633, 2004.

[412] I.M. Longini Jr, A. Nizam, S. Xu, K. Ungchusak, W. Hanshaoworakul, D. Cummings, and M. Halloran. Containing pandemic influenza at the source. *Science*, 309:1083–1087, 2005.

[413] V. Colizza, A. Barrat, M. Barthélemy, A.-J. Valleron, and A. Vespignani. Modeling the world-wide spread of pandemic influenza: baseline case and containment interventions. *PLoS Med*, 4:e13, 2007.

[414] T. D. Hollingsworth, N.M. Ferguson, and R.M. Anderson. Will travel restrictions control the International spread of pandemic influenza? *Nature Med.*, 12:497–499, 2006.

[415] C.T. Bauch, J.O. Lloyd-Smith, M.P. Coffee, and A.P. Galvani. Dynamically modeling SARS and other newly emerging respiratory illnesses: past, present, and future. *Epidemiology*, 16:791–801, 2005.

[416] I. M. Hall, R. Gani, H.E. Hughes, and S. Leach. Real-time epidemic forecasting for pandemic influenza. *Epidemiology and Infection*, 135:372–385, 2007.

[417] M. Tizzoni, P. Bajardi, C. Poletto, J. J. Ramasco, D. Balcan, B. Gonçalves, N. Perra, V. Colizza, and A. Vespignani. Real-time numerical forecast of global epidemic spreading: case study of 2009 A/H1N1pdm. *BMC Medicine*, 10:165, 2012.

[418] P. Bajardi, et al. Human Mobility Networks, Travel Restrictions, and the Global Spread of 2009 H1N1 Pandemic. *PLoS ONE*, 6:e16591, 2011.

[419] P. Bajardi, C. Poletto, D. Balcan, H. Hu, B. Gonçalves, J. J. Ramasco, D. Paolotti, N. Perra, M. Tizzoni, W. Van den Broeck, V. Colizza, and A. Vespignani. Modeling vaccination campaigns and the Fall/Winter 2009 activity of the new A/H1N1 influenza in the Northern Hemisphere. *EHT Journal*, 2:e11, 2009.

[420] M.E. Halloran, N.M. Ferguson, S. Eubank, I.M. Longini, D.A.T. Cummings, B. Lewis, S. Xu, C. Fraser, A. Vullikanti, T.C. Germann, D. Wagener, R. Beckman, K. Kadau, C. Macken, D.S. Burke, and P. Cooley. Modeling targeted layered containment of an influenza pandemic in the United States. *PNAS*, 105:4639–44, 2008.

[421] G. M. Leung, A. Nicoll. Reflections on Pandemic (H1N1) 2009 and the international response. *PLoS Med*, 7: e1000346, 2010.

[422] A.C. Singer, et al. Meeting report: risk assessment of Tamiflu use under pandemic conditions. Environmental Health Perspectives, 116:1563–1567, 2008.

[423] R. Fisher. The wave of advance of advantageous genes. Annals of Eugenics, 7:355–369, 1937.

[424] J. V. Noble. Geographic and temporal development of plagues. *Nature*, 250:726–729, 1974.

[425] D. Brockmann and D. Helbing. The Hidden Geometry of Complex, Network-Driven Contagion Phenomena. *Science*, 342:1337–1342, 2014.

[426] J. S. Brownstein, C. J. Wolfe, and K. D. Mandl. Empirical evidence for the effect of airline travel on inter-regional influenza spread in the United States. *PLoS Med*, 3:e40, 2006.

[427] D. Shah and T. Zaman, in SIGMETRICS'10, Proceedings of the ACM SIGMETRICS international conference on Measurement and modeling of computer systems, pp. 203–214, 2010.

[428] A. Y. Lokhov, M. Mezard, H. Ohta, L. Zdeborová. Inferring the origin of an epidemy with dynamic message-passing algorithm. *Physical Review E*, 90:012801, 2014.

[429] P. C. Pinto, P. Thiran, M. Vetterli. Locating the Source of Diffusion in Large-Scale Networks. *Physical Review Letters*, 109:068702, 2012.

[430] C. H. Comin and L. da Fontoura Costa. Identiying the starting point of a spreading process in complex networks. *Physical Review E*, 84:056105, 2011.

[431] D. Shah and T. Zaman. Rumors in a Network: Who's the Culprit? *IEEE Transactions on Information Theory*, 57:5163, 2011.

[432] K. Zhu and L. Ying. Information source detection in the SIR model: A sample path based approach. Information Theory and Applications Workshop (ITA); 1-9, 2013.

[433] B. A. Prakash, J. Vreeken, and C. Faloutsos. Spotting culprits in epidemics: How many and which ones? ICDM'12; Proceedings of the IEEE International Conference on Data Mining, 11:20, 2012.

[434] V. Fioriti and M. Chinnici. Predicting the sources of an outbreak with a spectral technique. *Applied Mathematical Sciences*, 8:6775–6782, 2012.

[435] W. Dong, W. Zhang and C.W. Tan. Rooting out the rumor culprit from suspects. Proceedings of CoRR, 2013.

[436] B. Barzel, and A.-L. Barabási. Universality in network dynamics. *Nature Physics*, 9:673, 2013.

[437] A. Barrat, M. Barthélemy and A. Vespignani. *Dynamical Processes on Complex Networks*. Cambridge: Cambridge University Press, 2012.

[438] S. N. Dorogovtsev, A.V. Goltsev, and J. F. F. Mendes. Critical phenomena in complex networks. *Reviews of Modern Physics* 80, 1275, 2008.

[439] P. Grassberger. On the critical behavior of the general epidemic process and dynamical percolation. *Mathematical Biosciences*, 63:157, 1983.

[440] M. E. J. Newman. The spread of epidemic disease on networks. *Physical Review E*, 66:016128, 2002.

[441] C. P. Warren, L. M. Sander, and I. M. Sokolov. Firewalls, disorder, and percolation in networks. *Mathematical Biosciences*, 180:293, 2002.

Index